Climate Change Management

Series editor

W. Leal Filho, Hamburg, Germany

More information about this series at http://www.springer.com/series/8740

Walter Leal Filho • Haruna Musa • Gina Cavan •
Paul O'Hare • Julia Seixas
Editors

Climate Change Adaptation, Resilience and Hazards

Springer

Editors
Walter Leal Filho
Faculty of Life Science
Research and Transfer Center
Hamburg University of Applied Sciences
Hamburg, Germany

Haruna Musa
Food and Tourism Management
Manchester Metropolitan University
Manchester, United Kingdom

Gina Cavan
Manchester, United Kingdom

Paul O'Hare
Geography & Environmental Management
Manchester Metropolitan University
Manchester, United Kingdom

Julia Seixas
Faculdade de Ciências e Tecnologia
Universidade Nova de Lisboa
Caparica, Portugal

ISSN 1610-2010 ISSN 1610-2002 (electronic)
Climate Change Management
ISBN 978-3-319-39879-2 ISBN 978-3-319-39880-8 (eBook)
DOI 10.1007/978-3-319-39880-8

Library of Congress Control Number: 2016946351

Printed on acid-free paper

This Springer imprint is published by Springer Nature
The registered company is Springer International Publishing AG Switzerland

Preface

This book, titled "Climate Change Adaptation, Resiliance and Hazards", is one of the outcomes of the "World Symposium on Climate Change Adaptation", held in Manchester, UK, on 2–4 September 2015. The event was co-organised by Manchester Metropolitan University (UK) and the Research and Transfer Centre "Applications of Life Sciences" of the Hamburg University of Applied Sciences (Germany), in cooperation with the Baltic University Programme in Uppsala (Sweden), International Climate Change Information Programme (ICCIP) and the United Nations University initiative "Regional Centres of Expertise on Education for Sustainable Development" (RCE).

The aims of the "World Symposium on Climate Change Adaptation" were as follows:

(i) To provide research institutions, universities, NGOs and enterprises all round the world with an opportunity to display and present their works in the field of climate change adaptation
(ii) To foster the exchange of information, ideas and experiences acquired in the execution of climate change adaptation projects, especially successful initiatives and good practice
(iii) To discuss methodological approaches and experiences deriving from case studies and projects, which aim to show how the principles of climate change adaptation may be implemented in practice
(iv) To network the participants and provide a platform so they can explore possibilities for cooperation.

Last but not least, a further aim of the event was to document and disseminate the wealth of experiences on climate change adaptation available today across the world. To this purpose, this book, which contains the papers presented at session 4 (Climate Change Adaptation, Resilience and Hazards (including floods)), has been compiled. It addresses the perceived need for publications which look at the links between adaptation tools and methods on the one hand and the means to handle hazards and foster resilience on the other. This publication departs from the hypothesis that countries and communities need to strengthen their resilience to a variety of hazards, whilst tackling the underlying causes of vulnerability, which entail elements as different as geographical conditions, governance, poverty and inequality. As shown via some of the examples documented here, much can be gained by affording risk and vulnerability a greater level of attention, since they are instrumental in resilience efforts and in achieving resilience outcomes.

This book is divided into two parts. Part I, titled "Climate Change Adaptation Practices", encompasses Chaps. 1–10, all of which describe the theory and practice of climate change adaptation under different angles.

Part II, titled "Fostering Resilience and Handling Hazards", offers a comprehensive overview of approaches related to resilience and entails Chaps. 11–22.

We want to thank all authors for sharing their know-how, as well as for the support they have provided in producing this book. All in all, the experiences gathered in this book serve the purpose of documenting and disseminating a wide range of experiences on climate change and resilience available today.

Enjoy your reading!
The Editors

Hamburg, Germany — Walter Leal Filho
Manchester, UK — Haruna Musa
Caparica, Portugal — Gina Cavan
Spring 2016 — Paul O'Hare
Julia Seixas

Contents

Part I
Climate Change Adaptation Practices

Chapter 1
Monitoring and Evaluation of Climate Change Adaptation in Coastal Zones: Overview of the Indicators in Use

Moktar Lamari, Jessica Bouchard, Johann Jacob, and Line Poulin-Larivière

Abstract Recently, the number of public policies initiated within the framework of CCA has increased in several western countries. On the other hand, the evaluation and measurement of CCA progress remain embryonic and inadequately charted by indicators quantifying the changes and impacts attributable to these policies that engender new stakes and concepts that are difficult to measure and do not always produce a consensus.

Our paper presents the results of a scoping review examining the extent, scope and nature of the literature dealing with CCA indicators and metrics. Using scoping-review protocol, bibliographic databases were examined (for the years 2005–2015), using key words, in both English and French, dealing with the measurement of progress in public CCA efforts in coastal areas. In all, 165 documents were selected and analyzed and more than 200 indicators were looked at.

Our analyses point out that the conceptual framework for CCA remains fragmented given the different scientific approaches and disciplines. The lack of consensus about CCA and about the indicators designed to assess CCA initiatives is a major limitation in coastal zone management. One way is to harmonize the practices for analyzing human and environmental systems respectively, both in the fields of the social and the natural sciences. In terms of governance, network management appears to be the most effective method in the context of social and environmental change. In terms of indicators, the WorldRiskIndex provides a useful estimate of the vulnerability of countries with respect to the effects of climate change in a context of rapid urbanization. In a regional context or a smaller geographical area, GIS stands out for being able to incorporate a lot of data, and to ensure their continuous update. Moreover, the ICZM is an effective approach at the national level that nevertheless takes into account the regional differences in coastal zones. The studies included in this analysis also demonstrate that the

M. Lamari (✉) • J. Bouchard • J. Jacob • L. Poulin-Larivière
Centre de recherche et d'expertise en évaluation (CREXE), University of Québec, École nationale d'administration publique, Québec, QC, Canada
e-mail: moktar.lamari@enap.ca

W.L. Filho et al. (eds.), *Climate Change Adaptation, Resilience and Hazards*, Climate Change Management, DOI 10.1007/978-3-319-39880-8_1

process of adopting CCA measures must be transparent and participatory since they aim for both short-term and long-term objectives.

Keywords Climate change adaptation • Monitoring and evaluation • Coastal zone • Indicators • Scoping review

Introduction

Since and throughout the last decade, efforts to mobilize collective action to advance climate change adaptation (CCA) have proliferated, as reflected in public policies and in activities and initiatives undertaken at various levels of government (Cooper and Pile 2014). These efforts, which can be quite costly and complex, involve a multitude of fields of intervention and stakeholders who are affected by the many facets and consequences of climate change.

In Canada, like in all Western countries, governments must support and structure their efforts in CCA, particularly in coastal areas, where shores are increasingly exposed to erosion as well as other phenomena associated with climate change. The need for collective action with regard to CCA is especially important and urgent in coastal zones and in wetland, river and sea ecosystems. In these ecosystems, climate change-related disturbances and unpredictabilities manifest primarily in rising sea levels, an increased frequency of storms, coastal erosion and loss of cliffs, the recurrence of devastating floods, the salinization of fresh water sources, and other phenomena that put densely populated coastal zones to the test. All of these phenomena greatly affect critical infrastructures (energy, transport, etc.) and even critical decision-making centers for governments and communities (IPCC 2014a).

As Governments are getting more aware of the risks associated with climate change, they are in the forefront of the development and implementation of innovative adaptation strategies. Indeed, several OECD countries have come up CCA initiatives and programs (OECD 2014; Mullan et al. 2013). At the same time, and considering the innovative nature of the actions to implement, governments and collective action as a whole must distinguish between effective actions from those that are not. In this context, reliable indicators are required to measure progress in CCA and support decision making. Significant amounts are often involved and policymakers and taxpayers want to ensure that the budgets allocated and the defined regulations give conclusive results, for example in reducing communities' vulnerability to extreme climate events. Several OECD countries have come up CCA initiatives and programs (OECD 2014; Mullan et al. 2013). Indeed, some governments and communities have also developed indicators to measure progress in CCA to support coastal management (Torresan et al. 2008; Harvey and Woodroffe 2008; Hanak and Moreno 2012).

Yet, despite the importance of the issues, concerns and actions involved, and the proliferation of research on initiatives and programs pertaining to CCA, we still know very little about the progress and results of the policies and initiatives implemented to advance CCA. Because no systematic approach to monitoring and evaluating CCA has emerged, the capacity to conduct assessments and to incorporate them in adaptation policies is limited. Consequently, although adaptation is occasionally subject to critical examination, the evaluation of CCA is still far from being an institutionalized practice (Preston et al. 2011). This is an important limitation especially for public administrators who cannot rely on a set of and validated indicators to support decision-making in adaptation.

On the one hand, this limitation could be attributed to the scarcity of research on the measurement and evaluation (M&E) of CCA efforts (Lemmen et al. 2008; MMM 2005; de Bruin et al. 2009). Most studies undertaken in the context of climate change also have a limited scope, their focus being more on documenting and measuring the efforts to mitigate climate change and less on analyzing and measuring the effects of CCA measures (Dupuis and Biesbroek 2013). M&E of CCA is also an emerging field of practice, albeit still riddled with a number of complicating factors. Existing assessment frameworks often focus on different and varied aspects of planning, making it difficult to identify an appropriate planning process, as well as generic indicators (Brooks et al. 2011; Hedger et al. 2008).

But the real challenge with M&E of CCA lies in its embryonic nature, insufficiently marked by indicators quantifying the changes and impacts due to different adaptation strategies and measures. Furthermore, adaptation refers to new issues and concepts which are not easily measurable and not always without controversies. The resulting confusion complicates the measurement of progress in CCA achieved through collective action on climate change (Ostrom 2010; Magnan 2009) and calls for improvements in M&E (Preston et al. 2011).

"Measuring" is understood as the systematic process of assigning a value to a phenomenon (and therefore something that can be observed) (Hinkel 2011). Because the different concepts and variables associated with a phenomena can take on different values, measuring will consist in establishing their state, in order to quantify and compare. Thus, the design and selection of indicators will be understood as the effort to operationalize a concept of a theoretical nature, meaning that the quality of an indicator can be assessed on the basis of ontological criterias. The quality of an indicator is also dependent on many methodological prerequisites. The most cited writings emphasize three types of attributes: reliability, accessibility and relevance of the data and measures used. To be reliable, an indicator must reflect in a relevant and robust way what the evaluator, the researcher or planner actually want to measure and observe.

Like any emerging area of research, the scientific literature on CCA in coastal zones is not mature, even if it is continually progressing to better define its theoretical contours and empirical indicators (Magnan 2009). Adaptation is a polysemic neologism which is theoretically hard to apprehend. The very nature of what constitutes adaptation is not constant (Hedger et al. 2008; OECD 2006) which complicates the determination of generic indicators. Is it a series of specific activities or decisions, or rather a set of processes evolving concurrently (Brooks and Frankel-Reed 2008; UNDP 2005) or a result to be achieved (UKCIP 2003)? The lack of consensus on what constitutes "good" adaptation or "maladaptation" is also a source of difficulty, preventing the application of universal criteria and standards to appreciate an adaptation response (Hedger et al. 2008).

Vague concepts and their various corollaries, which in some cases refer to undefined concepts, are also a source of difficulties. For a central concept such as vulnerability, Thywissen (2006) identified as many as 35 definitions, plus related terms referring to similar ideas (risk, sensitivity, fragility) or conversely similar (resilience, adaptability, adaptive capacity, stability) (Brooks 2003). This lack of consensus make it difficult to establish metrics on concepts whose definitions are shared by a large number of actors (Hinkel 2011). Furthermore, the inherent normative dimension in the appreciation of various concepts (adverse effect; significant climate variation) adds to the complexity to the operationalization of key CCA concepts. For one, a concept such as vulnerability do not refer to a single observable phenomenon (Moss et al. 2001; Patt et al. 2009) but rather to a theoretical concept, contingent to a given system, place or context, which can be measured on multiple spatial and temporal scales. Vulnerability will be perceived differently in different geographic region and will depend on various scenarios regarding climate change and socio-economic changes. It will also varies between sectors and social groups (Dunford et al. 2015), leading to very little certainty about the object around which to develop indicators (Cutter et al. 2003).

The numerous categories of CCA measures are pursuing various and ambitious goals (reduce vulnerability, develop adaptive capacity, reduce climate risk, address climate change, etc.). Several of these interventions have intangible nature, which complicates their evaluation in the light of a single paradigm or theory. Coming with indicators measuring the "invisible" and reliably quantifying all the benefits from a CCA measure are hard to come by. Indicators mostly account for the "visible part of the iceberg", but lose in relevance when one whishes to learn more on the underlying dynamics, currents, mechanisms, etc. by which the impacts associated with CCA interventions are happening.

The design and implementation of CCA measures are also associated with the intertwined of numerous interrelated and complementary initiatives. In many cases, indicators are struggling to "sort out" the various interventions and to isolate the impacts associated to each intervention. Frequent shifts (temporal and intergenerational) between the moment a CCA measure is implemented and its impacts become visible limit the relevance of certain indicators, as well as complicate the use of analytical methods and the definition of indicators, too oftently developed under static and ad hoc approaches. The emergence of certain impacts often depends on interactions between results and economic and social context, and this reality is hard to capture with a set of generic indicators.

There is also an equity dimension in CCA, and if the efficiency of a measure can be easily assess from the perspective of its beneficiaries, this demonstration is much more difficult when we adopt the perspective of the whole society bearing all the costs associated with the investment required by an adaptation measure. Very often, indicators address the beneficiaries of public intervention but fail to reflect all the costs that are diffuse and unevenly borne by taxpayers. From a methodological perspective, this distinction between the economic impacts and the social impacts of a CCA measure is not always easy to operate. The strong interdependence between social and economic factors makes it difficult to assess all the benefits for the society and so is their quantification as indicators.

Understanding the complexity in the course of designing M&E indicators can thus be realized in two directions: (i) according to the dimensions of the performance of an intervention (i.e., the elements of the logic model: inputs, processes, outputs, effects, impacts); (ii) whether this complexity is solely related to methodological issues, or is also weighted down by ontological difficulties. So, the more the purpose for performance monitoring concerns those responsible for implementing an intervention, or their direct beneficiaries, the lower the level of complexity. In this situation, chances are that indicators will refer to simple metrics, and that their collection will not present significant methodological challenges. As we move along in the chain of results, the identification process might become trickier, as it will require the analyst to reflect on specific outcomes and impacts theoretically associated to the intervention. Indicators will also have to be time and context specific, which reduces the potential use of generic indicators. Furthermore, as we adopt a societal perspective, more robust methods will be required to assess the benefits and costs collectively supported for one intervention. But the conceptualization of the concepts covered by the measure will be the real challenge, as M&E will still present different levels of complexity from a methodological standpoint, as well as ontological problems. Before even thinking for a method to gather data in order to assess the level of vulnerability of a community, one has to state what the concepts even means and how it can be apprehended if it is not directly observable.

With all the aforementioned difficulties associated with CCA indicators said, one can postulate the following hypothesis: Unlike with mitigation to climate change, where establish metrics are identified and used without controversies, existing CCA indicators are more controversial than mitigation indicators, and CCA indicators in coastal zone are even more complex. Our contribution aims to innovate on that subject, as these hypothesis have not been adequately treated in the literature. It also aims to reduce the knowledge gap on indicators to measure progress in CCA. It develops an analytical portrait of the indicators currently used to measure various dimensions relating to CCA in coastal zones. This portrait is built on a synthesis of knowledge (scoping review) from the scientific literature and from the most cited official documents in the context of the OECD countries. It revolves around three sub-questions that can be formulated as follows:

- What are the CCA indicators used in the coastal zones of the most affected OECD countries?
- How are CCA indicators designed and measured to assist in decision-making?
- What are the most common indicators and how are they valued in the development or implementation of a system of indicators?

The article is divided into four sections. In the first section, we present the methodological guidelines established as part of this scoping review. In the second section, we discuss the conceptual framework for CCA as well as indicators for measuring progress in CCA in coastal zones. In the third section, the results of this portrait are highlighted with a view to developing a set of indicators that are useful and usable in decision-making. This section also puts into perspective the impact of CCA monitoring indicators on decision-making. The fourth and final section discusses the governance of indicators for measuring CCA.

Methodology

This article adopts an investigative approach inspired by a method for synthesizing knowledge known as the scoping review method. Our investigations aim to generate empirical results from different types of research in order to draw valuable and useful findings for decision-making in matters concerning the monitoring and evaluation of CCA in coastal zones (Anderson et al. 2008). A scoping review is a method that is particularly relevant in a context where the research domain under study is relatively recent or in which the scientific writings are compartmentalized, contradictory or controversial. It allows to take stock of and better organize complex and relatively new issues (Levac et al. 2010; Davis et al. 2009). Unlike for a systematic review, the protocol of the scoping review does not lead to identify and consult systematically all the literature on a subject but rather to identify publications that meet certain criteria. This is a limitation to the study, as one might think of some other documents that would have been relevant to consult.

The literature mobilized for our scoping review includes documents from the OECD context, with a particular focus on Australia, the United States, France, New Zealand, the Netherlands and the United Kingdom—all countries that have made, to varying degrees, interesting experiences and initiatives in CCA in coastal zones. This is another limitation to the study, as other countries might have been interesting cases, but for time considerations, we had to focus on these countries. Two types of documentary sources are used: articles published in scientific journals with evaluation procedures that are rigorous and based on anonymity and official reports from government agencies or international organizations active in the field of CCA in coastal zones. The selected texts have been published during the last decade (2006–2015) and were written in either English or French.

This research used relevant databases from the social sciences as well as from multidisciplinary and environment-oriented fields. Eleven databases were consulted: (1) Arts & Humanities Citation Index, (2) Érudit, (3) JSTOR, (4) Sage, (5) Science Citation Index Expanded, (6) ScienceDirect (Elsevier), (7) Social Sciences Citation Index, (8) International Bibliography of the Social Sciences (IBSS), (9) Web of Science Core Collection, (10) Wiley, and (11) WorldWide Political Science Abstract.

The keywords were identified using the thesaurus of the databases presented above and by performing a preliminary documentary research. Thus, the term "climate change adaptation" was combined with synonyms of: coastal zone; indicators, or indexes; monitoring and evaluation; and governance, decision-making, management, public policy or institution. This allowed to identified 165 articles. After a first screening stage, based on the titles and abstracts, 50 articles were retained for a more comprehensive and detailed reading. Subsequently, eight of the 50 studies were excluded for non-compliance with the inclusion criteria established beforehand. Moreover, another four studies were rejected in a final validation because their methodology and results were insufficiently described and explained. This resulted in a total of 38 articles to proceed with. Detailed information on the methodology used is provided upon request by the authors.

A Conceptual Framework for CCA Issues

Despite the consensus on the need for CCA in the international arena, the term has many definitions. Cooper and Pile (2014: 91) presented a compilation of early definitions of CCA: "[t]he process through which people reduce the adverse effects of climate on their health and well-being, and take advantage of the opportunities that their climatic environment provides" (Burton 1992); "any adjustment, whether passive, reactive or anticipatory, that is proposed as a means for ameliorating the anticipated adverse consequences associated with climate change," (Stakhiv 1993), and "all adjustments in behaviour or economic structure that reduce the vulnerability of society to changes in the climate system" (Smith 1996).

In parallel, in the United Nations Framework Convention on Climate Change (UNFCCC), CCA is defined as "adjustments in ecological, social, or economic systems in response to actual or expected climatic stimuli and their effects or impacts. It refers to changes in processes, practices, and structures to moderate potential damages or to benefit from opportunities associated with climate change." According to the Organization for Economic Cooperation and Development (OECD), "various types of adaptation can be distinguished, including anticipatory and reactive adaptation, private and public adaptation, and autonomous and planned adaptation." In addition, the European Commission considers that CCA "aims at reducing the risk and damage from current and future harmful impacts cost-effectively or exploiting potential benefits [...] Adaptation can encompass national or regional strategies as well as practical steps taken at community level or by individuals" (Cooper and Pile 2014: 91).

Some of the above-mentioned definitions show commonalities and differences in their understanding of CCA. Some emphasize the anthropogenic causes of climate change, thereby highlighting the importance of human behavior in CCA, whereas other definitions focus on the natural and physical determinants, in which case collective action implicitly becomes somewhat exogenous. These nuances reflect the various levels of responsibility assumed by social actors in their efforts to advance CCA. In addition, social actors defend divergent interests, due to which their support of government actions and policies in CCA is not always unanimous (Cooper and Pile 2014).

Coastal areas have the particularity of being densely populated, holding the bulk of productive infrastructure and economic wealth. At the same time, coastal areas are also zones where the impacts of climate change are especially strong, as has been demonstrated over the last 20 years [rising sea levels; the extinction or migration of species, which has a significant impact on the fishing industry; coastal erosion, acidification of the oceans; changes in the distribution of marine species; changes in precipitation; and the recurrence of storms, ocean temperatures and marine winds and currents—all of which can cause extensive flooding and damages (McClatchey et al. 2014: 14)]. Adaptation measures can take many forms, whereby a dichotomy exists between interventions modifying the environment and those concerning human behavior. In the end, CCA measures undertaken in coastal zones are highly dependent on anticipated risks as well as on the public opinion regarding climate change-related risks and harms. Cooper and Pile (2014) highlight that CCA measures in these areas tend to focus on the short term and the search for cost-effective results allowing to solve immediate problems.

The scientific literature on CCA clearly identifies the need to monitor and evaluate adaptation efforts with a view to determining the effectiveness and the success of interventions. OECD outlines the contours of monitoring by giving it the status of a "continuing function that uses systematic collection of data on specified indicators to provide management and the main stakeholders of an ongoing development intervention with indications of the extent of progress and achievement of objectives and progress in the use of allocated funds." (OECD 2010: 27). On the other hand, evaluation is conceived as the "systematic and objective assessment of an on-going or completed project, programme or policy, its design, implementation and results. The aim is to determine the relevance and fulfillment of objectives, development efficiency, effectiveness, impact and sustainability" (OECD 2010: 21). Ultimately, "successful adaptation will be measured by how well different measures contribute to effectively reducing vulnerability and building resilience. Lessons learned, good practices, gaps and needs identified during the monitoring and evaluation of ongoing and completed projects, policies and programmes will inform future measures, creating an iterative and evolutionary adaptation process" (Sanahuja 2011: 15).

Because the scientific literature on these different concepts remains vastly fragmented, the recognized impacts of climate change and the adaptation measures and vulnerability indicators for the regions studied are not usually presented in a quantitative and systematic manner. This situation is not without generating a lack of empirical evidence to support the information and indicators presented (Hofmann et al. 2011; Nicholls et al. 2008). Sanahuja (2011) argues that the concept of CCA is polysemous and difficult to measure since it encompasses a multitude of disparate elements from sectors, disciplines, stakeholders, processes, diverse ecosystems and often divergent interests. As a result, the conceptual organization and structure surrounding the definition of CCA are fluid, which puts great stress on M&E efforts in adaptation. Given that the CCA challenge is marked by great complexity and requires a multidisciplinary analysis, Hofmann et al. (2011) recommend that we examine our notion of CCA more rigorously and resolve the confusion surrounding the terminology relating to CCA. This brings us to the concept of CCA indicators.

CCA Indicators for Coastal Zones

Our review of the literature confirms the absence of a standardized and consensual acceptance of notions dealing with metrics, measures or indicators in CCA. This absence also explains the overlaps, confusion and knowledge gaps on the subject. The IPCC (IPCC 2014a, b) defines measure as the quantity or degree of development of an observed object, in view of its supposed present condition; metrics as a group of values that, taken together, gives a broader indication of the state or the degree of progress towards the desired state; and indicator as a sign or estimate of the state of something, and often of the evolution of an observed object or phenomenon (IPCC 2014b).

Assessing the vulnerability or scope of an adaptation action cannot be done without the use of metrics or indicators to describe these components. The IPCC determines three types of use of indicators: an instrumental use, to determine the static state of a need to adapt (the measure of vulnerability); a dynamic use, to measure the dynamics and monitor the implementation of adaptation actions; and finally, an evaluative use, to measure the effectiveness or efficiency of a CCA initiative (IPCC 2014b).

The Organisation for Economic Cooperation and Development (OECD) defines an indicator as a "quantitative or qualitative factor or variable that provides a simple and reliable means to measure achievement, to reflect the changes connected to an intervention, or to help assess the performance of a development actor." (OECD 2010). As part of CCA, the indicators aim to "simplify, quantify, standardize and communicate complex and often disparate data and information." In addition, appropriate indicators allow building a solid foundation for evaluating the effectiveness and output of CCA measures undertaken (Sanahuja 2011).

At present, the measure of vulnerability to climate change in coastal zones is the most commonly used indicator at the international level. The IPCC defines vulnerability according to three central themes: "the magnitude and rate of climate variations to which a system is exposed (i.e., exposure); the degree to which a system could be affected by climate-related stimuli (i.e., sensitivity); the ability of a system to adjust or to cope with climate-change consequences (i.e., adaptive capacity)" (Torresan et al. 2008). Thus, contrary to the notion of sensitivity, vulnerability refers to a number of biophysical, socio-economic, institutional and political concepts.

The objective behind the design of vulnerability indicators is to assist in the decision-making, problem-solving and evaluation of business strategic performance measures in CCA. Yet, according to Hinkel (2011), in terms of public policy, the way in which such indicators are used remains unclear. An analysis of official documents produced by the UN and the European Union showed that public entities do not clearly state how these indicators should be taken into account. An analysis of the scientific literature showed that a large number of case studies do not state how vulnerability is estimated. According to Hinkel, the conceptual literature provides some answers by highlighting that vulnerability indicators serve to expose the needs with regard to CCA; to identify populations and the most vulnerable sectors; to alert the public and decision-makers about climate change; to assign resources to combat climate change to the most vulnerable areas; and to assess the performance of CCA initiatives.

Tibbetts and van Proosdij (2013) argue that CCA indicators are effective tools that shed light on policy design for both the short term and the long term. According to them, these tools can be made available to managers and persons in charge of CCA action, allowing these to identify the populations and areas exposed to higher levels of risk and to allocate CCA resources more efficiently (Tibbetts and van Proosdij 2013).

Romieu et al. (2010) examined the literature on the evaluation of the vulnerability of coastal zones, dividing these writings into two categories: one deals with risks associated with climate change, and the other concerns natural hazards at large (floods, storms, rising sea levels). They found that the studies focused on climate change have more of a long-term vision when analyzing vulnerability. Dupuis and Biesbroek (2013) conducted a comparative study of writings on CCA and the evaluation of CCA measures. The conclusions suggest that the methods used to evaluate the effectiveness and effects of CCA initiatives using valid indicators are inconsistent in the literature and face conceptual limitations. They therefore argue that the measurement of the effect of CCA measures should not be limited only to ex post evaluations, since adaptation measures are also designed to prevent future risks, which is complex to evaluate when the anticipated negative impacts have not yet manifested. In this context, the authors suggest viewing ex post analyses in conjunction with ex ante analyses, and to compare countries or regions with similar characteristics to highlight the most effective policies (Dupuis and Biesbroek 2013).

Because of the inherent methodological difficulties in evaluating CCA measures, several studies point to the need to analyze the processes in which policies are developed, including their goals and success factors. However, these studies hardly mention the design of indicators for this type of analysis.

Multi-criteria analyses are drawing renewed interest for CCA in coastal zones. Clearly, these analyses require a battery of indicators in order to be able to yield a selection and evaluation of CCA options for coastal zones. de Bruin et al. (2009) use the multi-criteria method to examine a number of options retained to advance CCA efforts in coastal zones, and list five types of criteria based on measurable indicators: (i) the importance of the expected gross impacts; (ii) the urgency associated with the reviewed interventions, (iii) the no-regret features of the options considered (meaning that the options are useful even in the absence of climate risk), (iv) indirect beneficial impacts on other sectors and geographical areas, and (v) effects beneficial to climate change mitigation (e.g., reducing greenhouse gas emissions). Each criterion is supported by a multitude of measurable and often demanding empirical indicators. The process of selecting the options to be applied to these multi-criteria analyses is, in turn, subject to a review composed of five steps: (1) identification of options on the basis of consultation with stakeholders, (2) qualitative examination of the options; (3) definition by experts of the criteria used to rank the options; (4) creation of scores for the options based on the selected criteria, (5) determination of the relative weights of the criteria and their indicators for classification purposes; and (6) interpretation and classification for decision-making purposes.

Bosello and De Cian (2014) conducted an extensive review of the methods recommended in the literature for CCA indicators in coastal zones. They first identified the exposure and vulnerability indicators, which take into account the elements exposed to risks associated with the negative impacts of climate change. Overall, they identified two main methodological approaches for measuring vulnerability to climate change in coastal zones:

- The Global Vulnerability Assessment (GVA) introduced by the IPCC to assesses the vulnerability of coastal zones to rising sea levels. The indicator takes into account mainly the level of exposure and risk in the study area and vulnerability has three components: exposure, sensitivity and adaptive capacity. It takes into account the population at risk (PaR) on the basis of the frequency of flooding due to climate change in the study area, protection standards in place, and population density. This indicator led to the creation of a global database for calculating the exposure and protection of coastal zones. Nicholls et al. (1999) improved the indicator by creating GVA1, which takes into account the evolution of the population, GDP and its exposure to climate change. In 2014, the authors extended the forecast of this database to the year 2080 (Bosello and De Cian 2014).
- The Dynamic Interactive Vulnerability Assessment (DIVA) measures exposure and vulnerability by adding an algorithm allowing to determine the optimal adaptation on the basis of the cost–benefit ratio. It is based on an extensive database of climate and socio-economic scenarios (Romieu et al. 2010). Hinkel (2010) developed this database that gathers the following types of information on the coastal topography: altitude, type of geomorphology, tidal range, type of terrain; as well as the population, protection status, and wetlands. It allows for the creation of a model structured into modules for performing vulnerability analyses according to different levels of adaptation (Bosello and De Cian 2014). This innovative and interactive database thus allows to integrate knowledge from different disciplines on coastal systems.

DIVA has the advantage of being flexible and including a vast range of data. Nevertheless, it is limited in that it is poorly suited to measuring vulnerability in local or regional contexts. Given its accessibility, DIVA is now among the most used estimation indicators of vulnerability in political, academic and scientific research.

Torresan et al. (2008) examined DIVA and confirmed its limited effectiveness in the analysis at the regional level. For this, they conducted an extensive investigation to identify indicators that are relevant for regional studies or smaller territories. They first identified the indicator *Aera X*, which allows to assess the vulnerability of coastal zones to flooding and other risks due to rising sea levels. The indicator is particularly useful for identifying the needs for protecting infrastructures in areas at risk. It takes into account the total area of square kilometers of a coastal territory under study. The authors also identified the coastal slope indicator, which measures the topographic slope of an area in degrees. This indicator also aims to estimate the risk of flooding of an area. Furthermore, the indicator of the wetland migratory potential estimates the vulnerability of ecosystems to rising sea levels. It is especially useful for decision-making in the context of urban development and regional coastal management. It includes geographical and geological elements (Torresan et al. 2008).

Leaving GVA and DIVA, Brenner (2008) discussed ways to assess the effectiveness and vulnerability of coastal zones due to climate change in relation to socio-economic variables. The author presents the indicator Geographic Information Systems (GIS) which, given the complexity of coastal regions, takes into account data from the geological, physical, chemical, biological, social, economic and political domains (Brenner et al. 2008).

United Nations University (UNU) published in 2014 the World Risk Report (WRR), which aims to systematically estimate the vulnerability of countries by studying their exposure to risks related to climate change and classifying their sensitivity to natural disasters. The report, which covers 171 countries, built its ranking from many external and internal factors. The WorldRiskIndex is calculated using 28 indicators for which the data are available worldwide and accessible to the public. Overall, this index is constructed from four elements: exposure, susceptibility, coping capacities and adaptive capacities. The WorldRiskIndex is calculated by multiplying the vulnerability index by the susceptibility index, the latter of which takes into account the coping and adaptive capacities.

Acosta et al. (2013) developed an interesting model for assessing CCA capacity in Europe and exposing the evolution of these adaptive capacities over time. The model is based on the conceptualization of vulnerability of the IPCC and combines forecasts and scenarios of climate change impacts in order to build a quantitative and static index of the vulnerability to climate change of European states. To determine the indicators to be considered in their model, the authors based themselves on the literature on adaptive capacities to climate change and on the socio-economic variables associated with them.

First, their adaptive capacities model by country is based on three levels of aggregation. The first level concerns the determinants of adaptation and has six components: equity, knowledge, technology, infrastructure, flexibility and economic power. The second level concerns the components awareness, ability and action. The third level of aggregation is adaptive capacity. On the basis of these aggregate levels, the authors integrated 12 indicators into their model.

This model of indicators has the advantage of including a wide range of socio-economic data and is therefore a useful tool for identifying the regions that are the most vulnerable to climate change. Thus, the vulnerability index generated by this model allows decision-makers to target the areas and regions that are the most at risk when implementing adaptation measures and channeling resources. By contrast, as the model estimates are based on long-term forecasts, the results generated do not target specific adaptation measures to be implemented. In sum, these authors emphasize the need to develop a better theoretical understanding of adaptive capacity in a regional context, which would involve conducting empirical research and meta-analyses targeting different regions and economic sectors (Acosta et al. 2013).

Governance and CCA Indicators in Coastal Zones

The fourth part of this analysis focuses on the governance in CCA, and more specifically on the concept of Integrated Coastal Zone Management (ICZM) and the role of stakeholders in this process. ICZM is a process that initially emerged from the 1992 Rio Summit and that was later taken up by the European Commission in a set of policy recommendations and opinions. The integrated management approach aims to take into account the environmental, economic and social aspects of a coastal territory. ICZM is a governance tool allowing coastal territories to deal with climate change and become more oriented towards sustainable development (Tang et al. 2011).

According to Hewett and Fletcher (2010), ICZM is a dominant paradigm for the management of coastal zones internationally. The authors conducted a case study of the United Kingdom, where ICZM actions have been taken since the early 1990s. Overall, the authors demonstrated clear benefits of ICZM, including: its role of consultation between the different sectors and decision-making levels; its ability to promote the participation of stakeholders and citizens; the fact that it promotes information exchange as well as conflict resolution and consensus building; its positive effect on the promotion of adaptation projects and their financing; and the creation of channels for exchanges between local and higher levels of governments (Hewett and Fletcher 2010).

However, several studies show that ICZM, despite its positive effects on partnership-building and decision-making processes, does have a significant limit, namely a fragility and dependence on shorter-term funding, making it vulnerable to contingencies. Indeed, partnerships initiated under ICZM are generally informal and therefore face difficulties in obtaining the necessary resources to remain viable (Hewett and Fletcher 2010).

Tang et al. (2011) also addressed this issue in the context of a larger study of the performance of ICZM strategies at the national and local levels. Their research evaluates the quality and results of 53 counties on the Pacific coast of the United States. The results of this study provide interesting details about the pitfalls of ICZM and also about the factors contributing to the success of this management approach. First, management plans and planning of coastal zones should clearly and consistently identify all the standards concerning this type of environment, allowing to integrate ICZM in a strong legal framework. This plan should also include a review of all the resources and environmentally sensitive areas. In addition, it must take into account the socio-economic context of the zone, and the elements considered essential for citizens, such as infrastructure, economic development projects and water quality. Finally, it must include extremely rigorous forecasts of the anticipated impacts of climate change (Tang et al. 2011).

Recent research shows that network governance is the approach that has the most success when it comes to CCA. Indeed, network governance promotes an efficient framework in a context like CCA that is characterized by a high level of risk and complexity. This mode of governance allows to manage the constraints brought by a large number of stakeholders at different levels.

In a survey conducted with 138 respondents in charge of CCA plans in the United States, Kettle and Dow (2014) showed that CCA plans adopted at the local and national levels identify potential risks associated with climate change but fail to identify strategies for the implementation and evaluation of the results of the CCA measures. These results confirm the findings of other studies showing that ICZM initiatives have limitations in terms of M&E. This, together with a lack of human and material resources for their implementation, impedes the adoption of adaptation strategies. In addition, entities often find themselves at an impasse due to substantial differences between the levels of government, representing a barrier to adaptation. The authors recommend the adoption of strategies favoring a better synergy between the various government agencies and the stakeholders in order to better allocate resources and risks. Network governance is presented as a possible solution to promote strategies that respect the priorities established at the national level while respecting the needs of different coastal regions. The researchers also call for more leadership from national governments to establish criteria for ranking priorities in each milieu and to accompany the various entities in the design, implementation and evaluation of their CCA initiatives (Kettle and Dow 2014).

Lemieux et al.'s (2013) conclusions, pointing in this same direction, also underline the need for measures to increase the transparency and accountability of agencies or organizations in charge of CCA strategies. This process of adopting such measures should be institutionalized, while also being dynamic, by offering guidelines, strategies and follow-ups in a sporadic manner. Obviously, the effectiveness of adaptation depends on additional factors, such as population density of the affected coastal region, and on the resources available to persons in charge of CCA to develop and implement strategies (Bradley et al. 2015).

Conclusions

This scoping review has highlighted the fact that the conceptual framework for CCA remains fragmented given the different scientific approaches and disciplines. The studies included in this analysis demonstrate that the lack of consensus about CCA and about the indicators designed to assess CCA initiatives is a major limitation in coastal zone management. A number of authors working in this relatively recent area of scientific research have proposed ways to address these limitations.

One way is to harmonize the practices for analyzing human and environmental systems respectively, both in the fields of the social and the natural sciences. The creation of better channels of exchange within the scientific community should also be prioritized, be it with regard to the transfer of data, knowledge or methodological guidelines. As each of these domains are facing complex issues, with anticipated impacts that are uncertain and extended over a long time horizon, they should collaborate to facilitate interaction and synergy. In that sense, vulnerability is a common concept that can serve as a basis for designing better CCA policies through a knowledge transfer process (Romieu et al. 2010).

The meta-analysis conducted by Hofmann et al. (2011) highlights the limits of the scientific literature on CCA, including the question of the quantification of impacts and the design of vulnerability indicators. As CCA initiatives require significant resources, the production of reliable data and recognized indicators is essential to allow for the *ex ante* and *ex post* evaluation of these initiatives.

In terms of governance, it appears that network management is the most effective method in the context of social and environmental change. Woodland and Hutton (2012) have discussed the importance of inter-organizational collaboration and have reiterated its expediency for solving complex public issues such as climate change. Indeed, their work has demonstrated that when an issue is at once political, social and economic, a collaborative approach is more effective. They developed a model outlining the criteria to be taken into account when assessing the action of a network: "(1) Operationalizing the construct of collaboration; (2) Identifying and mapping alliance teams and groups; (3) Monitoring stage/stages of development; (4) Assessing levels of integration; and (5) Assessing cycles of inquiry in high-leverage teams" (Woodland and Hutton 2012: 381). Jørgensen (2006) likewise addressed the importance of fostering partnerships between public, private and non-governmental organizations (NGOs) to address the challenges of climate change, and also assessed the effectiveness of this network collaboration for developing best practices.

In terms of indicators, this study shows that in a national and international context, the WorldRiskIndex provides a useful estimate of the vulnerability of countries with respect to the effects of climate change in a context of rapid urbanization. In addition, the DIVA has the advantage of being accessible and flexible and of including a considerable amount of data. Its integrated model allows to assess the impacts of climate change as well as the costs of adaptation measures. The simulations that can be produced with this tool also have the advantage of including different scenarios and inputs. Together with the fact that it is free of charge and easily accessible, DIVA is thus an essential indicator. The Global Vulnerability Assessment (GVA), introduced by the IPCC, can also be used in a national context to measure the vulnerability of coastal zones with regard to the level of exposure and risk, namely by calculating the exposure, sensitivity and adaptive capacity. The GVA1 developed by Nicholls et al. (1999) moreover allows to take into account the evolution of the population and the GDP.

In a regional context or a smaller geographical area, GIS stands out for being able to incorporate a lot of data, and to ensure the continuous update of that data.

Torresan et al. (2008), for their part, discuss other indicators used exclusively in a regional context. These provide a limited estimate of the vulnerability because they are more oriented towards the evaluation of flood risk or risks associated with the rise of water levels.

Moreover, the ICZM is an effective approach at the national level that nevertheless takes into account the regional differences in coastal zones. The studies included in this analysis also demonstrate that the process of adopting CCA measures must be transparent and participatory since they aim for both short-term and long-term objectives. As short-term risks tend to generate more concern among the public and stakeholders, governments must take swift action to implement CCA initiatives focusing on such risks. Nevertheless, the anticipated effects that extend over the long term should not be ignored either.

Thus, overall this study has shown that the systematic and consensual use of indicators in the scientific community is crucial for advancing the work of the evaluation of CCA initiatives. In a highly complex field such as climate change, the analyses of contributions associated with the realistic evaluation approach (Pawson and Tilley 2004) emerge as a possible solution in the literature. The realistic approach has the advantage of being based on the methodology of the natural sciences and allows to analyze programs in their particular context and according to the mechanisms in place. It provides a coherent framework that takes into account the role of stakeholders in the development of a program all the while maintaining enough flexibility to integrate the diversity of factors that influence the development and implementation of a policy (Pawson and Tilley 2004).

References

Acosta L, Klein RJ, Reidsma P, Metzger MJ, Rounsevell MD, Leemans R, Schröter D (2013) A spatially explicit scenario-driven model of adaptive capacity to global change in Europe. Glob Environ Chang 23(5):1211–1224

Anderson S, Allen P, Peckham S et al (2008) Asking the right questions: scoping studies in the commissioning of research on the organisation and delivery of health services. Health Res Policy Syst 6(7):12

Bosello F, De Cian E (2014) Climate change, sea level rise, and coastal disasters. A review of modeling practices. Energy Econ 46:593–605

Bradley M, Putten I, Sheaves M (2015) The pace and progress of adaptation: marine climate change preparedness in Australia's coastal communities. Mar Policy 53:13–20

Brenner J, Jiménez JA, Sardá R (2008) Environmental indicators GIS of the Catalan Coast. J Coast Conserv 11(4):185–191

Brooks N (2003) Vulnerability, risk and adaptation: a conceptual framework. Tyndall Centre Climate Change Res Work Pap 38:1–16

Brooks N, Anderson S, Ayers J et al (2011) Tracking adaptation and measuring development. IIED, London/Edinburgh

Burton I (1992) Adapt and thrive. Canadian Climate Centre, Downsview, ON, unpublished manuscript

Cooper J, Pile J (2014) The adaptation-resistance spectrum: a classification of contemporary adaptation approaches to climate-related coastal change. Ocean Coast Manag 94:90–98

Cutter SL, Boruff BJ, Shirley WL (2003) Social vulnerability to environmental hazards. Soc Sci Q 84(2):242–261

Davis K, Drey N, Gould D (2009) What are scoping studies? A review of the nursing literature. Int J Nurs Stud 46(10):1386–1400

de Bruin K, Dellink RB, Ruijs A, Bolwidt L, van Buuren A, Graveland J, Tassone VC (2009) Adapting to climate change in The Netherlands: an inventory of climate adaptation options and ranking of alternatives. Clim Change 95(1–2):23–45

Dunford R, Harrison PA, Jäger J, Rounsevell MDA, Tinch R (2015) Exploring climate change vulnerability across sectors and scenarios using indicators of impacts and coping capacity. Clim Change 128(3–4):339–354

Dupuis J, Biesbroek R (2013) Comparing apples and oranges: the dependent variable problem in comparing and evaluating climate change adaptation policies. Glob Environ Chang 23 (6):1476–1487

Finland. Ministry of Agriculture and Forestry (MMM) (2005) Finland's national strategy for climate change adaptation. MMM, Helsinki, p 281

Frankel-Reed J, Brooks N (2008) Proposed framework for monitoring adaptation to climate change adaptation. London

Hanak E, Moreno G (2012) California coastal management with a changing climate. Clim Change 111:45–73

Harvey N, Woodroffe CD (2008) Australian approaches to coastal vulnerability assessment. Sustain Sci 3:67–87, Springer

Hedger MM, Mitchell T, Leavy J, Greeley M, Downie A (2008) Desk review: Evaluation of adaptation to climate change from a development perspective. Institute of Development Studies, Lisa Horrocks, AEA Group, Brighton, p 60, August

Hewett T, Fletcher S (2010) The emergence of service-based integrated coastal management in the UK. Area 42:313–327

Hinkel J (2011) Indicators of vulnerability and adaptive capacity: towards a clarification of the science–policy interface. Glob Environ Chang 21(1):198–208

Hinkel J, Klein R (2009) Integrating knowledge to assess coastal vulnerability to sea-level rise: the development of the DIVA tool. Glob Environ Chang 19:384–395

Hinkel J, Nicholls RJ, Vafeidis AT, Tol RS, Avagianou T (2010) Assessing risk of and adaptation to sea-level rise in the European Union: an application of DIVA. Mitig Adapt Strat Glob Chang 15(7):703–719

Hofmann ME et al (2011) Classifying knowledge on climate change impacts, adaptation, and vulnerability in Europe for informing adaptation research and decision-making: a conceptual meta-analysis. Glob Environ Chang 21(3):1106–1116

IPCC (2014a) Coastal systems and low-lying areas [en ligne], ch. 5. p 32. http://ipcc-wg2.gov/AR5/images/uploads/WGIIAR5-Chap5_FGDall.pdf. Accessed 1 Sept 2015

IPCC (2014b) Glossary, rapport final, [online]. http://ipcc-wg2.gov/AR5/images/uploads/WGIIAR5-Glossary_FGD.pdf. Accessed 29 Jan 2015

Jørgensen M (2006) Evaluating cross-sector partnerships. Working paper presented at the conference public–private partnerships in the post WSSD context, Copenhagen Business School, 14 August

Kettle NP, Dow K (2014) Cross-level differences and similarities in coastal climate change adaptation planning. Environ Sci Pol 44:279–290

Lemieux C et al (2013) Natural resource manager perceptions of agency performance on climate change. J Environ Manage 114:189–198

Lemmen DS, Warren FJ, Lacroix J, Bush E (eds) (2008) From impacts to adaptation: Canada in a changing climate 2007. Government of Canada, Ottawa, ON, p 448

Levac D, Colquhoun H, O'Brien KK (2010) Scoping studies: advancing the methodology. Implement Sci 5(1):1–9

Magnan A (2009) Proposition d'une trame de recherche pour appréhender la capacité d'adaptation au changement climatique. VertigO, la revue électronique en sciences de l'environnement 9:3

McClatchey J, Devoy R, Woolf D, Bremner B, James N (2014) Climate change and adaptation in the coastal areas of Europe's Northern Periphery Region. Ocean Coast Manag 94:9–21
Moss RH, Brenkert AL, Malone EL (2001) Vulnerability to climate change: a quantitative approach. Pacific Northwest National Laboratory (PNNL-SA-33642). Prepared for the US Department of Energy, pp 155–167
Mullan M et al (2013) Planification nationale de l'adaptation: L'expérience des pays de l'OCDE. Éditions OCDE. doi:10.1787/5k483jp21xg0-fr
Nicholls RJ, Hoozemans FM, Marchand M (1999) Increasing flood risk and wetland losses due to global sea-level rise: regional and global analyses. Glob Environ Chang 9:S69–S87
Nicholls RJ, Wong PP, Burkett V et al (2008) Climate change and coastal vulnerability assessment: scenarios for integrated assessment. Sustain Sci 3(1):89–102
OECD (2006) Adaptation to climate change: key terms. OECD, Paris
OECD (2010) Glossaire des principaux termes relatifs à l'évaluation et la gestion axée sur les résultats [online]. http://www.oecd.org/dac/evaluation/2754804.pdf. Accessed 9 Feb 2015
OECD (2014) L'eau et l'adaptation au changement climatique : Des politiques pour naviguer en eaux inconnues. Études de l'OCDE sur l'eau. Éditions OCDE. doi:10.1787/9789264200647-fr
Ostrom E (2010) Polycentric systems for coping with collective action and global environmental change. Glob Environ Chang 20(4):550–557
Patt AG, Schröter D, de la Vega-Leinert AC, Klein RJ (2009) Vulnerability research and assessment to support adaptation and mitigation: common themes from the diversity of approaches. In: Assessing vulnerability to global environmental change: making research useful for adaptation decision making and policy. pp 1–25
Pawson R, Tilley N (2004) Realist evaluation. Sage, New York, NY
Preston B, Westaway M, Yuen EJ (2011) Climate adaptation planning in practice: an evaluation of adaptation plans from three developed nations. Mitig Adapt Strat Glob Chang 16(4):407–438
Romieu E, Welle T, Schneiderbauer S et al (2010) Vulnerability assessment within climate change and natural hazard contexts: revealing gaps and synergies through coastal applications. Sustain Sci 5(2):159–170, Springer
Sanahuja HE (2011) A framework for monitoring and evaluating adaptation to climate change. Community of Practice, Global Environment Facility, Washington, DC, p 78
Smith JB (1996) Development of adaptation measures for water resources. Int J Water Resour Dev 12(2):151–164
Stakhiv E (1993) Evaluation of IPCC adaptation strategies. Institute for Water Resources, US Army Corps of Engineers, Fort Belvoir, VA, draft report
Tang Z, Lindell MK, Prater C et al (2011) Examining local coastal zone management capacity in U.S. Pacific Coastal counties. Coast Manag 39(2):105–132
Thywissen K (2006) Components of risk: a comparative glossary. UNU-EHS, Bonn
Tibbetts J, van Proosdij D (2013) Development of a relative coastal vulnerability index in a macro-tidal environment for climate change adaptation. J Coast Conserv 17(4):775–797
Torresan S, Critto A, Dalla Valle M (2008) Assessing coastal vulnerability to climate change: comparing segmentation at global and regional scales. Sustain Sci 3(1):45–65
UNDP (2005) Adaptation policy framework for climate change: developing policies strategies and measures. In: Lim B, Spanger-Siegfred E (eds). Cambridge University Press
Woodland RH, Hutton MS (2012) Evaluating organizational collaborations: suggested entry points and strategies. Am J Eval 33(3):366–383

Chapter 2
An Evaluation of the Community Land Model (Version 3.5) and Noah Land Surface Models for Temperature and Precipitation Over Nebraska (Central Great Plains): Implications for Agriculture in Simulations of Future Climate Change and Adaptation

Jane A. Okalebo, Robert J. Oglesby, Song Feng, Kenneth Hubbard, Ayse Kilic, Michael Hayes, and Cynthia Hays

Abstract With increasing evidence of climate change, future decision-making among crop modelers and agronomists will require the inclusion of high-resolution climate predictions from regional climate models as input into agricultural system simulation models to assess the impacts of projected ambient CO_2 increases, temperature and general climatic change on crop production. Before they can be implemented in climate adaption studies and decision-support systems, weather variables must be reliable and accurate. This study evaluated weather variables generated from computer simulations using two land surface models, (LSMs) coupled to a regional climate model, namely, Weather Research Forecasting (WRF 3.2). The land surface models tested are the Community Land Surface Model CLM 3.5 and the Noah Land surface model. Ground truth observations

J.A. Okalebo (✉)
School of Natural Resources, University of Nebraska-Lincoln, 249 Hardin Hall, 3310 Holdrege Street, Lincoln, NE 68583, USA
e-mail: jane.okalebo@gmail.com

R.J. Oglesby
School of Natural Resources, University of Nebraska-Lincoln, 802 Hardin Hall, 3310 Holdrege Street, Lincoln, NE 68583, USA

Earth and Atmospheric Sciences, University of Nebraska-Lincoln, 221 Bessey Hall, Lincoln, NE 68588-0340, USA

S. Feng
Department of Geosciences, University of Arkansas, 228 Ozark Hall, Fayetteville, AR 72701, USA

K. Hubbard • A. Kilic • M. Hayes • C. Hays
School of Natural Resources, University of Nebraska-Lincoln, 802 Hardin Hall, 3310 Holdrege Street, Lincoln, NE 68583, USA

W.L. Filho et al. (eds.), *Climate Change Adaptation, Resilience and Hazards*, Climate Change Management, DOI 10.1007/978-3-319-39880-8_2

from 7 stations in Nebraska from a dry year, a normal year and a wet year (2002, 2005 and 2008 respectively) were used to evaluate the model results. Model results were also compared for their spatial ability to mimic distance-standard error weather variables. Both LSMs performed well in predicting the maximum and minimum temperatures in 2002, 2005 and 2008. Rainfall predictions by both models were not as reliable, based on evaluation for individual stations as well as spatially (state-wide).

Keywords Climate change • Land surface models • Regional climate models

Introduction

With ever increasing evidence of climate change, future decision-making among crop modelers and agronomists will require the inclusion of climate predictions in agricultural system simulation models to assess the impacts of projected ambient CO_2 increments and attendant climatic changes on crop production. These agricultural simulation models rely on predictions from Global Circulation Models (GCMs) to provide useful climatic and weather data to simulate crop responses. Water resource planners require accurate runoff estimates to develop safe and secure structural designs that incorporate the effects of climate change and variability. They also need to make informed decisions on energy production levels, instream flows, water supplies and water quality.

Several researchers (e.g., Brown et al. 2000; Mearns et al. 2001; Easterling et al. 2001; Niu et al. 2009; Ko et al. 2010; among many others) have used agricultural system simulation models to assess the impacts of projected ambient CO_2 increases and resultant climatic change on crop production. These crop models require weather data as inputs, and the sources of future weather data are predicted weather patterns from General Circulation Models. However, there are concerns about the "input-data-induced uncertainties" (Niu et al. 2009, p. 268) that reduce the confidence in results and thus, threaten the usefulness of the output generated from crop simulation models.

Concerns about the reliability of the output data from GCMs, especially at the 100 km spatial scale typically used in them. Of particular interest, GCMs rely on Land Surface Models (LSMs) to estimate surface gas exchange fluxes. LSMs utilize algorithms to estimate energy fluxes such as Latent Heat (LE), Sensible Heat (SH) and Soil Heat Flux (G). Clearly both agriculture and water resources will benefit from improved predictions of future climate. Land Surface Models are used to compute the hydrological, biogeophysical and biogeochemical processes involved in latent, sensible and soil heat land surface-atmospheric fluxes (Wei et al. 2009). A wide range of LSMs are currently in use today, each varying in their temporal and spatial scales and especially in their degree and type of physical parameterization. Unfortunately, even with the same forcings from the atmosphere; latent, sensible and ground surface fluxes can vary considerably from one LSM to another because they differ in their varied levels of complexity and their description of relevant processes; thereby introducing differences in simulated weather variables (e.g., PILPS, Pitman et al. 1999; Wei et al. 2009; Evans et al. 2005).

This purpose of the study was to evaluate weather variables generated from computer simulations using two land surface models, (LSMs) coupled to a regional climate model, namely, Weather Research Forecasting (WRF 3.2). The land surface models tested are the Community Land Surface Model CLM 3.5 and the Noah Land surface model. Ground truth observations from 7 stations in Nebraska from a dry year, a normal year and a wet year (2002, 2005 and 2008 respectively) were used to evaluate the model results. Additionally, spatial and temporal precipitation predictions were evaluated using the Precipitation-elevation Regressions on Independent Slopes Model (PRISM) daily estimates (Daly et al. 1994). The better LSM would be recommended for future weather variable predictions.

Expected Climate Trends for Nebraska

Throughout history the Earth's climate has seen changes at various scales; local, regional and global. It is expected to continue changing and this changes are being exuberated by anthropogenic activities such as burning of fossil fuels which have been documented to result in global warming (HPRCC 2013; Bathke et al. 2014). Nebraska with its continental climate, experiences a lot of variability in its climate from year to year. Long term historical records for Nebraska, prove that average annual temperatures have been changing over time and that the annual temperature has risen by about 0.6 °C (HPRCC 2013).

It is projected that by the end of this century with the range of representative concentration pathway scenarios (low to high), Nebraska's temperature is projected to increase from 2.22–2.78 to 4.44–5.00 °C. Additionally, the number of days above 55.6 °C (temperature stress days) is projected to increase by 13–15 to 22–25 additional days over the lower to higher spectrum of emissions (Bathke et al. 2014;Wilhite 2014). Occurrences of high temperatures will "become typical" (Wilhite 2014) by the middle of this century and the "number of warm nights" (Bathke et al. 2014; Wilhite 2014) is to be expected. The probability of frost during the growing season will reduce and the growing season will increase by approximately 2 weeks (Bathke et al. 2014; Wilhite 2014). With regard to rainfall, annual precipitation is projected to remain unchanged however, in the summer, rainfall is expected to decrease. The frequency and severity of droughts is expected to increase with increasing temperatures. For instance, in 2003 and 2012, Nebraska experienced drought during the growing season (April–October). During those years, increased water abstraction from the Ogallala aquifer for the purposes of irrigation, increased (Hornbeck and Keskin 2014). With these climatic projections and trends in mind, pragmatic decision-making among food producers will require the inclusion of the aforementioned climate predictions to assess the impacts of projected ambient CO_2 increments. Reliable weather predictions using both regional and land surface models, are therefore very essential in a bid to adapt to climate variability and change.

Case Study of Nebraska

In order to compare and evaluate the two Land Surface Models (LSMs) coupled to a regional climate model, a region centered on the state of Nebraska was selected (Fig. 2.1). Seven of Nebraska's weather stations with long historical records of ground truth data were used for point weather data evaluations. These stations are shown in Fig. 2.1. The 3 years selected for the LSM comparison studies included: 2002, 2005 and 2008 which were dry, average and wet respectively. The level of wetness was based on statistical long-term historical HPRCC weather data.

The Precipitation-elevation Regressions on Independent Slopes Model (PRISM) daily estimated rainfall amounts (Daly et al. 1994) were utilized to evaluate spatial and temporal rainfall patterns. These datasets are provided at approximately 4.4 km spatial resolution gridded datasets and have been developed by scientists at the Spatial Climate Analysis Service of Oregon State University. They are available online at http:www.ocs.orst.edu/prism/docs/meta/. Daly et al. (1994) employed a statistical topographic-precipitation relationship to interpolate station observations and fill in rainfall distribution data for areas whose terrain is intricate.

Weather Research Forecast (WRF) runs were conducted for April through October for each of the individual 3 years. A horizontal grid size resolution of

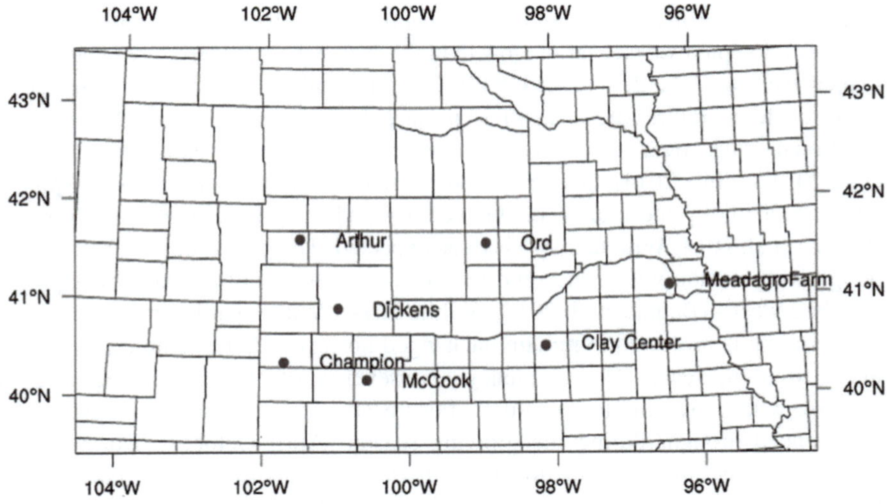

Fig. 2.1 Seven automated weather data network stations selected for evaluation of WRF-Noah and WRF-CLM3.5 weather prediction capabilities. *Source*: Author's figure developed using National Center for Atmospheric Research (NCAR) Command Language (NCL)

12 km and 27 vertical sigma levels were used in the runs. NCEP North American Regional Reanalysis (NARR) ds608.0 (https://rda.ucar.edu/) data, at 32 km horizontal resolution, were used for both lateral and lower boundary and initial conditions. The physics options that were applied for both the LSMs; CLM3.5 and Noah runs were similar apart from the number of soil layers and the surface layer option. For the CLM3.5 land surface model, 10 soil layers were included in the simulation while in the Noah runs 4 soil layers were simulated. The Noah land-surface model was represented using option 2 or the unified Noah land-surface model while option 5 was used to represent the CLM3.5 land surface model.

Both models used the WSM 5-class scheme (Hong et al. 2004) as the preferred microphysics option to estimate surface rainfall employing both its atmospheric moisture and heat tendencies. The shortwave radiation option chosen was that developed by Dudhia (1989) to estimate amount of energy absorbed, scattered and reflected from the surface relative to the cloud cover, vegetation, land surface characteristics such as albedo. The Rapid Radiative Transfer Model (RRTM) described longwave radiation transfer in the atmosphere to and from the earth's surface (Mlawer et al. 1997). The Monin–Obukhov surface layer scheme with its universal stability correction was selected for momentum, heat and moisture flux estimates. It was linked to the Yonsei University (YSU) boundary layer scheme that has an explicit entrainment layer that estimates transportation of mass, moisture, and energy. The new version of the Kain–Fritsch Scheme (tested in the Eta model) was selected for estimations in cloud formation, heat redistribution and precipitation estimations.

WRF Model

The Weather Research and Forecasting (WRF) Model, a mesoscale numerical weather prediction system, provides both operational forecasts and atmospheric research requirements (Skamarock et al. 2008). It shares several features with global climate models with respect to parameterizations of physics and dynamics. The main difference between GCMs and Regional Climate Models (RCMs) is the spatial and temporal resolutions at which they operate (smaller time steps and smaller grid point spacing for RCM). RCMs need to assimilate initial conditions and lateral boundary from reanalysis and/or GCMs (Evans et al. 2005). An essential feature of a regional climate model is the need to simulate land surface—atmosphere fluxes of energy, moisture, and momentum. This is typically handled via a Land Surface Model (LSM) component. WRF provides several LSM options. Available LSMs differ in their degree of complexity in estimating moisture and heat fluxes in various layers of the soil and in their "vegetation, root, and canopy effects and surface snow-cover predictions" (Skamarock et al. 2008, p. 73). The two specific ones evaluated in this study are described below.

Noah Land Surface Model

The Noah Scheme is one of the 'second generation' LSMs of the Advanced Research WRF (ARW) GCM that relies on both soil and vegetation processes for water budgets and surface energy closures (Wei et al. 2009). The model has evolved from the original Oregon State University (OSU) Land Model that was created in the 1980s (Mahrt and Pan 1984). It can simulate soil and land surface temperature, snow depth and snow water equivalent, both water and energy fluxes among others (Chen and Dudhia 2001; Ek et al. 2003; Feng et al. 2008). The model has four distinct soil layers (0.1, 0.3, 0.6 and 1.0 m) that reach a total depth of 2 m and one vegetation canopy layer. The Noah Scheme, which is commonly incorporated in WRF, utilizes the Penman equation to estimate potential evapotranspiration (PET). It has 16 soil and vegetation parameters that are employed to estimate soil temperature, soil moisture, snow cover and atmospheric feedbacks (Evans et al. 2005). In Noah; snow, vegetation and soil are all modeled as a single unit (Slater et al. 2007) over the whole grid box.

Community Land Model (Version 3.5): CLM3.5

The CLM3.5 is a sub-global vegetation land surface model (Collins et al. 2006) developed by the National Center for Atmospheric Research (NCAR) to serve as its Community Climate System Model (CCSM). It is a 'third generation' model and incorporates the influence of both nitrogen and carbon in the computations of water and energy fluxes. It was improved from the NCAR Community Land Model version 3 (CLM3) by adopting a sophisticated surface albedo scheme (Dickinson et al. 2006; Jin and Miller 2010) and enhancing its terrestrial water cycle (Oleson et al. 2008; Stöckli et al. 2008). The CLM3.5 improves the characterization of the land surface by subdividing each CLM3 cell into 8 sub-cells, thereby improving the accuracy of water and energy flux estimations between the land surface and atmosphere. Twenty-four land cover types and 10 soil layers are employed within the CLM3.5. Additionally cropped lands are characterized by their leaf area index, vegetation fraction and roughness height (Kueppers et al. 2008). The current vegetation dataset applied in CLM3.5 is based on a remotely sensed fractional vegetation cover dataset which is comprised of seven primary plant functional types (Bonan et al. 2002).

As this paper goes into publication, it is important to note that a new 'official' release of WRF3.5 is coupled to the newly released CLM4.0 (Kluzek 2013).

Results

Maximum and Minimum Temperature

The highest average temperature over the 2002, 2005 and 2008 Growing Seasons (GS) occurred during 2005 (Table 2.1). The lowest average GS temperatures recorded over the three study years 2002, 2005 and 2008 occurred in 2002. Minimum temperatures ranged between 280.6 and 285.2 K over the duration of the study (April–October) for all seven stations. In 2005, McCook, located in the south-western part of the state recorded the highest average GS temperature of 300.0 K while Arthur at the highest elevation recorded the lowest average maximum temperature (296.9 K) among the seven stations. During the year 2008; Arthur, Champion, Dickens, MeadagroFarm, Ord and Clay Center reported lower temperatures (0.56–1.29 K) than the 30-year climatological temperature recorded (source: hprcc.unl.edu Accessed 27th June 2013).

Precipitation

The year 2002 was a drought year in Nebraska, especially in the western parts. The average growing season (GS) rainfall for the seven stations was 318 mm in 2002. The year 2005 was moderate GS precipitation (467 mm) (Table 2.1) while the year 2008 received the highest amounts of GS precipitation (above normal—611 mm). The only CLM prediction that stood out conspicuously was in Champion in 2005 where the WRF-CLM3.5 prediction was about 260 mm above the actual observation while the WRF-Noah prediction stood at about 141 mm above the ground truth measurements. Apart from this incidence, WRF-CLM35 predicted rainfall totals compare much better to station observations than WRF-Noah. The largest over predictions by the Noah-WRF model occurred in 2005 for Clay Center (471 mm), Meadagrofarm (813 mm) and McCook (331 mm). WRF-CLM performed better with total rainfall predictions for Clay Center (+354 mm), Meadagrofarm (+492 mm) and McCook (+236 mm) above the observed values. The only significant rainfall total under-prediction by CLM and Noah LSMs occurred at Dickens Station in 2008.

Grid point precipitation estimate totals of the June, July and August [JJA] totals from both the WRF-CLM3.5 and WRF-Noah coupled models were compared to those from PRISM seasonal totals. Figure 2.2 illustrates the relative differences between WRF-Land Surface Model and PRISM observations for the years under study. WRF-CLM3.5 total GS rainfall predictions were lower than those of the WRF-Noah predictions. Over-predictions of about 2.5-fold, were generally common in the southeastern lower-elevation areas of Nebraska. However, the level of over-prediction was both quantitatively larger and spatially extended for the

Table 2.1 Precipitation, minimum and maximum temperature observations over the 2002, 2005, 2008 growing seasons (1st April–31st October) for Arthur, Champion, Clay Center, Dickens, McCook, MeadagroFarm, and Ord, Nebraska

		Actual observations								
		Precipitation (mm)			Minimum temperature degrees Kelvin			Maximum temperature degrees Kelvin		
Station		2002	2005	2008	2002	2005	2008	2002	2005	2008
Arthur	Sum	262.000	407.714	412.289						
	Average	1.224	1.905	1.927	281.480	281.645	280.807	296.540	296.885	295.937
Champion	Sum	193.534	420.520	446.278						
	Average	0.904	1.965	2.085	281.575	281.279	280.623	298.786	298.865	297.376
ClayCenter	Sum	393.000	381.496	698.902						
	Average	1.836	1.783	3.266	283.957	284.511	283.198	297.821	298.987	296.626
Dickens	Sum	191.754	560.846	624.080						
	Average	0.896	2.621	2.916	282.296	282.170	281.227	298.767	298.338	296.849
McCook	Sum	300.485	428.996	571.244						
	Average	1.404	2.005	2.669	284.046	283.937	282.716	299.770	300.019	298.355
Me adagro Farm	Sum	491.992	467.553	865.937						
	Average	2.299	2.185	4.046	284.712	285.220	284.023	298.447	299.317	297.291
Ord	Sum	393.000	578.152	659.156						
	Average	1.836	2.702	3.080	283.365	283.557	282.346	297.481	298.134	296.389
Average		317.966	463.611	611.127	283.062	283.189	282.134	298.230	298.648	296.975

Notes: *Data source*: Daily precipitation, minimum and maximum temperatures for each station were obtained from the High Plains Regional Climate Center (HPRCC) www.hprcc.unl.edu. *Table source*: Sum and average station weather values for the growing season (April–October) were calculated by author

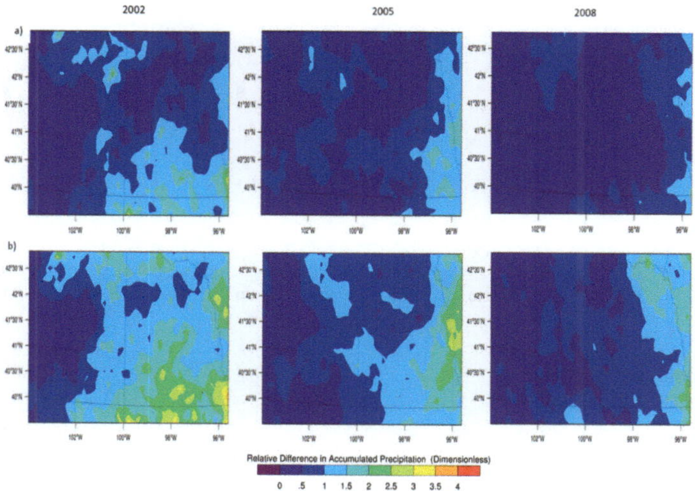

Fig. 2.2 Relative difference of (**a**) WRF-CLM and (**b**) WRF-Noah to Precipitation-elevation Regressions on Independent Slopes Model (PRISM) seasonal growing total precipitation (mm) over Nebraska during a dry year (2002), moderate year (2005) and wet year (2008). *Source* of observed data: Observed precipitation values from the Oregon State University, PRISM www.prism.oregonstate.edu were downloaded and summed for the growing season (April–October). *Source of graphics*: Relative differences were calculated and graphic visualizations were conducted by author using National Center for Atmospheric Research (NCAR) NCL

WRF-Noah precipitation model prediction results as compared to WRF-CLM3.5 precipitation totals and daily station observations.

Verification of Temporal and Spatial Distribution of WRF-LSM Coupled Temperature and Precipitation

The standard error of estimate (STEYX) associated with utilizing weather variables from a reference site to estimate data for adjacent sites was used to compare the corresponding ability of the WRF-CLM3.5 and CLM-Noah models ability to mimic the spatial structure of observed weather data. Using Champion as the reference site, the STEYX for precipitation, maximum and minimum temperature for the adjacent stations were calculated and plotted against the distance between Champion and the each of the other six stations (Fig. 2.3). STEYX increased with distance as expected. However, both the model results exhibited lower STEYX for maximum and minimum temperatures and did not adequately mimic the observed spatial variability resulting from non-uniform terrain, varied land use types and management; that may have resulted in complex atmospheric conditions on the ground.

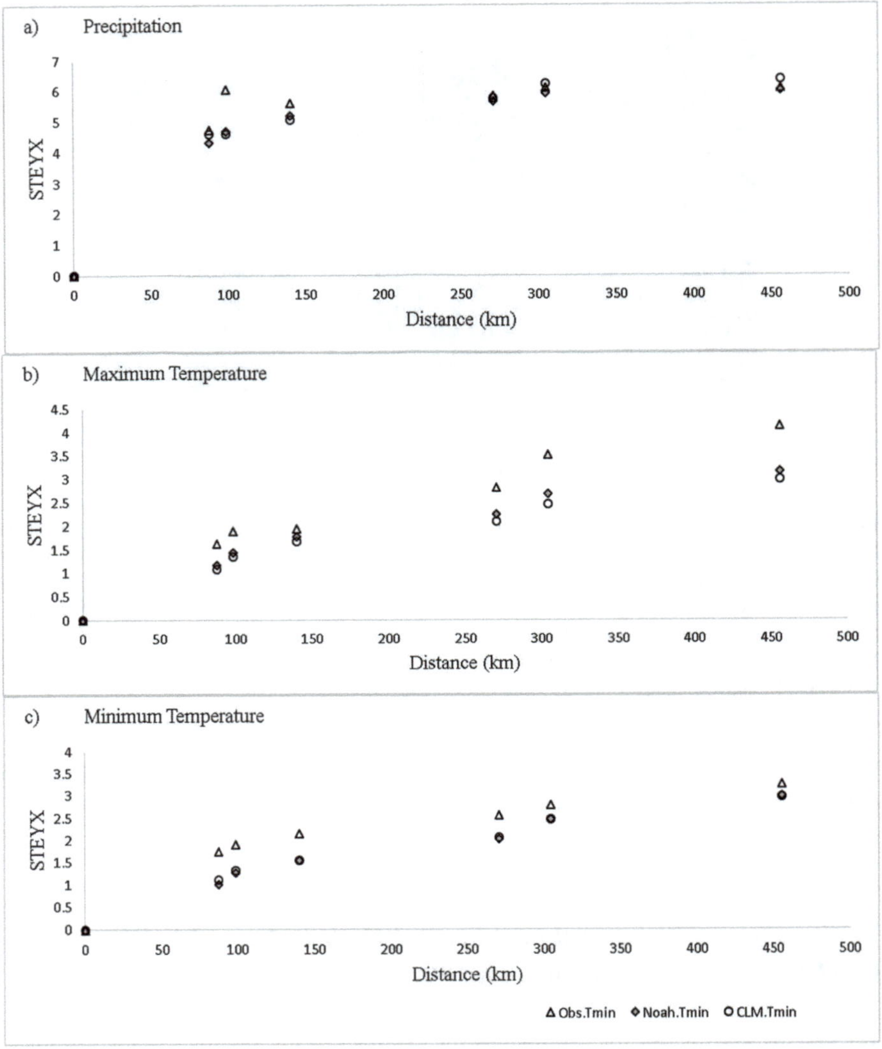

Fig. 2.3 Comparison of errogram for observed, WRF-CLM3.5 and WRF-Noah's precipitation, minimum and maximum temperature observations over the 2002, 2005, 2008 growing seasons (1st April–31st October) for Arthur, Champion, Clay Center, Dickens, McCook, MeadagroFarm, and Ord, Nebraska. Source: Authors' calculations

Discussion

Generally, there was a high correlation (>0.88) between the observed historical recorded temperature values and modeled predictions from both WRF-Noah and WRF-CLM3.5 for all the seven sites. However, WRF-CLM3.5 was always superior in predicting both daily maximum and minimum temperatures over the entire growing season (GS) for all the weather stations with an average root mean square difference (RMSD) of 3.55 K as compared to RMSD of 4.14 K for WRF-Noah.

Model predictions of maximum temperature tended to be more accurate during the summer months of June, July and August when the atmosphere is more homogenous, with minimal occurrences of cold fronts. It was also noticeable when comparing monthly averages, that model predictions of minimum temperature were noticeably most accurate (for both models) in the months of May and October (data not shown here). However the WRF-Noah minimum temperature estimates were consistently higher than WRF-CLM3.5 and observed values for all weather stations. Overall, the models performed better at predicting maximum temperatures than minimum temperatures. However, WRF-CLM3.5 was more accurate than WRF-Noah in both minimum and maximum temperature predictions as depicted by higher correlations and lower RMSD values when compared to actual values.

Generally, both LSMs over-predicted rainfall. WRF-CLM3.5 rainfall predictions, however, were closer to actual ground truth observations and PRISM estimates. Nevertheless, better rainfall predictions were realized during the months of April and May when convective (parameterized) precipitation is less important. Duffy et al. (2003), *as cited in* Caldwell (2010), likewise noted that during the fall and winter precipitation, predictions improved when convective precipitation was of less importance. According to other regional climate model studies [such as Done et al. (2005)], predicting warm season rainfall in continental regions is much harder over the summer than during cooler times of the year. Done et al. (2005) simulated warm season rainfall using WRF and determined that "*the longer-timescale feedback mechanisms are not being represented accurately in climate simulations*". Among candidate mechanisms that they recommended for further testing was convective cloud-radiation feedback (Done et al. 2005). The results of this study likewise demonstrate that precipitation estimates became more variable for both land surface models during the months of June, July and August (data not shown here).

The methods used by Regional Climate Models (RCMs) to generate precipitation are affected by boundary conditions and the model physics are very complicated and far from perfect. For example, other studies such as that conducted by Davis et al. (2006), concluded that WRF rain errors "suffer from a positive size bias that maximizes during the later afternoon". Additionally, WRF-land surface models "dramatically overestimated" precipitation (Jin et al. 2010) in the western United States. The usefulness or utility of precipitation estimates from (RCMs) within crop growth models is hampered by the unrealistic intensity and frequency distributions of precipitation. In order to utilize data from RCMs, rainfall predictions need to be adjusted or corrected for biases. If corrected values are as close to reality as possible, there is promise for applying data from RCMs in crop yield simulation runs to make predictions into the future of agricultural production. The daily variations of rainfall affect crop growth significantly and crop growth simulations will only be as accurate as the input weather variables that drive the crop growth models.

This study also highlights the fact that even with perfect models, the nature of nonlinear atmospheric processes and initial boundary conditions have a large part to play in the data generated by the climate model. Inherent systematic biases exist within WRF model. The complexity of the land surfaces and changes in land use; are not adequately represented at the coarse spatial resolution of the models. The computations conducted by the models do not give accurate estimates of complex biophysical processes.

Conclusion

The study herein examined two land surface models (Noah and CLM3.5) coupled to a regional climate model, namely, WRF. Initial, lateral and boundary conditions were similar. What followed was the selection of an LSM scheme. The study did not examine any internal errors or biases that the regional climate model may have through its model physics.

Both LSMs performed well in predicting the maximum and minimum temperatures in 2002, 2005 and 2008. Generally, there was a high correlation (>0.88) between the observed historical temperature values and modeled predictions from both WRF-Noah and WRF-CLM3.5 for all the seven stations. However, WRF-CLM3.5 was always superior in predicting temperature as demonstrated by the lower standard errors over the entire growing season (GS) for all the weather stations. WRF-Noah minimum temperature estimates in particular were consistently higher than WRF-CLM3.5. Rainfall predictions by both models were not as reliable, based on evaluation for individual stations as well as spatially (state-wide). Both WRF-Noah and WRF-CLM3.5 models over predicted rainfall spatially and temporally. Generally, WRF-Land Surface model precipitation prediction skills tended to be lower in the south-eastern parts of the state. The systematic errors within the WRF model's convective schemes require more research.

From the overall comparisons of temperature and rainfall weather variables (results above), we are able to determine that coupling WRF to the CLM3.5 produces results or predictions that are more accurate than those of the WRF-Noah combination which is attributed to better soil moisture parameterizations within CLM3.5. Closer observations at specific monthly standard errors may help pinpoint areas of weakness within model computations, internal WRF model error biases and sensitivities of model parameterizations. It is envisioned that further comparisons with surface and atmospheric observations will guide the formation and revision of algorithms that reduce biases thereby improving the quality of global and regional climate models in the future.

Acknowledgments We would like to acknowledge the use datasets from the High Plains Regional Climate Center weather, NCEP North American Regional Reanalysis datasets and Oregon's State University's PRISM dataset. Appreciation is also accorded to all the technical staff and support personnel at NCAR who revise and respond to questions regarding NCL. The authors would also like to thank the anonymous reviewers who critiqued the manuscript to improve it.

References

Bathke DJ, Oglesby RJ, Rowe CM, Wilhite DA (2014) Understanding and assessing climate change University of Nebraska–Lincoln implications for Nebraska. A synthesis report to support decision making and natural resource management in a changing climate. Available at http://snr.unl.edu/download/research/projects/climateimpacts/2014ClimateChange.pdf. Accessed 22 Mar 2014

Bonan GB, Levis S, Kergoat L, Oleson KW (2002) Landscapes as patches of plant functional types: an integrating concept for climate and ecosystem models. Glob Biogeochem Cycles 16. doi:10.1029/2000GB001360

Brown RA, Rosenberg NJ, Hays CJ, Easterling WE, Mearns LO (2000) Potential production and environmental effects of switchgrass and traditional crops under current and greenhouse-altered climate in the central United States: a simulation study. Agric Ecosyst Environ 78:31–47

Caldwell P (2010) California wintertime precipitation bias in regional and global climate models. J Appl Meteorol Climatol 49(10):2147–2158

Chen F, Dudhia J (2001) Coupling an advanced land-surface/hydrology model with the Penn State/ NCAR MM5 modeling system. Part I: model description and implementation. Mon Weather Rev 129:569–585

Collins WD, Bitz CM, Blackmon ML, Bonan GB, Bretherton CS, Carton JA, Smith RD (2006) The community climate system model version 3 (CCSM3). J Climate 19(11):2122–2143

Daly C, Neilson RP, Phillips DL (1994) A statistical-topographic model for mapping climatological precipitation over mountainous terrain. J Appl Meteorol 33:140–158

Davis C, Brown B, Bullock R (2006) Object-based verification of precipitation forecasts. Part I: Methodology and application to mesoscale rain areas. Mon Weather Rev 134:1772–1784

Dickinson RE, Oleson KW, Bonan G, Hoffman F, Thornton P, Vertenstein M, Yang Z-L, Zeng X (2006) The community land model and its climate statistics as a component of the community climate system model. J Climate 19:2302–2324

Done JM, Leung LR, Davis CA, Kuo B (2005) Simulation of warm season rainfall using WRF regional climate model. In: 6th WRF/15th MM5 users' workshop, Boulder, CO, USA, June 2005

Dudhia J (1989) Numerical study of convection observed during the winter monsoon experiment using a mesoscale two-dimensional model. J Atmos Sci 46(20):3077–3107

Duffy PB, Govindasamy B, Taylor K, Wehner M, Lamont A, Thompson S (2003) High resolution simulations of global climate. Part I: Present climate. Clim Dyn 21:371–390

Easterling WE, Mearns LO, Hays CJ, Marx D (2001) Comparison of agricultural impacts of climate change calculated from high and low resolution climate change scenarios: part II. Accounting for adaptation and CO_2 direct effects. Clim Change 51(2):173–197

Ek MB, Mitchell KE, Lin Y, Rogers E, Grunmann P, Koren V, Gayno G, Tarpley JD (2003) Implementation of Noah land surface model advances in the National Centers for Environmental Prediction operational mesoscale Eta model. J Geophys Res 108(D22):8851. doi:10.1029/2002JD003296

Evans JP, Oglesby RJ, Lapenta WM (2005) Time series analysis of regional climate model performance. J Geophys Res 110(D4):DO4104. doi:10.1029/2004JD005406

Feng X, Sahoo A, Arsenault K, Houser P, Luo Y, Troy T (2008) The impact of snow model complexity at three CLPX sites. J Hydrometeorol 9:1464–1481

High Plains Regional Climate Center (HPRCC) (2013) Climate change on the Prairie: a basic guide to climate change in the high plains region—UPDATE. Available at http://www.hprcc.unl.edu/publications/files/HighPlainsClimateChangeGuide-2013.pdf. Accessed 16 Jun 2014

Hong SY, Dudhia J, Chen SH (2004) A revised approach to ice microphysical processes for the bulk parameterization of clouds and precipitation. Mon Weather Rev 132(1):103–120

Hornbeck R, Keskin P (2014) The historically evolving impact of the Ogallala aquifer: agricultural adaptation to groundwater and drought. Am Econ J Appl Econ 6(1):190–219

Jin J, Miller NL (2010) Improvement of snowpack simulations in a regional climate model. Hydrol Process. doi:10.1002/hyp.7975

Jin J, Miller NL, Schlegel N (2010) Sensitivity study of four land surface schemes in the WRF model. Adv Meteorol

Kluzek E (2013) CCSM research tools: CLM4.0 user's guide documentation. Available at http://www.cesm.ucar.edu/models/cesm1.0/clm/models/lnd/clm/doc/UsersGuide/book1.html. Accessed 15 Feb 2014

Ko J, Ahuja LR, Kimball BA, Anapalli S, Ma L, Green TR, Ruane A, Wall GW, Pinter PJ Jr, Bader D (2010) Simulation of free air CO_2 enriched wheat growth and interaction with water, nitrogen, and temperature. Agric For Meteorol 150:1331–1346

Kueppers LM, Snyder MA, Sloan LC, Cayan D, Jin J, Kanamaru H, Kanamitsu M, Miller NL, Tyree M, Du H, Weare B (2008) Seasonal temperature responses to land-use change in the western United States. Global Planet Change 60:250–264

Mahrt L, Pan H (1984) A two-layer model of soil hydrology. Bound-Lay Meteorol 29(1):1–20

Mearns LO, Easterling W, Hays C, Marx D (2001) Comparison of agricultural impacts of climate change calculated from high and low resolution climate change scenarios: part I. the uncertainty due to spatial scale. Clim Change 51(2):131–172

Mlawer EJ, Taubman SJ, Brown PD, Iacono MJ, Clough SA (1997) Radiative transfer for inhomogeneous atmospheres: RRTM, a validated correlated-k model for the longwave. J Geophys Res Atmos 102(D14):16663–16682

Niu X, Easterling W, Hays CJ, Jacobs A, Mearns L (2009) Reliability and input-data induced uncertainty of the EPIC model to estimate climate change on sorghum yields in the U.W. Great Plains. Agric Ecosyst Environ 129(1–3):268–276

Oleson KW, Niu GY, Yang ZL, Lawrence DM, Thornton PE, Lawrence PJ et al (2008) Improvements to the community land model and their impact on the hydrological cycle. J Geophys Res Biogeosci (2005–2012) 113(G1)

Pitman AJ, Henderson-Sellers A, Desborough CE, Yang Z-L, Abramopoulos F, Boone A, Dickinson RE, Gedney N, Koster R, Kowalczyk E, Lettenmaier D, Liang X, Mahfouf J-F, Noilhan J, Polcher J, Qu W, Robock A, Rosenzweig C, Schlosser CA, Shmakin AB, Smith J, Suarez M, Verseghy D, Wetzel P, Wood E, Xue Y (1999) Key results and implications from phase 1(c) of the project for intercomparison of land-surface parameterization schemes. Climate Dynam 15:673–684

Skamarock WC, Klemp JB, Dudhia J, Gill DO, Barker DM, Duda MG, Huang X-Y, Wang W, Powers JG (2008) A description of the advanced research WRF version 3. NCAR Technical note NCAR/TN-475 + STR, June 2008

Slater AG, Bohn TJ, McCreight JL, Serreze MC, Lettenmaier DP (2007) A multimodel simulation of pan-Arctic hydrology. J Geophys Res 112:G04S45. doi:10.1029/2006JG000303

Stöckli R, Lawrence DM, Niu G-Y, Oleson KW, Thornton PE, Yang Z-L, Bonan GB, Denning AS, Running SW (2008) Use of FLUXNET in the community land model development. J Geophys Res 113:G01025. doi:10.1029/2007JG000562

Wei J, Dirmeyer PA, Guo Z, Zhang L, Misra V (2009) How much do different land models matter for climate simulation? Part 1: climatology and variability. J Climate 23:3120–3134

Wilhite DA (2014) Understanding and assessing climate change: implications for Nebraska. Heurmann Lecture. September 25, 3:30 p.m. Available at http://snr.unl.edu/download/research/projects/climateimpacts/HeuermannLecturePowerPoint_ClimateChangeImplicationsforNebraska.pdf. Accessed 22 Mar 2015

Chapter 3
Climate Change Frames in Public Health and Water Resource Management: Towards Intersectoral Climate Change Adaptation

Lindsay P. Galway, Margot W. Parkes, Kitty K. Corbett, Diana M. Allen, and Timothy K. Takaro

Abstract Effective and appropriate climate change adaptation requires a greater understanding and appreciation of the diverse ways in which the issue of climate change is constructed and understood. The ways in which an issue is framed should not be overlooked in interdisciplinary and intersectoral efforts given that implicit and divergent frames often impede the processes of knowledge integration and collaboration and therefore, can hinder adaptation processes. This study used frame analysis to identify and summarize the climate change frames in public health and water resource management texts. Five frames emerged from the analysis of the public health texts: *Preventing direct and indirect health impacts, promoting health and sustainability, climate change as a complex problem, strengthening the evidence base, and health equity in a changing climate.* Three frames emerged from the analysis of water resource management texts: *planning and decision-making under uncertainty, managing multiple drivers of water insecurity, and understanding impacts on complex systems.* Drawing on insights from this work, we assert that the notion of frames and the process of frame-reflection are useful tools to foster integration and intersectoral collaboration and an opportunity to foster enabling conditions for climate change adaptation.

L.P. Galway (✉)
Department of Health Sciences, Lakehead University, 955 Oliver Rd., Thunder Bay, ON, Canada
e-mail: lgalway@lakeheadu.ca

M.W. Parkes
School of Health Sciences, University of Northern British Columbia, Prince George, BC, Canada

K.K. Corbett
School of Public Health and Health Systems, University of Waterloo, Waterloo, ON, Canada

D.M. Allen
Department of Earth Sciences, Simon Fraser University, Burnaby, BC, Canada

T.K. Takaro
Faculty of Health Sciences, Simon Fraser University, 8888 University Dr., Burnaby, BC, Canada

W.L. Filho et al. (eds.), *Climate Change Adaptation, Resilience and Hazards*, Climate Change Management, DOI 10.1007/978-3-319-39880-8_3

Keywords Climate change • Adaptation • Framing • Public health • Water resource management • Interdisciplinary • Intersectoral collaboration

Introduction

There is now widespread consensus that adaptation to the current and future impacts of climate change is necessary (Field et al. 2014). Despite many compelling reasons for climate change adaptation, we currently face a disconcerting "adaptation deficit" (Burton 2004); a vast gap between needed action and the extent to which we are actually taking action to adapt to climate change. Over the last decade, a body of literature on the constraints for climate change policy and action has emerged. To date, most research has focused on informational and technological constraints, and our understanding of enabling conditions for climate change adaptation is generally lacking (Kiem and Austin 2013). Recognizing that adaptation is a social process over and above being a technical challenge underscores the need to address constraints with social and governance dimensions and focus on enabling conditions (Adger 2003; Field et al. 2014; Moser et al. 2012).

We argue that the challenge of "learning and working together" (Parkes et al. 2012), i.e., interdisciplinary research and intersectoral action, is an often overlooked yet important adaptation constraint. To develop and implement adaptive responses, we must consider how to engage a diversity of actors, integrate knowledge, and foster intersectoral collaboration (Warren and Lemmen 2014). More explicit attention to these interconnected processes is an opportunity to foster enabling conditions for adaptation. This in turn calls for a greater understanding of the diverse ways in which the problem of climate change itself is conceptualised and communicated across various divides. Actors from different sectors, disciplines, and perspectives construct and understand issues in different ways; they frame issues differently (Dewulf et al. 2007; Schön and Rein 1994). For example, public health professionals understand issues through public health lenses, and resource management professionals understand the same issues using their own theories, methods, and vocabularies. By examining the various ways in which the issue of climate change is framed across disciplinary and sectoral divides, we can develop tools and processes that promote knowledge integration and intersectoral collaboration and contribute towards enabling conditions for adaptation.

Against this backdrop, this research is guided by the question: '*How is the issue of climate change framed in the public health and water resource management sectors?*' Using the method of frame analysis, we analyze and summarize the various ways in which climate change is constructed and understood within two sectors that are central to climate change adaptation. Our intention is to contribute to the emerging literature and dialogue on enabling conditions for adaptation while also highlighting the potential utility of frames and frame-reflection for learning and working together to address the current adaptation deficit.

A Brief Background on Frames and Frame Analysis

There are various interpretations of the notion of frame and different approaches to frame analysis. In his influential paper *Framing: Towards Clarification of a Fractured Paradigm,* Entman (1993) aimed to "identify and make explicit common tendencies among the various uses of the terms [frames and framing] and to suggest a more precise and universal understanding of them" (Entman 1993). Entman offers the following definition; "to frame is *to select some aspects of a perceived reality and make them more salient in a communicating text, in such a way as to promote a particular problem definition, causal interpretation, moral evaluation, and/or treatment recommendation* for an item described (original italics)" (Entman 1993). Adapting Entman's work, Benford and Snow (2000) present core framing tasks: diagnostic framing, prognostic framing, and motivational framing. Diagnostic framing refers to problem identification, prognostic framing answers the question, 'what can and should be done?' and articulates solutions while motivational framing refers to the moral appeals and the rationale for addressing the issue. In addition to framing tasks, Van Gorp and Vercruysse (2012) highlight the role of framing devices including metaphors and exemplars. Frames thus are distinct combinations of diagnostic, prognostic, and motivational framing and the use of particular framing devices that together depict the distinct ways in which an issue is constructed and understood within written or spoken text. Since frames are not made explicit in communication, they must be interpreted through the analysis of language using frame analysis (O'Brien et al. 2007; Schön and Rein 1994). We draw primarily on the theoretical work of Entman (1993, 2000) and Benford and Snow (2000) to inform an analytical procedure utilizing prognostic framing, diagnostic framing, motivational framing, and framing devices as frame-signifying dimensions (herein referred to as frame dimensions).

Methods

The public health and water resource management texts were selected using a purposive sampling strategy (public health N = 30; water resource management N = 25). The first phase of the sampling strategy consisted of a search of the Web of Science database. The search consisted of the following terms within the "title" search field: "climat* change" or "climat* variability". The search results were then limited to the "public environmental occupational health" research area to select from public health literature and "water resources" research area to select from the water management literature. Next, the search was refined to include only texts published in the English language between 2007 and 2013 and editorial texts. The period of 2007 and 2013 was selected because we aimed to identify and described current frames of climate change rather than to document changes in frames over time. We focused on editorial texts in the scholarly literature because these texts are more likely to contain elaborated discussions of the issue of climate change, and thus depict the frames of the issue more fully, compared to research articles that tend to be heavily focused on methods (Huttenen and Hilden 2013). The initial selection of texts was screened and those that (a) were not explicitly focused on climate change and health or on climate change and water resources; (b) were short responses to other published work; or (c) were introductions to special journal issues primarily outlining the content of the journal issue were excluded. Although the Web of Science database includes a wide range of journals, the authors decided that the journal EcoHealth was missing from the "public environmental occupational health" research area. Consequently, literature from the journal EcoHealth was manually searched using the same process described above.

In the second phase of the sampling strategy, we identified relevant policy-oriented documents by reviewing the reference list of all texts selected from the scholarly literature. Policy documents from 2003 through to 2007 were selected, again to identify recent frames of climate change. The initial selection of policy-oriented texts was screened and those that were not explicitly focused on climate change and health or on climate change and water resources were excluded.

The frame analysis involved two main steps: immersion and frame identification. Immersion involved "obtaining a sense of the whole" (Tesch 1990) by carefully reading all of the text to become familiar with the climate change-related discourse and the data. To identify frames, NVivo software was used to generate a database containing statements from each text in the sample that illustrated frame dimensions. Sensitizing questions (Verloo and Maloutas 2005) were used to guide the identification of frame dimensions (see Table 3.1). This process was carried out separately for the water resource management and public health sample.

Frames evolved as patterns in the frame dimensions emerged within and across individual texts. Candidate frames were therefore "refined, combined and differentiated" iteratively (Porter and Hulme 2013). Frame summaries were generated to describe the identified frames. Original quotes from the data were used to illustrate frame dimensions. Text between 'quotation marks' is illustrative excerpts taken directly from the data.

Table 3.1 Frame dimensions and sensitizing questions

Frame dimensions			
Diagnostic framing	Prognostic framing	Motivational framing	Framing devices
– What is the nature of the climate change? – What is the climate change problem about? – What aspects of climate change are the primary foci?	– How can/should we respond to climate change? – What should be the outcome of policy and action? – Who should be responsible for responding to climate change?	– Why should we respond to climate change? – What is the rationale/motivation for action?	– What are common metaphors, terms, exemplars etc. used in relation to climate change? – What language characterises the frame?

Findings

Frames outline the nature of a given problem, what should be done, and rationale for action. Our frame analysis identified five climate change frames from the public health sample and three from the water resource management sample (see Figs. 3.1 and 3.2). Frame summaries are presented below.

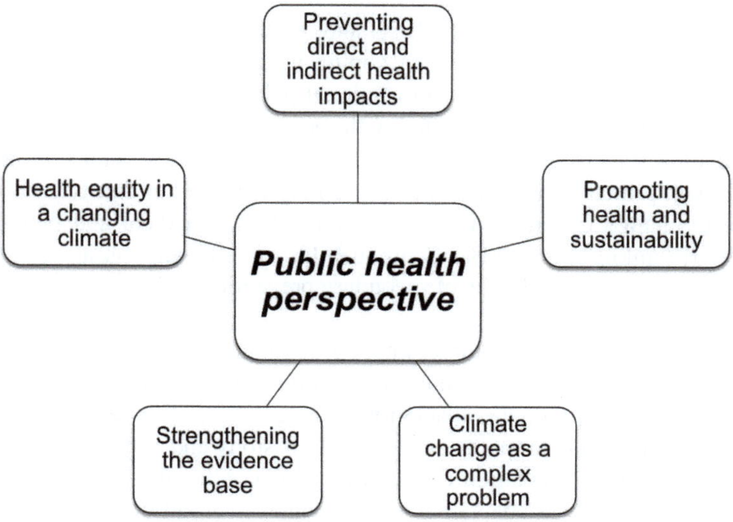

Fig. 3.1 Climate change frames from a public health perspective

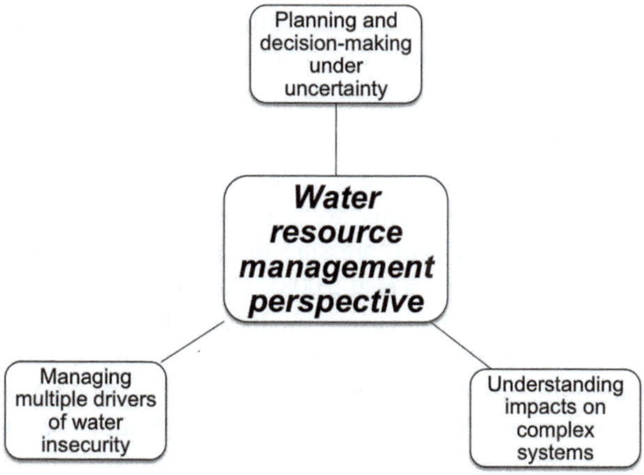

Fig. 3.2 Climate change frames from a water resource management perspective

Climate Change Frames from a Public Health Perspective

Preventing Direct and Indirect Health Impacts Texts depicting this frame emphasize that climate change is making us sick and will have increasingly serious direct and indirect impacts on health. Climate change is viewed as a major public health problem because it exacerbates the burden of disease, particularly among vulnerable populations like children, the elderly, and the chronically ill. The ways in which climate change will impact specific health outcomes and disease is the primary concern. The language of direct and indirect impacts features heavily.

Prognostic framing is focused on coping with the direct and indirect health impacts of climate change using conventional public health prevention strategies. Climate change adaptation is equated with secondary and tertiary prevention (i.e., early diagnosis and controlling the progression of disease), while mitigation is equated with primary prevention (i.e., preventing the onset of disease). Adaptation to climate change is conceptualized as 'conventional medical and public health practice' (Frumkin et al. 2008). Consequently, surveillance, disease control, and monitoring, the traditional suite of secondary prevention activities, are emphasized as adaptation strategies.

Predicting the direct and indirect impacts of climate change on health outcomes is emphasised as the primary means for informing adaptation strategies. The assumption that appropriate adaptation strategies can be identified based on predicted future health impacts underlies discussions and reflects a predict-and-provide approach to adaptation (Wise et al. 2013). Although mitigation does not feature as prominently as adaptation when it comes to prognostic framing, the links between mitigation and health are identified because of the view that mitigation strategies may improve health in the long run via health co-benefits. Finally, the need for 'integrated action… and multisectoral collaboration' (Campbell-Lendrum et al. 2007), in terms of both adaptation and mitigation, is emphasized in this frame.

The primary motivation for action is to protect vulnerable populations from the direct and indirect health impacts of climate change.

Promoting Health and Sustainability Texts representing this frame emphasize that climate change is a problem because it affects the fundamental drivers of health. Healthy and productive ecosystems are seen as critical to overall health and wellness and climate change puts 'pressure on the natural, economic, and social systems that sustain health' (Neira et al. 2008). The consequences of climate change are viewed more broadly than focusing on specific health outcomes. Prognostic framing therefore stresses the need for policy that can 'improve the conditions in which people are born, grow, live and work' (Stott 2012), while protecting ecosystems that support health. Protecting and promoting health *and* sustainability is highlighted as the aim of adaptation. The importance of working across sectors to achieve this dual vision of health and sustainability is recognized. Also, given the view that '[h]ealth professionals and organisations are well placed to help generate a more informed debate and policy response' (Sweet 2011), advocacy and education about the importance of climate change and need to for action are commonly discussed in texts expressing this frame.

Motivational framing is based on the dual vision of human health and sustainability and, in part, a desire to shift climate change-health conversations away from a focus on specific health outcomes and disease prevention towards a broader understanding of climate–health relationships.

Climate Change as a Complex Problem The view that climate change is a complex problem with interrelated implications for ecosystems, health, and social systems characterizes this frame. Climate change is presented as a problem of 'unprecedented scale and complexity' (McMichael and Wilcox 2009) characterized by non-linearity, feedback, and a high degree of interconnectedness in terms of causal pathways. Given this complexity, prognostic framing highlights the need for new modes of thinking and novel approaches to research and action. This is clearly illustrated by Forastiere (2010), who contends we must 'apply new modes of approaching and studying the issue, while continuously searching for collaboration from other disciplines'. Interdisciplinary research and collaborative action are seen as essential tools for working with this complexity, and therefore, are highlighted as a major feature of the climate change response. McMichael and Wilcox (2009) argue that 'we must stop thinking in outmoded differentiated sectoral terms'. Aside from recommendations for systems approaches and collaboration, texts illustrating this frame provide few specific recommendations when it comes to the details and practical aspects of novel ways of thinking and approaches for research and action, raising the need to explore and further develop options for dealing specifically with the complex nature of climate change.

Motivational framing is driven by the limitations of traditional modes of thinking and tools, which are seen as unsuitable for complex issues. Climate change is therefore also presented as an opportunity to develop novel and more effective approaches.

Strengthening the Evidence Base This frame constructs climate change as a problem of insufficient knowledge and evidence regarding the climate change impacts on health. There is a focus on specific health outcomes and climate-sensitive diseases in general. The lack of evidence regarding climate change impacts on health is largely explained by insufficient data and our limited methods to predict future health consequences. Articles demonstrating this frame tend to conceptualize uncertainty as a lack of knowledge: the view that uncertainty can be reduced and eventually eliminated with more data and knowledge features prominently. Prognostic framing revolves around generating more data and evidence that can be delivered to decision makers to develop policy and action. Generating knowledge about extreme weather–disease relationship is presented as a particularly useful means of strengthening the evidence base. Similar to the *Preventing direct and indirect health impacts* frame, a predict-and-provide approach to policy development underlies discussions (Wise et al. 2013).

Motivational framing here is rooted in the belief that health researchers and professionals have a responsibility to 'develop a more comprehensive body of evidence to inform decision makers and policy makers' (Hrynkow 2008).

Health Equity in a Changing Climate In this frame, climate change is seen as a health equity problem because it generates, and will amplify, 'inequalities in health within and across populations' (Künzli et al. 2000). Further, both mitigation and adaptation policy and action 'pose particular challenges for health equity' (Walpole et al. 2009).

Prognostic framing emphasizes that climate change responses should be aimed at reducing inequalities in health and ensuring access to basic human rights such as the right to water and food. Health equity is prioritized as the central goals of climate change policy and action. The texts illustrating this frame do reflect the assumption that evidence about climate change impacts directly leads to appropriate policy and action. Rather, the idea that 'ethical principles' (Singh 2012) are needed to guide adaptation and mitigation emerges. Jensen (2009) calls for the application of a 'climate change health equity filter' to assess any proposed adaptation and mitigation strategies. The assumption that those populations most responsible for climate change (i.e., western industrialized nation-states) should be primarily responsible for the costs of responding to climate change underlies discussions—'justice demands it' (Kiang et al. 2013).

The rationale for action in texts presenting this frame is the inherent injustice of climate change coupled with the mandate of human rights. Motivation stems in part from claims that 'health ethics has been absent in climate change discourse' (Singh 2012) and that ethical considerations should be the central considerations, 'not at the periphery' (Singh 2012).

Climate Change Frames from a Water Resource Management Perspective

Planning and Decision-Making Under Uncertainty In this frame, climate change is understood as a problem of planning and decision-making given the uncertainties surrounding climate change. The main challenge of climate change is 'how to plan for the future under highly uncertain conditions' (Rogers 2008). As Dessai et al. (2013) explains, 'water managers have often planned under the assumption of a stationary climate. This assumption is no longer valid'. We can no longer make management decisions based on the premise that 'future hydrology will not significantly deviate from past hydrology' (Barsugli et al. 2012). Reference to the 'assumption of stationarity' and the fact that stationarity is violated in a changing climate is common and in part underlies the motivation and rationale for action. Additional motivation for action stems from the belief that we need to adapt and make changes 'despite the fact that we have little faith in climate model projections and impact studies' (Dessai et al. 2013). The notion that uncertainty should not be used as an excuse for inaction emerges.

The uncertain nature of climate change impacts on the hydrologic cycle and water resources, particularly at the regional and local level, is highlighted. However, uncertainty here is not only about impacts. Uncertainty is conceptualized as a characteristic of complex problems and is seen as 'irreducible' (Barsugli et al. 2012). Adaptation processes must therefore 'embrace principles of decision making under uncertainty' (Gober 2013). The view that '[t]oo much attention has been focused on reducing, clarifying, and representing climatic uncertainty and too little attention has been directed to building capacity to accommodate uncertainty and change' underlies discussions (Gober 2013). Within this frame, adaptation calls for a shift away from predicting impacts and towards novel approaches to decision-making and governance. Specific recommendations include interdisciplinarity and adaptive co-management.

Managing Multiple Drivers of Water Insecurity The view that water resources are increasingly under pressure from many factors, including but not limited to climate change, characterizes this frame. Drivers of water insecurity include climate change as well as 'land use, aging infrastructure, urbanization, and changing social values.' (Connor et al. 2009). Current technologies, infrastructure, and management practices are insufficient to ensure water security in a rapidly changing world. Discussions of supply and demand are common among the texts demonstrating this frame because 'the supply of and demand for water resources will be substantially affected by climate change' (GWP 2009). Concerns regarding water quantity are at the forefront; discussions about water quality are secondary. Since climate change is seen as one of the many drivers of water insecurity, and because fragmented development across sectors is viewed as part of the problem, integration and intersectoral collaboration are underscored as key features for achieving water security. Integrated water resource management (IWRM) emerges as a key adaptation strategy to manage climate change impacts on water resources. The long-term sustainability of water resources, a goal that can be shared across sectors, is underscored as the overarching goal of climate change policy and action. Additionally, prognostic framing underscores that the drivers of water security vary across settings such that adaptation will need to be 'diverse and locally specific' (IWA 2012).

Motivation for action stems largely from the view that 'the main impacts of climate change on humans and the environment occur through water' (Connor et al. 2009). The management of water resources must therefore feature prominently in the climate change adaptation agenda.

Understanding Impacts on Complex Systems The *Understanding impacts on complex systems* frame is focused on the challenge of understanding and predicting change in the context of complex systems. Texts using this frame argue that we currently have little reliable knowledge of climate change impacts and, perhaps more importantly, also lack adequate understanding of key hydrological processes and relationships under current conditions. There is a high degree of uncertainty with regards to our current understanding of climate change impacts and 'uncertainty tends to increase as one goes down in scale and as one moves to more extreme events' (Blöschl et al. 2007). 'Dependence on local conditions is a distinguishing feature of hydrology that can make the effect of climate change less predictable' (Blöschl and Montanari 2010). The belief that regional climate projections do not adequately capture variability characterises this frame. With regards to regional projections, Beven asks if 'any of this work is fit for the purpose of adapting to, or managing for, the future?' while Blöschl and Montanari (2010) compares impact predictions to 'throwing the dice'.

Prognostic framing therefore is focused, primarily, on improving our 'knowledge of connections among climate, weather, and hydrology under current conditions' (Blöschl and Montanari 2010) and key processes that characterize complex systems. Second, better uncertainty estimation and improvements in modeling to reduce uncertainty are called for. There are parallels with the *Planning and decision-making under uncertainty* frame, uncertainty and complexity feature prominently in both for example. However, the major focus of prognostic framing here is the challenge of generating knowledge rather than the challenge of making decisions due to uncertainty and complexity.

Motivational framing centers on the belief that decision-makers and water managers want more evidence regarding the impacts of climate change; specifically evidence that is not plagued by uncertainty. Finally, climate change is generally presented not only as a significant challenge, but also as an opportunity for change.

Discussion: Frames and Frame-Reflection as Innovative Tools for Climate Change Adaptation

This set of frames is illustrative of the diverse ways in which climate change is constructed and understood across two purposefully selected sectors at a particular point in time (2007–2013). We present it less as an authoritative description than as a heuristic tool, a conversation piece to stimulate intersectoral discussion and action. We recognize that this set of frames is in flux. Environmental problems are continuously re-defined and constructed and frames are not static (Hajer 1995). Finally, although efforts were made to capture a suitable sample of texts for frame identification, it is possible that certain frames have not been adequately captured in the sample analyzed.

De Boer et al. (2010) contends that climate change is an issue that "can be framed and reframed in several ways". Our results show that this is certainly the case with regards to the public health and water resource management sectors. The fact that we identified eight distinct frames, within and across these perspectives, underscores the importance of considering the role of frames in our efforts towards knowledge integration and intersectoral collaboration. The literature increasingly acknowledges that learning and working together in the context of global change necessitates the reconciliation of multiple perspectives (Brown et al. 2010). Tools, mechanisms, and processes that create synergies across disciplines, sectors, and perspectives are lacking and sorely needed. We argue that the construct of frames and the process of frame-reflection (Schön and Rein 1994) could be useful tools to promote learning and working together in the context of climate change adaptation processes.

A plurality of implicit frames can impede mutual understanding and decision-making with regards to complex problems like climate change (Dewulf 2013; Gray 2003; Schön and Rein 1994). However, doing away with framing differences, or establishing that one particular way of understanding climate change is better than another, should not be the goal if we wish to learn and work together toward effective climate change adaptation (Dupuis and Knoepfel 2013; Pahl-Wostl 2006). Instead, we should aim to explore and make explicit the myriad frames and framing differences to better understand and appreciate various perspectives while utilizing points of convergence to achieve integrated understanding and motivate collective action. Schön and Rein (1994) have convincingly argued for situated frame-reflection as a means of navigating the challenge of learning and working together and addressing complex policy problems. We echo Schön and Rein's (1994) call for frame-reflective research, decision-making, and action, and suggest that this may be particularly fruitful for fostering enabling conditions for climate change adaptation. This involves acknowledging, respecting, and valuing diversity in interdisciplinary and collaborative efforts within in a particular context or setting. Engaging in the process of frame-reflection requires a high degree of critical self-reflection. Self-reflection may not come easily to many actors involved in climate change adaptation processes. Cornell (2010) writes, "most physical scientists are not habituated to reflection so... enter interdisciplinary areas unequipped for critical reflection". Building capacity for reflection may need to be purposefully addressed by drawing on specific design tools and frameworks [e.g., (Kolb 1984; Rolfe et al. 2001)].

Conclusion

This paper contributes to the growing body of research addressing the challenge of knowledge integration and collaboration by examining the ways in which climate change is framed across two sectors that are central to addressing the climate change adaptation deficit. Using frame analysis, we summarized the ways in which climate change is constructed within water resource management and public health texts. We argue that frames and framing should not be ignored in interdisciplinary and intersectoral activities given that implicit and divergent frames often underlie the challenge of learning and working together. Acknowledging, appreciating, and reflecting on a diversity of climate change frames could be a simple yet effective means of promoting enabling conditions for climate change adaptation.

References

Adger WN (2003) Social capital, collective action, and adaptation to climate change. Econ Geogr 79:387–404. doi:10.1111/j.1944-8287.2003.tb00220.x

Barsugli J, Vogel J, Kaatz L (2012) Two faces of uncertainty: climate science and water utility planning methods. J Water Resour Plan Manag 138(5):389–395. doi:10.1061/(ASCE)WR.1943-5452.0000188

Benford RD, Snow DA (2000) Framing processes and social movements: an overview and assessment. Annu Rev Sociol 26(1):611–639. doi:10.1146/annurev.soc.26.1.611

Blöschl G, Montanari A (2010) Climate change impacts—throwing the dice? Hydrol Process 24 (3):374–381. doi:10.1002/hyp.7574

Blöschl G, Ardoin-Bardin S, Bonell M (2007) At what scales do climate variability and land cover change impact on flooding and low flows? Hydrol Process 21(9):1241–1247. doi:10.1002/hyp.6669

Brown VA, Harris JA, Russell J (eds) (2010) Tackling wicked problems through the transdisciplinary imagination. Earthscan, London

Burton I (2004) Climate change and the adaptation deficit. In: Schipper E, Burton I (eds) Earthscan reader on adaptation to climate change. Earthscan, London, pp 88–95

Campbell-Lendrum D, Corvalán C, Neira M (2007) Global climate change: implications for international public health policy. Bull World Health Organ 85(3):235–237. doi:10.2471/BLT.06.039503

Connor G, Gallopin G, Hellmuth M, Rast W (2009) Climate change and possible futures. World water development report—water in a changing world. UNESCO, Geneva, pp 68–76

Cornell S (2010) Brokering interdisciplinarity cross the physical and social sciences. In: Bhaskar R (ed) Interdisciplinarity and climate change: transforming knowledge and practice for our global future. Routledge, New York, NY, pp 116–134

De Boer J, Wardekker JA, van der Sluijs JP (2010) Frame-based guide to situated decision-making on climate change. Glob Environ Change 20(3):502–510

Dessai S, Browne A, Harou JJ (2013) Introduction to the special issue on "adaptation and resilience of water systems to an uncertain changing climate." Water Resour Manage 27:943–948

Dewulf A (2013) Contrasting frames in policy debates on climate change adaptation. WIREs Clim Change 4:321–330. doi:10.1002/wcc.277

Dewulf A, Francois G, Pahl-Wostl C (2007) A framing approach to cross-disciplinary research collaboration: experiences from a large-scale research project on adaptive water management. Ecol Soc 12(2):14. http://www.ecologyandsociety.org/vol12/iss2/art14/

Dupuis J, Knoepfel P (2013) The adaptation policy paradox: the implementation deficit of policies framed as climate change adaptation. Ecol Soc 18(4):31. doi:10.5751/ES-05965-180431

Entman R (1993) Framing: toward clarification of a fractured paradigm. J Commun 43(4):51–58. doi:10.1111/j.1460-2466.1993.tb01304.x

Field C, Barros V, Dokken D (2014) Climate change 2014: impacts, adaptation, and vulnerability. Volume I: global and sectoral aspects. Contribution of working group II to the fifth assessment report of the Intergovernmental Panel on Climate Change. Cambridge University Press, Cambridge

Forastiere F (2010) Climate change and health: a challenge for epidemiology and public health. Int J Public Health 55(2):83–84

Frumkin H, Hess J, Luber G, Malilay J, McGeehin M (2008) Climate change: the public health response. Am J Public Health 98(3):435–445. doi:10.2105/AJPH.2007.119362

Gober P (2013) Getting outside the water box: the need for new approaches to water planning and policy. Water Resour Manag 27(4):955–957. doi:10.1007/s11269-012-0222-y

Gray B (2003) Framing in environmental disputes. In: Lewicki R, Gray B, Elliot M (eds) Making sense of intractable environmental conflicts: concepts and cases. Island Press, Washington, DC, pp 11–34

GWP (2009) Perspectives on water and climate change adaptation: introduction, summaries and key messages. GWP, Stockholm

Hajer M (1995) The politics of environmental discourse: ecological modernization and the policy process. Oxford University Press, Oxford

Hrynkow S (2008) Climate change and health research: time for teamwork. Environ Health Perspect 116(11):A470. doi:10.1289/ehp.12150

Huttenen S, Hilden M (2013) Framing the controversial: geoengineering in academic literature. Sci Commun 36(1):3–29. doi:10.1177/0963662513475966

International Water Association (2012) Climate change and the water industry: practical responses and action. Report No.: 10. Co-operative programme on water and climate. International Water Association, London

Jensen G (2009) Taking the reins of the white horse of climate change. J Epidemiol Community Health 63(4):269–270. doi:10.1136/jech.2008.083691

Kiang K, Graham S, Farrant B (2013) Climate change, child health and the role of the paediatric profession in under-resourced settings. Trop Med Int Health 18(9):1053–1056. doi:10.1111/tmi.12153

Kiem A, Austin E (2013) Disconnect between science and end-users as a barrier to climate change adaptation. Climate Res 58:29–41. doi:10.3354/cr01181

Kolb D (1984) Experiential learning: experience as the source of learning and development. Prentice-Hall, London

Künzli N, Kaiser R, Medina S et al (2000) Public-health impact of outdoor and traffic-related air pollution: a European assessment. Lancet 356(9232):795–801. doi:10.1016/S0140-6736(00)02653-2

McMichael AJ, Wilcox B (2009) Climate change, human health, and integrative research: a transformative imperative. Ecohealth 6(2):163–164. doi:10.1073/pnas.1007887107

Moser S, Jeffress Williams S, Boesch D (2012) Wicked challenges at land's end: managing coastal vulnerability under climate change. Annu Rev Environ Resour 37:51–78. doi:10.1146/annurev-environ-021611-135158

Neira M, Bertollini R, Campbell-Lendrum D, Heymann D (2008) The year 2008: a breakthrough year for health protection from climate change? Am J Prev Med 35:424–425. doi:10.1016/j.amepre.2008.08.011

O'Brien K, Eriksen S, Nygaard L, Schjolden A (2007) Why different interpretations of vulnerability matter in climate change discourses. Clim Pol 7(1):73–88. doi:10.1080/14693062.2007.9685639

Pahl-Wostl C (2006) The importance of social learning in restoring the multifunctionality of rivers and floodplains. Ecol Soc 11(1):10. http://www.ecologyandsociety.org/vol11/iss1/art10/

Parkes M, Charron D, Sánchez A (2012) Better together: field-building networks at the frontiers of ecohealth research. In: Charron D (ed) Ecohealth research in practice: innovative applications of an ecosystem approach to health. International Development Research Centre, Ottawa, pp 231–253

Porter K, Hulme M (2013) The emergence of the geoengineering debate in the UK print media: a frame analysis. Geogr J 179(9):342–355. doi:10.1111/geoj.12003

Rogers P (2008) Coping with global warming and climate change. J Water Resour Plan Manag 134 (3):203–204. doi:10.1061/(ASCE)0733-9496(2008)134:3(203)

Rolfe G, Freshwater D, Jasper M (2001) Critical reflection for nursing and the helping professions: a user's guide. Palgrave, New York, NY

Schön D, Rein M (1994) Frame reflection: toward the resolution of intractable policy controversies. Basic Books, New York, NY

Singh J (2012) Why human health and health ethics must be central to climate change deliberations. PLoS Med 9(6):e1001229. doi:10.1371/journal.pmed.1001229

Stott R (2012) Contraction and convergence: the best possible solution to the twin problems of climate change and inequity. BMJ 344:e1765–e1765. doi:10.1136/bmj.e1765

Sweet M (2011) Action on climate change requires strong leadership from the health sector. Health Promot J Austr 22(4):415–417

Tesch R (1990) Qualitative research: analysis types and software tools. Routledge, New York, NY

Van Gorp B, Vercruysse T (2012) Frames and counter-frames giving meaning to dementia: a framing analysis of media content. Soc Sci Med 74(8):1274–1281. doi:10.1016/j.socscimed.2011.12.045

Verloo M, Maloutas M (2005) Differences in the framing of gender inequality as a policy problem across Europe. Greek Re Soc Res 117(B1):3–10

Walpole S, Rasanathan K, Campbell-Lendrum D (2009) Natural and unnatural synergies: climate change policy and health equity. Bull World Health Organ 87(10):799–801. doi:10.2471/BLT.09.067116

Warren F, Lemmen D (eds) (2014) Canada in a changing climate: sector perspectives on impacts and adaptation. Government of Canada, Ottawa. http://www.nrcan.gc.ca/sites/www.nrcan.gc.ca/files/earthsciences/pdf/assess/2014/pdf/Full-Report_Eng.pdf. Accessed 27 Feb 2015

Wise R, Fazey I, Stafford Smith M, Park E, Eakin C, Campbell B (2013) Reconceptualising adaptation to climate change as part of pathways of change and response. Glob Environ Chang 28:325–336. doi:10.1016/j.gloenvcha.2013.12.002

Chapter 4
Rural Women Belief System and Attitude Toward Climate Change Mitigation and Adaptation Strategies in Nigeria

C.A.O. Akinbami, J.E. Olawoye, and F.A. Adesina

Abstract Climate change has affected both the natural and human systems, of which the women in the rural areas and their livelihood practices are the mostly affected. This study was conducted in some selected rural communities of Osun state in Southwest, Nigeria among women involved in different livelihood practices to find out issues about climate change impacts on the rural women such as: Are the rural women aware of climate change and its impacts? How prepared are they for climate change mitigation and adaptation strategies? Are there any socio-cultural barriers to combating climate change? The study therefore focused attention on their beliefs, attitude and perception about climate change. It also discussed the barriers their beliefs and attitude posed to the establishment and implementation of mitigation and adaptation strategies in the rural areas. Focus Group Discussions, in-depth interview and questionnaire were employed to capture awareness, actual beliefs and attitude, the effect of such attitude and beliefs on adopting mitigation and adaptation strategies. Data collected were analysed using Atlas.ti and SPSS. Most of the women in the rural areas are aware of the impacts of climate change in their environment, especially, on their livelihoods. However, the awareness level has not impacted on them positively to adopt any mitigation and adaptation strategies. This is due to their belief system that climate change is not a consequence of anthropogenic activities. Recommendations were made as to how these problems could be solved for the women in the rural areas to embrace mitigation and adaptation strategies.

C.A.O. Akinbami (✉)
Institute for Entrepreneurship and Development Studies, Obafemi Awolowo University, Ile Ife, Nigeria
e-mail: Abiolaakinbami1@gmail.com

J.E. Olawoye
Department of Agriculture Extension and Rural Sociology, University of Ibadan, Ibadan, Nigeria

F.A. Adesina
Department of Geography, Obafemi Awolowo University, Ile Ife, Nigeria

W.L. Filho et al. (eds.), *Climate Change Adaptation, Resilience and Hazards*, Climate Change Management, DOI 10.1007/978-3-319-39880-8_4

Keywords Climate change • Mitigation • Adaptation • Livelihood • Rural women • Socio-cultural barriers

Introduction

Climate change affects every aspect and sector of the socio-economic development, thereby affecting the well-being and economic growth in the human system. The vulnerable including the socially marginalised—the poor, children, women, the elderly and the indigenous people especially in the rural environment are at great risks because of their weak adaptive capacities. In Nigeria, women especially those from the rural areas, are particularly vulnerable to the effects of climate change because they constitute the majority of the country's poor and are more dependent for their livelihood on natural resources that are threatened by climate change.

Studies (IPCC 2013) have shown in unequivocal terms that global climate is accelerating with rapidly increasing evidences. The Federal Ministry of Environment (FME 2014) reported that the scale of climate change will increase with high anthropogenic emissions, greenhouse gas (GHG) concentration, and average global temperature. Climate models suggest that Africa's climate will generally become more variable, with high levels of uncertainty regarding climate projections in the Africa Sahel zone. Evidences indicate that the world has already warmed by 0.8 °C since the pre-industrial era. Temperatures in West Africa, and particularly the Sahel, have increased more sharply than the global trend, and the average predicted rise in temperature between 1980/1999 and 2080/2099 is between 3 and 4 °C, which is more than 1.5 times the average global trend (IPCC 2007).

Nigeria is like any other country, experiencing the impacts of climate change. According to an assessment of Nigeria's climate over the period of 1941–2000 done by the Nigeria Meteorological Agency (NIMET 2008), it was observed that significant changes were taking place. It has been shown that there is a possibility of Sea Level Rise from 1990 levels to 0.3 m by 2020 and 1 m by 2050, and rise in temperature of up to 3.2 °C by 2050 under a high climate change scenario has been predicted (DFID 2009). The low estimate predictions are for sea level rise of 0.1 and 0.2 m by 2020 and 2050 respectively, and a temperature increase of 0.4–1 °C over the same time periods. Sea level rise of 1 m could result in loss of about three-quarters of the land area of Niger Delta in the country. According to NASA 2015, year 2014 is the hottest in recent decades indicating that it is hotter than 1998. The ice caps and glaciers in the cold regions are also melting.

All of these have serious environmental and economic implications and call for urgent and well-targeted actions to minimize the potential disasters that could attend a full-scale climate change. The report of Nigeria's Second National Communication on Climate Change (FME 2014) underscore the fact that climate change continues to have increasingly negative impact on agricultural productivity in Nigeria, with greater water stress experienced particularly in the Sudano-Sahelian region of the country. The changes in climate are further putting pressure on fisheries and the rich biodiversity in the country impacting negatively on food security. It is obvious that these negative impacts are impeding the efforts of the Federal Government in striving to reduce poverty in the country.

The most affected are the women who according to 2006 census figures, make up 49 % of the total population in Nigeria (FRN 2007). At the global scale the pattern is the same Women who make up more than 60 % of the over one billion poorest in the world (UNDP 2006) are most affected by the impacts of Climate Change. In Nigeria, women have been known to be economic catalysts and that economic growth particularly in developing countries, can be stimulated through their involvement in various livelihood and enterprise practices, although, most women are involved in micro, small and medium scale enterprises (MSMEs) which incidentally contribute more than 97 % of all enterprises, 60 % of the nation's GDP and 94 % of the total share of national employment figure (Mayoux 2001; Ndubusi 2004).

Quite clearly, an effective response to the challenges of climate change must address the impact on all sectors particularly as it affects the women in the rural sector, who are among the most vulnerable groups. Their gendered divisions of labour often result in the overrepresentation of women in agricultural and informal sectors, and make them more vulnerable to disasters (WEDO 2008). These women are into various livelihood practices in their rural settings contributing to the economies of their communities, and as a result are considered agents of climate change response actions. In addition to their livelihood activities, they are also responsible for food production, water and energy supply for cooking to take care of their households (Enarson 2000). As climate change impact increases, these tasks become difficult to carry out (Patt et al. 2007).

The livelihoods of the rural indigenous people have become fragile because of their vulnerability not only derives from dependence on threatened ecosystems, but also on their comparatively limited access to infrastructure, services, and political representation to aid their preparedness for mitigation and adaptation strategies (Kronik and Verner 2010). Dependence of the rural people on cultural cohesion has also contributed to the vulnerability to climate change. To maintain their livelihood strategies, the indigenous people depend heavily on cultural, human, and social assets, including traditional knowledge systems and institutions that are now under increased stress (Salick and Byg 2007). Their knowledge systems which are based on experiments with nature and their ability to predict and interpret natural phenomena, including weather conditions, have been vital for their survival and well-being and have also been instrumental in the development of their cultural practices, social structures, trust, and authority.

The societal production of knowledge about nature's cycles has led to certain cultural practices. The practices, in turn, have resulted in the creation of cultural capital, which then is reproduced through practices and rituals. Cultural institutions are developed around these practices and rituals, serving to maintain, develop, and disseminate information. These cultural institutions thereby contribute to the social generation of knowledge. The cultural institutions strongly affect indigenous peoples' natural resource management, attitude, belief, health, and coping abilities. In many societies socio-cultural norms prevent women from migrating to look for shelter and work when a disaster hits in their environment (Patt et al. 2007)

In order to help the rural women, it becomes important to know if they are even aware of the changes in climatic condition and its impacts on their activities. Some of the questions begging for answers include: How prepared are the rural women for climate change mitigation and adaptation strategies? Are there any socio-cultural belief and barriers affecting their attitude towards mitigation of and adaptation to climate change in the rural areas?

The objectives of this paper are to assess the rural women awareness of climate change in the study areas; evaluate the perceived impact of the belief system of rural women on climate change in the areas, and examine the mitigation and adaptation strategies in place.

Women Development and Climate Change

Climate change refers to a change in the state of the climate that can be identified (e.g. using statistical tests) by changes in the mean and/or the variability of its properties, and that persists for an extended period, typically decades or longer (IPCC 2007). Climate change is one of the most serious environmental threats that can undermine the efforts to the achievement of both local and national governments as they strive to reduce poverty. Sustainable development goals negotiated at the UN in September 2015. It has impacts significant on agriculture, water resources and biodiversity as all of these are dependent on climate (e.g. Aydinalp and Cresser 2008; Speranza 2010). However, human activities through agriculture constitute a problem to the environment because about 20 % of the annual increase in anthropogenic greenhouse gas emissions through carbon dioxide (CO_2), methane (CH_4) and nitrous oxide (NO) gas emissions, with methane having the highest global warming potential, which is about 300 times the potential of carbon monoxide (CO) and about 20 times that of NO, are from the agriculture facilities which the women are involved in. Livestock rearing which is one of the common livelihood practices in the rural areas alone accounts for 40 % of the world's agriculture gross domestic product and would generate approximately between 5 and 10 % of global emission of greenhouse gases (FAO 2006; Aydinalp and Cresser 2008).

Women function as change agents in community natural resource management, innovation, farming and care giving and hold the key to adaptation to climate change. Responsibilities in households, communities and as stewards of natural resources position them well to developing strategies for adapting to changing environmental realities under the new, different weather patterns and extreme weather events produced by change in the climate (Enarson 2000). Research has also shown that climate change affects women differently from men because women are often in charge of growing and preparing food, gathering firewood for fuel, collecting water and caring for the ill in their families and communities, all of which tasks become more difficult and time consuming with the increased occurrence of floods and droughts associated with climate change (Ajani Onwubuya and Mgbenka 2013). But they lack means of dealing with climate change and more importantly, is the effect of the socio-cultural system on the women, which emphasis male dominancy in all respects. This also reflects in women under representation in policy and decision making processes around climate change at all levels, especially, at the local level (Brody et al. 2008; IUCN 2007).

This is a matter of concern not only because women comprise one of the most vulnerable groups of people, but also because they play an important role in the economy in terms of their contribution to the global and national economy. Most of the times, these roles are not fully acknowledged. For instance, women comprise 43 % of the agricultural labour force in developing countries, ranging from 20 % in Latin America to almost 50 % in some parts of Africa and Asia. In South and East Asia, the Middle East and North Africa, women's share of agricultural employment within total employment is higher than that of men. In rural areas with high levels of male outmigration, women's roles in agriculture are expanding, leading to dramatic changes in their responsibilities and tasks.

Sub-Saharan Africa is set to be one of the regions hardest hit by climate change for some reasons, which include the fact that 96 % of its population is dependent on rain-fed agriculture (World-Bank 2008) and the poor adaptive capacities of African countries which reflect in their general economic and technological under-development. Inhabitants in most African countries have poor or limited access to health services, they lack access to micro-finance support, have poorly developed transport systems and poor knowledge of the characteristics and dynamics of climate change especially among the women (IPCC 2007). The severity of climate change impacts on poor rural communities, whose incomes are mainly from subsistence agriculture, are not difficult to visualize. Successful adaptation actions are likely to be those that are finely tuned to the immediate needs of individual communities where local realities and social structures are taken into account. Adaptation to climate change may be described as activities that reduce the negative impacts of climate change and/or takes advantage of new opportunities that may be presented. In many cases, women and men have separate roles and different knowledge and a range of different coping strategies. Although various studies have focused on climate change impacts, gender relations and adaptation opportunities in Africa, there is the need to focus more on the perceptions, beliefs and attitude of the rural women towards climate change which determine their preparedness in terms of mitigation and adaptation strategies.

Theoretical Framework

The theoretical framework employed in this paper is the sustainable livelihoods approach (SLA). Livelihood studies were brought to the fore of development studies in the late 1990s and the beginning of the new millennium, when the

Sustainable Livelihood Framework was strongly promoted by the Department for International Development (DFID), the British state development cooperation agency. Chambers and Conway (1992) reported that "a livelihood comprises the capabilities, assets (stores, resources, claims and access) and activities required for a means of living: a livelihood is sustainable which can cope with or recover from stress and shocks, maintain or enhance its capabilities and assets and provide sustainable livelihood opportunities for the next generation; and which contribute to net benefits to other livelihood at the local and global levels and in the short and long term." However, Carney's definition of livelihood which has become general currency built on the one from Chambers and Conway states that "a livelihood system comprises the capabilities, assets (including both material and social resources) and activities required for a means of living. He further said that a livelihood is sustainable when it can cope with and recover from stresses and shocks and maintain or enhance its capabilities and assets both now and in the future, while not undermining the natural resource base" (Carney 1998). It is important to note that "livelihoods rarely refer to a single activity. It includes complex, contextual, diverse and dynamic strategies developed by households to meet their needs" (Gaillard et al. 2009). The SLA evolved within the context of the intentional development approach by which development practitioners were seeking to maximise the effectiveness of their interventions to help the disadvantaged (Morse and McNamara 2013). It is in effect a diagnostic tool which provides a framework for analysis leading to concrete suggestions for intervention (Allison and Horemans 2006; Tao and Wall 2009). It was typically applied in poorer countries as part of a planning phase for an intervention via policy, a development project or perhaps as the basis for more in-depth research. In that sense the SLA is an analysis of peoples' current livelihood and what is needed for an 'enhancement', and useful in avoiding the inappropriate interventions critiqued by the post-developmentalists.

The livelihood approach groups individuals into different livelihoods according to their access to assets (including both material and social resources) and their capabilities to combine them to livelihood strategies so that everyone will have the "opportunity to earn a sustainable livelihood". The sustainable framework has been illustrated with a model that makes it easier to understand the different components and their interrelatedness (Petersen and Pedersen 2010). Figure 4.1 depicts the SLA model. The vulnerability context describes the external environment that the poor people live in. This includes critical tends, such as technological trends or population trends. It also includes shocks such as natural disasters or economic inflation, and seasonality which refers to the way prices, employment opportunities and production might shift with the seasons. All of these factors will affect the assets that people have and thereby the sustainability of their livelihoods. The sustainable livelihoods framework is built on the belief that people need assets to achieve a positive livelihood outcome. Transforming structure and process includes the institutions, organisations and policies that frame the livelihoods of the poor, and they are found on all levels—from the household to the international level. These processes and structures determine the access that people have to different kinds of

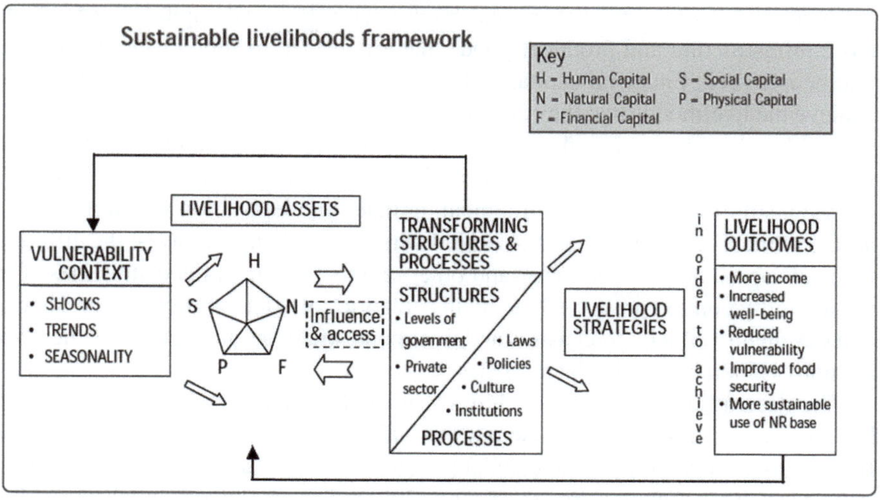

Fig. 4.1 An illustration of sustainable livelihoods framework (SLA) model. *Source*: (Petersen and Pedersen 2010; DFID 2000)

assets, and therefore the importance cannot be overemphasised. Examples of processes are international agreements, ownership rights and laws to secure the rights of the individuals, whereas structures might be the existence of ministries, banks that give credit to the farmers or self-help groups in the local community. Livelihoods strategies are the way that people act in order to achieve their desired livelihood. The access that people have to different kinds of assets affect the strategies that they employ, and the structures and processes in a given society also creates possibilities and constraints on the strategies that people are able to use. Finally Livelihood outcomes are the achievements of people's livelihood strategies. However, outcomes are to be described by the local people themselves, since these include much more than income. For outsiders it can be difficult to understand what people are seeking and why, because this is often influenced by culture, local norms and values (Petersen and Pedersen 2010).

People have different kind of assets that they combine to help them achieve the livelihoods that they seek. These assets have been broken into five 'capitals', which are: human capital (e.g., education, health); natural capital (e.g., land); financial capital (e.g., access to credit); social capital (e.g., community networks); and physical capital (e.g., infrastructure like markets and roads) (See Fig. 4.1).

Human capital is one of these assets, and refers to the skills, knowledge, ability to labour and good health that enable people to achieve their desired livelihoods. Human capital is essential in order to use the other kinds of capitals that exist. Social capital refers to the social resources that people can get help from in order to achieve their livelihoods—this could be through networking, membership of formalised groups or mere trust between people that make them help each another. Natural capital is to be understood in a very broad manner, since it both covers tangible factors, like natural resources such as trees, land etc., and more intangible products such as the atmosphere and biodiversity. Physical capital describes the basic infrastructure and producer goods that are needed to support the livelihoods that people seek. Financial capital is the financial resources that people can use to achieve the livelihoods that they are striving for. The ability to combine these assets to livelihood strategies is influenced by the prevailing transforming structures and institutions and the vulnerability context. The transforming structures and processes are the institutions, culture, environment, policies and legislation which determine access to the five different types of capital, terms of exchange between the different types of capital and the economic and other returns from livelihood strategies. The vulnerability context presents three main categories of vulnerability: trends, shocks and seasonality which affect assets and livelihood strategies and determine the level of (non) vulnerability (de Haan 2012).

The description of the model shows that the SL framework is a systemic and holistic way of describing the factors that affect the livelihoods of the poor. The framework is an attempt to understand poverty as a multifaceted concept, covering more than just economic growth (Krantz 2001). The framework emphasises that other aspects are important too, such as health, social status and natural resources. These factors have an impact on how people are able to take advantage of the economic opportunities, how they combine assets and thereby what livelihoods they can create. Moreover the description of the different factors show how important it is to include the poor, since they are the ones with the knowledge of the content associated with each factor, and of how the factors affect each other in positive or negative ways.

The livelihood approach became attractive because it has an open eye for the wider context in which the poor organise their livelihood strategies. The approach acknowledges that these strategies are embedded in structures and governed by institutions. This wider context is considered fundamental because an important part of the poverty alleviation policies and interventions is meant to aim at opportunities and constraints in these structures that would either enable or prevent the poor from organising effective livelihood strategies. If these policies and interventions could become more effective, it would bring the poor less vulnerability, more well-being and more sustainability. Therefore, notions like claims and access are considered key in the livelihood approach. These notions point at the possibility to call upon moral and practical assistance and to effectively use the resource in practice. Although the wider context or structure is not only regarded as a potential constraint to the livelihood strategies of the poor, the approach however also aims to stress the potential of livelihood strategies to influence and even to change structures. This attention for poor people's agency, as their capacity to integrate experiences into their livelihood strategies and to look for outlets of aspirations, ambition and solutions to problems, is prominent in the livelihood approach (Scoones 1998).

Applying Livelihood Profiles as Baseline for Vulnerability Analysis

The livelihood framework is increasingly influencing the approach of projects ranging from emergency response, to disaster mitigation to longer term. Livelihood Profiles are developed to serve as baseline information. The baseline information comprises a set of information dedicated to answering the fundamental question of how people survive in most years (Scoones 1998).

This study has used SLA to translate livelihoods analysis into practical, quantified information for decision makers with a practical geography attached. This model has also been used to link livelihood information to an analysis of the effects that a hazard will have on livelihoods and household income security especially as it relates to the rural women, their belief system and their economic empowerment activities. This study has also used the SLA as a holistic approach to get a comprehensive understanding of how people cope and how communities are internally differentiated in their response to climate variabilities.

Methodology

This study which was carried out in Osun state in the south-west of Nigeria. It precedes a bigger study which will cover several states in the various geopolitical zones of the country with different socio-cultural belief systems and vegetation zonation. Qualitative and quantitative methods were employed to gather the primary data. Both descriptive and inferential statistics were employed to analysis the quantitative data. The descriptive statistics involved the use of frequencies and percentages while the inferential statistics involved the use of bivariate level of analysis using chi square test at 0.01 level of significance. Furthermore, the Kendall Nonparametric bivariate correlations were employed to determine the significance and direction of association between the factors considered and awareness level. Atlas.ti software was employed to analysis the qualitative data. Atlas.ti was used to do the content analysis of responses by highlighting relevant quotes. The two communities were purposively chosen because of ease of entry into the areas. The communities were free of communal clashes as at the time of the study. However, the chosen communities have similar cultures and levels of development. The participatory research tools were employed to assess responses to climate change, awareness of and knowledge of climate change, and its impacts among the target groups in the communities.

Study Area

The study was conducted in two communities in Ile-Ife, Osun State. Ife Central, Ife East, Ife North, Ife South, are that make up Ile-Ife and its environ. Osun State is a landlocked state in South-western Nigeria and occupies a land mass of approximately 8602 km^2. It is known as the state of the living spring and is bounded on the west by Oyo State, in the east by Ondo and Ekiti States, in the north by Kwara State and in the south by Ogun State. The people of the state are Yoruba and trace their origin to Oduduwa and the town of Ile-Ife. The sub-ethnic groups comprise of Ife, Ijesha, Oyo and Igbomina. Osogbo is the capital of the state with a population of 2,203,016 (NPC 2010). The people of the state are mostly traders, artisans and farmers. The farmers produce food crops such as yam, maize, cassava, beans and cocoyam. The cash crops include cocoa and palm produce. The artisans make hand-woven textiles, tie and dye clothes, leather work, calabash carving and mat-weaving. For the purpose of this study Osun state was purposively selected due to ease of entry and suitability for this study. Osun state is dominated by rain forest climatic condition.

Sample and Sampling Techniques

The target population is the rural women who are into various livelihood practices: farming, food processing, and trading/artisans were sampled in the communities, because these are the prevailing business activities in the study areas. This study was conducted in two communities. In each community, one rural community was purposively chosen, that is, each from Ife Central and Ife East Local Government Areas (LGAs) of Osun state. The chosen communities serve as good representation where women experience the effect of climate change in the rural areas.

Research Instruments

Qualitative and quantitative approaches were employed to elicit responses from respondents. The qualitative approach involved the use of Focus Group Discussions among the women, in-depth interviews with community leader.

Sample Population (Sample Frame)

The two communities purposively selected for the study are Koola in Ife East LGA and Ilode in Ife Central LGA. The community in Ife Central LGA consists of a group of 30 women who are into vegetable farming. They are gathered in the area because of the availability of a flowing river which is an asset to their occupation. Questionnaire was administered on 20 of them while the remaining 10 women participated in focus group discussion (FGD).

The second sample population from Ife East LGA was conducted among women who are into various livelihood practices such as palm oil processing, kolanut trading, plantain trading and fruit trading. Fifty women were involved in the quantitative method while another ten women participated in the FGD. The in-depth interview (IDI) was conducted on the community leader.

Results and Discussion

Socio-Economic Characteristics of Respondents

About half of the respondents (50 %) were within the age range of 36 and above (Table 4.1). The mean age of the respondents was 32 years. This implies that most of the women were in their productive years. Hence their involvement in more than one livelihood activities. About 30 % of the respondents are involved in other occupation for various reasons such as: the need to source for more income to diversify and the need to augment income from major activities. A greater proportion (70 %) of the respondents were married. This again explains their involvement in more than one occupation. As married women, they have great responsibilities to cater for, which income from one occupation may not be sufficient to handle. Oberhauser and Pratt (2004) noted that married people have responsibility for provision of household needs of their families hence greater involvement in occupational diversification for economic empowerment.

The analysis of the educational level of the women revealed that about 83.4 % of them went to school out of which 45.7 % had primary education, that is, basic education; 25.7 % possessed secondary while 1.4 % had tertiary education where advanced skill acquisition could be acquired. This explains why most of the women in the rural areas in the study areas are mostly into farming and food processing activities. In this study, the respondents were also involved in farming (64.3 %), vegetable farming (28.6 %), livestock farming (5.7 %) and artisan (1.4 %). This result agrees with the submission of Ranjan (2006) which stated that level of education increases participation rate in occupations for rural women and that educated rural women are likely to possess skills which facilitate successful involvement in non-farm activities.

More than half (52.9 %) of the women had over 10 years of work experience. This is one feature aiding diversification of activities among the women and this is of great importance in risk management. According to Ajani and Igbokwe (2012) rural women with many years of experience in farming are more likely to diversify

Table 4.1 Socio demographic characteristics

Age group	N = 70	%
16–20 years	4	6
21–25 years	9	13
26–30 years	11	16
31–35 years	11	16
36–40 years	18	26
41 years>	17	24
Marital status		
Single	7	10
Married	49	70
Widowed	6	8
Separated	4	6
Divorced	4	6
Educational status		
No primary education	19	27
Primary education	32	46
Secondary education	18	26
Tertiary education	1	1
Religion		
Christianity	54	77
Muslim	16	23
Major occupation		
Crop faming	46	66
Vegetable farming	20	29
Livestock rearing	4	5
Creation of business venture		
1–5 years	6	9
5–10 years	27	39
10 years>	37	52
Average monthly income		
<N2,500	1	1
₦2500–₦5,000	4	6
₦5000–₦10,000	25	36
₦10,000–₦15,000	23	33
₦15,000–₦20,000	9	13
₦20,000–₦30,000	5	7
₦30,000–₦50,000	3	4

Source: Field Survey (conducted in 2015)

into agricultural activities making use of wealth of experiences they have acquired over the years.

Over 35 % of the respondents reported to be generating average monthly income between ₦5000–₦10,000, 32 % fell between the range of ₦10,000–₦15,000 while about 8 % earned less than ₦5000 and the remaining 25 % claimed to earn more than ₦15,000. The mean income is about ₦8000. From their claim, it's obvious that majority (67 %) of the women earn less than the approved minimum wage of ₦18,000 in the formal sector in the country. This again could be the reason why most of the women's livelihood practices remain at the micro level, since there seems to be lack of sufficient capital for expansion. Moreover, the study also found out that 60 % of the respondents have a household size of 6–10. The mean household size is 7. It can thus be interpreted that 7 people in a household would have to depend on ₦8000 in a month. This would lead to poverty being more entrenched in the rural areas, particularly among the women.

Awareness of Climate Change

All the respondents reported that they are aware of the changes that are taking place in the weather around them. According to the report from the FGD participants, it was stated by the participants that the changes had been observed as far back as 30 years' ago. This finding contradicts the notion stated by Egbe et al. (2014) that rural women are not aware of climate change. Though, the sufficiency of the awareness level may be queried because it is objective and not based on scientific evidence.

Their awareness cut across the elements of weather vagaries such as distortion of rainfall pattern, drought, sea level rise, temperature rise, and increase in humidity. More than three-fifths (62.9 %) indicated that there is less rainfall to give a good farm yield at the appropriate time, 11.4 % said that there is too much rainfall, that is, the few rainfall events experienced were heavy and had adverse effect on their farm products. More than a tenths (15.7 %) reported that there was more drought experienced and 10 % experienced more erosion. In the study areas, the women were able to monitor climate change trend. Figures 4.2a and b reveal the level of awareness in elements of climate change. The study reveals that rainfall, either the shortage or excess of it, is the most climate change effect being experienced by the women in the study areas (Fig. 4.3) since about 94 % of the women depend on rainfall for their farming activities.

During the FGD sessions, the women at Koola claimed that they have started to experience the instability in weather since 30 years ago and as a result, their livelihood had been grossly affected. Some of the respondents said:

> 'There has been too much of sunshine and lack of sufficient rain, thereby reducing the quantity of palm oil because the fruits would not be juicy due to lack of rain and too much of heat'. *Koola community, palm oil producer, 67 years.*

> 'Farm products are not doing well now because of lack of rain. The bunch of plantain harvested now cannot be compared with that of 20 years back. Now, they are smaller in size'. *Farm products seller, 55 years, Koola community.*

> 'In the olden days, when the harmattan season came at the appropriate time, it aided Kolanut production and they would be fresh during preservation. But now, there is much drought, as a result, kolanut fruits get burnt even while preserving them. And this leads to great loss for us because we have to pour the fruits away'. *Kolanut seller, 70 years, Koola community.*

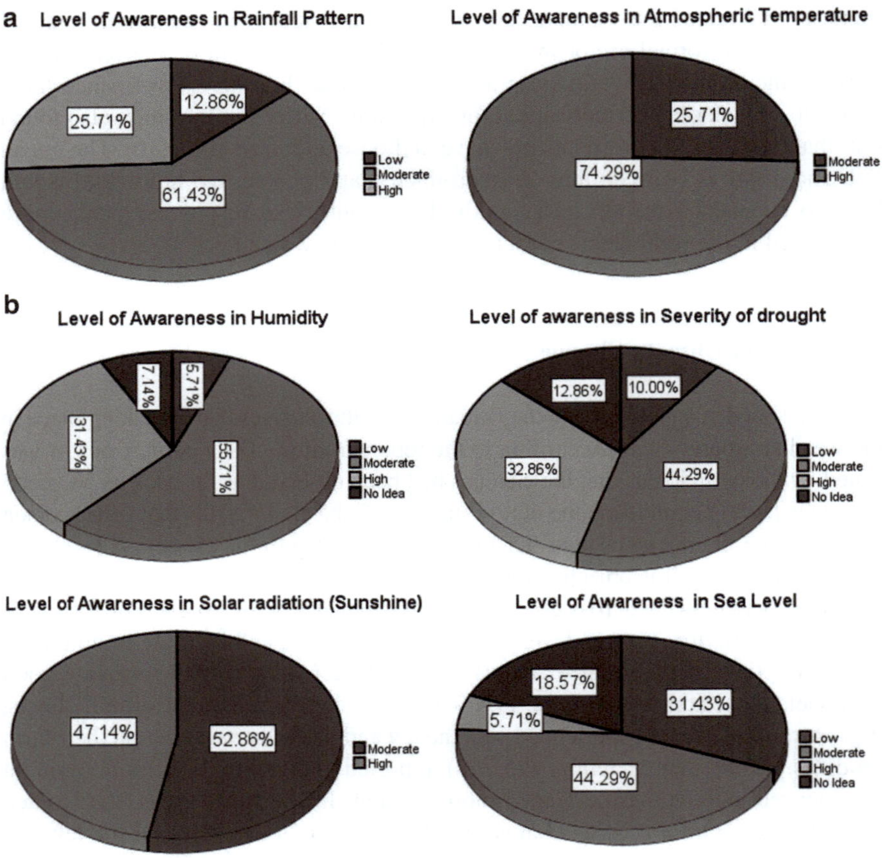

Fig. 4.2 **a** Level of awareness in elements of climate change. **b** Level of awareness in the elements of climate change

The situation was the same at Ilode, with the women who were into vegetable farming. They are aware of climate change and the effects on their livelihood practice. For them, there is too much of sunshine sometime and too much of rain some other time leading to flooding. They reported:

> 'We have not had it this bad before. About 3 years ago, the vegetable was getting rotten when it was to blossom due to the effect of too much of heat'. *Vegetable farmer, 51 years with over 10 years of farming experience. Ilode community'.*

> 'There has been too much of heat affecting the tender plant and sometimes we have to pour the vegetable away after harvesting and we have nothing to sell. Some other times, when it would rain, the rain would be too much causing most of the vegetable to be washed off by erosion'. *Vegetable farmer, 52 years with over 30 years of farming experience. Ilode community'.*

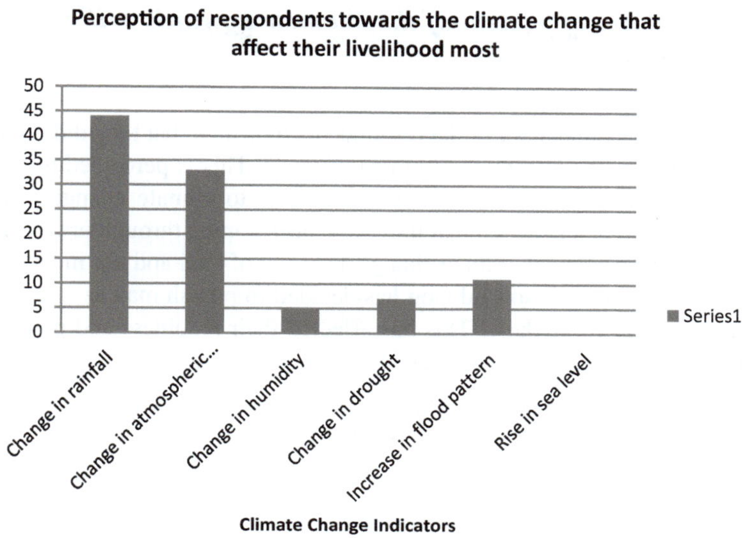

Fig. 4.3 Climate change element mostly affected by the women

Sources of Information

Information was sought from the respondents about their sources of information. Their responses revealed different sources of information, close to seven-tenths (65.7 %) got their climate change information from personal experience, 14.3 % form television, 11.4 % from radio and 8.6 % from fellow villagers. Though, personal experience may be taken as non-scientific and subjective, however, Salick and Anja (2007) brought out the importance of this type of information source. He stated that rural and indigenous people have developed knowledge systems which are based on experiments with nature (personal experience) and their ability to predict and interpret natural phenomena, including weather conditions. These have been vital for their survival and well-being and have also been instrumental to the development of their cultural practices, social structures, trust, and authority. They also possess the understanding of the relationship between society and nature and the notion that balances need to be maintained between the human, natural, and cosmological realms. So when changes occur, for example, in climatic conditions, people look to themselves and their social institutions and practices to see whether aspects of the way they lead their lives are causing imbalance and need to be rectified.

Attitude and Belief System of the Women Towards Mitigation and Adaptation Strategies

It is important to state that all of the respondents indicated that they did not receive any climate information from government officials. This is perceived to be a very serious issue among the rural women as it relates to climate change adaptation strategies. The women declared that they have been coping through prayers and use of charms. They perceived climate change as an act of God and not induced by the activities of man. They also said God has decided to punish man as a result of sin which is on the increase daily. This has affected their readiness and preparedness for mitigation and adaptation strategies.

> 'It is only the mercies of God that can help us out of this predicament. There is nothing that we can do'. *Chorus answer from the Koola women.*

From Ilode women:

> 'Our sins are too much and God is angry with us.'

The opinions of the community leader of Koola and the representative of the community leader in Ilode did not differ from the women'. Koola community leader said:

> 'Up to 10 years ago, the weather was normal, but things have changed so much since then and now. Rain does not fall as it used to, the sun is also too much. The climatic change has affected virtually all farm products in this community. No matter how hardworking and efficient the farmer may be, if there is no rain, there is nothing he can do. All these are as a result of our sins.' *Koola community leader. 72 years.*

Ilode community leader representative also said that:

> 'The women in this community are really suffering as a result of effect of erosion. When it rains, most of the houses are also flooded and the vegetable farms are badly affected. This is affecting the economic advancement of this community. We once called on the local government for help, our drainage system was widened but it has not totally helped. We pray that God will have mercy on us because we know that we have sinned'. *Ilode community leader representative. 50 years.*

At this point, this paper underscores the importance of training from the agricultural extension officers. If there has been continuous training programs organized to educate the rural women on the causes of climate change, how to control it, what activities they should stop carrying out, e.g., deforestation and activities to carry out e.g., afforestation and reforestation, the women would be well informed and have their belief and attitude changed positively. This attitude was seen to be having a negative effect on the mitigation and adaptation strategies they embark on. According to the women in Koola, the only form of mitigation strategy available in the community is ensuring that the drainage system in their individual compounds is always cleared of any debris for free flow of flood without affecting their houses. They remarked that:

'There is nothing to be done to control the effects of climate change. When it pleases God, He will stop it'. *Kolanut seller, 67 years, Koola community.*

The quote above represents the general opinion of the women. Also, at Ilode, the only mitigation strategy in place was the building of drainage system. The study found that there were no planned adaptation strategies in place in the communities in spite of their experiences. Even though, there may be autonomous adaptation strategies being practiced which they are not aware of. One of the women reported:

'Erosion disturbs us because this is river side area. It even gets to our houses often. We depend on the drainage as the saving grace. We also dug well in the farm to get water to irrigate our vegetable farm'. *Vegetable farmer, 36 years with over 9 years of farming experience. Ilode community'.*

The above statement revealed the autonomous adaptation strategy being practiced against drought but which they are not aware of.

Factors Influencing Level of Awareness

The study further determined the relationship between factors that can influence awareness level of the respondents. This is presented in Table 4.2. The table reveals that there is statistically significant association between Level of education of respondents and Awareness level ($p < 0.01$). The Source of information was also found to be statistically significant with awareness level of the respondents about climate change ($p < 0.01$). The Kendall's Non Parametric Correlations show that there is a significant positive association between level of education and awareness level ($r = 0.567$, $p = 0.001$) as well as between Source of information and awareness ($r = 0.391$, $p = 0.001$). This shows that increase in the level of education should enhance awareness level, however, it was discovered that respondents

Table 4.2 Factors influencing level of awareness

	Awareness Level				
Variables	High	Moderate	Low	χ^2-value	p-value
Level of education					
No formal education	18 (78.3)	1 (2.4)	–		
Primary	–	32 (100.0)	–		
Secondary	4 (17.4)	9 (21.4)	5 (100.0)		
Tertiary	1 (4.3)	–	–	68.014	0.01**
Source of information					
Television	10 (43.5)	–	–		
Fellow villagers	–	6 (14.3)	–		
Radio	3 (13.0)	5 (11.9)	–		
Personal experience	10 (43.5)	31 (73.8)	5 (100.0)	28.11	0.01**

Source: Survey (2015)
**Significant at 0.01 level of significance

without formal education have the greatest awareness level. This is so because of the heavy reliance on personal experience stated earlier on by the respondents.

If acquisition of formal level of education is not having the desired impact on awareness level of respondents, then an intervention in form of training from agricultural extension officers needs be encouraged. Since the scarcity of appropriate information to these women will always position them at the low ebb of development. According to Morse and McNamara (2013) in order for the women to be able to recover from "*stress and shocks*", low ebb development and poverty, they must be able to "*maintain and enhance*" capabilities and assets into the future. A central element in this 'resilience' to stress and shocks may well be the diversification of elements that comprise 'livelihood'. But this study suggests that no matter how well women are involved in livelihood diversification, as long there is lack of useful information, such diversification will not be fruitful.

This lack of relevant information could be one of the major factors that make the rural dwellers especially the women vulnerable. In order to help them, the institutional and cultural context within which these women operate must be understood. Once this is understood then interventions can be put in place to enhance livelihoods and their sustainability. According to Morse and McNamara (2013) the process of understanding their current situations so as to develop suggestions for improvement must be based upon a participatory approach. This is deem to be appropriate because it involves a bottom–top approach, where the rural dwellers are part of the decision process. This will make them to embrace whatever the outcomes of the process are.

Conclusion and Recommendation

Rural women were involved in various livelihood practices and these activities are being affected by the changes that are occurring in the global climate system. These women were not ignorant of the climate change and the effects on their livelihood practices. But the climate change has negative impacts on their economic activities, thus making them one of the mostly affected, since they depend on the natural resources for their survival. Furthermore, their belief system concerning climate change has affected their attitude negatively towards adopting mitigation and adaptation strategies. The study therefore concludes that for the fortune of these women to be changed positively which will enable them have sustainable livelihood practices and build their resilience to the adverse effects of climate change. It is recommended that there must be frantic effort on the part of the local government authorities in educating these women through the agricultural extension agents who are closer to the women in the rural areas, about the causes of climate change, activities to discontinue and the ones to imbibe. They should be educated on in-house mitigation and adaptation measures using locally available resources that address the climate change impacts in their communities and especially as they affect their livelihood practices. This will help them to develop and embrace any mitigation and adaptation strategies in their communities.

Limitations of the Study

The present study did not cover all the vegetation zones in Nigeria. Much more work is needed to capture the national picture of the subject of interest here to be able to study the climate change impact on the rural women in other zones. Also, the socio- cultural and belief systems of other regions of the country were not covered.

Acknowledgement "This research is supported by funding from the Department for International Development (DfID) under the Climate Impact Research Capacity and Leadership Enhancement (CIRCLE) programme".

References

Ajani EN, Igbokwe EM (2012) Promoting entrepreneurship and diversification as a strategy for climate change adaptation among rural women in Anambra State, Nigeria. J Agric Ext 16(2). ISSN 1119-944X

Ajani EN, Onwubuya EA, Mgbenka RN (2013) Approaches to economic empowerment of rural women for climate change mitigation and adaptation: implications for policy. J Agric Ext 17 (1):23–34. doi:10.4314/jae.v17i1.3

Allison EH, Horemans B (2006) Putting the principles of the sustainable livelihoods approach into fisheries development policy and practice. Mar Policy 30:757–766

Aydinalp C, Cresser MS (2008) The effect of global climate change on agriculture. Am Eurasian J Agric Environ Sci 3:672–676

Brody A, Demetriades J, Espen E (2008) Gender and climate change: mapping the linkages. BRIDGE-Institute of Development Studies, Sussex. Retrieved from http://www.bridge.ids.ac.uk/reports/Climate_Change_DFID_draft.pdf

Carney D (ed) (1998) Sustainable rural livelihoods: what contribution can we make? Department for International Development, London

Chambers R, Conway G (1992) Sustainable rural livelihoods: practical concepts for the 21st century. IDS Discussion paper 296. IDS, Brigton

De Haan LJ (2012) The livelihood approach: a critical exploration. Erdkunde 66(4):345–357. doi:10.3112/erdkunde, ISSN 0014-0015

DFID (2000) Framework introduction. Sustainable livelihoods guidance sheets. Available at http://www.eldis.org/go/topics/dossiers/livelihoods-connect/what-are-livelihoods-approaches/training-and-learning-materials

DFID (Department for International Development) (2009) Impact of climate change on Nigeria's economy. Final report

Egbe CA, Yaro MA, Okon AE, Bisong FE (2014) Rural peoples' perception to climate variability/change in cross river State-Nigeria. Sustain Dev 7(2). doi:10.5539/jsd.v7n2p25

Enarson E (2000) Gender issues in natural disasters: talking points and research needs in focus programme on crisis response and reconstruction workshop. IPCRR Working Paper no. 1. ILO, Geneva

FAO (2006) Livestock and major threat to the environment: remedies urgently needed. FAO, Rome. Available http://www.fao.org/newsroom/en/news/2006/10000448/index.html. Accessed 6 Oct 2011

FME (2014) Second national communication—Nigeria. Federal Ministry of Environment (FME), Abuja

FRN (Federal Republic of Nigeria) (2007) Official Gazette Extraordinary No 24, Government Notice No 21, vol 94. Lagos

Gaillard J, Maceda E, Stasiak E, Le Berre I, EspaLdon M (2009) Sustainable livelihoods and people's vulnerability in the face of coastal hazards. J Coast Conserv 13(2):119–129. doi:10.1007/s11852-009-0054-y

International Union for Conservation of Nature (IUCN) (2007) Gender aspects of climate change. Briefing paper. Retrieved from http://www.iucn.org/en/news/archive/2007/03/7_gender_climate_change.pdf

IPCC (2007) Summary for policymakers. In: Parry ML, Canzianni OF, Palutikof JP, Van der Linden PJ, Hanson CE (eds) Climate change 2007: impacts, adaptation and vulnerability. Contribution of working group II to the fourth assessment report of the IPCC. Cambridge University Press, Cambridge

IPCC (2013) Climate change 2013: the physical science basis. IPCC Working Group I contribution to AR5. https://www.ipcc.ch/pdf/assessment-report/ar5

Krantz L (2001) The sustainable livelihood approach to poverty reduction, an introduction. Swedish International Development Cooperation Agency, Stockholm

Kronik J, Verner D (2010) Indigenous peoples and climate change in Latin America and the Caribbean. World Bank, Washington, DC

Mayoux L (2001) Jobs, gender and small enterprises: getting the policy environment right. An ILO Working Paper on series on women's entrepreneurship development and gender in enterprises (WEDGE) No. 15

Morse S, McNamara N (2013) Sustainable livelihood approach: a critique of theory and practice, vol XVIII. Springer, Dordrecht, p 190, 36 illus. ISBN 978-94-007-6267-1

Ndubusi F (2004) Bankers list financing alternatives for SMEs. In: CBN seminar on small and medium industries equity investments scheme, Maritime

NIMET (2008) Nigeria climate review bulletin 2007. Nigerian Meteorological Agency. NIMET—P. 001, February 2008

NPC (2010) 2006 population and housing census, Priority table volume III population distribution by sex, state, LGA and Senatorial District. National Population Commission, Abuja

Oberhauser AM, Pratt A (2004) Women's collective economic strategies and transformation in rural South Africa, gender place and culture. Fem Geogr 2:209–228. doi:10.1080/0966369042000218464

Patt A, Angie D, Pablo S (2007) Gender and climate change vulnerability: what's the problem, what's the solution? Paper presented at the International women leaders global security summit, New York, 15–17 November 2007

Petersen EK, Pedersen ML (2010) The sustainable livelihoods approach: from a psychological perspective: approaches to development. June 2010 study report. Institute of Biology, University of Aarhus

Ranjan S (2006) Occupational diversification and access to rural employment: revisiting the non-farm employment debate *MPRA*. University Library of Munich, Germany, pp 4–8

Salick J, Anja B (2007) Indigenous peoples and climate change. Tyndall Centre for Climate Change Research, Oxford

Salick J, Byg A (2007) Indigenous peoples and climate change. Tyndall Centre for Climate Change Research, Oxford

Scoones I (1998) Sustainable rural livelihoods: a framework for analysis. IDS Working Paper 72

Speranza CI (2010) Resilient adaptation to climate change in African agriculture. German Development Institute (DIE), Bonn

Tao T, Wall G (2009) Tourism as a sustainable livelihood strategy. Tour Manag 30(1):90–98

UNDP (2006) Taking gender equality seriously—making progress. Meeting new challenges. http://www.undp.org/gender/. Accessed 7 Mar 2008

WEDO (2008) Gender, climate change and human security: lessons from Bangladesh, Ghana and Senegal. Report prepared for ELIAMEP. http://www.gdnonline.org/resources/WEDO_Gender_CC_Human_Security.pdf

World Bank (2008) World development report agriculture for development. The World Bank, Washington, DC, https://siteresources.worldbank.org/INTWDRS/Resources/477365-1327599046334/8394679-1327614067045/WDROver2008-ENG.pdf

Chapter 5
Climate Change and Human Security in a Regulatory Multilevel and Multidisciplinary Dimension: The Case of the Arctic Environmental Ocean

Sandra Cassotta, Kamrul Hossain, Jingzheng Ren, and Michael Evan Goodsite

Abstract Climate change determines the retreat of ice. This has created a huge access to petroleum, attracting strong interest by some states, especially energy hungry-countries and increasing competition between states, resulting in tension and threats, even military ones. Climate change has, therefore, to be perceived as a threat to international peace and security. However, recognition of the *nexus* between climate change, human security and conflicts in the prism of international law and politics is weak, leading to a difficulty institutions have for regulating and governing this *nexus*. Climate change can thus be considered as threat-multiplier, something that can exacerbate existing tensions, and the resolution of this threat will be the most difficult task to achieve where adaptation takes place in fragile and vulnerable areas, such as the Arctic, an area which is highly exposed to

S. Cassotta (✉)
Department of Law, University of Aalborg, Niels Jernes Vej 6b, 9220 Aalborg Ø, Denmark

Institue For Security and Development Policies (ISDP), Sino-Arctic Research Programme (SNAPP), Stockholm, Sweden
e-mail: sacp@law.au.dk

K. Hossain
Northen Institute for Environmental and Minority Law, Arctic Centre, University of Lapland, PO Box 122, University of Lapland, Rovaniemi, Finland

J. Ren
SDU Engineering Operations Management, Institute of Technology and Innovation, Campus Vej 55 DK - 5230, Odense, Denmark

Institue For Security and Development Policies (ISDP), Sino-Arctic Research Programme (SNAPP), Stockholm, Sweden

M.E. Goodsite
Institute for Technology and Innovation, SDU Civil Engineering, Odense University, Niels Bohrs Allé 1, 5230 Odense M, Denmark

Institue For Security and Development Policies (ISDP), Sino-Arctic Research Programme (SNAPP), Stockholm, Sweden

W.L. Filho et al. (eds.), *Climate Change Adaptation, Resilience and Hazards*, Climate Change Management, DOI 10.1007/978-3-319-39880-8_5

environmental risks and uncertainty. The region is populated by one of the most vulnerable groups, the indigenous people, such as the Inuit of the Arctic with low adaptive capacity compared to the pace of change. In the Arctic Environmental Ocean governance, access to natural resources, the potential of navigability, extension of maritime borders, and the desire of some states to extend their jurisdiction, all depict a situation of criss-crossing potential conflicts that could escalate and should, therefore, be perceived as "tinderboxes". This article relatedly explores the existing legal framework in the case of the Arctic environmental ocean to provide effective and legitimate governance for a peaceful and "stable space" to prevent threats from both Arctic and non-Arctic states. It will be shown that Arctic Environmental Ocean activities need multi-level governance (global, regional, national and local) and that Arctic environmental security challenges cannot be addressed without a broader holistic vision. The article treat the United Nations regulatory level and how it could support many issues which have an impact on Arctic Environmental Ocean governance, and the Law on the Sea. Methodologically, the way to increase effectiveness to maintain and stabilize the Arctic environmental ocean governance entails that "stability" is achieved by integrating elements of climatology, international law, political science and agent based modelling to act in a preventative way to protect the Arctic environmental ocean and its societies and formulate effective policies. The conclusion led on how the current Arctic environmental ocean framework could be changed in order to increase effectiveness by incorporating risk analysis into a universal equation based model to redesign a new regulatory package at United Nations level and recommend changes at institutional level.

Keywords Climate change • Human security • Arctic Environmental Ocean • Arctic security • United Nations (UN) law • United Nations Convention on the Law of the Sea (UNCLOS)

Introduction

Attention on climate change is continuously progressing at global, regional and national levels. The consequences of this environmental problem are not only potentially globally catastrophic, but also quite difficult to address and regulate due to scientific uncertainties and political disagreements. Climate change is hitting more intensively, and more than ever, the Arctic area. The present thinning and retreat of Arctic ice is one of the most serious geophysical consequences of global warming and is causing a major change to the face of the planet (Wadhams 2013). Most importantly, what takes place in the Arctic in the next decade will have consequence for the entire globe as the changing "Albedo Effect" alters the global climate, disrupting many equilibria both in the ecosystem and social sphere in what the climatologists defines as "cascading effect" (International Panel on Climate Change, IPCC 2015). Anthropogenic activities such as natural resources extractions are important drivers that generate rapid unpredictable changes both in the Arctic and in the rest of the planet (DPSIR model, European Environmental Agency, EEA 2009 and Omann et al. 2009). Polar Regions are the most vulnerable due to less resilience and biodiversity. The IPCC estimated global sea level rise by 2100 to be between 0.18 and 0.59 m (IPPC Fourth Assessment Report data and 2009 Copenhagen Climate Conference). Also the Arctic Monitoring and Assessment Programme (AMAP) estimated global sea level rise to be at the same period (2100) between 0.9 and 1.6 m (AMAP 2011). Climate Change is also facilitating new economic opportunities in the Arctic because of the opening up of the Arctic Ocean. The impact of climate change, such as ice melting, facilitates natural resource extraction and increased navigation through several of the Arctic sea routes. Arctic fossil fuels, for example, are expected to become drivers of geopolitical changes because climate change impact facilitates vast resource extraction, increases competition between states and results in tensions and threats, even military ones. This has literally changed the concept of security. Indeed, "*the threat from the skies today (are) not so much nuclear missiles as ozone-layer depletion and global warming*" as expressed by Mikhail Gorbatchev (Meyers 2007).

This article focuses on the climate change and human security and conflict *nexus* in the Arctic Environmental Ocean which until now, has been neither well developed nor well documented, especially from the prism of international law in the dynamics of an interdisciplinary and multi-level regulatory perspective. The term "Arctic Environmental Ocean" (hereinafter "AEO") refers, here to a system bounded by the sea floor, permanent sea-ice cap and surrounding land areas with inflow and outflow from North Pacific as well as the North Atlantic (Berkam and Vylegzhanin 2010). The interest in the AEO extends also from the Arctic states, indigenous people, stakeholders and business sectors to the non-Arctic States and global civil society. The difficulty consists to reconcile these interests between climate challenges, environmental security and new business opportunities.

The purpose of the paper is built upon the idea that governance of the AEO requires a holistic approach integrating different levels of sources of law and policy (i.e., local, national, regional, and global) and a systematic combination of different genres of disciplines as a method to tackle climate change effects in this region. The main hypothesis here, then, is whether the present international legal framework for AEO governance is adequate to avoid conflicts over underlying interests. In particular, it is asked whether the *nexus* poses societal challenges. Hence, the question of research for this paper is *how the nexus can be regulated in the most effective way in order to maintain and avoid conflicts over resources given the height level of exposure of environmental risks and unpredictable countries' assertive behaviours*. Unpredictability and "variability" to climate change rather than "changes", will be considered two of the problems of the *nexus* between climate change and human security, both of which can lead to an unpredictable proliferation of a myriad of potential "tinderboxes" that could develop "suddenly" and unexpectedly given the strategies of assertive "energy-hungry" countries, such as China or Russia. The new idea is to reframe and manage the AEO security with a global proactive preventative vision rather than reactive to face climate change.

Thus, the innovative elements of this paper is to propose a universal preventative global holistic approach and by reflective consequence to better protect and manage the international community as climate change and mitigation/adaptation are closely related to all aspect of the sustainable development agenda. For the first time, this paper advocate that increasing security of environmental concerns of the Arctic Environmental Ocean should be linked to global change because of the existence of the "cascading effect" of climate change on the Arctic and from the Arctic to the rest of the world. This requires a new regulatory approach and management to institutional and structural issues, including risk assessment analysis using probabilistic equation based modelling and agent based approach.

Commencing by enquiring as to what degree the effects of climate change can be tackled at the global level of the United Nations Security Council (UNSC), an assessment will be made of the extent to which improvements at the United Nations (UN) level can reverberates down to the regional level of AEO governance. The United Nations Convention on the Law of the Sea (UNCLOS) and the Arctic Council (AC) will be examined regarding how institutions can deal with uncertainty and potential threats associated with specified non-Arctic energy-hungry countries.

Finally, the article presents some recommendations and future prospects of interdisciplinary nature about the possibility for how changes at the UN level can be reflected at the AEO governance level by incorporating risks into the legal framework, assesses if these changes at the UN are realistically possible, and if not what could be the alternative at institutional level to increase AEO governance stability through the beneficial effect of reverberations from the global to the regional level.

Conceptual Background: Climate Change as a Threat to International Peace and Security

Climate change has proved to be an intractable problem, even now, twenty years after the impotent United Nations Framework Convention on Climate Change (UNFCCC 1992) and the weak enforcement of its Kyoto Protocol (1997). Not even such a protocol, the only binding instrument existing as a result of exhausting UN negotiations between developed and developing countries could block cataclysmic scenarios. It was an attempt to address climate change by aiming to stabilize greenhouse gas emissions (GHG) in the atmosphere at a level that would avoid "*dangerous anthropogenic interference with the climate system*". Another source of impotence is the fact that the United States (US), China and India, the world's largest GHG polluters, have not ratified the treaty. Through the Principle of "Common But Differentiated Responsibilities and Respective Capabilities (CBDR-RC)" the Convention imposed greater commitments on developed countries and lesser commitments and non-binding obligations on developing countries, thus leaving the responsibility of this global environmental problem to developed countries.

Against this scenario, during the last years, there has been an increasing perception by international legal scholars not only at national level but also at regional and international levels that climate change is seen as a security threat (Koivurova 2014; Scott and Andrade 2012).

An escalating persisting attention to this face of the phenomenon derives from the scientific community, especially from the IPCC's Fifth assessment report (IPPC 2015). This report notes an escalation of the level of probability of the consequences of climate change, such as the worsening of the severity of drought, the acceleration of land degradation and desertification, the intensifying of floods and tropical cyclones, the increase of malaria and devastating effects in vulnerable and fragile key areas of the earth, such as Sub-Saharan Africa or the Arctic. Such areas are highly exposed to environmental risks and uncertainty. They are populated by the most vulnerable people and communities, the indigenous people, such as, the Inuit of the Arctic, and in these areas, adaptive capacity is very low.

Once the critical thresholds are exceeded, the risks of climate change turn into catastrophic existential threats, pandemics such as AIDS, which when aggravated by the current economic global crisis will inexorably lead to a seeming "evaporation of the concept of security", a concept that is not solely a military one, no longer one we traditionally have known.

Today, the concept of security is no longer read through geostrategic eyes but through a *multi-faceted* optic. Indeed, the *nexus* between security, conflicts and climate change, has come to a point of non-return. The *nexus* conceptualize that climate change, associated with the current global economic crisis, aggravated by the crisis of getting as much as possible access, to natural resources, catalyzed by energy-deficit states running to violently exploit fossil fuels to the last drop, or find fossil fuel substitutes vital for some countries, such as the US and China, is a reality.

This will inevitably lead to the risk of igniting violent global struggles over dwindling reserves, today located in what has become one of the most "attractive but vulnerable easy to extract areas" on earth due to higher temperatures and melting ice: the Arctic ocean.

The environment, natural resource extraction in a world where natural resources are limited (water, earth, and energy), population migratory movements, together with possible energy resource conflicts and human rights are all factors affecting security, factors that have to be taken into account in the global concept of security.

In recent years, the debate on whether climate change should be perceived and treated as a security issue has been taken up by political scientists with the so-called "Copenhagen School of Security Studies" and to some very few extents by international legal scholars treating the law of the UN (Scheffran et al. 2012).

The Copenhagen School has been analyzing the process by which an issue comes to be perceived as an existential threat, a process of "securization". This is seen in the work of Wæver, who politicized climate change and framed it as a threat to national, international and human security, and was the first to coin the term "securization" referring to the process by which an issue comes to be represented not only as a political problem, but as an existential threat.

At the UN level, the discussion is about what the securization of climate change would signify in the global world dimension and from the prism of international law shifting the debate on what role the United Nations Security Council (hereinafter UNSC or the "Council" *tout-court*) might assume, and if this role could help achieve “full securization” and thus benefit the global response to climate change or in a way to contribute to reinforce the climate change regime.

The UNSC has held two debates on climate change: the first in April 2007 discussing the relationship between climate, energy and security; and the second in July 2011 (UNSC Debates 2007 and 2011).

In both debates, discussions focused on whether to make climate change a “security threat”. The discussions were unresolved and the Council did not pass a resolution. It is worth noticing that a Security Council’s resolution may give binding force, if adopted, but legitimacy of such measure would be questionable, unless law making is done by established sources of international law. Moreover, realistically speaking, any decision at the Council takes political dominance. Nevertheless, idealistically, the Council would act in a preventative way and even if the Council is a political institution, it would be a strong force in tackling security implications of climate change and its impact on resource availability into account in conflicts analysis, mission planning and mission monitoring. It remains to be observed, along this article, if an action in this sense from such a political institution could be realistically possible. The three main obstacles were (1) whether climate change fit the appropriate scope of Council interests, (2) the type of action the Council should take and (3) if a Council leadership role could help strengthen the effectiveness of climate change’s regime by reducing international security threats.

The obstacles were attributed to the political attitude of certain states supportive or not supportive of a UNSC role in climate change, and also around the CBDR-RC principle. The legal ambitions were limited by matters of *realpolitik* including the veto power at the Council of the five permanent members, amongst which were the US and China, also both non-signatories of the Kyoto Protocol. Neither China nor the US took part in the open debate of the Council during the first debate of April 2007, confirming the US reputation as environmental “laggard” (Mehling et al. 2013).

The debate at the UN level, even if very controversial, reflected the strong attention of the international community on a global level of climate change, perceived as an undeniable threat to international security.

This conceptualization is when climate change can exacerbate already existing tensions resulting in migration or potential armed conflicts (i.e., desertification and land degradation were the major factors in the conflict in Darfur, or competition for water resources or fuels competition over depleting natural resources). In these cases, the threat comes from armed conflicts rather than climate change but in this conceptualization climate change is a “threat multiplier” or “stress multiplier” which means that exacerbates already existing tensions or instabilities. These types of effects pose the greater danger in fragile and vulnerable areas.

One way climate change affects international security is through direct impact on the survival and well-being of populations by a rising sea level, extreme weather conditions, the spread of diseases, for example.

The main challenge to manage effectively the effects of climate change as a threat multiplier is the ability to prepare and adapt adequately. Adaptation is the most difficult concern and will require new thinking about future threats and responses. Adaptation responses to climate change, including frequent extreme events, will have to include many types of human reactions. Hence, adaptation will not be possible if climatologists studying the global phenomenon do not work in extremely close connection with social scientists, in particular international lawyers, and do not consider scenarios that include, for example, military planners who incorporate climate change threats into their strategic plans including predictive modelling or agent based modelling.

In that sense, adaptation must take into account the specific peculiarities of the phenomenon of climate change a nonlinear phenomenon that determines non-linear changes or occasional shifts that potentially produce nonlinear political events. Human existence worldwide will come under major stress, and "massive nonlinear events" in the global environment, will give rise to massive nonlinear societal events. It is not known how people, states or organizations will react and this could lead to an infinite series of reactions and surprises, very difficult to control.

The legal framework of international law existing at global level, in particular, the UN global level of response, should be modified to account for climate change' as a threat to international peace and security, and to protect environmentally vulnerable countries and those more exposed to risk conflicts, such as for example, the Middle East where many countries rely on the external sources, or the Arctic, where the retreat of ice paves the way for competition over resources, affecting indigenous people such as the Inuit, a totally unprotected people.

We advocate that environmental security framed as a collective security problem, can render climate change regime more effective. By incorporating security risk in this adaptation and giving it a central role especially in areas with a high level of exposure to environmental risks and uncertainty, and by incorporating these risks assessments into laws, potential armed conflicts and strategies to prevent conflicts occurring or to minimize global damages can be identified and prevented.

The AEO is the perfect case study to show that regardless of national, cultural or ethics differences, "we" humans are forced to cooperate because of the "cascade effect". Hence, cooperation is aimed for the benefits and protection, not only for Arctic communities but for the security and protection of all of us.

The Concept of Security in the Arctic and From Whom

In the current AEO "we" are the threat and it is really up to us to play a role. In the AEO, security of humans, environment and societies can be seen as a deal with the same level as traditional political, military and economic security. This is the reason why in the AEOG the question "what security, and from whom" receive a universal and global anwear: this is a comprehensive and collective security, "a security for all".

The new paradigmatic shift from national security, to regional security and finally, to global security in a multi-level dynamic's vision, implies a change in reference objects, the value to be protected and the security dangers. It concerns humans that have to cope with new policies and strategies where hazards and climate change' effects must be included as climate change is a threat to global security that all military planners must include in their calculation. In particular, attention should be given to the AEO case because of the "cascade effect" that starts in the Arctic. This led us to understand why we need to focus on the UN level and why is this level entagled with the Arctic level. "We" the people of the UN Charter, have the global responsibility to protect not only our self but also the "other" (states, indigenous people, groups, etc.) from "us", as the main cause of anthropogenic source of pollution as "we" are polluting and contributing to global warming and global environmental change including the "cascade effect". This means that we are both the author of the environmental damage and the victims at the same time. We are in fact, contributing to hydro-metereological hazards and we have the moral responsibility to protect the Arctic vulnerable people against ecocide of people who did not contribute to climate change at all, such as indigenous people. We have the ethical responsibility to permit adaptation to take place. This is why a global coordinated action of the role of military people, would be vital to protect in a preventative way, societal collective challenges.

The UNSC can tackle climate change on the basis of Article 39 of the UN Charter, stating that the Council can take action against any "threat to peace, breach of the peace, or act of aggression". The same article determines what the UNSC will do if it determines that such a situation will occur. If the situation occurs, the UNSC can make recommendations or decide which measures should be taken in accordance with Article 41 and Article 42 to maintain and restore peace and security (Conforti and Focarelli 2010).

Articles 39, 41 and 42 are contained in Chapter VII of the UN Charter. Chapter VII works in a way that for situations not involving the use of armed forces, it is Article 41 that authorizes the Council to take non-forcible measures. But if the Council deems necessary the use of force [which is normally prohibited under Article 2(4)], the Council can do so under Article 42.

In order to fight climate change and deal with environmental threats and emergencies, the Council could use the mechanism of Chapter VII. This means, theoretically, that the Council could pass (if a veto is not applied) a resolution under its Chapter VII powers but this would require compliance by all states. The Council could use Chapter VII and enact a resolution to assume the role of the peak body at the international level, addressing mitigation and adaptation, but the Council would have first to identify a "threat to the peace, breach of the peace or act of aggression" as a prerequisite to pass such a resolution. Here lies the heart of the problem: accepting climate change as security threat.

Full securization would be a stage that could be beneficial for a global policy response and could reflect positively also at the regional level and thus enhance adaptation. Nevertheless, the global response is far from that stage.

Despite the Council is not formally a "law-making body", it actually already started to use its mandatory powers to adopt binding resolutions and antiterrorist measures laying down general rules for all states. The Council could treat climate change with anti-terrorism measures with what is referred as the "1373/1540 model" (SCR 1373, 2001 and SCR 1540, 2004). This would help overcome the ineffectiveness of multilateral treaty governance and avoid the problem of the veto at the UNSC, adopt a resolution that imposes monitoring of compliance establishing, for example, even "green helmets" under "environmental peacekeeping operations", to intervene in conflicts caused by rising sea levels, deal with emergencies resources disputes, and even decide that all states participate in the climate change regime and ratify the treaty. The result would be an improved adaptation, which is the mayor problem challenge, but also one that most importantly turns to be much more problematic in adaptation: "variation" or "variability". This is why there is an urgent need to have a security approach in the AEO. But what does it mean to have a security approach in the AEO? It means a human security approach that link the environment security dangers to humans that causes "hazards" and aim to stabilize and prevent new security environmental and social challenges. This would both stabilize conflicts and address sources of uncertainity, given the easy access to natural resources. This can only be achieved with a public security approach at the UN level, based on the existing system of rule of law and by guaranteeng an effective legal framework that expands the international presence and establish new security forces, like the above mentioned "green helmets" and a multi-regulatory framework that shows evident interactions and synergies among the levels of source of law and policies. A security approach in the AEO means to stabilize and prevent degenerations of conflicts and it is a collective task of planetary interests that unify environmental, human and societal need in a sole concept. The individual and societies have much more difficulties managing variability in climate change. Variability includes for example changes in patterns or statistics of climate change, both of which produce much more uncertainty because adaptation to variability, not just change, requires far more information and resources for detection such as, for example, planning, decision-making and implementations.

Variation is much more stressful because stress much more humans and societies when we are talking about variability in access to natural resources and not by the level of access alone (Revilla et al. 2010).

The US and China, and recently Russia blocked the sole possibility for the UNSC to take a leadership role on climate change (King 2013). If this deadlock situation could be unblocked, this would probably determine a spill-over positive effects from climate change security threats responses over climate change regime effectiveness in general and improve adaptation.

Method

The assessment of the AEO from a regulatory point of view needs a multilevel regulatory evaluation of the different sources of law and policy, taking into account the role of institutions involved. This would entail a re-evaluation of different sources, such as the law of the UN and the law of the sea with the United Nation Convention of the Law of the Sea (UNCLOS), the AC and North Atlantic Treaty Organization (NATO) but also foreign policy concerns of both Arctic and non-Arctic states.

Risk and uncertainty of adaptation in AEO security, and variability rather than change in temperature, is analysed by combining different genres of disciplines, such as legal analysis, policy analysis, scenario analysis both theoretically and empirically, to change existing international law and to incorporate risk assessment reasoning based on disaster risk reduction (DRR).

Variability, not change is the most difficult factor to treat in adaptation to climate change because risk is observed at all the levels, and in all the sectors and fields. Tackling variability and permitting adaptation requires working with an agent-based modelling approach (Revilla et al. 2010).

The difficulties in changing law and the functioning of the institutions at the UN level and recognizing climate change as a threat to international peace and security, or even treating climate change with antiterrorist measures, are the same as the difficulties faced when the Conference of the Parties (COPs) of the UNFCCC was negotiated: obstacles in the decision-making process between big power relations, in particular the US and China which renders the linkage between climate change and geopolitics undeniable and thus the *nexus* between climate change, human security and conflict, evident. This set up aggravates the problems that may arise in the Arctic as ice retreat becomes even more dramatic and will exacerbate the behaviour of states. The suggested multi-level, interdisciplinary and comparative method between the different levels of sources of law and policy and the role of institutional actors involved in AEOG is necessary to understand how adaptation should be treated in fragile vulnerable areas, how climate change as threat multiplier is manifesting, how institution frameworks and structure should be changed, how legal frameworks should be redesigned, how the role of science could be used, in each cases, how priorities and structures should be undertaken in emergencies, and how the strength of the human component is in each areas. With regards to the study's delimitation of time and spatial dimension, for the first dimension, the paper covers the period of time from the post-Cold War ahead. For the second dimension, it connects what is happening after the Cold War with climate change in the Arctic environmental basin. The study considers the regulation, policy and management of resources located in the coastal states and in the continental platforms but also what is located beyond the continental platform since most of the resources are situated beyond the continental platform and refers to the narrative on the existing conflicts on the maritime delimitations. Also, since the term "resources" includes both "living" (fish and other sea/terrestrial living resources) and "non-living resources" (i.e., oil, gas and minerals), the resources treated in this study refer to the non-living resources solely.

Climate Change as a Threat to the International Regulation of the Arctic Environmental Ocean

From an Arctic security development point of view, the relationship between climate change and strategic interests related to sovereign claims must be emphasized. Nevertheless, due to the high level of institutionalization, including the collaboration among the Arctic Council (AC), the Nordic Council of Ministers (NCM) and the European Union (EU), the existence of the "rule of law", the cooperation among Nordic states plus the signature of the Ilulisat Declaration (a soft law instrument, not binding), the use of force in the Arctic is unlikely, realistically speaking. However, this situation is not static and could easily change due to unpredictable assertive behaviour of Arctic or non-Arctic energy-hungry countries acting in conflagration with the existing "tinderboxes", which calls for continuous re-evaluation and surveillance of the existing legal framework for the AEO.

Arctic Ocean Security

The Arctic Ocean can be considered a system bounded by the sea floor, a permanent sea-ice cap and surrounding land areas with inflow and outflow from the North Pacific as well as the North Atlantic. Seasonal solar forging is influenced by the Earth's axis, which is why the Arctic Circle is at 66.5° North latitude, providing an approximate limit for the Arctic Ocean (Berkam and Vylegzhanin 2010).

Within this Arctic system as a whole, meteorology and oceanography have a direct impact on the ecosystem and the populations of indigenous peoples of the surrounding "five costal states": Norway, Denmark, Canada, US, and Russia as well as the three non-costal Arctic states, Sweden, Finland, and Iceland which all make together, defined the "eight Arctic states".

In the last years, the Arctic Ocean has received renewed global interests from several parties and the challenge is to harmoniously reconcile resource exploitations and new geo-political and economic opportunities in a manner that promotes sustainable development. Environment must be integrated into the resolution of these challenges and must be protected in a sustainable way, merging these environmental changes and the risk that these activities entails using a peaceful and secure management of the Arctic Ocean, both for the civil society and investing stakeholders, and the Arctic states. This suggests the following precise definition of Arctic Ocean environmental security: "*an integrated approach for assessing and responding to the risks as well as the opportunities by an environmental state-change*" (Environmental and Security Initiative, UNDP 2011). The concept of security in the Arctic Ocean is therefore, implicitly and strictly linked to the environment and has to be understood in the same vein as the new concept of security explained at the beginning of this paper. From this point forward, the term "environment" will be implicitly and intrinsically encapsulated in the notion of Arctic security, as it is not possible to disentangle them.

These potentials represent, today a strong source of interests from both investors and states and also a latent underlying rumble of source of international conflicts and assertive behaviors not always easy to control and solve with diplomatic compromise or international agreements. Fossil fuels are, therefore, a potential factor of variability for Arctic environmental Ocean security, as it will be shown in the following paragraphs.

There are several "tinderboxes" that could lead to different scenarios in the Arctic. First, is the US-Canadian dispute over the maritime boundary in the Beaufort Sea; the US also rejects Canada's claim to the Northwest Passage. Second is Russia's claim to the Lomonosov Ridge, which is contested by Denmark and Canada, which have been working together to find evidence of a connection between the Greenland-Canada continental shelf and the Lomonosov Ridge. Third is Norway's claim to their position vis-à-vis Svalbard (the 200-nautical-mile "fishery protection zone" declared around the archipelago). This is not accepted by Russia and many other states including Norway's allies, such as Iceland. Fourth is the Canadian-Danish disagreement over the tiny Hans Island.

Despite the effects of climate change on these scenarios, political stability still prevails as there is an applicable international legal framework governing the Arctic Ocean based on the United Nations Convention of the law of the Sea (UNCLOS 1982). This obliges states to respond individually or jointly to possible new challenges. Nevertheless, future problems will consist in how to keep this stability under control given the high level of environmental risk which climate change brings about and unpredictable assertive attitudes by affected or involved countries. These potential conflicts could be shaken and exacerbated by the uncertain legal *status* of international law representing a challenge for the five Arctic coastal states and for the institutions and international organizations involved in Arctic Ocean governance.

This *status* of uncertainty on how to keep possible threats under control and maintain a certain Arctic peace and security also concerns non-Arctic states, such as China and southern European country members of the EU and which are also observers of the Arctic Council (AC) and classified as "energy-hungry" countries extremely interested in Arctic energy sources and new navigable routes.

The Arctic Ocean needs protection at the global level. In that sense, as noted by the United Nations Environment and Security Initiative: "*Peacefully resolving the overriding political, economic and social concerns of our time requires a multifaceted approach, including mechanism to address the links between the natural environment and human security*" (Environmental and Security Initiative UNDP 2011).

The United Nations Convention of the Law of the Sea (UNCLOS)

The Arctic region is mostly water surrounded by continents, meaning that there as also landmasses, unlike the Antarctic. It is a sea area where international law of the sea applies. Sovereignty depends on a country's territorial waters: this issue was very much debated at the UN level from 1960 to 1980 without any result. The Arctic region found a solution to its sovereignty issues only with the enactment of

UNCLOS a crucial treaty that define the rules for the use of world's sea for each nation around the global. UNCLOS is an "umbrella convention" which means that it globally covers several activities and agreements, such as fisheries agreements, and other sectorial and regional activities. Even with gaps in the formulation of its provisions, the convention is extremely important for peace and security issues as it is the only convention providing a comprehensive binding system for peaceful settlement disputes. In addition, UNCLOS present some shortcomings, for example, indigenous people which are very important for the Arctic, are not addressed trough UNCLOS.

The legality of territorial claims is determined by UNCLOS. This means that each state that wants to establish outer limits of its continental shelf beyond 200 nautical miles (NM) from its coast must make a submission to the Commission on the Limits of Continental Shelf (CLCS). The maximum is either 350 NM from the baseline (coast) or 100 NM beyond the 2500-m depth line (the letter rule does not apply to ridges, like the mid-Atlantic spreading ridge), whichever is better for the coastal state. The CLCS is set up under Article 76 and Annex II of UNCLOS. The CLCS is not composed of lawyers but scientists and work based on scientific evidence Even though it seems that that the provisions of the UNCLOS convention are clear, in reality there are gaps of clarity which open up too much room for interpretation with the risk it can be manipulated by states, especially non-Arctic states. The assessments often rely too much on scientific expertise as the CLCS is a technical body and has no competence in case of conflicts. Conflicts on sovereignty issues could be solved with agreements among states, or with an arbitral tribunal. If states cannot reach agreements, Art. 83 of UNCLOS refer and expressly make the link to Art. 38 of the Statute of the International Court of Justice to reach an "equitable solution" and in "good faith". This could stimulate cooperation between states, but denotes also the absence of binding mechanisms to solve conflicts and even if such mechanisms would exist, they could have no effect at all, since not all states are part of UNCLOS (i.e., the US has not ratified UNCLOS).

The United Nations Security Council (UNSC)

It has been noted previously how difficult it has been for the UNSC to include climate change as a primary responsibility for maintaining peace and security, as was raised in two proposals. Strong resistance came from two permanent members, the US and China, followed by Russia. Despite resistances, it is undeniable that climate change is now recognized by the UN system as mayor threat to both human and national security and is perceived as threat-multiplier, exacerbating already existing sources of conflicts (Report UNGA 2009).

The difficulties and resistances resulting from political disagreements in the UNSC *arena* impede a vote on a resolution that is very much needed and can have strong benefits in the Arctic in order to tackle the gaps and ineffectiveness for an effective and peaceful future in the Arctic Ocean governance as described previously.

Hence, a UNSC resolution is therefore, profoundly needed. Concretely, it should be an "organizational resolution which should be adopted, which is to say, establishing organs, or committees (UN Charter, Arts. 22 and 29)". It seems that the UNSC is neglecting the importance of the Arctic and not recognizing the cascading effect, this partly due to political disagreements, the lack of exchange between science and social scientists, the lack of information and the lack of knowledge as to the risks in climate change adaptation in the Arctic Ocean, which mixed all together, underestimate the relevance for establishing the *nexus* between Arctic issues, climate change and human security. Such a *nexus* should be considered crucial for the global level and also for the regional level because global progress reflects on regional and national normation processes.

The United Nations General Assembly (UNGA)

At the UN level, attention regarding the Arctic Ocean is missing, but not totally since there is one resolution from the United Nations General Assembly (UNGA) which very weakly mentions the importance of preserving the area of the Arctic Ocean. The resolution is not a specific resolution dedicated to the Arctic Ocean, but it is a general resolution about oceans titled "Ocean and the Law of the Sea" of the 2007 (Resolution UNGA 2007) in which it is stated under the formulation of "expressions" only a weak reference to the Arctic Ocean. It is enraging that the Arctic Ocean is given such little importance in a resolution like that, but even more unacceptable is that there is no resolution directed at all to the Arctic Ocean at this crucial *moment in time* where there is urgency and maximum level of alert on the effects of climate change in the Arctic Ocean. The gravest gap in this resolution is coming from treating the Arctic Ocean, as if it were like any other ocean and therefore not in need of a special legal framework. Thus, the minimum should at least be a modification of the UNGA resolution, as this would have an impact and be reflected at global level, which would in turn reverberate at the regional level and *vice-versa* as the synergies between the global and the regional will be beneficial for the regional and domestic level of states interested or involved in the Arctic Ocean regime. An ideal situation to improve the Arctic Ocean regime could be achieved by setting up a new organ, specifically a *liaison committee* at the UN level, working together with UNCLOS and the AC, all in cooperation with climatologists. This would recognize that Arctic Ocean governance and its system is strictly interconnected with the global dimension.

The Arctic Council (AC)

The AC is a central multilateral instrument of scientific cooperation in the Arctic region and crucial for the Arctic Ocean zone. The AC is characterized by its *soft* law, and its flexible and intergovernmental nature. Although the AC has recently established a secretariat, it still has no budget of its own, and does not enact binding laws but uses and creates *soft* law mechanisms. The AC is not equipped to deal with aspects of military security as it was originally created as a *forum* of cooperation and discussion regarding issues of scientific nature. Its work has been strictly connected to the work of the International Arctic Science Committee (IASC), the predecessor of the Arctic Monitoring Assessment (ACIA).

The main role of this institution is, in fact, to protect the Arctic environment by establishing cooperation among states when setting up scientific projects with the specific intent to involve indigenous people as "permanent participants".

It is certain that the AC is relevant for the protection of the Arctic Ocean environmental security especially in cases of natural resources use or industries such as shipping, oil and gas, and climate change.

Findings and Discussions: High Level of Exposure to Environmental Risks in Adaptation of the Arctic Environmental Ocean and the Need to Incorporate an Equation in Future Legal Framework

Climate change and human security are linked by uncertainty and unpredictability because climate change is a nonlinear phenomenon occurring in a multi-level dimension with scale effects at the global, regional and domestic level where discourses, policies and norms are generally translated in a top-down process for the "common goods" as observed with the attempt of the UN to play a role to tackle climate change and security from the prism of international law. But as observed, it was just an attempt, as the UN did not do this and was powerless to do so. A successful role of the UN in regulating the *nexus* between climate change and human security would have positive reverberation at the regional level in the case of the Arctic ocean case and at the domestic level of both Arctic and non-Arctic states, improving adaptation at all the levels.

As remarked previously, the *phenomenon* of adaptation is a very important element for the climate change and human security *nexus*.

In adaptation, risk plays a crucial role as it is the problem common to all the multi-regulatory levels, sectors and fields. Uncertainty and risks relevant to adaptation law and policy can be of two kinds: (1) uncertainties caused by our lack of knowledge, described as "epistemic risks" where the risks due to lack of knowledge gradually decreases as knowledge increases (but it is not easy because the effects of climate change do not occurs linearly) as explained above, and (2) uncertainties caused by randomness inherent in the phenomenon at hand, described as "aleatory risks". The risks inherent in randomness, for example of floods, will never disappear. These two kinds of uncertainty and risks are problematic not only for the business sector (i.e., for those who have to draft insurance policies) but also for law and policy as it requires long term approach of policy makers and regulators.

Hence, both environmental and humans risks, including possible war conflicts, must be treated together in order to permit adaptation to take place as climate change and damages does overlap with natural disasters. The issue is strictly linked with the studies related to disaster risk reduction (DDR) and with the frequency of extreme events.

For all that to proceed, the effects of climate change as drivers of insecurity need to be made very clear. An author, David Simon (Simon 2013) has proposed how the primary classical equation of the definition of disaster can be modified.

The primary classical equation is the following:

$$\mathrm{DR} = \mathrm{H} \times [(\mathrm{V}/\mathrm{C}) - \mathrm{M}]$$

were Disaster Risk (DR) is the product of Hazard (H) multiplied by the ratio of between Vulnerability (V) to Capacity to Protect oneself (C), minus Risk Mitigation Through Collective Action (M).

Simon's proposed modifying the above primary equation in order to include the importance of taking into account the Intensity of the hazards (I), the Duration (D) and the extent of asset/livelihood Exposure (E) when considering Climate Risks (CR). These modify the above equation in the following:

$$\mathrm{CR} = [(\mathrm{H} \times \mathrm{E})]/\mathrm{D}\ [(\mathrm{V}/\mathrm{C}) - \mathrm{M}]$$

This second version would be extremely suitable for application to the case of the Arctic Ocean environment because it considers the geographic location as an important variable affecting the likelihood and degree of exposure.

A place such as the Arctic Ocean environment where the level of exposure to environmental risks and uncertainty is high should be treated according to this second version equation. Nevertheless, if applied to the Arctic Ocean environmental governance, the equation would need to be modified and to be integrated with social science elements and problems in order to prevent and manage peace and security in this vulnerable zone. Simon's equation should include the interaction of other factors such as, possible risk of conflicts and interactions with other unpredictable variables, which could the unpredictable assertive behaviours of states, leading to possible conflicts.

Hence, Simon's second version equation could be modified once again whereby Climate Risk (CR) could be replaced by Arctic Ocean Governance Risk (AOGR), and the inclusion of Hazard (H), Intensity of a conflict (I) and Exposure Level During a Conflict (E) lead to a third version:

$$\mathrm{AOGR} = (\mathrm{H} \times \mathrm{I} \times \mathrm{E})/\mathrm{D}(\mathrm{V}/\mathrm{C} - \mathrm{M})$$

Finally, since AOGR cannot be observed detached from the global dimension, and since environmental climate change in the Arctic determines change at the planetary global level, Total World Climate Risks (TWCR) are given by the Climate Risks (CR) multiplied by AOGR to obtain a forth equation:

$$\text{TWCR} = \text{CR} \times \text{OAGR}$$

The proposed modified equations could be incorporated into the legal framework for the maintenance of peace and security as an attempt to prevent uncertainties and instabilities with an attempt to achieve adaptation and manage adaptation. But two factors make it difficult for humans and societies to achieve and manage adaptation, and most importantly, to design a legal and policy framework of "adaptive law" that simultaneously prevents the negative effects of climate change and preserve security. The first is the presence of invisible factors. The second is the existence of "variability" and unpredictability. This is problematic for adaptation to climate variability, not just to change, as highlighted in a research programme, taking precisely the Arctic as a case study (Revilla et al. 2010) because it requires far more information and resources than we have for detection, planning decision-making and successful implementation.

The study explains that variability is more challenging than the changes themselves because the greater stress on humans and societies is caused by variability in access to natural resources, not by the level of access alone.

A possible venue to alleviate this latter difficulty could be to apply Agent-based modelling to climate change, ecosystem and security in the special case of the Arctic Ocean environment which would be suitable to satisfy the need to take into account the "variability" of the phenomenon of adaptation, as depicted in Fig. 5.1 below:

Figure 5.1 shows how possible conflicts due to unpredictable assertive behaviours of states could be considered and incorporated into adaptation processes using agent-based modelling.

The Fig. 5.1 explains that the processes of adaptation (and the laws and societies which must include the risk in order to be adaptive and updated to climate change effects) have different phases (5) which requires actions to be taken. If during these phases there is a possible conflict these actions for adaptation can be interrupted or delayed and this will have negative reverberations of the process of adaptation.

The integration of environmental variability into risk analysis requires a predictable and probabilistic approach. Estimation of risk requires the applied methodology to be predictably and probabilistically based as for example, seen in stochastic computer simulations. Stochastic simulation models can also accommodate various global change scenarios, which may not be readily accomplished by mathematical analysis.

We therefore, propose combining some of these techniques to give more elaborate risk assessments that include the effects of climate changes and the risks of conflicts. In order to do that, it would be a possibility to build spatial and spatial-temporal frameworks, notably in the context of the complex, large-scale ecological dynamics driven by global change could be built.

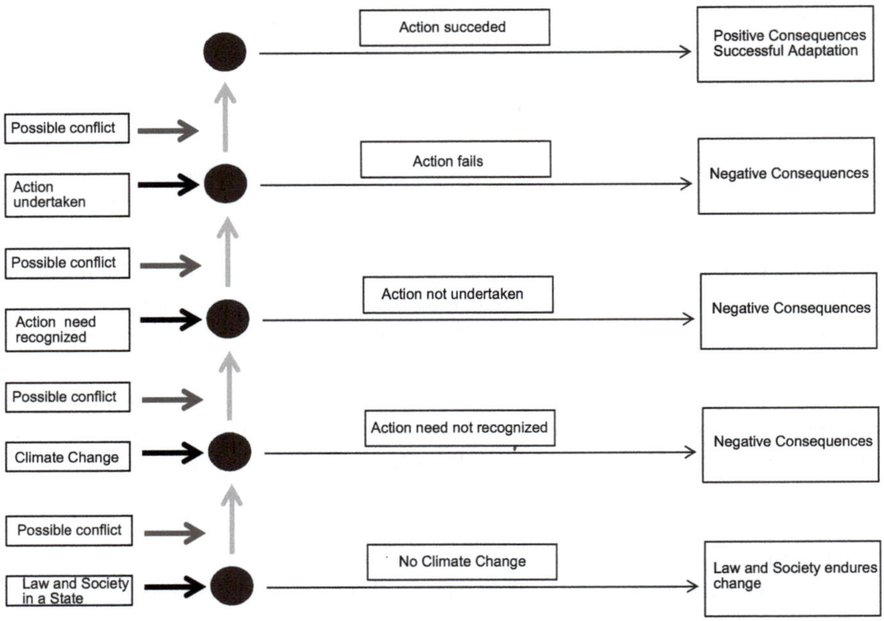

Fig. 5.1 Agent based model on variability of adaptation's conflict predictions into adaptation plans for vulnerable areas

The development of theoretical models and the use of computer simulations could contribute significantly to risk assessment field through the development of predictive models that incorporate both environmental and geopolitical data-set. These models, which could include stochastic environmental effects due to climate change, could allow us to make probabilistic predictions that can be reasonably precise when we consider averages over large scales.

We should bear in mind that deterministic simulations (those which do not take into account variation) are based on algebraic equations that predict the likely outcome of sampling, while stochastic simulation models mimic random processes. Although being transparent and analytically tractable, deterministic predictions cannot deal with the same level of complexity over many generations as stochastic simulations. One advantage of combining these approaches is apparent from simulation used to verify the accuracy when prediction equations are developed.

Therefore, the data utilised could be integrated into a modelling framework by using, for example an individual-based modelling (IBM). In such cases the IBM or agent-based approaches can be appropriate ways to allow variation in many environmental and geopolitical parameters as well as variable and complicated conditional interactions. Likewise the geospatial implementations of IBM can account for specific spatial effects. Empirical data can be entered in the models at several levels, but the real strength of this methodology is the bottom-up design. Here data is typically included on the lower levels, and patterns on higher levels can be observed as emergent properties of the system. As factors or estimates can be manipulated at almost all levels of the modelling, effects from ecological changes can be predicted, especially those related to spatially and temporally dynamic environments. If the information obtained can be combined with empirical data, the model will provide a powerful tool for understanding real-world dynamics.

Conclusion

From all the previous interdisciplinary analysis combining international law, elements of climatology, political science and agent based modelling this paper set some recommendations of interdisciplinary nature about future prospects for the global management and regulation of climate change in the AEO case and at global level. In particular:

- Adaptation to climate change as a consequence of security threats should take into account the "nonlinearity of climate change" as this same nonlinearity can determine a nonlinearity of social events, behaviors, and totally unpredictable social conflicts that will mean that countries with unpredictable strategies and behaviors have a greater impact on security. This means that there is a correlation between the nonlinearity phenomenon and its repercussions on societies that should be taken into account when adopting plans and designing legal framework and policies.
- Because of the previous finding, much attention at the UN level should be dedicated to the "cascading effect". In that sense, AEO security should be perceived not as a regional issue but as a global one and as a consequence regulated through a global and systematic regulatory structure. The UN level could support many issues which have an impact on AEO governance. Nevertheless, the UN level needs to be reformed at the institutional level and needs to enact new kinds of legal acts, something which is not easy to be achieved at least in a short term.
- It would be recommended for the UN to work out on the identification of general principles of international law that could be applicable to the case of the Arctic Ocean governance as this would be extremely necessary, considering the urgency of the catastrophic climate change effects in the AEO.
- There are a number of questions that have not been solved by the Council (and that the Council could solve) that are extremely relevant, especially for their reverberating effects for the maintenance of peace and security in the Arctic ocean: How should dispute about sovereignty and exploitation rights be settled in order to create more energy security; How can environmental security be achieved in this region under the impact of climate change; How can the military use of the Arctic be regulated in a way that it does not endanger the global peace and security; How can military security be used to create a nuclear weapon free zone.

- The starting point to implement the previous recommendations could be to take into consideration and apply both the proposed revised equations and the agent-based modelling analysis could represent the universal basic reasoning for a framework package of adaptive interdisciplinary regulatory proposal which could be integrated in a future resolution at the UN level in order to design the *nexus* between climate change and human security. This could be crystallized in a resolution/s using a new design to incorporate risks and uncertainties in a global international law framework, creating a more pro-active climate change law rather than a reactive.
- From the institutional point of view, to implement the recommendations sets in this paper, it would be auspicious to create a *liaison committee* linking the work of the UNSC and the AC because the UNSC lacks scientific data and information, a gap that could be complemented by the AC. In the *liaison committee*, lawyers and climatologists should work hand in hand on the risk, especially to set thresholds, considered alarming or in warning. This has to be done again mainly through an agent-based modeling analysis, as suggested in this paper.

Acknowledgments The authors would like to thanks Vito Todeschini, PhD Fellow at the Department of Law of Aarhus University for reading and commenting the first section.

References

Arctic Monitoring and Assessment Program (AMAP) (2011) Snow, water and permafrost in the Arctic. Oslo. www.amap.no/swipa/SWIPA2011ExecutiveSummaryV2.pdf

Berkam PA, Vylegzhanin AN (2010) Environmental security in the Arctic Ocean, The NATO science for peace and security series—C: environmental security. Springer, New York, NY, pp xx–xxi

Charter of the United Nations (UN), Arts. 22 and 29

Conforti B, Focarelli C (2010) The law and practice of the United Nations, Fourth Revised Edition. Martinus Nijhoff Publishers, pp 75–145

Copenhagen Climate Conference (2009)

DPSIR model, (Driving Forces, Pressures, States, Impacts, Responses) (2009) European Environmental Agency (EEA)

First-Debate on Impact of Climate Change on Peace, Security, Hearing over 50 Speakers, 566rd meeting, UN Doc. SC/9000, April (2007) Department of Public Information, Security Council Holds and Debates Climate Change and Security, July 20 (2011) Institute for Environmental Security, UN Security Council

Intergovernmental Panel on Climate Change (IPCC) (2015) Fifth assessment report. http://www.ipcc.ch/report/ar5/index.shtml

King E (2013) China and Russia block UN Security Council climate change action, responding to climate change (RTCC). 19 February 2013

Koivurova T (2014) Climate change and international security. Koninklijke Brill NV, Leiden

Kyoto Protocol to the United Nations Framework Convention on Climate Change (1997) Kyoto, 10 December

Mehling M, Frenkil D (2013) Climate law in the United States: facing structural and procedural limitations, Chapter 18. In: Hollo EJ et al (eds) Climate change and the law. Springer, New York, NY, p 474

Meyers N (2007) Environmental security concerns: sources. NATO workshop on energy and environmental challenges to security, Budapest, 21–23 November 2007, at 1–11

Omann I et al (2009) Climate change as a threat to biodiversity: an application of the DPIR approach. Ecol Econ 1–7. Elsevier

Report, United Nations, General Assembly (2009) Climate change and its possible security implications. September 2009, document A/64/350, at 1–20

Resolution adopted by the General Assembly on 22 December 2007, 62/215, Ocean and the Law of the Sea, General Assembly, A/RES/62/215, Sixty-second session, Agenda item 77(a)

Revilla CC et al (2010) Agent-based modelling of climate change, ecosystem, and security: a research programme. Prepared for the 250th Anniversary conference on climate change and security, The Royal Norwegian Society of Science and Letters (Det Kongelige Norske Videnskabers Selskab, DKNVS), Trondheim, Norway, 21–24 June 2010, p 4

Scheffran J et al (2012) Climate change, human security and violent conflict—challenges for societal stability, vol 8, Hexagon series on human and environment security and peace. Springer, New York, NY, pp 3–29

Scott VS, Andrade RCD (2012) The global response to climate change: can the security council assume a lead role? Brown J World Aff 8(Spring/Summer):215–226

Environment and Security Initiative (2011) United Nations Development Programme (UNDP), United Nations Environmental Programme (UNDP), Organization for Security and Cooperation in Europe (OSCE), North Atlantic Treaty Organization (NATO), United Nations Economic Commission for Europe (UNECE) and Regional Environmental Centre for Central and Eastern Europe (REC). See http://www.envsec.org

Simon D (2013) The environmental determinants of human security in the context of climate change. In: Redclift RM, Grasso M (eds) Handbook on climate change and human security. Edward Elgar, Cheltenham, Northampton, MA, pp 113–120

United Nations Convention of the Law of the Sea, Dec. 10, 1982, reprinted in 21 ILM 1261 (1982)

United Nations Framework Convention on Climate Change (UNFCCC) (1992) New York, 9 May

Wadhams P (2013) Diminishing Sea-ice extent and thickness in the Arctic Ocean. In: Berkman A, Vylegzhanin N (eds) Environmental security in the Arctic Ocean. Springer, New York, NY, pp 15–57

Chapter 6
Citrus and Tomatoes Response to Climate Change: Survey of Farmers' Perception and Adaptation Strategies in Northern Nigeria

O Adebisi-Adelani

Abstract Climate Change is a major challenge to agricultural development in Africa, Nigeria in particular and the world at large. Agriculture, as one of the most weather-dependent of all human activities, is highly vulnerable to climate change. The study was carried out to examine farmers' perception and adaptation strategies to citrus and tomatoes response to climate change in Northern Nigeria.

Multi-stage sampling procedure was used to select the study area. Purposely, North central and North-eastern zones were selected due to distinct comparative advantage in horticultural crop production. Benue (Central zone) and Gombe (North-east) agricultural zones were selected using random sampling technique, while Agricultural Development Programme (ADP) zones that are known for mass production of selected horticultural crops within the state were purposively selected. However, 271 farmers were proportionately sampled.

Secondary (weather data) as well as primary sources were used for data collection. Primary sources of data used include structured questionnaire and Focus Group Discussions (FGDs) which were subjected to both descriptive and inferential statistics. Result of analysis shows that majority of the respondents were between the ages of 35 and 45 years of age, married (89.7 %), and had formal education (84.5 %). Fifty-seven percent (57 %) were in high adaptation category, while 52.0 % were favorably disposed to the effect of climate change on production. Binomial logit regression shows that soil fertility, farmers' financial capacity and access to inputs to a large extent determine their choice of adaptation strategy. However, the inferential statistics result shows a significant difference in the adaptation strategies ($p = 0.000$, $t = -24.106$) and farmers' perception of citrus and tomatoes response to climate change ($p = 0.004$, $t = -182.269$).

O. Adebisi-Adelani (✉)
Farming Systems and Extension Research, National Horticultural Research Institute, P. M. B. 5432, Jericho, Ibadan, Nigeria
e-mail: adelanidotol@yahoo.com; adelanidotol@gmail.com

W.L. Filho et al. (eds.), *Climate Change Adaptation, Resilience and Hazards*, Climate Change Management, DOI 10.1007/978-3-319-39880-8_6

Intervention by relevant government and nongovernmental agencies toward adaptation strategies to mitigate the effect to this response will ultimately promote food security in Nigeria.

Keywords Climate change • Adaptation • Perception • Nigeria • Food security

Introduction

Changing climate had been accredited directly or indirectly to human activity that alters the composition of the global atmosphere and which led to the natural variability observed over comparable time periods. As implied from the definition therefore, the climate system can vary naturally, and when augmented, becomes a change (Adejuwon 2004).

Climate change is a major challenge to agricultural development in Africa and the world at large. Agriculture (being one of the most weather-dependent of all human activities) is highly vulnerable to climate change. Climate change threatens agriculture production through rising temperatures, changes in rainfall patterns or the increase of drought. This is directly linked to reduced soil productivity and to a higher incidence of pests and diseases (LEISA 2008). Horticultural crops such as mango, citrus, pepper and tomatoes are dependent on climatic factors for their growth. Thus, they are not left behind in the effect of climate change, simply because vegetables are very sensitive to water availability and minor stress of temperature outside the optimal range. Also flowering stage of fruit trees are susceptible to heavy rainfall. The fact that agricultural production in Nigeria is primarily rain–fed further reinforces the importance of climate to agriculture.

Some areas in the south western Nigeria that used to be termed rain forest are becoming derived savannah. Crops like water melon, cucumber and carrot that are usually grown in the savannah in those days are now being grown in the areas leading to the disappearance and extinction of tree crops which are supposed to be cash crops in these areas. Thus, the existence of such crops is being endangered. Farmers had to live with the realities of climate change to be able to manage the situation and to maintain their enterprise.

Odjugo (2009) reveals that the ways Nigerians are responding, perceiving and adapting to the changing climate have not been well investigated. Furthermore, while research efforts have focused on climate change at global scale, regional climatic patterns in Nigeria, have received limited attention. Worst still, the impact of climate change on agriculture especially horticulture has not received the desired attention (Nyelong 2004; Ati and Iguisi 2007).

The foregoing suggests the need for studies on the various ways farmers perceive the effect of climate change on horticultural crop production in Nigeria. This is in order to provide direction for the intervention that is currently emerging and to minimize the adverse effect of climate change on agricultural production at large.

Specific Objectives of the Study

1. Identify the socioeconomic characteristics of citrus and tomato farmers in the study area.
2. Assess the respondents' awareness of changes in major climate parameters.
3. Ascertain farmers' perception of the effect of climate change on citrus and tomato.
4. Determine the adaptation strategies of citrus and tomato farmers due to climate change and identify factors that influence the strategies used.

Research Hypotheses

Ho1: There is no significant difference in the perception of citrus and tomato farmers about climate change within the zones.

Ho2: There is no significant difference in the adaptation strategies used by citrus and tomato farmers within the zones.

Methodology

Study Area

Northern Nigeria is predominantly occupied by Hausa, Fulani, Gwari, Borim, Kanuri, Tiv, Jukun and many other tribal groups. Nigeria is the most populous nation in West Africa, with a population of about 200 million. In northern Nigeria, there are two distinct seasons; wet season and prolonged dry season. Temperatures during the day remain constantly high while humidity is relatively low throughout the year, with little or no cloud cover. The mean monthly temperatures during the day exceed 36 °C while the mean monthly temperature at night falls below 22 °C.

Presently, Northern Nigeria is made up of the following 19 Nigerian states:

Adamawa, Bauchi, Benue, Borno, Gombe, Jigawa, Kaduna, Kano, Katsina, Kebbi, Kogi, Kwara, Nasarawa, Niger, Plateau, Sokoto, Taraba, Yobe and Zamfara (northernnigeriatourism 2009). Figure 6.1 shows the map of the study area.

Sampling Techniques

Multi-stage sampling procedure was used to draw sample for the study. Purposely, North central and North-eastern zones were selected due to distinct comparative advantage in horticultural crop production. Benue (North central zone) and Gombe

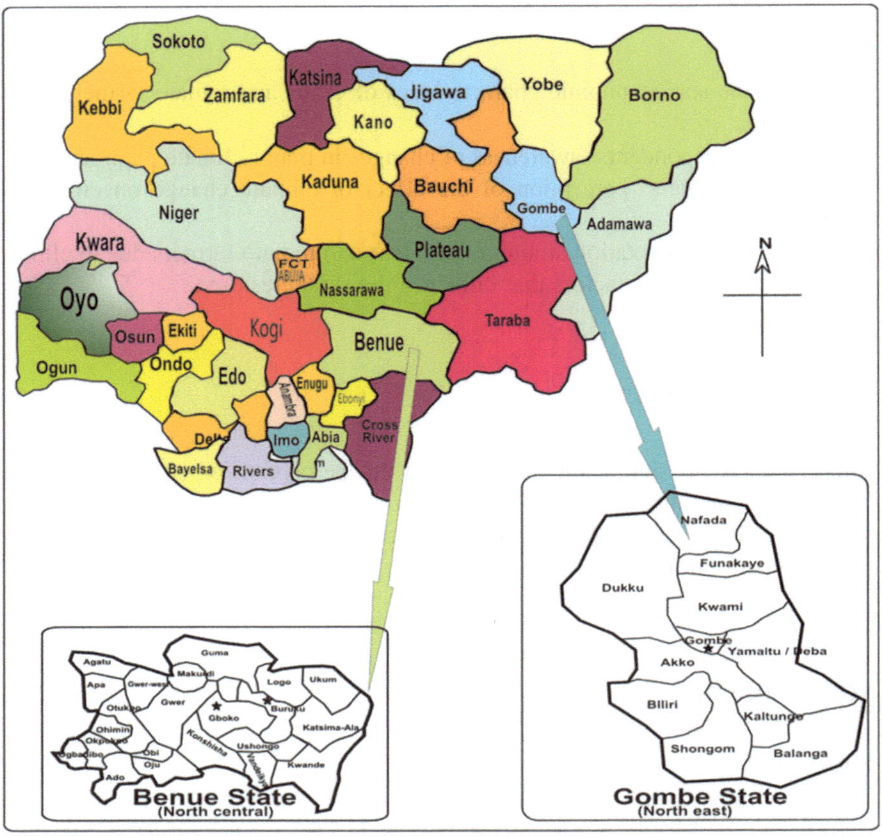

Fig. 6.1 Area of study (Gombe and Benue states). *Source*: 'Original', 2014

(North-east) agricultural zones were selected using random sampling technique, while Agricultural Development Programme (ADP) zones that are known for mass production of selected horticultural crops within the state were purposively selected. However 25 % of ADP blocks and 50 % of cells within the selected blocks and zones were randomly selected. Consequently, from the list of fruits and vegetable producers association, a proportionate sample of 271 farmers comprising of 157 and 114 Citrus and Tomatoes farmers respectively were selected.

Instrument for Data Collection

a. Qualitative method of data collection—Data was collected through the use of Focus Group Discussions (FGDs).
b. Quantitative method—Administration of validated interview schedule.

c. Secondary data: Rainfall and temperature data collected from the two major cities in each of the zones.

Limitation of the Study

The limitation of this study is that it focused only on rain-fed agriculture.

Results and Discussion

Socioeconomic Characteristics of the Respondents

The result (Table 6.1) reveals that the respondents had mean age of 46.0 years $\pm$ 7.6. This is an indication that most of the respondents are still in their active years and they contribute significantly to agricultural production of the country. This finding is consistent with that of Yekinni (2010) and Salimonu (2007) who reported a mean age of 43.2 and 48.1 years for farmers in different studies carried out across agricultural zones of Nigeria. Distribution of respondents by sex shows that 84.9 % were male, while 15.1 % were female as shown in the Table. This finding corroborates Oyedele (2005) who reported that rural women in Nigeria do not have direct access to land ownership and inheritance. Results reveal that 10.3 % had no formal education, 19.2 % completed primary education, and 44.3 % had secondary education, while respondents with tertiary education were 21.0 %. This implies that majority of the respondents were literates. Discussants during the FGDs corroborated this when they stated that *"they can read and write"*. This is consistent with Oladeji (2011) who stated that farmers have one form of education or the other in a related study.

The result further reveals that 52.0 % of respondents had between 10 and 20 years in horticultural production. This implies that majority of the respondents had been into horticultural production for over a decade; meaning that they are highly knowledgeable and experienced in horticultural farming this will help them to discuss better on the effect of climate change on their production over the years and be better informed about its general effects on livelihood. On family size, the modal family size class was between 5 and 8 persons, while the mean family size was 6 $\pm$ 3.4. This depicts a fairly large family size in Nigeria. The implication of this finding is that if the family size is big, the family responsibilities for poor women become enormous in the time of natural disaster like flood or drought and thus women becomes more vulnerable. This big family size will in turn make the woman and girl child more prone to climate change fury (Aaditya 2011). Majority (59.8 %) of the horticultural farmers have between 1 and 3 ha, 24.1 % had $<$1 ha, 11.1 % had between 3.1 and 5.0 ha and 2.2 % (5.1–7.0 ha). This implies that farmers are smallholders and it is due to the fact that Spencer (1990) put the upper limit of

Table 6.1 Frequency distribution of socioeconomic characteristics of respondents

Variable description	Frequency	%	Parameters
Age (years)			
35–45	192	70.8	Mean = 46.0 ± 6.7
46–55	64	23.6	
56–65	10	3.7	
66–75	3	1.1	
>75	2	0.7	
Sex			
Male	230	84.9	Mode = male
Female	41	15.1	
Educational level attained			
No formal	28	10.3	
Primary education	52	19.2	Mode = secondary
Secondary education	120	44.3	
Tertiary education	57	21.0	
Others	14	5.2	
Number of years in horticultural production			
0.00	3	1.1	
<10	47	17.3	
10–20	141	52.0	
21–30	60	22.1	Mode = 10–20
31–40	8	3.0	
41–50	11	4.1	
>50	1	0.4	
Family size			
0.00	18	6.6	
1–4	59	21.8	
5–8	100	36.9	Mode = 5–8
9–12	73	26.9	
>12	21	7.7	
Income			
0.00	9	3.3	
<10,000	1	0.4	
10,000–30,000	27	10.0	Mode=> 90,000
31,000–50,000	55	20.3	
51,000–70,000	22	8.1	
71,000–90,000	15	5.5	
>90,000	142	52.4	
Farm size			
<1.0	67	24.7	
1.0–3.0	162	59.8	
3.1–5.0	30	11.1	Mode = 1.0–3.0
5.1–7.0	6	2.2	
7.1–9.0	4	1.5	
>9.0	2	0.7	

small scale farming at three hectares. This shows the dominance of small farm size holdings in the study area.

Respondents' Awareness of Changing Features of Climate Parameters

The result in Table 6.2 show that respondents were aware of the general decrease in yearly amount of rainfall ($\overline{x}=0.92$), reduction in rainfall days ($\overline{x}=0.81$), increase possibility of loss of soil nutrients ($\overline{x}=0.70$), prolonged dry season ($\overline{x}=0.67$), increased frequency of drought in recent decades ($\overline{x}=0.69$) and increased intensity of drought in recent decades ($\overline{x}=0.65$). The result implies that citrus and tomato farmers in Nigeria were aware of the changes that occur in climate parameters in recent years.

To corroborate these results, farmers in Gombe stated during the FGD that "we are aware that there has been less rainfall in the past 2 years when compared to 3–5 years ago when rainfall normally starts in June. Although it is less in terms of duration, the intensity has been high which then leads to flooding".

The findings were consistent with Oyekale et al. (2009) who reported that 58.6 % of cocoa farmers in Nigeria were aware that there is low rainfall in recent years with other climatic parameters.

Figures 6.1 and 6.2 on rainfall and temperature distribution from secondary data also supported the findings that in using the lines of best fit there is general decrease in yearly amounts of rainfall and slight increase in temperature. Farmers are also

Table 6.2 Distribution of respondents' awareness of changing features of climate parameters

	NE		NC		Total			
Statement	F	%	F	%	F	%	WMS	SD
General decrease in yearly amounts of rainfall	91	33.6	158	58.3	249	91.9	0.92	0.27
Intense Harmattan period	90	33.2	55	20.3	145	53.5	0.53	0.49
Reduction in rainfall days	73	26.9	147	54.2	220	81.2	0.81	0.53
Increased possibility of loss of soil nutrients	63	23.2	128	47.2	191	70.5	0.70	0.45
Prolonged dry season	85	31.4	96	35.4	181	66.8	0.67	0.47
Incidence of sand dunes	60	22.1	47	17.3	107	39.4	0.39	0.49
Wind dryness	81	29.9	59	21.8	140	51.7	0.52	0.50
Increased rainfall intensity	62	22.9	86	31.7	148	54.6	0.55	0.50
High humidity	50	18.7	109	54.6	159	58.7	0.59	0.49
Increased frequency of drought in recent decades	62	22.9	126	46.5	188	69.4	0.69	0.46
Increased intensity of drought in recent decades	76	28.0	100	36.9	176	64.9	0.65	0.48

WMS weighted mean score; *SD* standard deviation

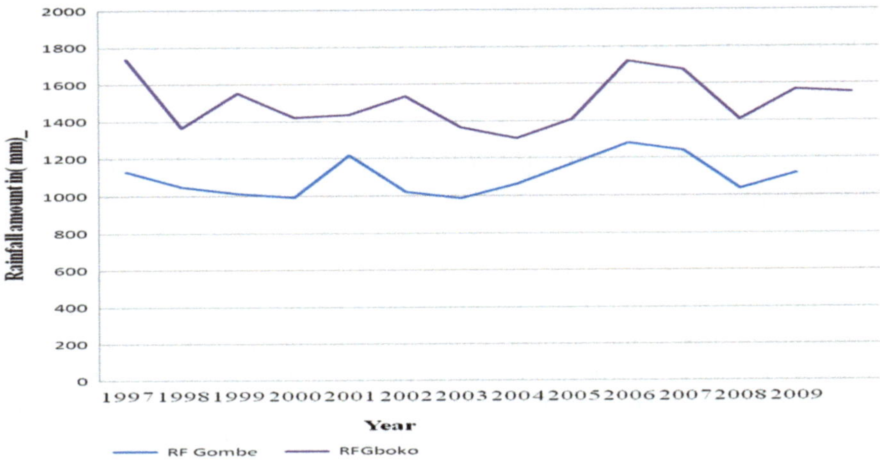

Fig. 6.2 Rainfall distribution pattern in Gombe and Gboko (Nigeria). *Source*: Stackhouse (2010)

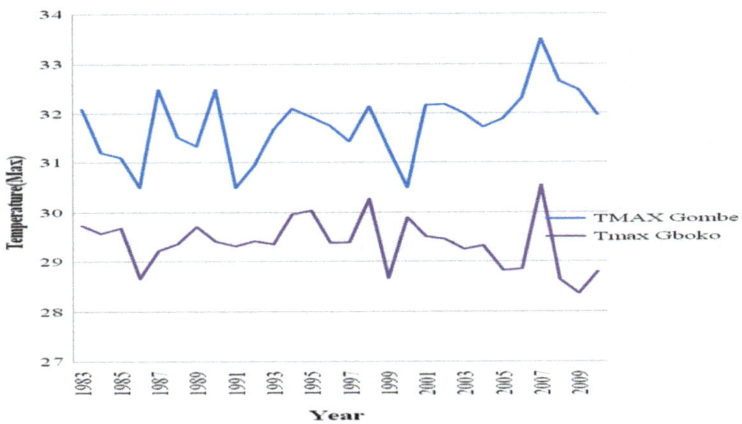

Fig. 6.3 Temperature distribution pattern in Gombe and Gboko (Nigeria). *Source*: Stackhouse (2010)

increasingly aware of climate change which could make them to be vulnerable (Fig. 6.3).

Perceived Effect of Climate Change on Citrus and Tomato Farmers' Production

The summary of the response on the perceived effect of climate change on selected horticultural crops production is as shown on Table 6.3. The mean of total response was 3.3, therefore statements with mean score below 3.3 were considered to be of

Table 6.3 Frequency distributions of respondents according to their perceived effect of climate change

	Perception Statement	SD	D	UD	A	SA	Mean
1.	Irregular rainfall increases the prevalence of pest infestation on Citrus crops	7.4	2.6	1.1	25.5	63.5	4.4
2.	Irregular rainfall reduces disease infection on fruits and vegetables	20.7	18.1	5.9	35.4	19.9	3.2
3.	Drought results in early droppings of fruits of horticultural crops	4.8	6.6	8.5	30.3	49.8	4.1
4.	Harmattan causes plants to flourish	11.4	21.4	13.3	34.3	19.6	3.3
5.	Curling of leaves is a sign of climatic variability	4.4	4.8	7.0	45.0	38.7	4.1
6.	Increased temperature means climate has changed and thus leads to high yield of horticultural crops	38.4	18.8	13.7	18.5	10.7	2.4
7	Dropping of flowers which prevents further fruiting is an indicator of climatic change	17.0	8.9	12.5	24.0	37.6	3.6
8.	Weather has become hotter and drier resulting in yield increment	7.7	13.3	18.5	31.7	28.8	3.6
9.	Change in frequency of droughts has led to improve yield	11.4	14.4	18.8	28.8	26.6	3.4
10.	Incidence of flood leads to opportunities for growers	16.6	15.1	4.1	34.7	29.5	3.5
11.	Irregular temperature is an indication that climate has changed thus reduction in yield	7.7	5.9	10.0	27.3	49.1	4.0
12.	Longer dry season increases weed infestation	21.8	30.3	12.9	22.1	12.9	2.7
13.	Flooding is a threats to growers	11.1	7.7	7.4	23.2	50.6	3.9
14.	Less clearly defined seasons helps horticultural farmers to increase their rate production	20.7	25. 5	12.9	23.2	17.7	2.9
15.	Flooding generally encourage rapid horticultural production	19.6	14.0	8.9	32.1	25.5	3.3
16.	Cost associated with damaging weather events increases cost of production	14.0	21.4	20.7	28.0	15.9	3.1
17.	Change in timing of rains can affect production negatively	12.5	11.8	10.7	29.2	35.8	3.6
18.	High humidity reduces the quality of horticultural crops	16.6	28.0	16.6	15.9	22.9	3.0
19.	Taste and nutritional value of fruits and vegetables can be better off as a result of run-off	25.8	13.7	19.9	27.3	13.3	2.9
20.	Delayed rainfall alters production pattern of horticultural crops	9.6	15.5	14.0	25.8	35.1	3.6

negative effect while statements with mean and above were considered to be of positive effect. Respondents positively perceived that irregular rainfall increases the prevalence of pest infestation on Citrus crops ($\overline{x} = 4.4$). Also, drought results in early droppings of fruits of horticultural crops ($\overline{x} = 4.1$) and dropping of flowers which prevents further fruiting is an indicator of climatic change ($\overline{x} = 3.6$). Delayed

Table 6.4 Frequency distribution of perceived effect category of respondents

Perceived effect category	Frequency	Percentage
Negative perceived effect	130	48.0
Positive perceived effect	141	52.0
Total	271	100.0

rainfall alters production pattern of horticultural crops ($\bar{x} = 3.68$) and dry season (FADAMA) vegetable farming is encouraged as a result of drier season ($\bar{x} = 4.0$).

Perceived Effect Categories of Respondents in the Study Area

The result (Table 6.4) shows that 48.0 % of respondents were unfavourably disposed to the effect of climate change on production in the study area while 52.0 % were favourably disposed to the effect of climate change on production. This implies a positive impact on their production, indicating that climate change leads to low yield of the crops in question. In a related study carried out by Marhjan et al. (2011) in Nepal they found out that flooding at the time of harvesting swept out the consumable agricultural produces impacting huge agricultural loss. In the same vein there was a general agreement by the respondents on increase in temperature, drastic change in weather, and generally reduction in yield in a study on farmer's perception of the effects of climate change and coping strategies in the three ecological zones of Nigeria (Tologbonse et al. 2010).

Adaptation Strategies Used by Respondents

The result shows (Table 6.5) that considering farmers adaptation strategies to climate change under **Crop management**, (56.5)% of farmers always used altering input such as varietal/species followed by use of different planting days (48.7 %). Adaptation under **Soil fertility management** results shows that respondents always used barrier hedges along contour to the soil erosion 59.4 %, followed by soil protection through tree planting 53.5 %. In **water management technique used** respondents always use the adaptation strategies as follows: managing water to prevent water logging erosion and run off (51.3 %), wider use of technologies to harvest water (41.0 %). Adaptation under **insect and pest management** include: wider use of integrated pest and pathogen management, (46.9 %) and planting pest and diseases resistant varieties (34.7 %). **Diversification** as an adaptation strategy in the study reveals diversifying income through altering integration with other farming activities (61.1 %) as well as moving to different site (52.0 %). The result implies that farmers have been using one form of adaptation strategy or the other though at a minimal rate and this depends on location. During the FGD, discussants in *North-central stated that they alter planting dates and build barriers along*

Table 6.5 Frequency distribution of respondents' adaptation strategies to climate change

Adaptation strategies	Use		Frequency of use		
	Yes	No	Rarely	Occasionally	Always
Crop management					
Altering inputs such as varieties/species	56.5	43.5	8.5	20.3	27.7
Monitoring or improving quarantine capabilities	35.1	64.9	3.0	9.2	22.9
Use of varieties and species resistant to pest and diseases	41.7	58.3	4.4	17.0	20.3
Altering the timing or location of cropping activities	55.0	45.0	5.5	21.4	18.1
Different planting dates	48.7	51.3	7.0	17.0	24.7
Shorten length of growing period	33.9	66.1	5.9	18.1	10.0
Crop relocation	43.9	56.1	10.7	20.7	12.5
Planting drought resistant varieties	38.0	62.0	9.2	15.1	13.7
Soil fertility management					
Barriers hedges along contours to the soil erosion	59.4	40.6	8.4	25.5	25.5
Change amount of land	26.9	73.1	4.8	14.4	7.7
Soil protection through tree planting	53.5	46.5	10.3	21.0	22.1
Soil conservation	39.9	60.1	8.9	16.6	14.4
Water management					
Expansion of rainwater harvesting	32.5	67.5	4.4	11.4	16.6
Water storage and conservation technique	33.6	66.4	4.4	14.4	14.8
Water re-use	29.2	70.8	8.1	10.0	11.1
Desalination	25.8	74.2	10.0	6.3	9.5
Wider use of technologies to harvest water, conserve soil moisture	41.0	59.0	4.1	11.1	25.8
Managing water to prevent water logging, erosion and run-off	51.3	48.7	3.3	22.1	25.8
Increase irrigation	43.5	56.5	7.0	17.3	19.2
Planting flood resistant varieties	36.3	63.5	7.0	19.2	10.3
Increase water conservation	42.4	57.6	8.5	19.9	14.0
Pest and Insect management					
Planting pest and diseases resistant varieties	34.7	65.3	5.9	11.8	17.0
Wider use of integrated pest and pathogen management, development	46.9	53.1	12.2	16.6	18.1
Diversification					
Move to different site	52.0	48.0	13.7	22.1	16.2
Changes from crops to livestock	38.7	61.3	9.2	17.7	11.8
Farming to non-farming	33.6	66.4	4.4	14.0	15.1
Diversifying income through altering integration with other farming activities	61.6	38.4	7.0	18.5	36.2

(continued)

Table 6.5 (continued)

Adaptation strategies	Use		Frequency of use		
	Yes	No	Rarely	Occasionally	Always
Others					
Prayer	65.7	34.3	8.5	11.4	45.8
Change use of chemicals, fertilizers and pesticides	50.9	49.1	4.4	23.6	22.9
Use of weather insurance	29.5	70.5	4.1	15.9	9.6

Table 6.6 Frequency distribution of categorization of adaptation strategies

Adaptation categories	Frequency	Percentage
High adaptation (mean and above)	116	42.8
Low adaptation (Below the mean)	155	57.2
Total	271	100

SD = 16.6, Max = 71.0, Mean = 31.023, Min = 0.00

contour, while those in the North-east maintain the use of irrigation. Findings of Ayanwuyi et al. (2011) on farmers' perception of impact of climate change on food crop production corroborated this finding which indicated that increased water conservation, planting of different crops and change row orientation are common adaptation strategies employed by farmers.

Categorization of Respondents According to Adaptation Strategies

The study went further to compute adaptation scores for the area under study; findings revealed that 57.2 % of respondents in the study area have low adaptation strategy score while 42.8 % (Table 6.6) have high adaptation strategy score. It therefore implies that generally farmers have low adaptation strategies. This is in line with Salau et al. (2012), in his study who stated that farmers have low adaptive capacity to climate change.

Result of Binomial Logit Regression Showing Factors that Affect Respondents' Choice of Adaptation Strategies

The result in Table 6.7 reveals that soil fertility negatively and significantly ($t = -3.790$; $p = 0.002$) influenced respondents choice of adaptation; implying that the fertile the soil, the less the choice of adaptation strategies being used. The table also reveals that finance at the same time significantly ($t = 2.610$;

Table 6.7 Binomial logit regression showing factors that affect respondents' choice of adaptation strategies

Factors	Coefficient (β)	t value	p value	Decision
Tenure right	0.373	1.540	0.124	Not significant
Household size	0.108	0.432	0.666	Not significant
Soil fertility	−1.161	−3.790	0.002	Significant
Off farm activities	−0.127	−0.529	0.597	Not significant
Access to inputs	1.385	4.836	0.000	Significant
Access to fertiliser	0.169	0.635	0.525	Not significant
Wealth	0.615	2.610	0.009	Significant
Access to extension	0.411	1.805	0.071	Not significant
Access to credit	0.100	0.426	0.670	Not significant

Table 6.8 T-test difference in the adaptation strategies between the North-east and North-central zones of Nigeria

Adaptation strategies	F-value	t-value	df	Mean difference	P-value	Decision
Assumed	0.795	24.106	270	27.974	0.000	Significant
						Level of sig 0.05

p = 0.009) influenced respondents choice of adaptation strategies; implying that finance is a major factor that affect farmers' choice of adaptation strategies. Finally, the table shows that access to inputs significantly influenced (t = 4.836; p = 0.000) farmers' choice of adaptation strategies; implying that the more access to inputs the more the choice of adaptation strategies used by respondents. This is an indication that level of soil fertility, farmers' financial capacity and access to inputs greatly determine their choice of adaptation strategy.

Hypotheses Testing 1

Adaptation Strategies Between the North-East and North-Central Zones of Nigeria

The result shows (Table 6.8) a significant difference in the adaptation strategies in the two zones under study (P = 0.000, t = −24.106). This result implies that in the two zones considered in this study the rate at which both zones make use of adaptation strategies is quite different, which may be as a result of the difference in time at which they started noticing change in climate or as a result of the differences in the intensity of change and impact.

Table 6.9 T-test difference in the perceived effect strategies between the North-east and North-central zones of Nigeria

Perceived effect	F-value	t-value	df	Mean difference	P-value	Decision
Assumed	0.177	182.269	270	134.5978	0.004	Significant
						Level of sig 0.05

Hypotheses 2

Perceived Effect Strategies Between the North-East and North-Central Zones of Nigeria

The result (Table 6.9) shows a significant difference in the perceived effect of citrus and tomato farmers about climate change in the two zones under study ($P = 0.004$, $t = -182.269$). The implication of this is that in the two zones under study, farmers perceived the effect of climate change on the production of tomato and citrus differently, this may still be due to the period they started noticing change in climate.

Conclusions

Arising from the study, respondents are aware of change in climate parameters such as rainfall, temperature and wind. The adaptation strategies of respondents to climate change are still a little above average despite their high level of awareness and knowledge. The adaptation strategies across the North central and North-east are significantly different. It suffices therefore that, any type of adaptation strategy introduced is likely to be adopted to improve on their present form of adaptation strategies.

In the same vein, the perceived effect of climate change on production across the zones are also different, while it was found out that farmers had favourable perception to the effect of climate change, in this wise climate change has a lot of negative impact on their production. Finally adaptation choice by farmers is mainly determined by soil fertility, wealth and input availability.

Recommendations

The following recommendations are put forward based on the findings of this study for the development of horticulture industry in Nigeria.

- Bulletins on awareness creation on climate change, preferably in local languages should be made available to farmers so as to further create more awareness.

- Varieties of horticultural crops that are resistant to drought and certain pests and diseases that are as a result of climate change can be introduced to farmers in the study area by the research Institute.
- Several projects like shelter belt project, afforestation projects climate change programmes should be funded by the government of each zones and states.
- Efforts on mitigation and adaptation strategies should be based on religious and gender bias. In this wise farmer should be made to realize that though God is all in all, but they are responsible and at the same time liable to some of the actions they partake in for example when they cut trees and refuse to re-plant. It is therefore pertinent to introduce mitigating and adaptation strategies that will safe guard livelihood of the respondents and salvage them from poverty and other problems like harsh or extreme weather conditions on agricultural produce particularly tomatoes and other horticultural crops that are classified as crops that are highly sensitive to weather conditions. The resultant effect was described as low quality and poor quantity of the produce which make them to attract low market price.

References

Aaditya S (2011) Climate change, developing country, development, family planning fertility, overpopulation, population growth, poverty, women. 12 August 2011. Posted in Wikinut > News > Environment. http://nut.bz/1u3rs_iy/. Downloaded Sept 2011

Adejuwon SA (2004) Impacts of climate variability and climate change on crop yield in Nigeria. Being the lead paper presented at the stakeholders' workshop on assessment of impacts and adaptation to climate change (AIACC), Conference Centres Obafemi Awolowo University, Ile-Ife, 20–21 September, p 14

Ati OF, Iguisi OE (2007) Recent trends in the duration of rainy season in some station north of latitude 9^{0N} of Nigeria. Paper presented at the 49th annual conference of the association of Nigerian geographers held at University of Abuja, Abuja—Nigeria, 16–18 October 2007

Ayanwuyi E, Kuponiyi F, Ogunlade A, Oyetoro JO (2011) Farmers' perception of impact of climate change on food crop production in Ogbomosho Agricultural zone of Oyo state Nigeria. Glob J Hum Soc Sci 10(7):34–39

LEISA Magazine (2008) Living soils desertification. Posted on September 23, 2008, LEISA Magazine 24:2. wordpress.com/2008/09/23/leisa-magazine

Maharjan K, Sigdel ER, Sthapit BR, Regmi BR (2011) Tharu community's perception on climate changes and their adaptive initiations to withstand its impacts. West Terai Nepal Int NGO J 6(2):035–042

northernnigeriatourism (2009) REF—Northern Nigeria tourism and trade copyright @2009 NNT. www.northernnigeriatourism.com/nnt/about-northern-nigeria.html. Retrieved 22 Jan 2015

Nyelong PN (2004) Global warming and global waters. J Energy Environ 17(1):79–90

Odjugo PAO (2009) Shift in crop production as a means of adaptation to climate change in semi-arid region of Nigeria. Paper presented at the international conference of Nigeria meteorological society, Osun State University, 6–11 December 2009, p 211

Oladeji JO (2011) Farmers' perception of agricultural advertisements in Nigerian newspapers in Ibadan municipality, Oyo State. Niger J Media Commun Stud 3(3):97–101

Oyedele OO (2005) Information and training needs for improved technique of citrus production among farmers in Southwestern Nigeria. An unpublished PhD Thesis submitted to the department of Agricultural Extension and Rural Development University of Ibadan, p 211

Oyekale AS, Bolaji MB, Olowa OW (2009) The effect of climate change on cocoa production and vulnerability assessment in Nigeria. Medwell Agric J 4(2):77–85

Salau ES, Onuk EG, Ibrahim A (2012) Knowledge, perception and adaptation strategies to climate change among farmers in Southern Agricultural Zone in Nasarawa State. Niger J Agric Ext 16 (2):199

Salimonu KK (2007) Attitude to risk in resource allocation among food crop farmers in Osun State, Nigeria. PhD Thesis submitted to Agricultural Economics Department, University of Ibadan, Ibadan

Spencer DSC (1990) Research priorities for sustainable agriculture in sub-Saharan African. In: Lawani SM, Odemwingie TE (eds) Communication for sustainable agriculture. IITA, Ibadan

Stackhouse PW (2010) NASA climatology resource for agro climatology daily average data (evaluation version). National Aeronautics and space. Administration available at http://power.larc.nasa.gov

Tologbonse EB, Auta SJ, Bidoli TD, Jaliya MM, Onu RO (2010) Farmers perception of the effects of climate change and coping strategies in three agro-ecological zones of Nigeria. Paper presented at the agricultural extension society of Nigeria annual conference held in Obafemi Awolowo University, Ile-Ife, Osun State, Nigeria, between 10 and 14 May, 2010

Yekinni OT (2010) Determinants of utilization of information of information communication technologies for agricultural extension delivery in Nigeria. A PhD Thesis submitted to Agricultural Extension and Rural Development Department, University of Ibadan, Ibadan, p 226

Chapter 7
Local Networks of Resilience and Climate Adaptation: The Case of Istanbul

Rana I. Connelly and Pınar G. Bal

Abstract Large coastal cities are often the engines of national growth but also tend to be areas with high sensitivity to the impacts of climate change. Therefore, integrated adaptation plans are essential for turning them into resilient cities. National competitiveness strategies are, however, increasingly at odds with the very idea of resilient cities, either forcing urban sprawl to its limits or transforming green spaces into grey spaces within the city. In the midst of heated conflict about how to use public land, the role of networks of local initiatives to protect green spaces and residential rights of poor and marginal groups becomes pivotal to achieve equity and urban resilience. The aim of this paper is to explore the dynamics of such networks in Istanbul and investigate how to integrate them into local climate change adaptation plans. Conflicts over Istanbul's historical urban vegetable gardens (bostan) and the construction of the third bridge are good examples of sites of contestation which, unless resolved, seriously hinder any possibility of agreement and action on climate adaptation plans.

Keywords Resilience • Urban farming • Green spaces • Local adaptation • Climate change

Introduction: Protests and Conflicts

Many cities form regional and global alliances to mitigate climate change. They share their experiences and commit to support innovative ideas to reduce greenhouse gas emissions from urban activities. Unquestionably, the scale and pace of climate change reconstitute global interconnectedness and local innovations (Bulkeley 2005; Taylor et al. 2012).

R.I. Connelly (✉)
Marmara University-EU Institute, Istanbul, Turkey
e-mail: ranaizci@marmara.edu.tr; ranaizci@gmail.com

P.G. Bal
Faculty of Economics and Administrative Sciences, Beykent University, Ayazağa–Maslak Campus, Istanbul, Turkey

W.L. Filho et al. (eds.), *Climate Change Adaptation, Resilience and Hazards*,
Climate Change Management, DOI 10.1007/978-3-319-39880-8_7

However, local governments often face difficulties in integrating hard (engineering based interventions/innovations) and soft adaptation (legal, institutional, social and economic incentives/ innovations) measures into everyday urban life. Hence local governments seek to establish stakeholder partnerships from different parts of society to implement climate change plans and strategies at the local level. Building and promoting public engagement in adaptation plans offer more flexible solutions for cities to improve their resilience. According to the IPCC (Intergovernmental Panel on Climate Change) definition, resilience is

> The capacity of social, economic, and environmental systems to cope with a hazardous event or trend or disturbance, responding or reorganizing in ways that maintain their essential function, identity, and structure, while also maintaining the capacity for adaptation, learning, and transformation (IPCC 2014, p. 5).

The term resilience might have different meanings for social and physical sciences, but in either case it is linked to reducing vulnerability which has three components: exposure, sensitivity and adaptive capacity of natural and/or social systems (IPCC 2014; Beichler et al. 2014). And, as Adger has suggested, "...vulnerability to environmental change does not exist in isolation from the wider political economy of resource use" (Adger 2006, p. 270). Cultural, technological, economic and institutional structures can shape vulnerability. Nevertheless, resilient ecosystems can reduce social and ecological vulnerability. Green areas are particularly important for socio-ecological systems to adapt to challenges and to reduce their sensitivity to any hazard. Therefore, land use patterns have a great impact on strengthening resilience in urban areas (Haq 2011; Cavan and Kazmierczak 2011).

At this point, two important questions arise:

- How can the adaptation agenda be more efficiently integrated into land use planning in urban areas?
- What is a local action, who are the local actors, and to what extent can non state actors affect urban social and ecological resilience?

Recent protests showed that citizens in Istanbul demanded more citizen involvement in urban planning. They expressed their disappointments about decisions of local and national authorities to shape their small neighbourhoods as well as Istanbul's landscape by implementing mega plans with strong economic growth arguments. Decisions that caused wide spread discomfort included construction of the third bridge and the third airport around the Northern Forests of Istanbul (Kuzey Ormanları), redesigning the green land for other uses in the Gezi Park/Taksim, the Validebağ grove (Validebağ korusu), the Kuzguncuk vegetable garden (Kuzguncuk bostanı) as well as urban renewal projects about Sulukule and the Yedikule vegetable gardens (Yedikule bostanı) (Gerçek 2014; Akçalı and Korkut 2015; Özkaynak et al. 2015).

Local initiatives were evolved to protect historical neighbourhoods, green spaces, protected sites and to encourage alternative life styles. Neighbourhood communities started to take legal actions and organised public events. Civil society organisations also took part in this process. They supported local urban stewardship attempts providing expertise and resources. The emergence of strong local initiatives demonstrated citizens' growing demand for living in harmony with nature, not against it, in their cities. Collective memories of landscape, demands for equity and participation in city life also shaped these initiatives.

Even though these protests did not start with the aim of affecting local adaptation strategies and plans, their aims are apparently in line with the *raison d'être* of adaptation. Local initiatives often aim to maintain urban ecosystems and to support diversity of interests and necessities in urban planning. There is a growing body of research trying to understand co-management and social networks in climate change adaptation (Tompkins and Adger 2004; Newman and Dale 2005; Bodin et al. 2006). Although most of this research focuses on communities heavily dependent on natural resources, the role of local networks on urban ecosystems has started to receive more attention (Enqvist et al. 2014; Ernstson et al. 2010; Haq 2011). This study claims that local environmental initiatives have an increasing role to improve urban resilience in Istanbul whose socio-ecological integrity is under threat due to its status as the engine of Turkish economy. A number of mega projects and local plans have been particularly challenging the ecology, social justice and quality of life in the city. This study will firstly refer to the recent developments in Istanbul which have led to protests and the emergence of local initiatives. The case of the third bridge demonstrates how mega projects can easily disturb the socio-ecological resilience of a mega city. Then conflicts over the Yedikule and Kuzguncuk vegetable gardens will also be examined within the framework of urban/peri urban agriculture, which is now regarded as one of the main elements of urban resilience in adapting to the impacts of climate change. The study, finally, analyses the approaches of each local initiative to the specific area of contention. It is difficult to define such new and dynamic networks, initiatives and groups. Nevertheless, they all prefer loose and non-hierarchical structures and formulate innovative responses to problems. At this point, it is premature to draw precise conclusions about the impact of such networks on urban resilience: but we hope to stimulate new lines of inquiry on the development of effective communication for local adaptation plans in Turkey.

Redesigning the City: Resilient or Dispossessed and Vulnerable?

There have always been protests demanding social justice and alternative urban life styles in big cities. Since the early 2000s, however, all around the world ardent protestors have taken the streets to claim their right to the city (Mayer 2009; Özkaynak et al. 2015). Intra generational equity, sustainable land use practices and participation are at the centre of these claims (Özkaynak et al. 2015).

Various economic reasons as well as opportunities of different lifestyles have attracted people to cities. Although better quality of life is the motto for urban life, cities are often vulnerable human agglomerations. They depend on large quantities of long distance supplies to meet the demands of ever increasing population. Moreover, urban land expansion causes severe ecological degradation in suburban areas, leading to continuous decline in inner city areas. Traffic congestion, poor air quality and social inequality are, *inter alia*, the most detrimental effects of unfettered urban growth. Earthquakes and other natural disasters, such as hurricanes, aggravate existing socio-ecological problems and inequalities. Local governments often resort to urban renewal projects to address these problems. However, since most urban renewal projects depend on property led or project based approaches, they often cause the loss of urban green space and lead to gentrification, pushing the locals out of the neighbourhood (Aksoy 2014). Thus, urban fabric might be severely damaged and social exclusion—particularly in historical but economically deprived urban areas—becomes a new urban normality. For instance, the Sulukule urban renewal project in Istanbul was criticized for not implementing the original plan for supporting the revitalisation of economic and socio-cultural development of the area (Eren 2014). As a historic Roma neighbourhood dating back to the Byzantine era, Sulukule needed to renew its building stock and improve its infrastructure. However, it ended up with displacement of its inhabitants, demolishing registered historic buildings and damaging civic culture (Eren 2014).

Since housing, post-earthquake/disaster reconstruction and economic revitalisation are the key priorities of urban renewal projects, built environment in and around the target areas often increases. However, climate friendly urban development requires a balance between built and natural environments. So far, there is not any particular sign that urban renewal projects have paid enough attention to the impacts of climate change in Istanbul. On the contrary, environmental concerns and public participation seem rather marginalized within these projects. Ideological views on the future of urban life also affect land use patterns in the city. Accordingly, other renewal or regeneration projects in Istanbul have caused even greater social unrest and resistance as in the cases of Gezi Park (2013) and of Validebağ grove (2014).

Natural forests, gardens, recreational green areas and public parks have always been the landmarks of Istanbul because Istanbul has a rich cultural and ecological heritage under the influence of a transitional climate between the Mediterranean and the Black sea. The topography of the city also engenders micro climatic zones and rich biodiversity making urban and peri-urban agriculture possible throughout centuries. However the status of vegetable gardens was also challenged by recent urban renewal projects.

Despite increasing complaints about environmental governance in Istanbul, the Metropolitan Municipality claims that green spaces and parks are being improved and enlarged during their terms of governance. However, green space per capita in Istanbul is 6.5 m^2 in 2010 (IBB 2010). This percentage is far below the recommended minimum standard per citizen (9 m^2) by the United Nations Food and Agriculture Organisation (Singh et al. 2010). The Municipality has recently pledged to sow 115,000 trees of 40 different types including fruit trees in 300 areas of Istanbul. Water catchment areas and highways are the priority areas within this plan (IBB 2015). Local governors, through new afforestation and reforestation projects, also aim to achieve further reductions in Istanbul's green house gas emissions (IBB 2015). Istanbul Metropolitan Municipality has also launched extensive transport projects (Marmaray) and waste management plans stating that these plans would reduce Istanbul's greenhouse gas emissions. It is also important to note that Istanbul is a member of C40 Cities.

Nevertheless, new transportation and urban renewal projects threaten the existence of natural systems which has enabled the city to survive to date. New projects clearly indicate that Istanbul, as a coastal mega city, is spreading into undisturbed and semi-disturbed forest and coastal systems. Construction of the third bridge on the Bosphorus as part of the Northern Marmara Motorway is one of these mega projects.

The Northern Forests and the Third Bridge: Water Resources, Wildlife and Livelihoods

The Northern Forests are of vital importance for urban sustainability in Istanbul (and beyond Istanbul). The Northern Forests include city forests, coastal areas, sand dunes, rivers, becks, springs, historical water catchment areas, dams, lakes, natural parks, an arboretum and natural forests on both the European and Asian sides of Istanbul. These areas also provide valuable refugee for birds. Istanbul's forests are registered as one of the 200 most important ecological sites in the world. They also absorb and stock significant amounts of CO_2 from urban activities (Tolunay 2014).

The third bridge is a part of the Northern Marmara Motorway Project and is located in the northern part of Istanbul facing the Black Sea. Construction of the bridge started in 2013 but, before and during its construction, various legal cases were brought to the court. A number of civil society organisations and associations strongly argued that the construction of the third bridge was illegal, since it violated 1/100,000 scaled Istanbul Provincial Environmental Plan. Nevertheless, according to the third bridge construction consortium, the bridge symbolizes the modern face of Turkey and "is going to be the widest suspension bridge in the world" with "8 lanes for motorway and 2 lanes for railway" (ICA 2013). The Consortium claims that with the construction of this bridge, inner city traffic is expected to be reduced significantly. It has been also argued that the bridge aims to reduce the costs for cargo traffic and to increase employment (ICA 2013). On this point, a number of experts warn that the main function of the third bridge would be to serve mostly to intercity and cargo traffic causing even a greater regional environmental damage rather than to reduce inner-city traffic (Şahin 2013; Çalışkan 2010).

The third bridge was not subject to the Environmental Impact Assessment (EIA); however a private company conducted an Environmental and Social Impact assessment. According to the court cases, the original route of the bridge has been revised (Gülersoy and Gökmen 2014; Hürriyet Daily News 2015). Various nationwide and Istanbul based organisations (Green Party, Istanbul Chamber of Architects and Engineers, Doğa Derneği/Doğa, TEMA/The Turkish Foundation for Combating Soil Erosion, For Reforestation and the Protection for Natural Habitats) and local groups have expressed their strong opposition to the third bridge and have taken legal action. More than 30 cases were brought to court. A movement was also formed (Kuzey Ormanları Savunması/Northern Forests Defence) to protect these forests and support the sustainability of the socio-ecosystems in the area. Northern Forests Defence describes itself as a grassroots movement dedicated to stop any plans which would harm urban and rural environment (Northern Forests Defence 2015). It includes sociologists, urban planners, students and other volunteers from different parts of the society. The movement states that they reject any hierarchy and are open to any ideas and participation. The movement functions on the basis of rotating responsibility and moderation.

While legal actions and protests on the cancellation of the third bridge plans still continue, construction of the third bridge has already threatened the wetlands, sand dunes, historical water catchment areas and endemic species at an alarming level. Wild boar tried to escape from their damaged environment and appeared in the city (Weise 2015). Some boar also tried to swim to cross the Bosphorus. The villages around these areas have also faced the threat to lose their livelihoods. Due to the construction of an express road between the Trans-European Motorway (TEM) and the Northern Marmara Motorway, a huge picnic area in the Asian side was almost totally destroyed (Güvemli 2014; Ocak and Sönmez 2014).

Urban planners argue that most of the mega transportation projects essentially provide infrastructure for other projects and make investments in those areas possible (Yalçıntan et al. 2014). The first and the second bridges caused unplanned urbanisation and expansion of industrial zones in areas between the core and periphery as well as in the outskirts of the city (Terzi and Bölen 2012). The second bridge on the Bosphorus paved the way for the construction of the Sabiha Gökçen Airport, Istanbul Racing circuit/Formula 1 (Istanbul Park) and big shopping outlets (Yalçıntan et al. 2014). According to many specialists, a new city project in the north is the main factor behind the third bridge and the motorway plan. They also argue that the third bridge would not solve Istanbul's traffic problem but create a new source of congestion in different centres (Yalçıntan et al. 2014; Gerçek 2014).

At this stage, it is also difficult to predict the micro-climatic changes (such as in humidity, temperature, energy flows, evaporation, local winds) which might occur due to these mega projects—and their impacts on various ecosystems—in and around Istanbul (Türkeş 2014). This uncertainty might engender further difficulties to cope with and adapt to the impacts of climate change in Istanbul and neighbouring cities. Yet the third bridge is not the only threat to Northern Forests; the third airport and Kanal Istanbul are two other recent mega projects which also threaten Istanbul's ecosystems (Gerçek 2014; Gülersoy and Gökmen 2014; Northern Forests 2015). The Northern Forests Defence, other local initiatives, NGOs and experts groups have organized campaigns and meetings against the construction of the third bridge, the third airport and against the housing projects in some parts of the historical city forests.

Urban Food as Part of Adaptation Measures and Istanbul Vegetable Gardens

Urban -and peri-urban- agriculture is listed as one of the prerequisites for climate change adaptation (IPCC 2014). Urban agriculture might broaden and raise awareness necessary to respond to the impacts of climate change. It can also expand collaborative relationships between different actors which in turn might encourage effective public involvement in the local adaptation plans (Aylett 2014). Vegetable gardens (bostan), orchards and farms have always been essential parts of Istanbul city life both within and outside the city fortifications for centuries.

The Historic Vegetable Gardens of Yedikule

Yedikule historic vegetable gardens have been used as urban agricultural land for more than 1500 years. Historical evidence and documents confirm the existence of farming areas around the city walls (Theodosian Landwalls) during the Byzantine era (Barthel et al. 2010; Kaldjian 2004). Ottoman documents also provide significant amount of detailed evidence about the management of historic vegetable gardens. This area is the one and only remaining example of both Ottoman and Byzantine urban farming practices (Başer and Tunçay 2010). However, the Marmaray project (railway tunnel underneath the Bosphorus was opened in 2013), recent restoration plans for city land walls and the construction of Yenikapı meeting area challenge the status of historic vegetable gardens. In 2013, the Fatih Municipality decided to run a recreation project from Belgrade gate to Yedikule gate for the preservation of inner land walls. The Fatih Municipality has also announced its plans for a park and recreational area along the walls. This plan included creating cycling routes and an artificial river, providing more security and preventing crime in deserted parts around the walls and building a playground for children (Koca 2014; Çorakbaş et al. 2014). The Fatih Municipality's plan to transform some part of historic vegetable gardens as a public space in this impoverished neighbourhood was warmly welcomed by many inhabitants. However, recent luxury housing development in the area has raised considerable concerns about the status of historic vegetable gardens (Çorakbaş et al. 2014).

At this point, local networks started to emerge. The locals, city planners, environmental groups, architects have formed a number of small networks to protest this decision. In addition to legal processes, a petition campaign was organised by Slow Food/Fikir Sahibi Damaklar, the School of Yedikule Historic Vegetable Gardens (Yedikule Bostan okulu) was established, several summer courses and art performances were held. The Initiative to Protect Yedikule Gardens was formed; a report was prepared by experts with the support of the Association of Archaeologists, Istanbul Branch and submitted to the UNESCO Istanbul (Çorakbaş et al. 2014; İnce 2014; Koca 2014). They have demanded more transparency about the decision and asked whether it would be possible to integrate the park project into vegetable gardens without obscuring farming activities (Koca 2014). After long protests and court cases, the park project was halted (Vardar 2014).

This was not the first attempt by local governments to transform vegetable gardens into a built environment. However, this is the first large scale project about the status of vegetable gardens as Yedikule is now at the centre stage of one of the new gentrification processes in Istanbul (Koca 2014). Local governors clearly stated their desire to protect the built environment (the walls, wooden house and stables in the vegetable gardens) but they did not acknowledge vegetable gardens as valuable to be included in the preservation project. However, protesters reminded the local government that under UNESCO guidelines, gardens were also to be protected. Networks involved in this case have encouraged urban agricultural activities in different parts of the city. They continue their model urban framing activities and festivals in the vegetable gardens.

The Kuzguncuk Vegetable Garden

Kuzguncuk is, in contrast to the Yedikule vegetable gardens, located in a more protected natural environment. It is a well known district with its natural beauty and traditional neighbourhood features. Due to its characteristic architecture, some parts of the district are often used as a background for many movies and TV dramas. There were three vegetable gardens in Kuzguncuk. However, only one of them still exists—known as Ilya (Ilia) Garden by residents—is named after its last tender (Koca 2014). This area has been registered as a vegetable garden for 700 years and provided fresh food to the neighbourhood. The Directorate General of Foundations, Istanbul 2nd Regional Directorate (Vakıflar Genel Müdürlüğü-Istanbul Bölge Müdürlüğü) obtained the garden's ownership in 1977. In 1986, the status of the garden was amended to include a public school project (Koca 2014). The neighbourhood did not need a new school and this school project was never put into practice. However, due to this small change in master plans, this area had to face new challenges. In 1992, the garden was rented to another foundation chaired by a famous businessman for 10 years to build a hospital. Locals protested immediately and, after extended effort, their resistance stopped the project. In order to prevent future threats to the existence of their only green land, Kuzguncuk residents through their local association (Kuzguncuk Derneği) tried to rent the garden by themselves. It was very costly; they could not manage to get the right to use. But a garden nursery rented the vegetable gardens for 10 years and kept the land as it was. In 2011, the Üsküdar Municipality rented the land and a private school project was announced in the area (Koca 2014). Subsequently, public protests took place. Locals managed to halt the project and offered their plan about the future of garden to the Üsküdar Municipality. According to their project, the land was going to be divided into small allotments and priority would be given to Kuzguncuk residents (Aksu 2014). The Üsküdar Municipality accepted this plan and as of mid-2015, the Kuzguncuk garden keeps its status as an urban farming area and as a place for shelter in case of a disaster.

Local Networks and Grassroots Groups for a Socio-Ecological Resilient City

Urban environmental degradation either creates new actors or challenges the long established governance structures in urban politics. Some local issues often initiate *ad hoc* movements which have loose non-hierarchical features in contrast to non-governmental organisations. They organise demonstrations, petitions, and sit-ins to raise public awareness and to pressure decision makers to change their decisions about a specific issue. They often depend on the actions of local people who would be most affected by the decision taken at the local/or national level and who cease their activism once they reach their aim. Community based partnerships, grassroots organisations, social networks and local networks share certain features with this kind of movement. Even if they have nation wide support they are geographically local (Young 1997). It is very difficult to delineate the differences among them and between *ad hoc* movements. However, community based partnerships, grassroots groups, social networks and local networks not only pressurize local/national governments but also try to empower locals and to promote alternative life styles or innovative solutions to the problems. They also aim to achieve long term co-operative action in contrast to the short term aims of *ad hoc* movements. Thus some argue that citizen groups or networks might provide the most appropriate form of environmental stewardship to eliminate urban hotspots and protect urban green land (Enqvist et al. 2014). Even though there are certain limitations to their possible success, networks and partnerships provide more opportunities for marginalised groups (mostly women, students, poor, disabled and elderly), locals in the area and those who envisage alternative urban life styles (supporters of urban permaculture, social volunteering, reducing car dependency) to take action.

Local initiatives are not new in Turkey. However, local initiatives to protect urban green space are becoming increasingly important as urban areas in Turkey come under increasing pressure from new competitiveness targets, financial crises and rapid urbanisation. Local networks might emerge due to a forced change in the historical usage of a green space for the interest of a small group and/or touristic recreation project as in the case of the Yedikule vegetable gardens. In this case, not only a historical urban area - a medium of interaction between humans and ecosystems - but also the source of income of a community was threatened. What is more, the landscape and social fabric of the area faced the risk of being altered. Short term impacts of such an urban regeneration plan would have included displacement of locals, creation of urban heat islands and traffic congestion in the area. Since the Yedikule vegetable gardens are close to a coastal zone, extreme meteorological events would have hit new built areas with great density. Responses to this regeneration plan were therefore organised around the idea of maintaining urban farming and recreating the human-urban ecosystem interaction. Social networks and local initiatives act, in this case, as a moderator between locals who see some parts of the garden as a desolate land to be regenerated and locals who want to maintain their farming activity or conserve the land as it is. The volunteers who belong to these networks have demonstrated that those who support urban farming are not disillusioned environmentalists or idealists but ordinary citizens of Istanbul.

A different citizen engagement has taken place to protect the Kuzguncuk vegetable garden. In this case, a small scale recreation project threatened another historical green space. Neighbourhood based communities played the central role in the fight against the municipality decision to change land use practice in the area. If the plans of the Üsküdar Municipality had been realised, the collective identity of an old neighbourhood and integrity of a very old ecosystem would have been irrevocably damaged. Through their efforts, this green space retains its status as vegetable gardens and continues to function as a gathering point in case of a disaster. Locals also obtained the right to farm through the allotment system. Since the construction plans in the area were halted, this area continues to serve as a critical ecosystem in a partially protected environment. In this case, the Kuzguncuk neighbourhood community has acted as entrepreneurs of change and offered innovative solution for a conflict (Bulkeley and Betsill 2013).

Taking local practice as well as knowledge into consideration is one of the essential elements of adaptation strategies. However, this is not a sufficient condition to achieve sustainable climate adaptation. "Local power differences and divergent interests in the community" should also be integrated into strategies (Taylor et al. 2012, p. 108). In both cases of urban farming, local initiatives in Istanbul have aimed to improve socio-ecological conditions and to promote alternative ways of development in order to overcome intergenerational inequality and environmental degradation (Young 1997).

Urban resilience is a multifaceted concept which "refers to the ability of urban systems to withstand, adapt to, and recover from climate related hazard" (Aylett 2014, p. 9). The third bridge project is still posing a great risk to resilience of the city. The damage imposed upon ecosystems in the area weakens their capacity to adapt to the climate change and recover from climate hazards. Impacts of the construction are visible. Some locals face the risk of being displaced; others might lose their source of income while some of them have to cope with air and noise pollution. A great variety of forest vegetation, coastal ecosystems, inland water ecosystems and animal species are under threat. The overall impact is not, however, limited to local losses. Once this project is completed and integrated into the northern Marmara motorway and the third airport project, serious damage to the entire ecosystem on both sides of Istanbul appears inescapable. Water scarcity is the most expected and short term outcome of these developments. In this case, the problem might be regarded as a regional issue since its impacts would be extended throughout Northern Marmara. Another mega project, construction of the third airport, exacerbates the environmental destruction caused by the third bridge project (Gülersoy and Gökmen 2014; Northern Forests Defence 2015). At present, attempts to protect the Northern Forests have taken place mostly at the local level. However, the variety of actors involved in this issue and its wide scale impacts are likely to create not only a new social movement for urban resilience but also to contribute to more inclusive environmental activism which has already started with the protests against various mining and energy projects in Turkey (Özkaynak et al. 2015).

How climate change action is defined at national level frames the action at the local level (Schreurs 2008). Yet, local governments as the main actors in urban politics can produce a wide variety of mitigation and adaptation initiatives (Fünfgeld 2015; Schreurs 2008). Typical interpretations of sustainable urban development define resilience as an essential part of urban planning (ICLEI 2014). However, imminent and projected disturbances by the third bridge to the ecosystems in Istanbul reveal that local authorities not only underestimate the impacts of climate hazards and ecosystem resilience but also overlook Turkey's international commitments to protect biodiversity and wildlife as well as to combat against desertification and to reduce pollution (Budak 2014).

Conclusion: Ecosystem Based Adaptation and Urban Green Spaces

Adaptation is a continuous process which needs to comply not only with changing climate but also changing priorities, life styles and values (Brown et al. 2011). Thus, sustainable adaptation should respond to temporal and spatial challenges. Unquestionably adaptation to climate change goes beyond one size fits all approaches; consists of not only country specific but city specific solutions. Grassroots actions might provide social connectivity necessary to address dynamic forces of adaptation. They simply offer new "or emergent forms of collaborative action" (Feola and Nunes 2014, p. 234). They can promote resilience through participation and innovations based on alternative systems of energy and food systems (Feola and Nunes 2014).

Urban green spaces provide ecological, economic and social benefits for locals. They might contain high biodiversity and can function as shelters and meeting points in case of disasters. Some areas are suitable for urban and peri-urban agriculture activities which provide local employment and enhance food security especially for deprived neighbourhoods. Expanding green spaces in urban areas also contribute to adaptation plans since they can reverse the impacts exacerbated by urban heat island effect (Cavan and Kazmierczak 2011). To put it differently, land use patterns in urban areas "regulate urban climate" (Haq 2011, p. 602).

Citizen involvement to conserve urban green spaces might fill an important gap in local adaptation plans of Istanbul. Their demands might force local governors of Istanbul to accept that the city is a socio-ecological system and to revise their adaptation plans according to ecosystem based adaptation strategies which reassess the links between use of land and built environment (e.g., green roofs, rainwater harvesting, urban agriculture, supporting drought tolerant gardens, restoration of coastal ecosystems, developing open spaces, creating permeable surfaces) (Colls et al. 2009).

Since 1960s, environmental movements in Turkey have been challenging various state decisions. Turkey tries to achieve the dual goals of economic growth and wealth creation. However, its market based regulations have so far created a significant number of tensions about natural richness and land use patterns in the country. In Turkey, inhabitants of many areas with high biological diversity have found themselves in the middle of heated conflicts due to energy projects. Istanbul is, on the other hand, facing with challenges of demand for land. Hitherto, political decisions and the dominant view on "nature" have evidently favoured techno-centric approaches towards green land in the city. And Istanbul becomes more risk prone to climate change related hazards as natural life in and around the city diminishes. While local and national authorities continue to expand and regenerate the city under smart city projects, local initiatives increasingly remind them that city belongs to its inhabitants and adaptation to impacts of climate change can only be achieved by collective action.

References

Adger WN (2006) Vulnerability. Glob Environ Chang 16:268–281

Akçalı E, Korkut U (2015) Urban transformation in Istanbul and Budapest: neoliberal governmentality in the EU's semi periphery and its limits. Polit Geogr 46:76–88

Aksoy A (2014) İstanbul'un Neoliberalizmle İmtihanı. In: Candan A, Özbay C (eds) Yeni İstanbul Çalışmaları Sınırlar, Mücadeleler, Açılımlar. Metis Yayınları, Istanbul

Aksu F (2014) Kuzguncuk'ta Ortak Karar Bostan Park, Hürriyet. 30 Ekim 2014. http://www.hurriyet.com.tr/gundem/27482376.asp. Accessed 01 Feb 2015

Aylett A (2014) Progress and challenges in the urban governance of climate change: results of a global survey. MIT, Cambridge, MA. http://observgo.uquebec.ca/observgo/fichiers/94238_GLR-1.pdf. Accessed 11 Feb 2015

Barthel S, Sörlin S, Ljungkvist J (2010) Innovative memory and resilient cities: echoes from ancient Constantinople. In: Sinclair PJJ, Nordquist G, Herschend F, Isendahl C (eds) Urban mind. Cultural and environmental dynamics. Uppsala Universitet, Uppsala

Başer B, Tunçay HE (2010) Understanding the spatial and historical characteristics of agricultural landscapes in Istanbul. ITU A/Z 7(2):106–120

Beichler SA, Hasibovic S, Davidse BJ, Deppisch S (2014) The role played by social-ecological resilience as a method of integration in interdisciplinary research. Ecol Soc 19(3):4. doi:10.5751/ES-06583-190304. Accessed 11 Jun 2015

Bodin Ö, Crona B, Ernstson H (2006) Social networks in natural resource management. What is there to learn from a structural perspective? Ecol Soc 11(2). http://www.ecologyandsociety.org/vol11/iss2/resp2/. Accessed 08 Feb 2015

Brown A, Gawith M, Lonsdale K, Pringle P (2011) Managing adaptation: linking theory and practice. UK Climate Impacts Programme, Oxford

Budak S (2014) Istanbul'da Yapılması Planlanan Projeler ve Taraf Olunan Uluslararası Sözleşmeler. In Istanbul'un Geleceğini Etkileyecek Üç Proje. 3. Köprü, 3. Havalimani ve Kanal Istanbul. TEMA Vakfi Uzman Görüşleri, TEMA Istanbul. http://www.tema.org.tr/folders/14966/categorial1docs/1244/BUYUKPROJELER20032014data.pdf. Accessed 11 Jun 2015

Bulkeley H (2005) Reconfiguring environmental governance: towards a politics of scales and network. Polit Geogr 24:875–902

Bulkeley H, Betsill MM (2013) Revisiting the urban politics of climate change. Environ Polit 22 (1):136–154

Cavan G, Kazmierczak A (2011) Urban greening to adapt urban areas to climate change. Oxford road corridor case study. Ecocities Project, University of Manchester, Manchester

Colls A, Ash N, Ikkala N (2009) Ecosystem based adaptation: a natural response to climate change. IUCN Gland, Switzerland. http://cmsdata.iucn.org/downloads/iucn_eba_brochure.pdf. Accessed 24 Feb 2015

Çalışkan ÇO (2010) 3. Köprü Projesi Değerlendirme Raporu, TMMOB Şehir Plancıları Odası, Istanbul. https://mutlukent.files.wordpress.com/2010/09/spoist_3-koprurapor.pdf. Accessed 07 Jun 2015

Çorakbaş FK, Aksoy A, Ricci A (2014) A report of concern on the conservation issues of the Istanbul Land Walls World Heritage Site. Istanbul, January 2014. https://oxfordbyzantinesociety.files.wordpress.com/2014/02/report_land_walls_whs.pdf. Accessed 07 Jun 2015

Enqvist J, Tengö M, Bodin Ö (2014) Citizen networks in the Garden City: protecting urban ecosystems in rapid urbanization. Landsc Urban Plan 130:24–35

Eren Öİ (2014) What is the threshold in urban generation in the context of urban identity? The case of Turkey. Spatium Int Rev 31:14–21

Ernstson H, Barthel S, Andersson E, Borgström ST (2010) Scale-crossing brokers and network governance of urban ecosystem services: the case of Stockholm. Ecol Soc 15(4):28. http://www.ecologyandsociety.org/vol15/iss4/art28/. Accessed 12 Feb 2015

Feola G, Nunes R (2014) Success and failure of grassroots innovations for addressing climate change: the case of the transition movement. Glob Environ Chang 24:232–250

Fünfgeld H (2015) Facilitating local climate change adaptation through transnational municipal network. Curr Opin Environ Sustain 12:67–73

Gerçek H (2014) 3. Köprü ve Kuzey Marmara Otoyolunun Istanbul'un Ulaşım ve Trafik Sorunlari Üzerindeki Etkileri. In Istanbul'un Geleceğini Etkileyecek Üç Proje. 3. Köprü, 3. Havalimani ve Kanal Istanbul. TEMA Vakfi Uzman Görüşleri, TEMA, Istanbul. http://www.tema.org.tr/folders/14966/categorial1docs/1244/BUYUKPROJELER20032014data.pdf. Accessed 24 Feb 2015

Gülersoy NZ, Gökmen EY (2014) Istanbul ve Tartışmalı Üç Proje: 3.Köprü, 3 Havalimanı ve Kanal Istanbul. In Istanbul'un Geleceğini Etkileyecek Üç Proje. 3. Köprü, 3. Havalimani ve Kanal Istanbul. TEMA Vakfi Uzman Görüşleri, TEMA, Istanbul. http://www.tema.org.tr/folders/14966/categorial1docs/1244/BUYUKPROJELER20032014data.pdf. Accessed 11 Jun 2015

Güvemli Ö (2014) 3. Köprü Evsizleri. Cumhuriyet. 06 Ekim 2014. http://www.cumhuriyet.com.tr/haber/turkiye/127623/3._kopru_evsizleri.html. Accessed 18 Feb 2015

Haq ASM (2011) Urban green spaces and an integrative approach to sustainable environment. J Environ Prot 2(5):601–608

Hürriyet Daily News (2015) Court cancels access road plans for Istanbul's third bridge. 11 June 2015 http://www.hurriyetdailynews.com/court-cancels-access-road-plans-for-Istanbuls-third-bridge.aspx?pageID=238&nID=83855&NewsCatID=340. Accessed 11 Jun 2015

İBB, Istanbul Büyükşehir Belediyesi Park ve Bahçeler Müdürlüğü (2010) Park ve Bahçeler Müdürlüğü 2004–2010. http://www.ibb.gov.tr/tr-TR/BilgiHizmetleri/Istatistikler/Documents/bldhizmetleri/2010/parkvebahceler_mud_2004-%202010.pdf. Accessed 21 Mar 2015

IBB, Istanbul Büyükşehir Belediyesi Park ve Bahçeler Müdürlüğü (2015) Istanbul'a 115 bin yeni ağaç. 23 January 2015. http://www.ibb.gov.tr/tr-TR/Pages/Haber.aspx?NewsID=22449#.VRqB7fysUho. Accessed 02 Mar 2015

ICA (2013) Proje Hakkında: İlklerin Köprüsü. http://www.3kopru.com/proje/PROJE-HAKKINDA/1. Accessed 12 Mar 2015

ICLEI (2014) Resilient cities 2014: congress report. http://resilient-cities.iclei.org/fileadmin/sites/resilient-cities/files/Full_papers/RC2014__Congress_Report__Final.pdf. Accessed 12 Feb 2015

İnce E (2014) Harvard Üniversitesi Yedikule Bostanları'nda Derste. Radikal. 16 Ağustos 2014. http://www.radikal.com.tr/cevre/harvard_universitesi_yedikule_bostanlarinda_derste-1207134. Accessed 02 Feb 2015

IPCC (2014) Climate change 2014: impacts, adaptation and vulnerability. Part A: global and sectoral aspects. In: Contribution of working group II to the fifth assessment report of the Intergovernmental Panel. Cambridge University Press, Cambridge and New York, NY

Kaldjian PJ (2004) Istanbul's Bostans. A millennium of market gardens. Geogr View 94 (3):284–304

Koca A (2014) Istanbul'un Yaşayan İki Bostanı Neden Yok Edilmek İsteniyor? Güncel Dosya: Bostanlar. YAPI 386:58–63. https://bogachandundaralp.files.wordpress.com/2014/01/guncel_dosya.pdf. Accessed 01 Feb 2015

Mayer M (2009) The right to the city in the context of shifting mottos of urban social movements. City: Anal Urban Trends Cult Theory Policy Action 13(2–3):362–374

Newman L, Dale A (2005) Network structure, diversity and proactive resilience building: a response to Tompkins and Adger. Ecol Soc 10(1). http://www.ecologyandsociety.org/vol10/iss1/resp2/. Accessed 14 Feb 2015

Northern Forests Defence (2015) The third airport project vis-a vis life, nature, environment, people and law, Istanbul. http://www.kuzeyormanlari.org/wp-content/uploads/2015/05/3rd_airport_project.pdf. Accessed 06 Jun 2015

Ocak S, Sönmez Y (2014) An aerial tour of Istanbul's disappearing forests. Hürriyet Daily News. 02 September 2014. http://www.hurriyetdailynews.com/an-aerial-tour-of-Istanbuls-disappearing-forests.aspx?pageID=238&nID=71201&NewsCatID=340. Accessed 01 Mar 2015

Özkaynak B, İskender AC, Ertör-Akyazı P, Ertör I (2015) The Gezi Park resistance from an environmental justice and social metabolism perspective. Capital Nat Socialism 26(1). doi:10.1080/10455752.2014.999102. Accessed 10 Jun 2015

Schreurs MA (2008) From the bottom up local and subnational climate change politics. J Environ Dev 17(4):343–355

Singh VS, Pandey DN, Chaudhry P (2010) Urban forests and open green spaces: lessons for Jaipur, Rajasthan India. RSPCB Occasional Paper No. 1/2010. http://www.indiaenvironmentportal.org.in/files/RSPCB-OP-1-2010.pdf. Accessed 05 Feb 2015

Şahin Ü (2013) Üçüncü Köprü'yü Neden Durdurmalıyız (2) 17 Temmuz 2013. Yeşil Gazete. http://yesilgazete.org/blog/2013/07/17/ucuncu-kopru%E2%80%99yu-neden-durdurmaliyiz-2/. Accessed 05 Jun 2015

Taylor RD, Olwig MF, Chettri N (2012) Adaptation as innovation, innovation as adaptation: an institutional approach to climate change. Appl Geogr 33:107–111. doi:10.1016/j.apgeog.2011.10.011

Terzi F, Bölen F (2012) The potential effects of spatial strategies on Urban Sprawl in Istanbul. Urban Stud 49(6):1229–1250

Tolunay D (2014) Istanbul'da Yapılması Planlanan Projelerin Orman Ekosistemi ve Endemik Türler Üzerindeki Etkileri. In Istanbul'un Geleceğini Etkileyecek Üç Proje. 3. Köprü, 3. Havalimani ve Kanal Istanbul. TEMA Vakfi Uzman Görüşleri, TEMA, Istanbul. http://www.tema.org.tr/folders/14966/categorial1docs/1244/BUYUKPROJELER20032014data.pdf. Accessed 24 Feb 2015

Tompkins EL, Adger NW (2004) Does management of natural resources enhance resilience to climate change? Ecol Soc 9(2):10. http://www.ecologyandsociety.org/vol9/iss2/art10/. Accessed 17 Feb 2015

Türkeş M (2014) Istanbul'da Yapılması Planlanan Projelerin Yerel İklim ve İklim Değişikliği Üzerindeki Etkileri. In Istanbul'un Geleceğini Etkileyecek Üç Proje. 3. Köprü, 3. Havalimani ve Kanal Istanbul. TEMA Vakfi Uzman Görüşleri, TEMA, Istanbul. http://www.tema.org.tr/folders/14966/categorial1docs/1244/BUYUKPROJELER20032014data.pdf. Accessed 08 Jun 2015

Vardar N (2014) Yedikule Bostanları Planı İade Edildi. Taraflar Temkinli. Istanbul Bia Haber. 25 Kasım 2014. http://www.bianet.org/bianet/toplum/160216-yedikule-bostanlari-plani-iade-edildi-taraflar-temkinli. Accessed 24 Feb 2015

Weise Z (2015) Wild boars arrive in Istanbul due to loss of habitat. The Guardian. 6 March 2015. http://www.theguardian.com/environment/2015/mar/06/wild-boars-arrive-in-Istanbul-due-to-loss-of-habitat. Accessed 08 Jun 2015

Yalçıntan MC, Çalışkan ÇO, Çılgın K, Dündar U (2014) İstanbul Dönüşüm Coğrafyası. In: Candan A, Özbay C (eds) Yeni İstanbul Çalışmaları Sınırlar, Mücadeleler, Açılımlar. Metis Yayınları, Istanbul

Young S (1997) Community-based partnerships and sustainable development: a third force in the social economy. In: Baker S, Kousis M, Richardson D, Young S (eds) The politics of sustainable development. Routledge, London, New York, NY

Chapter 8
Extreme Weather Events and the German Economy: The Potential for Climate Change Adaptation

Ulrike Lehr, Anne Nieters, and Thomas Drosdowski

Abstract Although climate change is a global challenge, its effects occur locally and differ by region. A feasible adaptation strategy needs to assess regional damages and their socio-economic effects. For Germany, the largest threat comes from extreme weather events, which will impact residential and commercial buildings, infrastructure and in the case of heat waves will limit labor productivity. This paper presents findings from a study of economic effects of climate change adaptation until the year 2050 in Germany on different scales. In particular, the authors have applied an input–output-based macroeconometric model, adjusting it to cope with the challenges of damages from heat waves, and river flood events, by integrating suitable adaptation measures to such events into the model. Infrastructure damages, shifts from domestic production to imports, and low levels of productivity due to heat waves, are some of the topics the paper deals with. Comparing scenarios with (a) integrated extreme weather events and (b) adaptation measures with a reference scenario without extreme weather or adaptation, the simulation results reveal slightly negative effects on economic sectors and Germany's economy as a whole. These effects intensify over time and hurt the economy. Adaptation measures reduce the damages and pay off, but the economy is still worse off with climate change.

Keywords Extreme weather events • Heat wave • River flood • Adaptation

Introduction

In the last 15 years, Germany experienced severe extreme weather events, particularly river floods and heat waves. The flood events of the rivers Elbe, Danube and Oder in the Eastern part of Germany caused massive damages in the years 2002,

U. Lehr (✉) • A. Nieters • T. Drosdowski
Institute for Economic Structure Research (GWS mbH), Heinrichstrasse 30, 49080 Osnabrueck, Germany
e-mail: lehr@gws-os.com

W.L. Filho et al. (eds.), *Climate Change Adaptation, Resilience and Hazards*, Climate Change Management, DOI 10.1007/978-3-319-39880-8_8

2006, 2010 and 2013 amounting to several bn. Euros (Muenchner Rueck 2003; Chorynski et al. 2012). The most severe heat wave in Germany occurred in the year 2003, followed by above average hot summers in 2006 and 2010 (UNEP 2004; Muenchner Rueck 2004).

Germany's industry contributes 30.8 % to the country's Gross Domestic Product (GDP); a high contribution in comparison with other European countries (France with 19.4 %, UK with 20.6 % or the US with 20.7 %).[1] The industry is situated close to waterways and suffers losses from floods, which in turn affect further upstream and downstream industries. Moreover, Germany is the fifth densely populated country in Europe (after Malta, Belgium, Netherlands and UK) and the damages from floods in densely populated areas are high. In addition, potential inundation areas are cultivated and populated. As the frequency of occurrence and the severity of extreme weather events likely increase in the future (IPCC 2013), adaptation measures will be taken. Estimates of the likely benefits from adaptation measures based on estimates of the prevented damages are relevant to support decision makers to take the appropriate steps. Our results contribute to this decision base.

Latest since the United Nations Conference on Environment and Development (UNCED) in Rio de Janeiro in the year 1992 economists developed different approaches to quantify economic effects of climate change. Integrated Assessment Models (IAM) extend economic general equilibrium models with damage functions driven by physical impacts, mostly focus on the increase of global average surface temperature. FUND (Climate Framework for Uncertainty, Negotiation and Distribution) (Tol 1997; Tol et al. 2014), Dynamic Integrated Climate-Economy Model (DICE) (Nordhaus 1992; Nordhaus and Sztorc 2013), The Policy Analysis for the Greenhouse Effect (PAGE) (Hope 2011) and MERGE (Model for Evaluating Regional and Global Effects of GHG reductions) (Manne et al. 1995) are the most prominent IAM models. Patt et al. (2010) find that IAM rarely are applied to extreme weather events, though examples can be found in the FUND 3.17 model (Anthoff and Tol 2013). Disaster Impact Models (DIM) focus on the latter, e.g., on floods, hurricanes or earthquakes on a regional level (Hallegatte et al. 2011; Rose and Liao 2005), but lack the overall economic perspective. Although the first reference to adaptation in scientific analyses can already be found in the First Assessment Report of the IPCC in 1991, only the Cancún summit in 2010 has brought this topic back on the UNFCCC framework agenda (Agrawala et al. 2011). In the last 15 years, adaptation to climate change has increasingly entered the literature on climate change policies. Agrawala et al. (2011) give a comprehensive overview on methodologies and adaptation costs at the sectoral, regional and global level. The focus is on adaptation to consequences of gradual temperature changes (e.g., sea-level rise, changes in the tourism sector or in agricultural production). Adaptation to extreme weather events (heat waves, floods, storms) is as well considered, albeit only in the context of their effects on human health. In creating

[1]All data for 2014 from Central Intelligence Agency (2015).

AD-DICE, as an extension of the DICE model, de Bruin et al. 2009 integrate adaptation and mitigation in one model.

Mitigation policy decisions have been supported by the results of a further modelling approach, called macroeconometric simulation models. Examples are E3E by Cambridge Econometrics (Barker et al. 2015; Pollitt et al. 2014; Barker et al. 2011) or the model described in this contribution, PANTA RHEI (Lutz et al. 2005; Lehr et al. 2008; Meyer et al. 2012). The basic idea behind this approach is that the economy is characterized by several non-equilibrium states, such as unemployment, inefficient use of resources etc. and therefore an alternative approach to equilibrium modelling was sought. The research question which will be answered in the remainder of this contribution is: Can this type of model be extended to simulate the effects of climate change and climate change adaptation?

The remainder of this paper is organized as follows: The next section describes the changes needed in the modelling framework and defines the scenarios for (a) frequently recurring extreme weather events (river floods and heat waves) and (b) adaptation measures. In the following section, results are presented in terms of effects on the German economy as a whole and on individual economic sectors until 2050. The last section concludes and gives an outlook.

The Modelling Challenge

This section introduces the model. However, since the model comprises several thousand equations, a closed analytical representation cannot be given and the reader is referred to the available model descriptions in the literature (Lehr et al. 2011; Grossmann et al. 2012).

The Model PANTA RHEI

PANTA RHEI (Lutz et al. 2005; Lehr et al. 2008; Meyer et al. 2012) is an environmentally extended version of the econometric simulation and forecasting model INFORGE (Ahlert et al. 2009; Meyer et al. 2007). A detailed description of the economic part of the model is presented in Maier et al. (2015). Among others, it has been used for economic evaluation of different energy scenarios that have been the basis for the German energy concept in 2010 (Lindenberger et al. 2010; Nagl et al. 2011).

The behavioral equations reflect bounded rationality rather than optimizing behavior of agents. All parameters are estimated econometrically from time series data (1991–2012). Producer prices are the result of mark-up calculations of firms. Output decisions follow observable historic developments, including observed inefficiencies rather than optimal choices. The use of econometrically estimated equations means that agents have only myopic expectations. They follow routines developed in the past. This implies in contrast to equilibrium models that markets will not necessarily be in an optimum and non-market (energy) policy interventions can have positive economic impacts.

Structural equations are usually modeled on the 59 sector level (according to the European 2 digit NACE classification of economic activities) of the input–output accounting framework of the official system of national accounts (SNA) and the corresponding macro variables are then endogenously calculated by explicit aggregation. In that sense the model has a bottom-up structure. The input–output part is consistently integrated into the SNA accounts, which fully reflect the circular flow of generation, distribution, redistribution and use of income.

The core of PANTA RHEI is the economic module, which calculates final demand (consumption, investment, exports) and intermediate demand (domestic and imported) for goods, capital stocks, and employment, wages, unit costs and producer as well as consumer prices in deep disaggregation of 59 industries. The disaggregated system also calculates taxes on goods and taxes on production. The corresponding equations are integrated into the balance equations of the input–output system.

Value added of the different branches is aggregated and gives the base for the SNA that calculates distribution and redistribution of income, use of disposable income, capital account and financial account for financial enterprises, non-financial enterprises, private households, the government and the rest of the world. Macro variables like disposable income of private households and disposable income of the government as well as demographic variables represent important determinants of sectoral final demand for goods. Another important outcome of the macro SNA system are net savings and governmental debt as its stock. Both are important indicators for the evaluation of policies. The demand side of the labor market is modeled in deep sectoral disaggregation. Wages per head are explained using Philips curve specifications. The aggregate labor supply is driven by demographic developments.

The model is empirically evaluated: The parameters of the structural equations are econometrically estimated. On the time consuming model-specification stage various sets of competing theoretical hypotheses are empirically tested. As the resulting structure is characterized by highly nonlinear and interdependent dynamics the economic core of the model has furthermore been tested in dynamic ex-post simulations. The model is solved by an iterative procedure year by year.

The energy module captures the dependence between economic development, energy input and CO_2 emissions. It contains the full energy balance with primary energy input, transformation and final energy consumption for 20 energy consumption sectors, 27 fossil energy carriers and the satellite balance for renewable energy (AGEB 2011). The energy module is fully integrated into the economic part of the model.

Input–output tables provide detailed insights into the flows of goods and services between all sectors of the economy and the interdependence of the economy of a country and with the rest of the world. They are closed accounting schemes where the identity of the sum of inputs and the sum of outputs has to hold in each sector.

PANTA RHEI covers effects for the German economy in total and for 59 economic sectors. Regional disaggregation exists, but it has not been part of the simulations described here.

Integration of Damages

In a first step, PANTA RHEI is extended to allow modelling damages resulting from extreme weather events, the second step is the simulation of adaptation. To effectively integrate extreme events and adaptation measures into the model, it is essential to firstly identify economic sectors most severely affected by extreme events in the past. Since no time series exist concerning extreme weather events in Germany, cost and damages of river flood events and heat waves are primarily derived from the impacts of the Oder/Elbe flood in 2002 and the heat wave in 2003, functioning as reference events or benchmark. Table 8.1 gives an overview of the observed physical impact, its translation into model variables and economic quantities and the literature used.

For instance: A flood event on the Rhine, where numerous production sites of the machinery industry are located, leads to damages on buildings and production sites in the machinery sector. Companies decrease output, because the capital stock is damaged, i.e., the production machinery is damaged, wet, spoiled etc. The industry will rebuild and claim reimbursement from insurances. The economy as a whole experiences a slowdown from the lacking output and growth from the repair measures. There is a time lag between the damage and the full recovery, which depends on the industry analyzed. Intermediate input production slows and is replaced by imports. Transport infrastructure damages impede the transport of intermediate goods. All this leads to an increase of imports in PANTA RHEI. Similar proceeding holds for heat waves. More detailed information on the approach of including extreme weather events in PANTA RHEI can be found in Nieters et al. (2015).

Scenarios: Exploring Consistent Future Developments

The next step involves the definition of future scenarios. The comparison of results from simulation runs under different scenario assumptions gives estimates of relative economic effects. The reference case is based on the reference projection for the German energy system (cf. GWS, Prognos, EWI 2013).

In scenario 1, extreme weather enters the model as shocks and economic interdependences are temporarily overridden. The "shocks" reflecting flood events were modeled in the form of 10-year events, which means that the processes described above take place every 10 years until 2050. The decade between 2041 and 2050 is an exception, since two flood events occur. The "shocks" reflecting heat

Table 8.1 Integration of extreme events in PANTA RHEI

Damages	Target variables	Expected main effects	Sources
Damages of a river flood on...			
...production sites	– Capital stock of: *Machinery* – Other current transfers	Increase in buildings investment and investment in plant and equipment, loss in production	Muenchner Rueck (2003) Braeuer et al. (2009)
...dwellings	– Capital stock of: *Real estate* – Other current transfers – Disposable income	Increase in buildings investment and investment in plant and equipment, decrease in consumption	Muenchner Rueck (2003) Braeuer et al. (2009)
...transport infrastructure	– Capital stock of: *Public administration* – Other current transfers – Production output	Increase in buildings investment, unit costs and depreciation, loss in production	Muenchner Rueck (2003)
...production	– Imported intermediate goods: *Chemicals* *Machinery* *Metals and semi-finished products* *Basic metals* *Automobiles and parts* *Agriculture*	Increase in imports and prices	Ludwig and Brautzsch (2002) BMI (2013)
...disaster management	– Government spending: *Defense* *Public order and safety* – Tax increases for private households	Lower government spending in fields other than defense and public order and safety (e.g., for education), lower disposable income for private households	BMI (2013)
Damages of a heat wave on...			
...agriculture	– Imported intermediate goods: *Agriculture*	Increase in imports and prices and decrease in production value	Braeuer et al. (2009) Fischer and Schaer (2010)
...energy sector	– Electricity imports	Increase in imports and prices and decrease in production value	Rademaekers et al. (2011)
...labor productivity	– Labor productivity	Decrease in average wages per hour, increase in employment	Huebler et al. (2008) PIK (2014)
...ship traffic	– Input coefficients of land and ship transport services – Imported intermediate goods: *All*	Increase in imports and prices	Jonkeren et al. (2011)

waves are modeled in the form of 4-year events. It has to be stressed that we do not make a point regarding the likelihood of occurrence of flood events or heat waves. The occurrence in 10 or 4-year intervals is merely an assumption in accordance with estimations of climate experts concerning an increase in frequency and intensity of extreme weather events in Europe (IPCC 2014; Braeuer et al. 2009).

Scenario 2 includes adaptation measures. Investing in adaptation measures leads to lesser future damages. Adequate adaptation to prevent damages from river floods events is the creation of additional retention areas, which are located next to running waters and used in case of flood discharge as flood-plain areas and the reinforcement of dikes, paid from public funds. Adaptation to heat waves in electric power generation requires cooling towers. The costs are borne by the utility companies. Companies prevent heat-related reductions in labor productivity by greening roofs and installing air conditioning and have to bear the costs for these investments. Air conditioning is regarded an adaptation strategy as long as environmentally friendly air conditioners are installed and energy from renewable sources is used. Table 8.2 shows the adaptation measures, the translation into model variables and the literature used, similar to Table 8.1.

Scenario comparison shows differences in a large set of economic quantities. Results are given as changes in the following indicators:

- Differences in overall GDP stand for differences in economic performance.
- Differences in employment reflects the social aspects of a future development.
- Private consumption reflects individual well-being.
- Production indicates the opportunities and activities of a country's industry.
- Sector specific production gives an indication of winners and losers.

Table 8.2 Modeling adaptation

Adaptation measures			
Creation of additional retention areas	– Increase in buildings investments – Compensation for farmers increases other current transfers	Assumption: All flood damages are avoided Described expected main effects (s. above) are reduced	Gruenig et al. (2013)
Reinforcement of dikes	– Increase in buildings investments		Gruenig et al. (2013)
Installation of cooling towers	– Increase in buildings investments	Increase in energy imports is avoided	Van Ierland et al. (2007) Weisz et al. (2013)
Roof greening	– Increase in buildings investments	Assumption: Reduction of labor productivity is avoided	Gruenig et al. (2013) Altvater et al. (2012)
Installation of air conditioning	– Increase in buildings investments		ZIA (2014), BKI (2013)

As has been pointed out above, the systems boundary is Germany. Note that the frequency of the extreme weather event cannot be a forecast and therefore the development over time of the economic results strongly hinges on the assumed frequency.

Results I: Extreme Weather Events Compared to Reference Scenario

Simulation results indicate that overall economic effects of recurring heat waves and river floods are moderate in Germany until 2050. Figure 8.1 shows the development of the four indicators described above. All results are produced as follows: the model is run under different scenarios and produces time series of the respective indicator. Subtraction of GDP under the reference scenario from GDP under the extreme weather scenario gives the GDP effect from extreme weather events. Figure 8.1 shows these differences for all for indicators. Negative values mean the indicator is smaller in the extreme weather event scenario compared to the reference.

Overall, the data show a negative effect of weather extremes on the economy, Private consumption is annually between 1 and 3 bn., production between 3.5 and 30 bn. and GDP between 1.8 and 19.5 bn. Euros lower than in the reference. Only employment is higher under extreme events. The positive development can be attributed to two effects: firstly, production shifts from capital-intensive sectors to labor-intensive sectors, as capital is destroyed by the flood and production goes down in some capital-intensive sectors, whereas employment in the labor intensive construction sector grows. Secondly, the increase is due to falling productivity, which is compensated by the model with businesses hiring more personnel, whose wage incomes have become smaller.

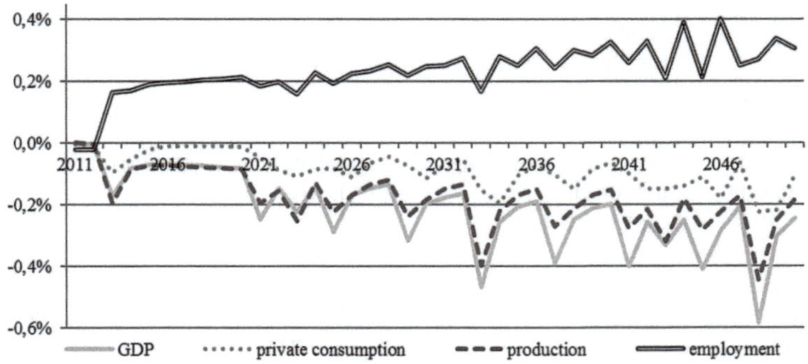

Fig. 8.1 Economic development, difference (in %) extreme weather versus reference scenario, price adjusted

The effect is intensifying over time: In the decade from 2031 to 2040, average effects are higher than in the following decade, even if less events occur, hence effects seem to be lasting, and the predicted increase in the number and intensity of extreme weather events in Germany may challenge its economic development in the future.

Results II: Adaptation to Extreme Weather Events Compared to (a) Extreme Weather and (b) Reference Scenario

Economic effects from adaptation are twofold in the model: they prevent or reduce extreme weather damages and they induce economic activity in the construction sector, in transportation, and in the production of adaptation goods. Figure 8.2 shows the simulation results as differences between two simulation runs. The economic impacts of adaptation are positive in comparison to the extreme weather scenario. The effect is increasing over time, which is due to the fact that firstly, the impacts of extreme weather events are as well rising. Secondly, adaptation measures are cumulative over the years. Hence, the more adaptation measures are undertaken, the more damages are avoided. However, the positive deviation is diminishing in the last decade. This development results from rising costs accompanying some of the adaptation measures.

Figure 8.3 shows the development of GDP in the extreme weather and the adaptation scenario until 2050 in comparison to the reference scenario. It can be seen that the selected adaptation measures reduce the negative effects of extreme weather events, however, in both scenarios the German economy is worse off in comparison to a scenario without extreme weather events.

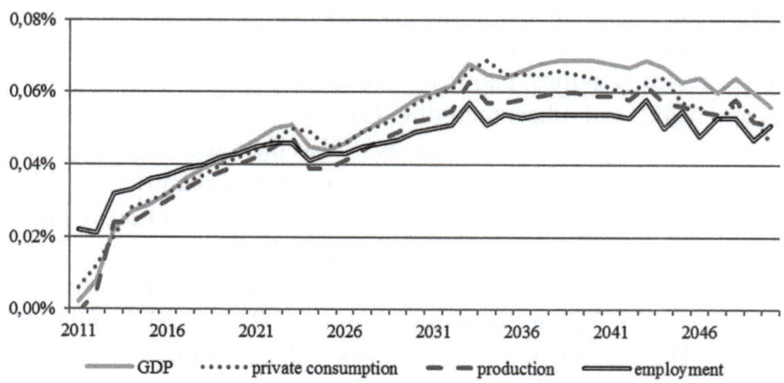

Fig. 8.2 Economic development, difference (in %) extreme weather versus adaptation scenario, price adjusted

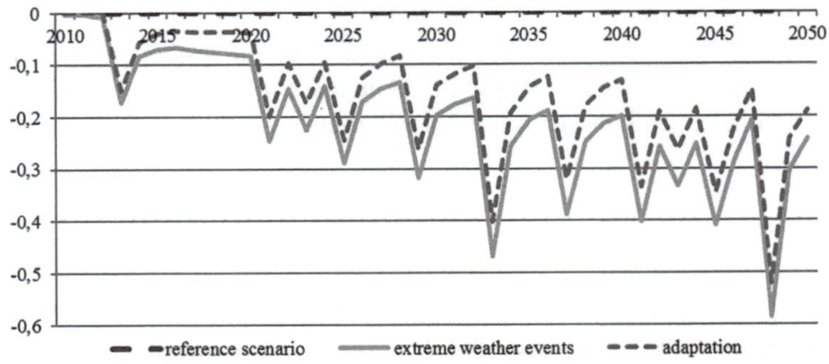

Fig. 8.3 GDP, differences reference versus extreme weather and adaptation scenario, price adjusted

A More Detailed Picture: Extreme Weather and Economic Sectors

How are individual sectors affected by extreme weather? Are there winners and losers in the economy? The contribution of sectors to the aggregate changes can differ markedly. Since the analysis of effects on the total economy may underestimate the effects on individual economic sectors, sector-specific effects have been included in the analysis. Instead of a detailed description of single effects of extreme weather events and adaptation measures on the economic sectors, an overview of the most and least affected industries is given.

For the sector specific analysis, we need one further indicator: Gross Value Added (GVA). Gross Value Added by sector is defined as the difference between turnover and costs by industry. Wages and profits are paid from Gross Value Added. It serves as a measure of sector specific success. In 2014, GVA in Germany amounted to roughly 2.5 tr. Euros. As presented in Table 8.3, the most important sectors with respect to economic performance are "other services" (administrative and support service activities, public administration and defense, education, social activities and health and other) recording GVA of 1.25 tr. Euros. The next largest sector is manufacturing (535 bn. Euros), as well as the trade sector (250 bn. Euros).

Comparing the reference and the extreme weather scenario (see Table 8.4) we find small deviations from the reference scenario. An exception is the transport sector with percentage changes between 0.4 and 0.5 per decade.

Transport and communication services show the largest effects with an average increase in GVA of 0.5 % in the period between 2021 and 2030 and 0.4 % in the other decades. The construction sector follows with deviations amounting to values between 0.1 and 0.3 %. Striking are the simulation results for the agricultural sector, indicating a slightly positive effect—as far as average deviations per decade are considered. Figure 8.4 presents annual deviations from the reference scenario in the sectors agriculture and trade (in percent) and reveals that the total effect on these sectors, positive in both cases, does not reflect the vulnerability of the sectors.

Table 8.3 Gross value added on the sectoral level in 2014, in bn. Euros

Economic sectors	GVA in 2014 in bn. Euros
Agriculture, forestry and fishery	20
Mining and quarrying	7
Manufacturing	535
Energy and water supply	71
Construction	114
Trade; maintenance and repair of motor vehicles	250
Hotel and restaurant industry	45
Transport and communication services	150
Financial intermediation	92
Other services	1252

Table 8.4 GVA, average deviations per decade, extreme weather versus reference scenario, in percent

Economic sectors	2011–2020 (%)	2021–2030 (%)	2031–2040 (%)	2041–2050 (%)
Agriculture, forestry and fishery	0.07	0.01	0.11	0.09
Mining and quarrying	−0.17	−0.31	−0.27	−0.28
Manufacturing	−0.07	−0.22	−0.19	−0.28
Energy and water supply	−0.04	−0.09	−0.13	−0.16
Construction	0.14	0.21	0.28	0.29
Trade; maintenance and repair of motor vehicles	0.08	0.12	0.18	0.20
Hotel and restaurant industry	−0.04	−0.14	−0.14	−0.17
Transport and communication services	0.43	0.48	0.43	0.38
Financial intermediation	0.06	0.07	0.10	0.09
Other services	−0.02	−0.08	−0.09	−0.11

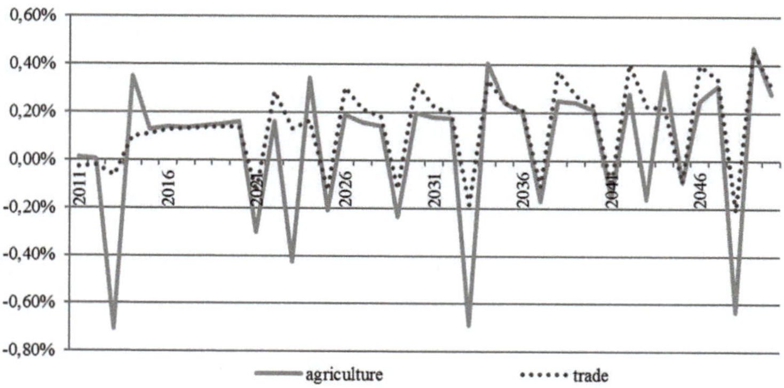

Fig. 8.4 GVA, deviations from the reference scenario of the sectors trade and agriculture, in percent

Simulation results indicate both years with negative and years with positive effects. Even if the negative effect on GVA in the agricultural sector is rather strong on an annual basis in comparison to the other sectors (−0.71 %), the overall effect on the sector is positive, meaning that negative impacts in the years in which extreme events occur are overcompensated by positive effects in the years in between. The same is true for the trade sector: negative effects are observable but positive effects predominate. This means, the stronger the impact, the stronger the counter-reaction in the years following extreme weather events. An increasing number of extreme events may pose a challenge particularly for the agricultural sector. The relatively strong counter-reactions can be attributed to price increases after an extreme weather event due to a likely considerable crop loss. Higher prices in years following an extreme event may compensate farmers for the losses in the previous one, thus leading to an increase in GVA.

Effects of Adaptation to Extreme Weather Events on Economic Sectors

In the adaptation scenario, governments and companies undertake investments to adapt to river floods by extending retention areas and reinforcing dikes. Impacts of heat waves on labor productivity and on power plants are mitigated by investments in green roofing, air conditioning and cooling towers. Table 8.5 shows the average deviations per decade of GVA on the sectoral basis, comparing the reference and the adaptation scenario. Simulation results of the latter reflect a combination of both, impacts of extreme weather events, and adaptation measures. This is because, firstly, undertaking adaptation measures does not mean that all impacts of extreme weathers are avoided at once. This is happening incrementally. Secondly, not all of

Table 8.5 GVA, average deviations per decade, adaptation versus reference scenario, in percent

Economic sectors	2011–2020 (%)	2021–2030 (%)	2031–2040 (%)	2041–2050 (%)
Agriculture, forestry and fishery	0.12	0.05	0.15	0.16
Mining and quarrying	−0.03	−0.17	−0.13	−0.16
Manufacturing	0.00	−0.15	−0.11	−0.19
Energy and water supply	0.00	−0.02	−0.03	−0.05
Construction	0.28	0.47	0.68	0.81
Trade; maintenance and repair of motor vehicles	0.11	0.16	0.23	0.24
Hotel and restaurant industry	0.01	−0.06	−0.05	−0.10
Transport and communication services	0.05	0.12	0.10	0.10
Financial intermediation	0.09	0.11	0.15	0.13
Other services	0.00	−0.04	−0.04	−0.06

the modeled impacts can be reduced by the selected adaptation measures: whereas it is assumed that all effects of river floods can be avoided by the end of the simulation period, the selected measures are for example not suitable to adapt to low river water levels during a massive heat wave. Also in the adaptation scenario ship transport companies have to shift to other means of transportation to deliver (intermediate) goods and raw materials.

The differences in GVA between the reference and the adaptation scenario are similar to the results obtained by comparing the extreme weather and the reference scenario. As discussed by the following examples, effects are small and some of the sectors may benefit from adaptation measures, whereas others may suffer reductions in GVA. In all sectors (an exception is the sector transport and communication services) the positive effect is intensifying in the adaptation scenario, whereas negative effects are diminishing (see, for comparison Table 8.4). In most of the sectors, the development can be attributed to avoided damages. The manufacturing sector, for example, faces lower damages on buildings and properties, since communities gradually invest in additional retention areas and dike reinforcement. Average deviations per decade from the reference scenario amount to between −0.11 and 0.19 %. Hence, the sector mining and quarrying records lower deviations as well. An increasing number of utility companies does not need to reduce or even cease energy production in case of heat waves due to the installation of cooling towers. Therefore, reductions in GVA decrease and amount to between −0.02 and −0.05.

The sectors already "benefitting" from extreme weather events (decade averages), show even higher values in terms of GVA in the adaptation scenario. This is because there is a combination of extreme events and adaptation measures in this scenario. While, for example, the agricultural sector is less affected by river flood events, simultaneously it benefits from compensation payments for the creation of retention areas. Average deviations of GVA in the adaptation scenario are up to 0.16 % higher than in the reference. Investments in dike reinforcement, green roofing or air conditioning lead to a rise in GVA in the construction sector in comparison to the reference (up to 0.81 %) and as well to the extreme weather scenario. The transport sector, in turn, benefits from the positive development in the trade and construction sector. However, in comparison to the extreme weather scenario, where the deviation from the reference amounts to up to 0.17 % in the last decade, the deviation of the adaptation scenario from the reference run is up to 0.10 % and thus decreasing. This is because only the demand for construction material increases in comparison to the extreme weather scenario and the reference, implicating an increased need for transport services. The demand for disposal of rubble, and for machinery and other equipment in affected regions is lower than in the extreme weather scenario, since damages are lower.

Overall, all sectors benefit from adaptation measures, even if this involves financial burdens for companies and the government. Some sectors (agriculture, construction, trade and others) benefit from extreme weather events when focusing on GVA and decade averages, however investments in adaptation measures intensify the positive development. At the same time adaptation to extreme events diminishes negative effects resulting from these events. The sector transport and communication services is the only sector not following a similar pattern. It benefits from adaptation measures in comparison to the reference scenario but compared to the development in the extreme weather scenario it performs worse.

Conclusions

The analysis has shown that extreme weather events have a small effect on the German economy. However, the analysis shows an intensification of the impacts on economic sectors and the economy as a whole in the future. Thus, recurring extreme weather events have the potential to weaken Germany's future economic performance. Adaptation measures not only reduce negative effects on the German economy, but also stimulate economic activity and might also lead to the development of innovations demanded on international markets. However, it has to be emphasized that the results do not take climate change effects on other countries into account. The decline or loss of production sites along the value chain in countries producing raw materials or intermediate goods for German production incurs losses in Germany. Further research into these effects is required.

On a sector-specific level, manufacturing as well as mining and quarrying are the most heavily affected industries with respect to GVA when comparing the extreme weather and the adaptation scenario with the reference scenario. The transport sector benefits the most from extreme events. However, its economic performance diminishes in the adaptation scenario. The construction sector responds above average positively to adaptation measures. For future research, this detailed sector specific analysis needs a detailed regional analysis, too. Not only industries differ in their vulnerability to extreme weather events, but also the individual regions in Germany. Reductions in agricultural production, for example, will have stronger impacts on the economy of Schleswig-Holstein than on North Rhine-Westphalia, since agricultural production in Schleswig-Holstein plays a more important role for the value added in that region than in North Rhine-Westphalia. Regarding this issue, further research is necessary to gain a more comprehensive insight into the impacts of extreme weather events and, thus, adaptation measures in Germany. Detailed knowledge about a region's vulnerability to extreme events is decisive with respect to adaptation strategies. Within a regional approach, the burden sharing process will gain more relevance. Although the effects spread over the whole economy, the local or regional effects on individual households might be much larger. These aspects escape a macro-modeling exercise. Moreover, some responses of the economic sectors to extreme events and adaptation measures reveal weaknesses underlying input–output-based macroeconometric models when analyzing effects of extreme weather events. Further adjustments are needed to capture modified economic relationships resulting from extreme events unhinging preliminary relationships and their interdependencies.

The access to more comprehensive and more detailed data on damages and costs resulting from extreme weather events and costs associated with concrete adaptation measures in Germany will improve simulation results. The program for further research therefore has to meet the threefold challenge of a wider regional focus, including trade and international value chains, a narrower regional focus within Germany and the need for more accurate, sector (and region) specific data on damages from all kinds of extreme weather and temperature increases.

References

AGEB (2011) Energiebilanzen für die Bundesrepublik Deutschland, Satellitenbilanz "Erneuerbare Energien". Arbeitsgemeinschaft Energiebilanzen e.V, Berlin

Agrawala S, Bosello F, Carraro C, de Cian E, Lanzi E (2011) Adapting to climate change: costs, benefits, and modelling approaches. Int Rev Environ Resour Econ 5(3):245–284

Ahlert G, Distelkamp M, Lutz C, Meyer B, Moennig A, Wolter MI (2009) Das IAB/INFORGE-Modell. In: Schnur P, Zika G (eds) Ein sektorales makrooekonometrisches Projektions- und Simulationsmodell zur Vorausschaetzung des laengerfristigen Arbeitskraeftebedarfs, vol 318. IAB-Bibliothek, Germany, pp 15–175

Altvater S, de Block D, Bouwma I, Dworak T, Frelih-Larsen A, Goerlach B, Hermeling C, Klostermann J, Koenig M, Leitner M, Marinova N, McCallum S, Naumann S, Osberghaus D, Prutsch A, Reif C, van de Sandt K, Swart R, Troeltzsch J (2012) Adaptation measures in the EU: policies, costs, and economic assessment, report as deliverable for the project "Climate Proofing" of key EU policies—short term actions. Tender CLIMA.C.3/SER/2010/0009, Berlin

Anthoff D, Tol RS (2013) The climate framework for uncertainty, negotiation and distribution (FUND). Technical description, version 3.7. http://www.fund-model.org/. Accessed 07.06.2016

Barker T, Lutz C, Meyer B, Pollitt H, Speck S (2011) Modelling an ETR for Europe. In: Ekins P, Speck S (eds) Environmental tax reform (ETR): a policy for green growth. Oxford University Press, Oxford

Barker T, Alexandri E, Chewpreecha U, Ogawa Y, Pollitt H (2015) GDP and employment effects of policies to close the 2020 emissions gap. Climate Policy. doi:10.1080/14693062.2014.1003774

BKI (2013) Kostenplanung, Baukosten, Bauelemente

BMI (2013) Bericht zur Flutkatastrophe 2013: Katastrophenhilfe, Entschaedigung, Wiederaufbau. Kabinettbericht des Bundesministeriums des Inneren

Braeuer I, Umpfenbach K, Blobel D, Gruenig M, Best A, Peter M, Lueckge H (2009) Klimawandel: Welche Belastungen entstehen für die Tragfaehigkeit der Oef-fentlichen Finanzen? Endbericht. Ecologic Institute, Berlin

Central Intelligence Agency (2015) Germany. France, UK, US. In: The World Factbook. Retrieved from https://www.cia.gov/library/publications/the-world-factbook/

Chorynski A, Pinskwar I, Kron W, Brakenridge G (2012) Catalogue of large floods in Europe in the 20th century. In: Kundzewicz ZW (ed) Changes in flood risk in Europe, special publication 10. IAHS Press, Wallington, Oxfordshire

De Bruin K, Dellink R, Agrawala S (2009) Economic aspects of adaptation to climate change: integrated assessment modelling of adaptation costs and benefits. OECD Environment Working Paper, 6. OECD Publishing. doi:10.1787/225282538105

Fischer EM, Schaer C (2010) Consistent geographical patterns of changes in high-impact European heatwaves. Nat Geosci 3(6):398–403

Grossmann A, Lehr U, Wiebe KS, Wolter MI, Fleissner P (2012) Feasible futures for the common good. Energy transition paths in a period of increasing resource scarcities. Progress report 5b: E3.at and HYBRIO 57—model comparison

Gruenig M, Troeltzsch J, Schulze S (2013) Gutachten zu den oekonomischen Folgen des Klimawandels und Kosten der Anpassung fuer Hamburg, Hamburg

GWS, Prognos, EWI (2013) Gesamtwirtschaftliche Effekte der Energiewende. Endbericht. Projekt Nr. 31/13. Im Auftrag für das Bundesministerium für Wirtschaft und Energie

Hallegatte S, Ranger N, Mestre O, Dumas P, Corfee-Morlot J, Herweijer C, Wood RM (2011) Assessing climate change impacts, sea level rise and storm surge risk in port cities: a case study on Copenhagen. Climate Change 104(1):113–137

Hope C (2011) The PAGE09 integrated assessment model: a technical description. Cambridge Judge Business School Working Paper 4 (11)

Huebler M, Klepper G, Peterson S (2008) Costs of climate change. The effects of rising temperatures on health and productivity in Germany. Ecol Econ 68:381–391

IPCC (2013) Climate change 2013. The physical science basis. In: Contribution of working group I to the fifth assessment report of the Intergovernmental panel on climate change, Geneva

IPCC (2014) Climate change 2014: impacts, adaptation, and vulnerability. Part B: regional aspects. In: Contribution of working group II to the fifth assessment report of the Intergovernmental panel on climate change, Geneva

Jonkeren O, Jourquin B, Rietveld P (2011) Modal-split effects of climate change: the effect of low water levels on the competitive position of inland waterways transport in the river Rhine area. Transp Res Part A Policy Pract 45(10):1007–1019

Lehr U, Nitsch J, Kratzat M, Lutz C, Edler D (2008) Renewable energy and employment in Germany. Energy Policy 36:108–117

Lehr U, Mönnig A, Wolter MI, Lutz C, Schade W, Krail M (2011) Die Modelle ASTRA und PANTA RHEI zur Abschätzung gesamtwirtschaftlicher Wirkungen umweltpolitischer Instrumente—ein Vergleich. GWS Discussion Paper 11/4

Lindenberger D, Lutz C, Schlesinger M (2010) Szenarien für ein Energiekonzept der Bundesregierung. Energiewirtschaftliche Tagesfragen 60:32–35

Ludwig U, Brautzsch H-U (2002) Die Hochwasserkatastrophe und das Sozialprodukt in Deutschland. Wirtschaft im Wandel 12:353–356

Lutz C, Meyer B, Nathani C, Schleich J (2005) Endogenous technological change and emissions: the case of the German steel industry. Energy Policy 33:1143–1154

Maier T, Mönnig A, Zika G (2015) Labour demand in Germany by industrial sector, occupational field and qualification until 2025—model calculations using the IAB/INFORGE model. Econ Sys Res 27(1):19–42

Manne A, Mendelsohn R, Richels R (1995) MERGE: a model for evaluating regional and global effects of GHG reduction policies. Energy Policy 23(1):17–34

Meyer B, Lutz C, Schnur P, Zika G (2007) Economic policy simulations with global interdependencies: a sensitivity analysis for Germany. Econ Syst Res 19:37–55

Meyer B, Meyer M, Distelkamp M (2012) Modeling green growth and resource efficiency: new results. Miner Econ 24(2):145–154

Muenchner Rueck (2003) Topics. Jahresrueckblick Naturkatastrophen 2002. www.munich-re.com. Accessed 7.06.2016

Muenchner Rueck (2004) Topics geo. Jahresrueckblick Naturkatastrophen 2003. www.munich-re.com. Accessed 7.06.2016

Nagl S, Fuersch M, Paulus M, Richter J, Trüby J, Lindenberger D (2011) Energy policy scenarios to reach challenging climate protection targets in the German electricity sector until 2050. Util Policy 19:185–192

Nieters A, Drosdowski T, Lehr U (2015) Do extreme weather events damage the german economy? GWS Discussion Paper 15 (2)

Nordhaus WD (1992) An optimal transition path for controlling greenhouse gases. Science 258:1315

Nordhaus W, Sztorc P (2013) DICE 2013R: introduction and user's manual, 2nd edn. Yale University Press, New Haven, CT

Patt AG, van Vuuren DP, Berkhout F, Aaheim A, Hof AF, Isaac M, Mechler R (2010) Adaptation in integrated assessment modelling: where do we stand? Clim Change 99(3):383–402

PIK (2014) Klimafolgen online. http://www.klimafolgenonline.com/. Accessed 15.4.2014

Pollitt H, Alexandri E, Chewpreecha U, Klaassen G (2014) Macroeconomic analysis of the employment impacts of future EU climate policies. Climate Policy. doi:10.1080/14693062.2014.953907

Rademaekers K, van der Laan J, Boeve S, Lise W (2011) Investment needs for future adaptation measures in EU nuclear power plants and other electricity generation technologies due to effects of climate change: final report. European Commission, Brussels

Rose A, Liao S (2005) Modeling regional economic resilience to disasters: a computable general equilibrium analysis of water service disruptions. J Reg Sci 45(1):75–112

Tol RS (1997) On the optimal control of carbon dioxide emissions: an application of FUND. Environ Model Assess 2(3):151–163

Tol RSJ, Rose S, Anthoff D, Waldhoff S (2014) The marginal damage costs of different greenhouse gases: an application of FUND. Econ Open Access Open Assess E J 8(2014–31):4–10

UNEP (2004) Impacts of summer 2003 heat wave in Europe. Environmental Alert Bulletin, United Nations Environmental Programme, Nairobi

Van Ierland EC, de Bruin K, Dellink RB, Ruijs A (2007) A qualitative assessment of climate adaptation options and some estimates of adaptation costs. Study performed within the framework of the Netherlands Policy Programme ARK as Routeplanner projects 3, 4 and 5

Weisz H, Koch H, Lasch P, Walkenhorst O, Peters V, Voegele S, Hattermann FF, Huang S, Aich V, Buechner M, Gutsch M, Pichler P-P, Suckow F (2013) Methode einer integrierten und erweiterten Vulnerabilitaetsbewertung: Konzeptuell-methodische Grundlagen und exemplarische Umsetzung für Wasserhaushalt, Stromerzeugung und energetische Nutzung von Holz unter Klimawandel. Bundesministerium für Umwelt, Naturschutz und Reaktorsicherheit, Berlin

ZIA (2014) Daten und Fakten zu Buero- und Verwaltungsimmobilien. Zentraler Immobilien Ausschuss

Chapter 9
A Conceptual Framework for Understanding Vulnerabilities to Extreme Climate Events

Harry Polo Diaz

Abstract Many regions of the world are experiencing impacts of climate change of increasing variability, including drought and flood events. Proactive adaptation to climate change builds resiliency and reduces vulnerability to extreme events, lessening their impact and also their classification as "disasters." Adaptive strategies need to address the changing climate, other exposures (i.e. globalization and neo-liberalism), and sensitivities (i.e. unequal access to economic capital or lack of human capital).

This paper presents a research framework used by an international and interdisciplinary research project for assessing and building resiliency to climate change and extreme events of drought and flood in five countries of the Americas. The paper discusses how past, present, and future vulnerabilities are integrated into the research process, the complexities and nuances of dealing with local vulnerabilities to extreme climate events, and the incorporation of an adaptive governance assessment.

Keywords Climate change • Vulnerability • Extreme events • Americas

Introduction

We are living a time of major social, economic and environmental changes that are already affecting our lives in different ways, changes that emerge from the increasingly complex interrelationships between social and ecological systems. The period has been characterized as, "the Anthropocene", a historical moment where social systems have become increasingly dysfunctional in their relations with nature causing serious disruptions to environmental stability that pose "increasing threats to human security for both present and future generations" (O'Brien 2013: 72; Hackmann and Moser 2013; Wheeler 2012).

Climate change is one of the multiple expressions of this global environmental change. An unprecedented concentration of greenhouse gases in the atmosphere is

H.P. Diaz (✉)
University of Regina, Regina, SK, Canada, S4S 0A2
e-mail: harry.diaz@uregina.ca

W.L. Filho et al. (eds.), *Climate Change Adaptation, Resilience and Hazards*,
Climate Change Management, DOI 10.1007/978-3-319-39880-8_9

linked to an overall warming of the planet, which has been interfering and affecting climate and weather patterns and presenting one of the most prominent sources of increased human vulnerability. The World Meteorological Organization has estimated that from 2001 to 2010, more than 370,000 lives were lost as a result of extreme climate conditions, including heat waves, cold spells, drought, storms and floods, marking a 20 % increase in deaths compared to the previous decade (WMO 2013). The magnitude and frequency of extreme climate events are projected to increase under climate change, potentially increasing people's vulnerabilities and associated risks.

This paper deals with the conceptual approach used by an international, comparative, and interdisciplinary research project that focuses on the present and future vulnerabilities of rural people—farmers and residents of small towns—to the increasing number of extreme climate events. The Vulnerability and Adaptation to Climate Extremes in the Americas (VACEA) project takes place in five American countries: Argentina, Brazil, Canada, Chile, and Colombia. The goal of the project is the understanding of present and future extreme climate events, not just in terms of climatic hazard parameters such as timing, duration, intensity and geographic scope, but relative to human exposure-sensitivity. Disasters are a spatial interaction between hazards and a social system that is sensitive to the event and likely to suffer human and economic loss as a result of this interaction (Wisner et al. 2004; Paul 2011). Thus, VACEA focuses on the nature of hazards that emerge in the context of climate change and their impacts on rural people, who are characterized by different degrees of vulnerability due to unequal social conditions.

A systematic understanding of the relationships between extreme climate events and the socio-economic conditions that contribute to climate vulnerabilities is fundamental in order to grasp the implications of climate change. These relationships, however, are difficult to grasp using traditional scientific approaches. They are a "wicked" problem, which "is a complex issue that defies complete definition, for which there can be no final solution, since any resolution generates further issues, and where solutions are not true or false or good or bad, but the best that can be done at the time" (Brown et al. 2010: 4; see also Rittel and Webber 1973; Batie 2008; Conklin 2006). Part of the wickedness of these is related to our attempts to define and explain them using traditional disciplinary modes of inquiry, which tend to overemphasize some aspects of these problems and ignore others. In this perspective, the VACEA project has developed and strengthened an interdisciplinary approach to understanding these climate-social events, one that combines the efforts of both natural and social scientists.

The central focus of the VACEA project is climate vulnerability. Vulnerability, in very general terms, "is the measure of an entity's inability to deal with a natural disaster" or any form of stress (Paul 2011: 68). There is a multiplicity of definitions of vulnerability (see Paul 2011; Birkmann 2006; Patt et al. 2009), although VACEA has emphasized the approach used by the IPCC, where vulnerability is a function of exposure, sensitivity and adaptive capacity to climate (a more systematic discussion of these terms is found in the next section of this paper).

The paper explains the conceptual and methodological framework that has provided direction to the work of the project. Initially, the paper explains the general perspective of the project to assessing both present and future vulnerabilities, an important aspect given that climate change is a temporal and spatial process without a clear end. The second section outlines the conceptual framework adopted by the project for an assessment of local vulnerabilities. This is followed by a section that provides a brief discussion on adaptive governance, an institutional capital that is important to rural people to reduce their vulnerabilities to climate. Finally the paper offers some insights learned in these assessments for future work in this area.

Dealing with Present and Future Vulnerabilities

Climate change, as a component of global environmental change, is expected to have a myriad of complex impacts upon our lives. Many of these of these impacts may be beneficial—such as an extension of the growing season in agriculture—but most of them will be problematic. As is argued by Feliciano and Berkhout, "contemporary analysis of the impacts of climate change and environmental change is concerned with the factors that underpin risk, vulnerability and human resilience, and how these are perceived, framed, and managed in different social contexts" (2013: 226).

In this context the project assumes climate change mainly as a risk issue. Following Smit and Pilisova, who argue that "the key adaptations are less of often those related to changes in longer-term average temperature and more often related to the frequency and magnitude of extremes such as droughts or floods" (2003: 11), the project emphasizes the point that at local and regional scales the major climate hazards are related more to variability and not to averages. Extreme forms of variability—such as drought and floods—are especially important because they escape the adaptive ranges that characterize local or regional systems. Based on a common experience, local people, communities and institutions learn to adapt to climate within a certain range of climatic conditions—the adaptive range—defined by the historical "ups and downs" of climate variability. Our studies in western Canada have demonstrated that farmers living in areas prone to drought are able to cope with this phenomenon for a period of 2 or 3 years, while those farming in areas that are not historically affected by water scarcities show a lower adaptive range that is usually restricted to a year (Warren and Diaz 2012). This plasticity of human response to the environment, based on experience and learning, is what allows people and systems to deal with different climatic conditions and with a range of historical variability (for an excellent discussion of human adaptation to environmental conditions see Moran 2008). Thus, climate hazards—or extreme climate events—are climate phenomena that escape what is considered to be the normal historical variability and are problematic to people due to their incapacity to deal with them. In this context, the main problem of climate change is an increasing variability that escape the existing coping ranges of people, a new variability that lead to more frequent and more severe climate extremes and to an increase of existing vulnerabilities or the emergence of new ones.

The VACEA project addresses the consequences of global climate change for regional climate variability and extremes and the associated vulnerabilities and adaptive strategies of rural people, who are highly vulnerable because their livelihoods makes them highly exposed and sensitive to climate variability and extremes. The project seeks to analyze the current vulnerabilities in the context of projected shifts in climate variability, including the frequency and intensity of extreme events, an analysis that should produce important insights into rural people's future risks and opportunities, informing the adoption of more appropriate local practices and adjustments to governance policies.

With this perspective in mind, the project adopted a model for assessing vulnerability that highlights the need to understand it within the context of past and present, as well as future climate conditions (see Smit and Wandel 2006: 288), a model presented in brown at the center of Fig. 9.1. It makes use of three sets of interrelated activities to realize this present and future vulnerability assessment. The first set involves the development of an understanding of the past and current degree of vulnerability of rural systems, where the effort involves an identification of how climate-related factors influence individuals, communities or economic sectors, as well as specific ecosystems, and what ability exists to manage changes in these. The second set of activities involves constructing future climate projections for the area where the system occurs, with an emphasis on the frequency and magnitude of extreme climate events. Finally, bringing together the insights produced by the first two set of activities in order to assess future vulnerabilities based on how the current vulnerabilities will be affected by the expected future conditions. The approach, undoubtedly, is characterized by some degree of uncertainty regarding future climate and social conditions, but the first two sets of

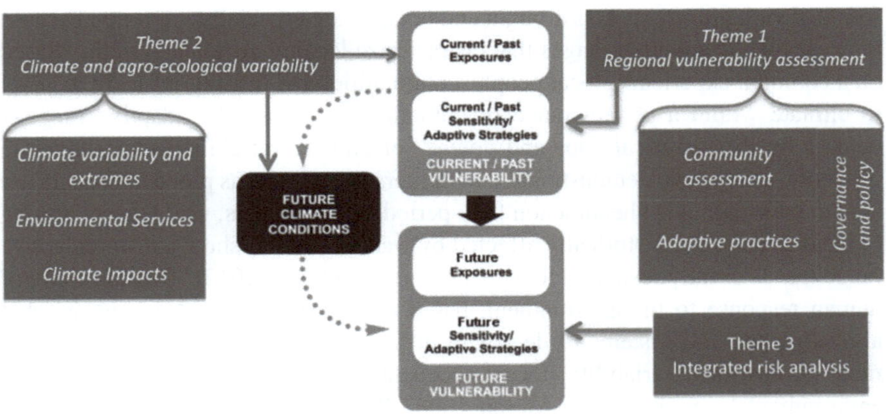

Fig. 9.1 Vulnerability assessment model and research themes

activities provide at least some degree of relatively secure knowledge that should satisfy the need for action in relation to the future climate risks (see Murphy 2014).

Figure 9.1 also illustrates in blue the sets of interrelated research activities used in the project. These activities fall under three major research themes: Regional Vulnerability Assessment (Theme 1), Climate and Agro-Ecological Variability (Theme 2), and Integrated Risk Analysis (Theme 3). The themes appear in Fig. 9.1 in relation to the vulnerability assessment model and the investigation of past, current and future exposure, sensitivity and adaptive capacity. As indicated before, the project integrates the work of both natural and social scientists in order to grasp the complexities of vulnerabilities n the context of the continuous interaction between climate and rural society. Social scientists are fundamentally involved in Theme 1, assessing the vulnerabilities of rural communities, the capacity of regional and national governance systems to reduce rural vulnerabilities, and the robustness of specific adaptive practices. Natural scientists' work is focused on Theme 2, dealing with existing climate variability and their impacts on ecosystems, as well as with future climate scenarios. Theme 3 is expected to integrate the insights from the natural and social disciplines produced in the contexts of the first two themes in order to construct an interdisciplinary understanding of the complexity of future extreme climate events and their impacts (Repko 2012).

Dealing with Local Vulnerabilities

Given the interest of the VACEA project on the consequences of climate change on regional climate variability and extremes and their associated risks for rural people, a central research component of the VACEA project has been a vulnerability assessment of rural social conditions (presented in the previous section as Theme 1, "Regional Vulnerability Assessment"), which is an internationally recognized approach for assessing and understanding the social dimensions of climate hazards (for a discussion of the approach see Smit and Wandel 2006). This assessment facilitates a comprehensive understanding of vulnerability in terms of (a) the magnitude of the threats that extreme climate events present, (b) determine priorities for adaptation, and (c) contribute to policy development.

This research component has been organized around a group of rural community vulnerability assessments, and the role that some other entities—governments and policy and adaptive practices—play in the reduction of rural vulnerability. This section discusses the conceptual framework that informs the community vulnerability assessment, which constitute the core of Theme 1.

As indicated before, vulnerability is defined in the VACEA project as the degree to which a system, such as a rural community or a farm, is susceptible to the adverse effects of stressors and change (Smit and Wandel 2006; Wisner et al. 2004). Following the definition of the IPCC (2001: 995), the project emphasizes the roles of climate variability and climate change as stressors that create risks (and opportunities) for rural people. In more precise terms, we define vulnerability as a function of two dimensions: first, exposure to climate hazards and their impacts; and, second, the social conditions that determine sensitivity—the degree to which a system is affected by its exposure to a climate-related stimuli—and adaptive capacity, the ability of a system to adjust to climate risks and opportunities by increasing its coping range. Figure 9.2 represents these two dimensions of vulnerability. Exposure is a characteristic of a climate system and it refers to the frequency of climate hazards—i.e. droughts, storms, and others—and their attributes—such as intensity, duration, and coverage—that define the magnitude of their impact on social systems. Sensitivity and adaptive capacity, on the other side, are characteristics of the social system and are mainly determined by people's access and control of essential resources (they are also called determinants as we could see in Fig. 9.2) that support their livelihoods. It is the existence of these resources that define, to a large extent, the coping range of individuals or local systems, such as a farm or a community. In this perspective, vulnerability emerges from the interactions between the human and the natural systems.

In very simple terms, a social system that is characterized by limited resources is more vulnerable and, consequently, more conditioned to be impacted by climate hazards. Figure 9.2 lists these resources based on what the IPCC calls "the determinants of adaptive capacity" (IPCC 2001: 893; for a similar list of resources see Department for International Development 2000). Access and control of these resources are important to reduce vulnerabilities, but it is the capabilities of actors to organize them into adaptive activities what defines the balance between

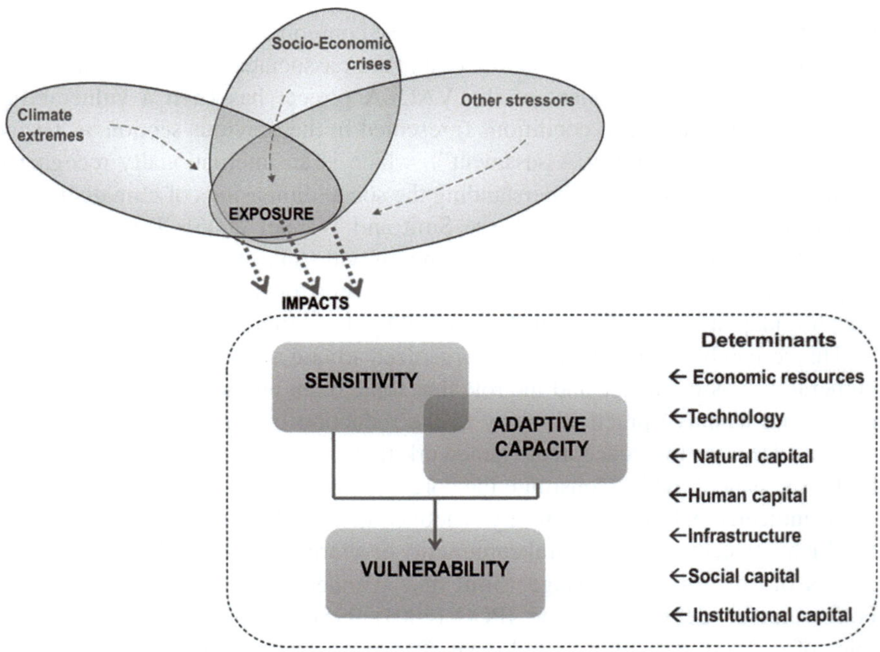

Fig. 9.2 The dimensions of vulnerability. *Source*: Wandel et al. (2016)

sensitivity (determined by lack of or limited resources), and adaptation (defined by the existence of resources that could be mobilized to reduce sensitivity).

These determinants of adaptive capacity, which are also called resources, are:

- **Economic resources**. The existence of monetary capital, financial means, wealth, productive resources, and others forms, which could contribute to the development of an adaptive capacity.
- **Technology**. The availability and access to technology—such as irrigation systems, flood control measures, warning systems, and others—as well the existence of a capacity to develop new technologies that could contribute to a more robust adaptive capacity.
- **Natural capital**. The availability and access to basic environmental services (water, soil, seeds), which are fundamental to the viability of rural livelihoods.
- **Human capital**. The educational and knowledge levels, as well as expertise, we find in a system. It includes traditional knowledges about nature, and especially climate and weather, and their relationships with agricultural practices. Systems with the capacity to produce, disseminate and store information (high educational levels or efficient communication among producers to disseminate successful practices) have a better ability to understand and predict climate hazards, reducing their vulnerability to climate and climate-related events.
- **Infrastructure**. The existing of proper housing conditions, drainage systems, weather-resistant roads, coastal defense, and others forms of allows regions and populations not only to cope with extreme weather events but also to recuperate faster from their impacts.
- **Social capital**. The existence of social networks characterized by trust and reciprocity that integrate individual resources to facilitate collective tasks (Putnam 1995; Coleman 1988).
- **Institutional capital**. Established institutions facilitate the management of climate-related risks—such as the existence and availability of insurance services, water conservation programs, and others—reinforcing (or debilitating) the adaptive capacity of the population.

This last form of capital is found both at the local and the provincial and national levels. The VACEA project, following findings from previous projects that indicate that the adaptive capacity of communities is always nested in larger institutional contexts (see Hurlbert et al. 2009), has included a governance assessment as part of its research activities, as it is indicated in the next section of the paper. No less relevant is gathering information about the role that social capital plays in the reduction of the vulnerability of communities. We have evidence from previous related projects that social capital—a local and informal institution that emerges around local institutions such as kinship, friendship—is an important resource in dealing with the impacts of disasters or having access to a larger number of other resources (Diaz et al. 2003).

In the process of carrying out the assessment we have given especial attention to the fact that vulnerability is not an unalterable condition but rather it is subjected to changes depending on the intensity of the stressor, the quality and quantity of resources that are available to rural people and the capacity of the rural people to manage the resources. As an example, resources could be limited and if they are used unwisely in a situation of vulnerability it could leave a family with the necessary resources to face future risks. Also important to consider in the community vulnerability assessment is the process of differentiation that characterizes the integration of rural people to economic and social processes in the five countries, a process where some actors are able to be better integrated than others. This difference in the process of integration is due both to a historical process of economic marginalization and to institutional failures, which result on an unequal distribution of resources essential to adaptive capacity. A clear example is that some rural actors have a better adaptive capacity than others due to a better economic situation. Landless peasants, small producers, and women are normally more vulnerable, a condition that could become worse with extreme climate events (Wisner et al. 2004).

It is also important to emphasize the point that vulnerabilities are not only related to access and control of the listed resources or determinants, but also to other conditions. The nature of productive systems creates specific conditions of vulnerability for different type of agricultural producers. As an example, the water demands vary between farmers and ranchers, as well as among different type of producers, during the year. No less relevant is the localization of the productive units within the basin. Non-existent or limited access to irrigation is a fundamental issue for agricultural producers in the context of increasing water scarcities (Warren 2016).

A final point of discussion in relation Fig. 9.2 are the sources of stress that affect local vulnerabilities (presented at the top left corner of the figure). We have identified "climate extremes" and "socio economic crises", but there are a variety of other non-climate stressors that could affect local people in rural localities, such as government policies or animal diseases. Thus, vulnerability to extreme climate events could be strongly interlocked to other types of vulnerability. Climate is not the single determinant of the communities' vulnerability. Rather, climate and water stresses are part of a suite of stresses that individuals and communities must manage on their everyday life. Our studies in Canada and Latin America indicate that rural people are exposed to several non-climatic stressors—such as market conditions, political processes, domestic catastrophes, and others—which are normally more relevant to them than climate (Wandel et al. 2010; Montana 2012). Particularly problematic for them is the interlocking of climatic and non-climatic vulnerabilities at a single moment in time, such as the case of a drought at a moment in which market crop prices are low. It is this interlocking of stressors that multiply the negative impacts of risks leading to double exposures (Leichenko and O'Brien 2008).

Adaptive Governance

One important resource or determinant of adaptive capacity is formal institutional capital. In a previous study we learned that local adaptive capacity is always nested in larger institutional frameworks that contribute either to make this capacity more robust or to debilitate it. A variety of practices, processes, systems, and infrastructure, are attempted and taken by rural people to reduce climate-related risk and to create new opportunities. Accumulating assets, relocating human resources, diversifying income sources and crops, redefining land use, adopting new technologies are some of the indicators of the existence of an adaptive capacity. This local adaptive capacity has been shaped to a large degree by a wider decision-making networks at different levels (IACC 2009). In this context we considered relevant to assess the capacity of governance to reduce local vulnerabilities through different programs and policies.

According to Mosser, governance can be "conceived as the set of decisions, actors, processes, institutional structures, and mechanisms, including the division of authority and underlying norms, involved in determining a course of action" (2009: 315). It is a term to be contrasted from the similar, but differentiated, terms of "government" and "management." "Management" refers to the processes of decision-making, coordination and resource deployment that occur within a given institutional setting (Hatfield-Dodds et al. 2007: 3) while "government" centers on the institutions and actions of the state. Governance is wider than both of these terms, encompassing non-state actors such as businesses and civil society, which are brought into the societal steering of natural resources and social actors. Governance involves the range of institutions through which government agencies, citizens and groups articulate their interests and mediate their differences, participating in some of the decision-making processes of governments (Armitage et al. 2009; Kooiman 1993). Thus governance, in relation to extreme events, refers to the range of political, social, economic, and administrative systems that respond to, manage, and anticipate extreme events. A systematic community vulnerability assessment requires not only an evaluation of the local adaptive capacity but also of the capacity of external institutional systems to contribute to a reduction of local vulnerabilities to a variety of stressors, including extreme climate events. Of particular importance in the community vulnerability assessment is the identification of the key organizations interconnecting with community members and, specifically, the community members' relationships with local governments. In these terms, we are referring to what Adger refers as synergistic social capital, where "local management and government intervention work together to reduce risks" (2003: 43).

Governments could have a limited or even negative role in reducing the vulnerability of rural communities, either because of a policy deficit (absence of specific policies, policy perspective, or just a simple urban bias) and/or a style of governance that limit the capacity of government agencies to provide the necessary resources to rural people (different agency priorities, lack of inter-agency integration and coordination, etc.) (Hulbert and Diaz 2013). Building on information obtained in the community vulnerability assessments, the VACEA's governance assessment has then as its purpose to provide information on the inter-linkage of government programs and policies and their contribution to governance and ultimately, community vulnerability.

Rural people mediate stressors and assets through local institutions, such as bonding and bridging social capital (Adger 2003), which are based on cultural practices, deep rooted lifestyles and ideological premises. This mediation may give rise to institutional capital or adaptive mechanisms, which relate in part to first, the assets which a community has at its disposal; and second, the interplay of government (federal, provincial and local governments) and civic institutions and the bridges and barriers to adaptation provided by these entities. Governance, thus, includes the local processes of decision making in relation to climatic events which is exercised by local institutions at the community level. Thus, an important research focus of VACEA's assessment is, accordingly, the local government which is mediating these community decisions through a combination of policy tools and policy processes as set by the federal and provincial governments.

In the context of the VACEA project we have oriented our effort to understand the extent to which multiple forms of governance could be understood as adaptive governance, which spans a range of political, social, economic, and administrative systems and develops, manages, and distributes a resource in a manner that promotes adaptive capacity through collaborative, flexible, and learning-based issue management across different scales. It is important to note that the governance assessment is not only an assessment of government or governance agencies. Rather, it is an exploration and assessment of the entire network of actors, institutions, relationships, organizations, and entities involved in managing and responding to climate variability, hazards, and extreme events.

Based on an increasing literature on governance and its specific dimension as adaptive governance (see among others, Berkes and Folke 1998; Folke et al. 2005; Olsson et al. 2006; Hatfield-Dodds et al. 2007; Burris et al. 2005; Lebel et al. 2006; Scholz 2005; Knieling and Leal 2013; Hill 2013; and Hulbert and Diaz 2013), the VACEA project has focused its governance evaluation on the following characteristics that exemplify adaptive governance:

- Responsiveness—the ability of governance networks, organizations and actors to respond in a timely manner to climate variability, hazards and extreme events, involving issues such as the capacity of the agency to respond to or account for ecosystem dynamics, climate variability, hazards, and extreme climate events and the existence of early warning systems.
- Reflexivity—the social learning aptitude of extreme climate events governance institutions, which include issues such as the capacity of the governance regime to assess or reassess practices for assisting adaptation to climate variability, hazards, and extreme climate events, its openness towards uncertainties, and the existence of constant monitoring and evaluation processes.

- Flexibility—the ability of the water and extreme climate events governance institutions to respond in a variety of manners as appropriate to the situation, context and particular needs of the community, dealing with issues such as the capacity to modify adaptation practices in response to unanticipated events or the adjustments of practices to take into account different needs and requirements.
- Capacity—the informational, human, and social capital in existence necessary to respond appropriately to climate variability, hazards, and extreme events, including the existence of leaders (government or significant social actors or networks in communities) that are capable of responding to climate variability, hazards, and extreme events, the availability and access to necessary and appropriate information, and
- Equity—the fairness of the extreme climate events governance regime in dealing with processes and impacts, including issues such as the existence of opportunities for multiple frames of reference, opinions, and problem definitions as well as the involvement of different actors, levels and sectors in the governance process, the implementation of responses to climate variability, hazards, and extreme events equitable to all community members.

Based on institutional profiles and data collected in the in-depth interviews and focus groups of the CVA these themes have been explored in the governance assessment. An assessment of local governance took place at the same time than the community assessments but the assessments of regional and national governance bodies was done post facto the community assessments.

Some Challenges

Conceptual frameworks are social constructions—a product of scientific deliberations—that help us to organize our approach to the understanding of reality. In these terms, they are not definitive. Rather they are subject to changes and modifications based on their confrontation with the empirical reality. Our conceptual framework, which we applied in the field research carried out in the five countries, was able to provide direction and develop a better understanding of the issue under consideration. Based on this experience and some new developments in the field of adaptive capacity, we think it is necessary to integrate new issues that could improve our understanding of the impacts of extreme climate events and people's capacity to reduce the risks associated to these impacts. Two aspects could be important here.

The first one is a very relevant issue that is necessary to consider in terms of the purpose of adaptation. Pelling (2012) identifies the need to frame the process of adaptation in the context of sustainable development, an argument that is also emphasized in the last IPCC report (Denton et al. 2014). Pelling advances the argument that adaptation should go beyond simple resiliency, which he defines as "a refinement of actions to improve performance without changing guiding assumptions or the questioning of established routines (2012: 53). In other words, he argues against adaptive strategies oriented to maintain what we have been doing in the past, an approach that takes us into "the sustainability of the unsustainable". In these terms he argues the need to redefine adaptation as a process of transformation in the context of sustainability. This is an important aspect to be considered in our analysis of the existing adaptive strategies and of their capacity to secure sustainability.

A second aspect, no less important, is the need to establish a difference between community vulnerability, which involves individuals, households, and local groups, from the vulnerability of larger systems (see Paul 2011: 76–83). As indicated before, community vulnerability is nested in larger social frameworks such as those imposed by governance. However, there is a need to move beyond an assessment of policies and programs. There is also the need to assess the existence and resilience of a variety of services that are fundamental to everyday life and which are normally provided by governance. Availability of potable water, electricity, and health services, among others, are essential to local people. Murphy (2009), provides an interesting example based on the case of the ice storm that affected eastern Canada in 2008, when most of the infrastructure that supported the distribution of electricity was destroyed by the weight of ice, leaving a large number of rural people, farms, and rural business without defense against the cold weather.

These aspects, among others, force us to assume the complexity of climate change. All social groups and societies are vulnerable in different degrees to disaster, and are likely to become even more vulnerable in the coming future. In this way we need to intensify our efforts to develop more comprehensive understanding of the different manifestations of global environmental change and its associated risks in order to reduce its wickedness.

References

Adger WN (2003) Social aspects of adaptive capacity. In: Smith J, Klein R, Huq S (eds) Climate change, adaptive capacity and development. Imperial College Press, London

Armitage DR et al (2009) Adaptive co-management for social-ecological complexity. Front Ecol Environ 7(2):95–102

Batie S (2008) Wicked problems and applied economics. Am J Agric Econ 5:1176–1191

Berkes F, Folke C (eds) (1998) Linking sociological and ecological systems: management practices and social mechanisms for building resilience. Cambridge University Press, New York

Birkmann J (2006) Measuring vulnerability to promote disaster-resilient societies: conceptual frameworks and definitions. In: Birkmann J (ed) Measuring vulnerability to natural hazards. Towards disaster-resilient societies. United Nations University Press, New York

Brown V et al (2010) Towards a just and sustainable future. In: Brown V, Harris J, Russell J (eds) Tackling wicked problems through the transdisciplinary action. Earthscan, London

Burris S, Drahos P, Shearing C (2005) Nodal governance. Aust J Leg Philos 30:30–58

Coleman J (1988) Social capital in the creation of human capital. Am J Soc 94:S95–S120

Conklin J (2006) Wicked problems and social complexity. In: Conklin J (ed) Dialogue mapping. Building shared understanding of wicked problems. Wiley, West Sussex (It is also found in the website of the CogNexus Institute: cognexus.org/wpf/wickedproblems.pdf)

Denton et al (2014) Climate resilient pathways: adaptation, mitigation, and sustainable development. In: IPCC (ed) Climate change 2014: impacts, adaptation, and vulnerability, Part A: Global and sectoral aspects. Cambridge University Press, Cambridge

Department for International Development, UK (2000) Sustainable livelihoods framework. In: Livelihoods connect: sustainable livelihoods guidance sheets from DFID. UK Department for International Development, London (Sec. 2)

Diaz H et al (2003) Community case studies of vulnerability to climate change: Cabri and Stewart Valley, Saskatchewan. In: Marchildon G (ed) A dry oasis. Institutional adaptation to climate on the Canadian plains. CPRC Press, Regina

Feliciano D, Berkhout F (2013) The consequences of global environmental change. Introduction to Part 3. In: ISSC, UNESCO (eds) World social science report 2013. OECD/UNESCO, Paris

Folke C, Hahn T, Olsson P, Norberg J (2005) Adaptive governance of social-ecological systems. Annu Rev Environ Resour 30:411–473

Hackmann H, Moser S (2013) Social sciences in a changing global environment. In: ISSC, UNESCO (eds) World social science report 2013. OECD/UNESCO, Paris

Hatfield-Dodds S, Nelson R, Cook DC (2007) Adaptive governance: an introduction, and implications for public policy. A paper presented at the ANZSEE conference, Noosa Australia, 4, 5 July 2007

Hill M (2013) Climate change and water governance. Adaptive capacity in Chile and Switzerland. Springer, Hamburg

Hulbert M, Diaz H (2013) Water governance in Chile and Canada: a comparison of adaptive characteristics. Ecol Soc 18(4):61

Hurlbert M, Corkal D, Diaz H (2009) Government and civil society: adaptive water management in the South Saskatchewan River Basin. In: Marchildon G (ed) A dry oasis: Institutional adaptation to climate on the Canadian plains. CPRC Press, Regina

IACC (Institutional Adaptation to Climate Change) (2009) Integration report: the case of the South Saskatchewan River Basin. CPRC Press, Regina. Available at www.parc.ca/mcri/pdfs/papers/into1.pdf

IPCC (Intergovernmental Panel on Climate Change) (2001) Climate change 2001: impacts, adaptation vulnerability. Contribution of working group II to the third assessment report of the intergovernmental panel on climate change. UNEP/WMO, Geneva

Knieling, Leal (2013) Climate change governance. Springer, Hamburg

Kooiman J (1993) Modern governance. Sage, London

Lebel L, Anderies JM, Cambell C, Folke S, Hatfield-Dodds S, Hughes TP, Wilson J (2006) Governance and the capacity to manage resilience in regional social-ecological systems. Ecol Soc 11(1):19

Leichenko R, O'Brien K (2008) Environmental change and globalization. Doubles exposures. Oxford University Press, New York

Montaña E (2012) Escenarios de cambio ambiental global, escenarios de pobreza rural. Una mirada desde el territorio. CLACSO-CROP, Buenos Aires, 331 pp

Moran E (2008) Human adaptability. An introduction to ecological anthropology. Westview Press, Boulder

Mosser S (2009) Whether our levers are long enough and the fulcrum strong? Exploring the soft underbelly of adaptation decisions and actions. In: Adger WN, Lorenzoni I, O'Brien K (eds) Adapting to climate change. Thresholds, values, governance. Cambridge University Press, Cambridge

Murphy (2009) Leadership in disaster. Learning for a future with global climate change. McGill-Queen's University Press, Montreal

Murphy R (2014) Uncertainty and claims of uncertainty as impediments to risk management. In: Lockie S, Sonnenfeld D, Fisher D (eds) Routledge international handbook of social and environmental change. Routledge, London
O'Brien K (2013) What is the problem? Putting global environmental change into perspective. In: ISSC (International Social Science Council) (ed) World science report 2013. UNESCO, Paris, pp 71–78
Olsson P, Gunderson LH, Carpenter SR, Ryan P, Lebel L, Folke C, Holling CS (2006) Shooting the rapids: navigating transitions to adaptive governance of socio-ecological systems. Ecol Soc 11(1):1–18
Patt A, Schoroter D, Klein R, de la Vega-Leiner D (2009) Vulnerability research and assessment to support adaptation and mitigation: common themes from the diversity of approaches. In: Patt A, Schroter D, Klein R, de la Vega-Leinert C (eds) Assessing vulnerability to global environmental change. Earthscan, London
Paul BK (2011) Environmental hazards and disasters. Contexts, perspectives and management. Wiley, Oxford
Pelling M (2012) Resilience and transformation. In: Pelling M, Manuel-Navarrete D, Redclift M (eds) Climate change and the crisis of capitalism. Routledge, London
Putnam R (1995) Bowling alone: America's declining social capital. J Democr 6(1):66–78
Repko A (2012) Interdisciplinary research. Process and theory. Sage, London
Rittel H, Webber M (1973) Dilemmas in a general theory of planning. Policy Sci 4:155–169
Scholz JT (ed) (2005) Adaptive governance and water conflict. New institutions for collaborative planning. Resources for the Future, Washington, DC
Smit B, Pilisova O (2003) From adaptation to adaptive capacity and vulnerability reduction. In: Smith J, Klein R, Huq S (eds) Climate change, adaptive capacity and development. Imperial College Press, London
Smit B, Wandel J (2006) Adaptation, adaptive capacity and vulnerability. Glob Environ Change 16:282–292
Wandel J, Pittman J, Prado S (2010) Rural vulnerability to climate change in the South Saskatchewan River Basin. In: Sauchyn D, Diaz H, Kulshreshtha S (eds) The new normal. The Canadian Prairies in a changing climate. CPRC Press, Regina
Wandel J et al (2016) Drought and vulnerability: a conceptual approach. In: Diaz H, Hurlbert M, Warren J (eds) Vulnerability and adaptation to drought: the Canadian Prairies and South America. University of Calgary Press, Calgary
Warren J (2016) The troubled state of irrigation in Southwestern Saskatchewan: the effects of climate variability and government offloading on a vulnerable community. In: Diaz H, Hurlbert M, Warren J (eds) Vulnerability and adaptation to drought: the Canadian Prairies and South America. University of Calgary Press, Calgary
Warren J, Diaz H (2012) Defying Palliser. Stories of resilience from the driest region of the Canadian Prairies. CPRC Press, Regina
Wheeler S (2012) Climate change and social ecology. A new perspective on the climate challenge. Routledge, London
Wisner B, Blaikie P, Cannon T, Davis I (2004) At risk. Routledge, London
World Meteorological Organization (2013) The global climate of 2001-2010—a decade of climate extremes. WMO-No-1103, Geneva

Chapter 10
Adaptation Through Climate Smart Agriculture: Status and Determinants in Coastal Bangladesh

M. Mustafa Saroar and Walter Leal Filho

Abstract The adoption of climate smart agriculture as an adaptation strategy is overwhelmingly stressed in recent times. The Bangladesh Climate Change Strategy and Action Plan 2009 has envisaged adaptation through adoption of climate smart agriculture. Although the literature on agricultural adaptation in general has noticeably increased in recent times, little is known about the adoption of climate smart agriculture as a strategy of adaptation against climate change and extreme events. This study is intended to achieve two objectives. First, to assess the status of the adoption of climate smart agriculture as an adaptation strategy and second to examine the factors that influence the adoption of climate smart agriculture in environmentally stressed areas of coastal Bangladesh. The empirical part of this study was conducted in Dacope Upazila (sub-district) of south-west coastal Bangladesh. Two hundred and thirty five randomly selected households were interviewed through a semi-structured questionnaire during March–June in 2011. From a literature review, 25 indicators which capture three dimensions of climate smart agriculture, such as sustainability of production, resilience to change and the potential for mitigation of emissions were used in the questionnaire. Responses against each of these were rated in a 5-point Likert scale. Information about various socio-demographic, economic, ecological, and adaptive behavioural characteristics of households and their farms were collected as well. By employing the Principle Component Analysis (PCA) technique, weak adopters and strong adopters of the three dimensions of climate smart agriculture are identified. Finally, by employing the Multivariate Probit model, the influence of various factors on the adoption of climate smart agriculture are assessed. The adoption of sustainable production is mostly related to the frequent exposure to rainfall (irrigation facility), sharecropping, and the age of the household head. The adoption of adaptive measures for resilient agriculture are related to frequent exposures to dry spells,

M.M. Saroar, PhD (✉)
Khulna University of Engineering & Technology (KUET), Khulna 9203, Bangladesh
e-mail: saroar.mustafa@yahoo.com

W.L. Filho
Faculty of Life Sciences, Hamburg University of Applied Sciences, Lohbrügger Kirchstraße 65, 21033 Hamburg, Germany

W.L. Filho et al. (eds.), *Climate Change Adaptation, Resilience and Hazards*,
Climate Change Management, DOI 10.1007/978-3-319-39880-8_10

the size of farm landholding and frequent exposures to saline water. On the other hand the adoption of measures for the reduction of emissions are related to land degradation by salt, membership of a social group, and the education of the head of household. Other factors have very limited influence in this regard. Finally this study came up with some policy suggestions, the implementation of which would help the coastal inhabitants to better adopt climate smart agriculture as an adaptation strategy to ensure livelihood and food security against climate change and extreme events.

Keywords Adaptation • Climate smart agriculture • Coastal livelihood • Food security • Bangladesh

Introduction

Evidence based research on the adoption of climate smart agriculture (CSA) could offer new possibilities for food security against the impacts of climate change and extreme events (Thornton and Lipper 2013; Campbell et al. 2014). This evidence-based research is aimed to showcase the adoption of CSA in coastal Bangladesh by answering two research questions. First, what is the status of the adoption of climate smart agriculture (CSA) in coastal Bangladesh? Second, what are the factors that influence the adoption of CSA? Agriculture in the southwest coast of Bangladesh, an area located at the lower edge of Ganges-Brahmaputra-Meghna (GBM) mega delta, is highly susceptible to climate change (Pervez and Henebry 2015; Rahman and Rahman 2015).

Climate change refers to a change in the state of the climate that can be identified by changes in the mean and/or the variability of its properties and persists for an extended period, typically decades or longer (IPCC 2007). Although the term refers to any change in climatic variables over time—whether due to natural variability or as a result of human activity—the United Nations Framework Convention on Climate Change (UNFCCC) conceptualizes climate change as a change in climate due to global warming which is attributed directly or indirectly to human activity (UN 1992). In fact, increasing concentrations of CO_2 and other greenhouse gases from anthropogenic activities have caused the warming of the global climate (Houghton et al. 2001). The impact of climate change is global. Every sector of economy would be affected, yet the agriculture, and subsistence agriculture in the tropics in particular would suffer the most (Challinor and Wheeler 2008; Niles et al. 2015).

Climate change affects agricultural productivity through five main processes: increasing temperatures, rainfall variability, frequency and severity of extreme events, and increasing incidence of pests and diseases (Reilly et al. 2003; World Bank 2008). Crop production is directly influenced by changes in precipitation and temperature. Precipitation is the main source of all fresh water resources and determines the level of soil moisture, which is a critical input for crop growth (Reilly et al. 2003; Calzadilla et al. 2014). But it may also have a negative impact if extreme precipitation causes flooding and waterlogging. Temperature and soil moisture determines the length of the growing season and controls the crop development and water requirements. In general, higher temperatures shorten the crop cycle and reduce crop yields, because higher temperature leads to increased crop water requirements (World Bank 2012a). Due to the increasing temperature, the developmental stage starts earlier and the growth period gets shortened. Warmer climate causes earlier seeding for spring crops, later seeding for autumn crops, and accelerated crop growth (FAO 2010).

Climate variability, especially changes in rainfall patterns, is particularly important for rainfed agriculture. Soil moisture limitations reduce crop productivity and increase production risk in rainfed farming systems (Calzadilla et al. 2014). Although the risk of climate variability is reduced by the use of irrigation, irrigated farming systems depend on surface runoff for ground water availability, which are subject to change under climate change (Pervez and Henebry 2015). Even short-term extremes in precipitation and temperature can be critical for crop growth, especially if they coincide with key stages of development. Increased salinity is another significant climatic variable that affects crop agriculture in the coastal region. In fact, due to increased temperature and evapotranspiration the salinity in soil and water increases which in turn degrades the soil quality by dwindling soil fertility, and increasing erosion (Thornton and Lipper 2013; Wood et al. 2014).

Agriculture is the main source of livelihood for 1.3 billion smallholder farmers worldwide (World Bank 2008) and is highly vulnerable to climate change and extreme events, particularly in the tropics (Salinger et al. 2005). While there is no universally accepted definition of 'smallholder farmers' most of them cultivate small areas of land (often less than 2 ha), use family labor, and depend on their farms as their main source of both food security and income generation (Nagayets 2005; Morton 2007; Singh et al. 2015). It is estimated that smallholder farmers represent 85 % of the world's farms (Nagayets 2005) and provide more than 80 % of the food consumed in the developing world (IFAD 2013; FAO 2013). Across the world, smallholder farmers are considered to be disproportionately vulnerable to climate change because changes in temperature, rainfall and the frequency or intensity of extreme weather events directly affect their crop productivity as well as their food security, income and well-being (Thornton and Gerber 2010; Bennett et al. 2014; Campbell et al. 2014; Singh et al. 2015).

In the tropics, and particularly in monsoon Asia, the majority of the population derives their livelihood from subsistence agriculture which is dominated by smallholder farmers. Unfortunately their production systems are at tremendous risk from climate variability and change as they depend on the complex interactions of monsoon systems and local heat and hydrological feedbacks (Rahman and Rahman 2015; Singh et al. 2015). In fact, a growing number of studies show that the productivity of many crops (e.g., rice, wheat, maize, etc.) owned by smallholder farmers in developing countries are expected to be significantly reduced in the coming decades due to increased climate variability and change (Challinor and Wheeler 2008; Thornton and Gerber 2010; Bennett et al. 2014; Calzadilla, et al. 2014; Campbell et al. 2014; Rahman and Rahman 2015). What happens to smallholder farmers in the future as the climate changes, will therefore have significant social, economic and environmental consequences globally, especially in mega deltas where agriculture is the key source of livelihood (Singh et al. 2015).

Coastal Bangladesh, located at the lower edge of Ganges-Brahmaputra-Meghna (GBM) basin is one of the world's most populous/mega deltas that supports the livelihoods of over 35 million people primarily through subsistence agriculture (Saroar and Routray 2010; Pervez and Henebry 2015; Rahman and Rahman 2015). This region is likely to experience higher temperatures and less rainfall in the dry winter months (November to January) as a result of climate change. At the same time climate change is predicted to lead to an intensification of the hydrological cycle (Huntington 2006; Pervez and Henebry 2015). The primary source of precipitation in the lower edge of the GBM delta is the Indian monsoon, which is projected to be impacted by global warming (Kripalani et al. 2007; Sabade et al. 2011; Pervez and Henebry 2014; Rahman and Rahman 2015). Average monsoon precipitation is projected to increase with a possible extension of the monsoon period (Kripalani et al. 2007), yet freshwater resources in dry season may be impacted adversely (Schuol et al. 2008; Cisneros et al. 2014; Pervez and Henebry 2014). The increased severity of localized droughts in some parts is already being reported in the southwest coast of Bangladesh (GoB 2009; Rahman and Rahman 2015). Therefore, resulting changes in regional water endowments and soil moisture will affect the productivity of cropland, leading to changes in food production and ultimately affecting food security (Pervez and Henebry 2015). Therefore, the issue of livelihood and food security has become a major challenge in coastal Bangladesh like many other parts of the Asia Pacific region (Thornton and Gerber 2010; Bennett et al. 2014).

Although the overwhelming majority of literature focuses on the impacts of climate change on agriculture, agricultural operation itself is a key contributor to the emission of greenhouse gases which in turn causes climate change. Food systems contribute significantly to global warming and are responsible for 19–29 % of global emissions, the bulk of which come directly from agricultural production activities (i.e. N_2O and CH_4) and indirectly from land cover change driven by agriculture (CO_2) (Vermeulen et al. 2012). Therefore about one-third of total emission of greenhouse gases are claimed to be attributed to agricultural operation (Chartzoulakis and Bertaki 2015; Rahman and Rahman 2015). Sixty percent more food production is estimated to be needed in 2050 compared to 2005/2007 (Alexandratos and Bruinsma 2012). Therefore, agricultural emission of greenhouse gases is likely to continue with projected increases in food demand. Therefore to address the overall impacts on climate change on subsistence agriculture there is a clear need to ensure (i) increased agricultural productivity despite climate being a limiting factor; (ii) increased resilience through anticipatory adaptation in agriculture; and increased mitigation efforts to reduce emission of carbon and other greenhouse gases.

To date most efforts to address the impacts of climate change on agriculture are centered on identifying coping and adaptation options to manage the risks of climate change and extreme events. Therefore, significant innovation is observed in the adaptation of agriculture against the impacts of cyclonic surges, heavy rainfall, waterlogging, flooding, and localized droughts. In this regard, various programs are undertaken at multiple levels and scales by numerous actors including the community. However little effort is observed to lower the amount of emission of greenhouse gases from agricultural activities or to sequester carbon. Therefore, there is a clear need to identify [adaptation] options that would make smallholder farmers resilient to climate change, minimize the emission of greenhouse gases and increase food production (Rosenzweig et al. 2013). The CSA is argued to be one of the most cost effective measures in this regards which will contribute to increased production, increased resiliency and decreased emissions (Burney et al. 2010; Wollenberg et al. 2011; Campbell et al. 2014).

While the adoption of CSA can help to achieve the triple benefits of increased productivity, increased resilience through adaptation, and the decreased emission (of greenhouse gases) through mitigation measures, it does not take place automatically (Howden et al. 2007; Niles et al. 2015). To popularize the adoption of CSA there is a clear need to identify the drivers that are more likely to influence the adoption of CSA. In this regard, the research done so far is fragmented and sporadic in nature. The overwhelming majority of research deals with the identification of adaptation options (Howden et al. 2007; Prokopy et al. 2008; Barnes and Toma 2012; Arbuckle et al. 2013; Wood et al. 2014). By contrast, many studies have only identified the amount of greenhouse gases that are released from agricultural operation. Some studies even go further by suggesting mitigation measures to reduce emission from commercial agricultural operation and food systems. The context of the last group of studies are developed economies; recommendations of these studies have little significance for rainfed coastal subsistence agriculture. Yet none has quantitatively assessed the influence of various factors on the adoption of CSA in a coastal context where livelihood largely centers on subsistence agriculture. Identifying the drivers and quantification of influence of those drivers on the adoption of CSA at a local level is the first step toward designing an elaborated program of intervention to popularize the adoption of CSA. This study, as the first of this kind, will fill this gap in both knowledge and practice.

Climate Change and Coastal Agriculture in Bangladesh: A Succinct Review

Bangladesh is located at the interface of two different environments. The Himalayas are in the north and the Bay of Bengal is in the south (Pervez and Henebry 2015). The climate of Bangladesh is characterised by heavy rainfall induced by the Indian monsoon and seasonal droughts in the dry season caused by irregular rainfall and high temperatures (Trenberth 2008; Ahasan et al. 2010). Typical summer temperatures range from 20 to 38 °C and winter temperature range from 10 to 20 °C. Due to global warming, many climatic variables are projected to change significantly by the middle of this century. Annual rainfall in Bangladesh varies from 1400 mm in the west to 4300 mm in the east, with 80 % occurring between May and September. The average annual rainfall in Bangladesh has increased at a rate of 5.52 mm per year during the period 1958–2007 (Shahid 2010; Dastagir 2015). It is predicted that monsoon rainfall will increase about 10–15 % by 2030 (Kripalani et al. 2007; Sabade et al. 2011; Rahman and Rahman 2015). Rainfall variation has increased the risk of both flood and drought in coastal Bangladesh. Heavy rainfall causes coastal flooding when high river discharge accompanied by low drainage capacity, and backwater effects with tidal surges from the Bay of Bengal (Banerjee 2010). By contrast drought at the coast is projected to increase because of a reduction in rainfall during the winter and an increase in temperature (Pervez and Henebry 2014). Although increases in temperature will not be very significant, the sea surface temperature in the Bay of Bengal is predicted to increase from 0.35 to 0.72 °C by 2050 (Chowdhury et al. 2012) which is much faster than the change in the global ocean temperature (Dailidiene et al. 2011). This higher sea surface temperature is attributed to Global warming (Dasgupta et al. 2011). With the increase of sea surface temperature the frequency and intensity of cyclonic storms and surges are projected to increase (Ahmed et al. 2013; Dastagir 2015). Coastal Bangladesh is historically prone to violent storms and tropical cyclones. Between 1877 and 1995, Bangladesh was hit by 154 cyclones, including 43 severe cyclones (Dasgupta et al. 2011). On average, a severe cyclone hits the country every three years (GoB 2009). Recently, two super cyclones (category 4) Sidr and Aila stuck the entire south-west coast in 2007 and 2009 respectively. One-third of the south-west coast was under salty water for about two years. Due to the accumulation of a thick layer of salt, the soil became unsuitable for crop agriculture and pasture production. Even after five years of rainfall only a partial recovery of soil productively has been reported. Around 14.6 million people in coastal Bangladesh are vulnerable to inundation due to cyclonic surges, and this number will increase to 18.5 million by 2050 (World Bank 2012b; Dastagir 2015). About 5690 km^2 of coastal Bangladesh has been identified as a high-risk zone where the depth of inundation might exceed 1 m for an extended period of time because of congestion in the natural drainage systems (Karim and Mimura 2008).

Being one of the largest deltas in the world, coastal region of Bangladesh lies just less than 2 m above sea-level. Sea-level appears to be rising by 15.9–17.2 mm each year in coastal Bangladesh (Schiermeier 2014), while global sea-level rises 2–3 mm each year (Pethick and Orford 2013). Sea-level rise will cause millions of people to be permanently displaced (Dasgupta et al. 2010). The sea-level could rise more than 1 m by 2100 (Rahmstorf 2007), and such a rise would affect a 710 km long coastline, with the global heritage Sundarbans mangrove forest totally lost (Agrawala et al. 2003). Bangladesh is a global hot spot for climate change. According to the global climate risk index, Bangladesh was ranked first in 2012 among countries vulnerable to climate change while it is ranked sixth in 2015 (Harmeling and Eckstein 2012; German Watch 2013; Kreft et al. 2014).

Anticipated climate change and extreme events such as coastal flooding, droughts, cyclonic surges, and sea level rises are likely to play a critical role in increasing the salinity in soil and water (Huq et al. 1998). Saltwater from the Bay of Bengal has already extended over 100 km in coastal Bangladesh (Ahmed and Diana 2015). The salinity has recently increased in coastal rivers to 4 ppt in the monsoon and 13 ppt in the dry season (Khan et al. 2011). About 1.05 million ha of land in coastal Bangladesh has been affected by varying degrees of soil and water salinity (Sikder 2013), which is predicted to increase to two million ha by 2050 (Conway and Waage 2010). Therefore, the intrusion of saltwater is now considered to be the most important single factor that may severely affect coastal agriculture and the livelihood and security of smallholder farmers as agriculture is and continues to be the most dominant occupation of smallholder farmers in the coastal region (GoB 2009). The presently practiced crop agriculture may not be able to withstand increased temperatures, drought and salinity unless coastal agriculture is successfully aligned with the adoption of CSA (Dastagir 2015; Iizumi and Ramankutty 2015). However, one needs to be mindful that adaptation through CSA might not take place automatically; certain factors govern people's adoption of CSA as adaptation strategies against climatic change and extreme events (Adger et al. 2003; Adger 2006; Rahman and Rahman 2015). This study examines the influence of various such factors on farming household's adoption of CSA as an adaptation strategy in coastal Bangladesh against the impacts of climate change and its associated extreme events.

Materials and Methods

Designing the Research Instruments

The research instruments include questionnaires for household surveys, and checklists for focus group discussion and guide questions for key informant interview (KII). The household survey questionnaire includes two important elements that relate to the basic research question. The first element is to capture the concept of CSA that suits the local agro-ecology. The second element is to identify a host of factors that affect the adoption of CSA at a local level. A good review of CSA is done in order to identify a possible range of attributes of CSA and an array of factors that determine the adoption of CSA at multiple scales. With regards to the first element, the following is extracted.

- Many studies conceptualize CSA as a more natural productive system, which provides improved ecological and social functions (Thornton and Lipper 2013). For example it can also include measures that help governments to achieve national food security and poverty reduction goals, while meeting adaptation needs and building resilient agro-ecological systems that actively sequester carbon (FAO 2010; Kumar and Nair 2012; Agrawal et al. 2014).
- At a micro level, the adoption of climate smart agriculture involves some processes, actions, and techniques in farming practices in order to sustainably

increase production, mitigate emission of greenhouse gases, particularly CO_2, and better cope with, manage or adjust to some changing condition, stress, hazard, risk or opportunity (Brooks et al. 2005; Smit and Wandel 2006).

- The core components of CSA includes practices and technologies that sustainably increase productivity, support farmers' adaptation to climate change, and reduce levels of greenhouse gases (Campbell et al. 2014).
- Sustainable production involves mostly low input agriculture practices and organic farming. This may include: use of bio-fertilizer, use of bio-pest management, avoiding onsite burning of crop residuals, inter cropping, planting legume, integrated aquaculture and agriculture, use of surface water for irrigation, and rainwater harvesting (FAO 2010; Koohafkan et al. 2012; Vermeulen 2014; Chartzoulakis and Bertaki 2015).
- Emission reduction involves mitigation measures in order to reduce the emission of CO_2 and other greenhouse gases and to increase carbon sequestration in soil, plants, and trees. This may include: planting trees to control land degradation, maintaining conservation buffer strips, alley cropping, conservation tillage, planting perennial plants on plot boundaries, etc. (FAO 2010; Kumar and Nair 2012; Agrawal et al. 2014; Mbow et al. 2014; Chartzoulakis and Bertaki 2015).
- Adaptation involves a wide range of activities to adjust the farming system. Adaptation for resilience may include: terracing to avoid nutrient loss, keeping fallow for nutrient recycling, cultivating stress resistant cultivars, and mulching to retain soil moisture (Lasco et al. 2011; Chartzoulakis and Bertaki 2015).
- Some other aspects that span across three dimensions of CSA have also emerged as indicators of CSA at the local level. These include- ecosystem based new techniques of soil, water and plant nutrient management, improved on-farm water storage and irrigation, use of crop varieties that are more tolerant of heat, droughts, floods and salinity, the diversification of farm enterprises (including mixed crop and tree systems), and the building of the capacity of institutions to enhance collective action, and the use of climate information services (Vermeulen 2014).

Based on the above literature, seven national and local level senior researchers who are experts in integrated coastal zone management, agro-ecological zoning, climate change adaptation, coastal agriculture and food security, were consulted to frame a questionnaire that would capture CSA in the context of coastal morphology, local hydrology, local vulnerability to climate change, and other socio-ecological systems. Table 10.1 presents the indicators that were developed after a literature review and the expert's consultation and are used to frame the questions to assess the levels of adoption of CSA by the smallholder farmers in the study area. Respondents were asked to rate each of the 25 indicators of CSA in a 5-point Likert scale (Very rarely adopt = 1 to Very often adopt = 5). While it was aimed to identify three separate sets of indicators for measuring each of the three dimensions of CSA (e.g. sustainable production, emission reduction and adaptation), some of the indicators are able to measure more than one dimension of CSA. The dimension that best suits a specific set of indicators is determined using Principal Component Analysis (PCA) which is discussed later in the appropriate section.

Table 10.1 Rotated factor loading matrix of factor-variable which constitutes the concept-Climate Smart Agriculture (CSA)[a]

Variable/statement of specific CSA practice	Major dimensions of CSA		
	1	2	3
Do you practice agroforestry?	0.356	***0.752***	
Do you keep land fallow to use as perineal pasture?	0.561	***0.605***	
Do you use bio-fertilizer?	***0.776***		
Do you use bio-pest management?	***0.837***		
Do you practice integrated aquaculture and agriculture?	***0.846***		
Do you practice conservation tillage?	***0.699***		
Do you allow growth of fodder in your land?	−0.352		
Do you use surface water for irrigation?			
Do you regularly drain out submersed land to cultivate?			
Do you use crop bi-products to feed cattle/poultry?			
Do you cultivate saline tolerant crops?			***0.817***
Do you harvest rainwater for use in micro-agriculture during winter?			***0.850***
Do you practice inter cropping?	***−0.618***	0.306	
Do you plant legume to get green manure?	***−0.689***		
Do you keep crop residuals as in the field as folder of animals?	***−0.646***		
Do you plant trees on plot boundary to control land degradation?		***−0.798***	
Do you plant perennial plant on plot boundary?		***−0.775***	
Do you maintain conservation buffer strips?		***−0.805***	
Do you avoid burning of agricultural residues on site?		***−0.761***	
Do you practice cover crops to control salinization?	***−0.563***		0.387
Do you practice aeroponics agriculture to avoid salinity problem?			***0.823***
Do you cover the seedbed by polythene sheets to protect from dense fog?			***0.797***
Do you practice mulching to retain soil moisture during winter cropping?			
Do you keep land fallow at a regular intervals for nutrient recycling?	***−0.462***	0.447	
Do you cultivate untraditional crops that demand less water?			***0.797***
Variance (%)	21.60	16.90	16.07
Cumulative variance (%)	19.60	38.50	54.57

[a]Extraction method: Principal Component Analysis; rotation method: Varimax with Kaiser Normalization (Rotation converged in 14 iterations). Italic values in each column constitute a dimension of CSA

With regards to the second element, the following is extracted.

- Micro-macro paradox is observed with regards to the factors that influence adoption of CSA at micro (local) and macro level (national). For instance, adoption of CSA at micro-level can be influenced by a local availability of farm-level techniques, whereas at the macro-level adoption of CSA may be influenced by even international policy and finance mechanisms (Berry et al. 2006). CSA practices are often required to be aligned with and supported by national policy and a legal and regulatory framework.

- However, most often the adoption of appropriate suite of CSA at the micro-level depends on the agro-ecological conditions. Many other factors as well may influence the adoption of CSA. These factors include- various bio-physical factors, social, economic, demographic, and psychological attributes of farming households and institutional settings where the farming system operates (Tanaka et al. 2006; Schlenker et al. 2007; Adger et al. 2009; Haden et al. 2012). Some of the specific factors are a limited capacity to adapt to climate change, low education levels, low income, limited land areas, poor access to technical assistance, market and credits, and often chronic dependence on external support (Howden et al. 2007; Morton 2007).

  Therefore, forgoing factors and attributes of farming households were included in the questionnaire (see Table 10.1) which later are used as explanatory variables in the Probit model to assess the influence of various factors on the adoption of CSA in coastal Bangladesh (Table 10.3).

Selection of the Study Site

Throughout the whole of coastal Bangladesh, agriculture is at risk from climate change and extreme events, yet it is the most affected in the south-west coastal Khulna region. The Dacope Upazila (sub-district) was selected as the study area. It is flanked in the south by the swampy mangrove forest '*Sundarbans*' and the bay of Bengal and is exposed to a multiplicity of climatic disasters including cyclones, surges, salinity intrusion, coastal flooding, and droughts. Primary data and information was collected from two villages of Sutarkhali Union Parishad (the lowest level of local government unit in Bangladesh) of Dacope (Fig. 10.1).

The typical agro-ecological characteristics of this area are as follows. The majority of the population depends on rainfed crop agriculture, mostly wet-season rice. The area under dry season rice has been expanding recently, though at a slower pace. Such an expansion of agricultural land has been taking place at the expense of increased emissions of greenhouse gases such as methane and carbon dioxide as the clearing of swampy forestland results in the release of those gases. Appropriate agricultural practices through the adoption of CSA could address the issue of emission reduction. Moreover as the expanded agricultural fields have always bordered the edge of Mangrove swamp Sundarbans, even normal tides inundate most lowlands in the entire study area. The chronic exposure to salinized tidal water has been decreasing the productivity of soil and agricultural production as well. The productivity of rice and other cereal crops such as wheat (maize has only recently been introduced) has been low due to other climate change induced constrains such as waterlogging, and seasonal (dry-season) droughts. Moreover, periodic but sustained exposure to cyclonic surges often brings huge quantities of salt water from the bay of Bengal and leaves thick layers of salt on the

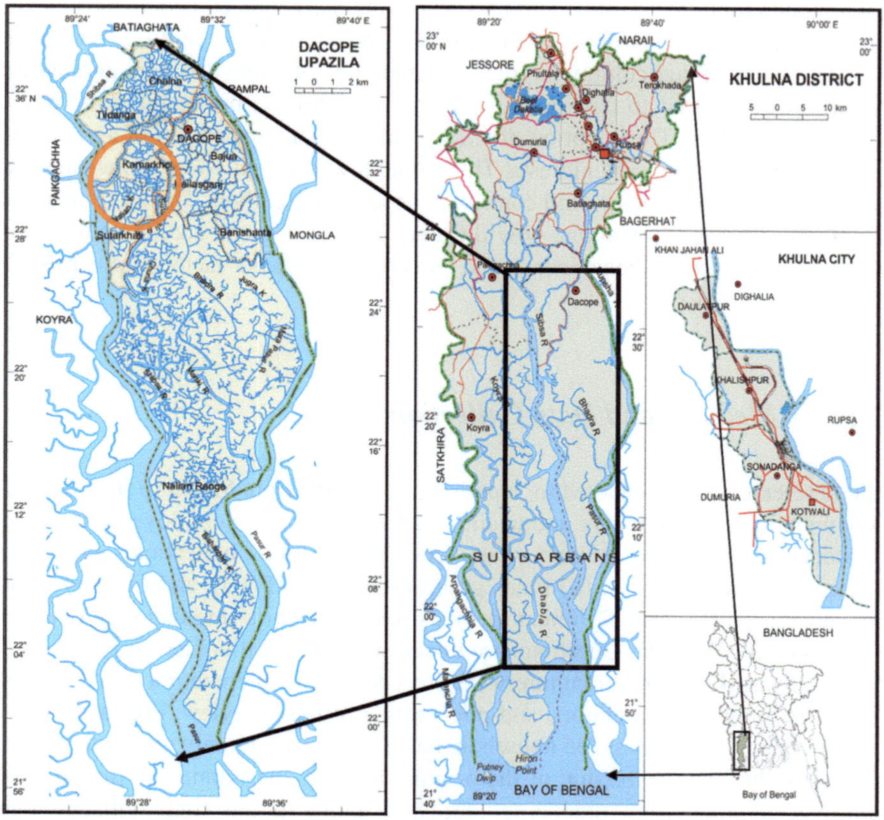

Fig. 10.1 Study area: Dacope Upazila

submerged agricultural land which seriously undermines the productivity of soil for crop cultivation. In the last 50 years the area has been exposed to more than 10 category-4 cyclones. Most recent of such extreme events were the category 4 cyclones-*Aila* and *Sidr*. The cyclone *Sidr* hit in the month of May 2007 and *Aila* struck in November 2009. Cyclone Nargis and Mahasen also hit the area. The death tolls were not so high for a number of reasons including good preparedness, yet the impacts on agriculture and livelihood were unprecedented.

After the super cyclones Sidr and Aila, the lush green agricultural landscape turned into a death valley in the entire area. Thousands of hectares of cultivable lands, often with standing crops, were underwater for many months as the trapped water could not pass away due to congestion in the natural drainage systems. The agricultural fields, horticulture, fish ponds and shrimp *ghers* (enclosures) were abandoned for the next 3 years as the salt layers that the surge water left overland were too thick to washout completely from normal rainfall. Hundreds of NGOs and grassroots organizations have been trying to enhance the smallholder farmers' resilience against climatic events through various adaptation measures. Agriculture and related jobs are the only activities where the marginalized people have developed their skills. Therefore, most efforts of adaptation are centred toward the revitalization of agriculture through the adoption of climate smart agriculture. While it is reported to have CSA by many smallholders but again many others yet to adopt CSA. The adoption of CSA would stop the expansion of agricultural field to swampy Sundarban forests. This would lower the emission of methane, and carbon dioxide and thus would contribute to sequester carbon. Therefore this area could be a model to examine how different agro-ecological, socio-cultural, economic, institutional, behavioural and governance related factors influence the adoption of CSA.

Survey Methods and Data Collection

Field research was conducted for a period of 3 months from March to June 2011. A combination of quantitative and qualitative techniques were employed in a participatory manner for primary data and information collection, including: (i) questionnaire surveys, (ii) focus group discussions (FGD), and (iii) key informant/expert interviews. Different methods were used for triangulation in data and information collection. Questionnaire interviews with smallholder farming households engaged in crop agriculture were preceded by the preparation of a questionnaire, training of enumerators, and pilot testing of the interview. For face-to-face questionnaire interviews, 235 household heads selected randomly were interviewed by using the Bengali version of the original English questionnaire. Around 1500 farming families were practicing crop agriculture in the study villages. The assumed proportion of smallholder farmers used for determining sample size is 0.75 with a confidence level of 95 % and an error limit of 6 %, i.e., a sample size 235 from 1500 population is acceptable. However, 225 questionnaires were considered complete responses and were used for this research. Several visits were made to surveyed farms to identify the impacts of climate change on crop agriculture. Agricultural farmers were interviewed at either farm sites or on their homestead, which lasted about an hour on average. The semi-structured questionnaire has captured information on surveyed family's socio-demographic attributes, various assets, agricultural practices, exposure to climatic variables and extreme events and adaptation practices to enhance resilience, mitigate emissions and increase production to secure their livelihood. Three FGD sessions were conducted with crop farming households to obtain information regarding their attitudes towards the adoption of practices that follow the science of CSA. A total of 28 farmers participated in three FGD sessions, where each group consisted of 8–12 people. FGD sessions lasted about 2 hours, and were held in farmers' houses and in their backyards. A total of seven key informants—who are believed to hold specialist knowledge about various methods of the adaptation of agriculture against climate change, and are expected to be able to answer questions about opportunities and challenges for CSA—were interviewed in their offices and work place. The data was analyzed using SPSS version 21 to get both descriptive and inferential statics which are discussed in the results section.

Results and Discussion

Adaptation Through Climate Smart Agriculture

In order to reduce a large number of variables (that relate to CSA practice) into few meaningful and manageable categories, all 25 statements/questions are entered in a factor analysis. Principal component analysis (PCA) method is employed to bring these 25 statements/questions under few major dimensions of CSA. Following the Kaiser criterion, i.e. only factors/components having Eigenvalue >1, initially six factors are extracted using a varimax rotation. Further processing is done to group those statements into the three distinct dimensions of CSA, i.e. Sustainable Production, Emission Reduction, and Resilience to change (Adaptation). The last processing left five statements and retained 20 statements under the three dimensions of CSA which is shown in Table 10.2 (load factor <0.300 is not shown).

Table 10.2 List of independent variables used in Probit model

Variables in details[a]	Variable (abbreviated)	Coding
Age of the respondent	Age (yr)	
Distance travel to sell products	Travel distance (km)	
Level of education of HHH	Education (yr schooling)	
Household size	Household size (No.)	
Income of family	Income of family (Taka)	
Farm landholding size in acre	Landholding (acre)	
No. of years living in the locality	Duration of living (yr)	
If access to credit is easy	Access to credit (d)	if yes: 1; otherwise: 0
If access to input market is easy	Access to market (d)	if yes: 1; otherwise: 0
If household head is a male	Sex of HHH (d)	if yes: 1; otherwise: 0
If land is suitable for more than one crops	Land suitability (d)	if yes: 1; otherwise: 0
If most land parcels are salt affected	Salt land (d)	if yes: 1; otherwise: 0
If there is other source of income	Other income (d)	if yes: 1; otherwise: 0
If member of social group	Social membership (d)	if yes: 1; otherwise: 0
If the family is a severe victim of recent cyclone Sidr	Victim of Sidr (d)	if yes: 1; otherwise: 0
If cultivate as share cropper as well	Sharecropper (d)	if yes: 1; otherwise: 0
If frequently exposed to rainfall	Exposure to rainfall (d)	if yes: 1; otherwise: 0
If frequently exposed to drier condition	Exposure to dry spell (d)	if yes: 1; otherwise: 0
If frequently exposed to water logging	Exposure to water (d)	if yes: 1; otherwise: 0
If frequently exposed to salinity	Exposure to salinity (d)	if yes: 1; otherwise: 0

[a](d) indicates that the variable is dummy/binary coded

The result shows that the value of determinant of correlation matrix is greater than 0, the Kaiser-Meyer-Olkin value for sampling adequacy is 0.87, the Bartlett's test of sphericity is significant at $p < 0.0001$ and the average communality is >0.500. Therefore, it can be claimed that reduction of 20 statements under three major dimensions (of CSA) is statistically valid (Field 2005; Hair et al. 2006). The remaining five statements do not signify CSA practice.

The first dimension of CSA can be termed as "sustainable production", which constitutes nine variables/statements that explain 21.6 % of the variances. The second dimension can be named as "emission reduction" which consists of six variables/statements that explain 16.9 % of the variances. The third dimension, which relates to "resilience to change (adaptation)", includes five variables/statements and explains 16.1 % of the variance of the concept-CSA.

Strong and Weak Adopter of Climate Smart Agriculture

First, by employing simple arithmetic methods, the mean score of each of the three dimensions of CSA, e.g. "sustainable production", "emission reduction" and "resilience to change (adaptation)" are computed. Second, based on the sample mean of each CSA dimension, each farming household is classified as either a weak adopter (if it scores less than sample mean) or a strong adopter (if it scores more than or equal to sample mean). In this way the strong and weak adopter of CSA as regards Sustainable Production, Emission Reduction, and Resilience to Change (Adaptation) are identified.

Greater variability is observed in adaptation through the adoption of climate smart agriculture (CSA). In regards to sustainable production, 58.2 % households are strong adopters of CSA while the remaining 42.8 % are weak adopters. Similarly, with regards to the resilience to climate change (adaptation), 61.3 % households are strong adopters while 38.7 % are weak adopters. However, in the case of emission reduction, only 45.8 % households are strong adopters and 54.2 % households are weak adopters of CSA. Therefore, it is found that the adoption of CSA is highest in the resilience to climate change (adaptation) dimension and lowest in the emission reduction dimension. A higher adoption of adaptation practices in agriculture is probably attributed to decade long efforts of NGOs and GOs to popularize community based adaptation.

Factors Influencing Adoption of CSA: Application of Probit Model

It was hypothesized that various agro-ecological, hydrogeological, socio-economic, demographic, behavioral, exposure to hazards, and access to climate information are the determinants of the adoption of climate smart agriculture. From a literature review a total of 20 such factors were identified and data and information about these variables was collected from the household surveyed (Table 10.2). These variables are used as independent variables. As dependent variables, each of the three dimensions of CSA such as "sustainable production", "emission reduction" and "resilience to change (adaptation)" are used in Probit model. Here the Probit model is built because there is inherent ordering in the dependent variables, for instance, weak adopters of CSA versus strong adopters of CSA. Three Probit models, one for each of "sustainable production", "emission reduction" and "resilience to change (adaptation)" are developed. In the next couple of sections the results of the three Probit models are discussed in sufficient detail. For employing Probit model the generalized linear (GLM) procedure is followed.

Before running the Probit model, bivariate correlation is performed and due to strong colinearity ($r > 0.80$) among the "size of farmland holding" and "yearly total income", only the size of a farm landholding is used in the Probit model. Similarly, cross-tabulation for each categorical/binary independent variable with each of the three dependent variables is performed. The independent variable which has a zero count/frequency or a count/frequency less than five in any cell of cross-tabulation matrix was excluded from the model to ensure robustness of the model's output. Finally, after running the model, factors/variables that significantly explain the variation in respondent's adoption of CSA are identified and their influences are assessed. These outputs are analysed to identify policy suggestions for the increased adoption of CSA in coastal settings.

To determine the factors that affect the adoption of CSA in the dimension of "sustainable production", "emission reduction" and "resilience to change (adaptation)" three separate Probit models were developed. The first Probit model predicts the influence of factors on farming households' probability to be strong adopters of "sustainable production". The second and third Probit models predict the influence of factors on farming households' probability to be strong adopters of "emission reduction" and "resilience to change (adaptation)" respectively.

Model outputs are presented in Table 10.3. For the first model, the omnibus test shows that the model fits well and the results are statistically significant (LR chi-square = 50.58, df = 19, $p < 001$). Among the 19 hypothesized variables, age of respondents ($p = 0.06$), land affected by salt ($p = 0.09$), share cropping ($p = 0.09$), frequent exposure to rainfall ($p = 0.008$), frequent exposure to dry spell ($p = 0.09$), and frequent exposure to salinity ($p = 0.03$) have appeared as significant predictors of the adoption of CSA in the dimension of "sustainable production". Among these 6 factors, respondents who are frequently exposed to rainfall are 79 % more likely to adopt CSA in the dimension of "sustainable production". Similarly, respondents who are sharecroppers have 72 % higher probability to adopt this. In the same way, aged heads of family have a 61 % higher probability to adopt this. However, respondents who are frequently exposed to drier conditions are 44 % less likely to adopt CSA in the dimension of "sustainable production". Similarly respondents whose lands are affected by varying degrees of salinity and are exposed to saline

Table 10.3 Probit model to predict the influences of various factors on adoption of CSA

Determinants	Model 1 Sustainable production B: Coefficient (Probability)	Model 2 Emission reduction B: Coefficient (Probability)	Model 3 Resilience to change B: Coefficient (Probability)
Age of the respondent	0.037 (0.61)**	−0.003 (0.63)	0.052 (0.52)**
Distance travel to sell products	0.009 (0.60)	0.013 (0.63)	−0.011 (0.50)
Level of education of HHH	0.003 (0.60)	−0.111 (0.59)***	0.107 (0.54)**
Household size	−0.107 (0.55)	0.062 (0.65)	−0.191 (0.43)
No. of years living in the locality	−0.007 (0.59)	−0.007 (0.63)	−0.022 (0.49)
If access to credit is easy	0.226 (0.68)	−0.164 (0.56)	−0.084 (0.47)
If access to input market is easy	−0.126 (0.55)	0.134 (0.68)	0.148 (0.56)
Landholding size in acre	0.054 (0.62)	−0.315 (0.50)**	0.811 (0.79)*
If household head is male	−0.267 (0.49)	0.287 (0.73)	−0.735 (0.23)***
If land is suitable for more than one crops	0.300 (0.71)	−0.521 (0.42)*	0.004 (0.50)
If most land parcels are salt affected	−0.421 (0.43)*	0.908 (0.89)***	−0.127 (0.45)
If there is other source of income	−0.206 (0.52)	0.216 (0.71)	−1.148 (0.13)***
If member of social group	−0.013 (0.59)	0.564 (0.81)*	−0.032 (0.49)
If severely affected by recent cyclone Sidr	−0.287 (0.48)	−0.116 (0.58)	−0.906 (0.18)***
If cultivate as share cropper as well	0.349 (0.72)*	0.345 (0.75)	−0.147 (0.44)
If frequently exposed to rainfall	0.566 (0.79)***	0.251 (0.72)	−0.638 (0.26)**
If frequently exposed to drier condition	−0.395 (0.44)*	−0.480 (0.44)*	0.883 (0.81)***
If frequently exposed to water logging	−0.067 (0.57)	−0.062 (0.60)	0.534 (0.71)**
If frequently exposed to salinity	−488 (0.40)**	−0.337 (0.50)	0.664 (0.75)**
Likelihood Ratio Chi-square	50.58***	112.84***	141.26***
N	225	225	225

1. "Frequent exposure to cyclone" is excluded from the models as the responses are constant (100 % same)
2. "Household yearly income" is also excluded because of its co-linearity with "Size of farm landholding" ($r > 0.80$, $p < 0.01$)
*Significant at 0.10; **Significant at 0.05; ***Significant at 0.01

water are 43 and 40 % less likely to adopt CSA in the dimension of "sustainable production".

For the second model, the omnibus test shows that the model fits well and results are statistically significant (LR chi-square = 112.844, df = 19, p < 001). Among the 19 hypothesized variables, education level of household's head (p = 0.005), size of landholding ($p = 0.04$), suitability of land for multiple crops ($p = 0.058$), salt affected land ($p = 0.003$), membership of social group ($p = 0.095$), and exposure to drier condition ($p = 0.063$) have appeared as significant predictors of adoption of CSA in the dimension of "emission reduction". Among these six factors, salt affected landholders are 89 % more likely to adopt CSA in the dimension of "emission reduction". This is probably attributed with the fact that salt affected land is often used for family forest as this land could not support crop agriculture. Respondents who are members of social groups have a 81 % higher probability to adopt this. Probably for the same reason, NGO's beneficiary households are more aware of the benefits of plantation in salt affected fallow land. However, respondents with higher education are 59 % less likely to adopt CSA in the dimension of "emission reduction". Respondents with large landholdings are 50 % less likely to adopt this. Similarly, respondents who experience frequent drier conditions and have land suitable for multiple cropping are 44 % and 42 % less likely to adopt CSA in the dimension of "emission reduction".

For the third model, the omnibus test shows that the model fits well and the results are statistically significant (LR chi-square = 141.259, df = 19, $p < 001$). Among the 19 hypothesized variables, education level of respondents ($p = 0.019$), size of handholding ($p = 0.079$), sex ($p = 0.008$), multiple sources of income ($p = 0.000$), severe exposure to recent cyclone sidr ($p = 0.001$), frequent exposure to rainfall ($p = 0.026$), frequent exposure to dry spell ($p = 0.004$), frequent exposure to water logging ($p = 0.035$), and frequent exposure to saline water ($p = 0.018$) have appeared as significant predictors of adoption of CSA in the dimension of "resilience to change". Among the nine factors, respondents who are frequently exposed to drier condition are 81 % more likely to adopt CSA in the dimension of "resilience to change". Similarly, large landholders are 79 % more likely to adopt CSA in resilient dimension. Respondents who are frequently exposed to saline water have a 75 % higher probability to adopt this. In the same way respondents who are frequently exposed to water logging and who are educated have a 71 % and 54 % higher likelihood to adopt this. However, respondents who are frequently exposed to rainfall are 26 % less likely to adopt CSA in the dimension of "resilience to change". Similarly, male headed households and severe victims of recent cyclones are 23 and 18 % less likely to adopt CSA in the dimension of "resilience to change".

Conclusion and Limitations of the Research

This study has tried first to identify a range of indicators to assess the adoption of climate smart agriculture in the coastal context of Bangladesh. Through a rigorous review of literature and with a consultation with relevant experts, it came up with a conceptual model of climate smart agriculture that suits the coastal context of monsoon Asia including Bangladesh. The CSA model is empirically tested in a coastal setting of south-west Bangladesh which is historically prone to various hydro-meteorological disasters and is now at risk from climate change and extreme events. People in the study area are more comfortable to adopt CSA in the realm of resilience through adaptation. Conversely, they are less likely to adopt measures that specifically target emission reductions for sequestration of carbon. It is unclear whether this is due to government policy. Because while the National Adaptation Program of Actions (NAPA) and the Bangladesh Climate Change Strategy and Action Plan put significant emphasis on adaptation, they put less importance on the issue of greenhouse gas emission from the agricultural sector (GOB 2005; 2009).

Adaptation through sustainable agriculture is more common among families who are frequently exposed to rainfall and possess solid experience in coastal agricultural. Although the adoption of measures to lower emission of greenhouse gases (i.e. methane and carbon dioxide) is rather less common, the practice is increasing among families who are member of NGOs and are educated as well. This probably happens because of the popularization of social/agro-forestry in coastal villages by NGOs. Although coastal smallholder farming communities could benefit from the adoption of climate smart agriculture, the adoption of CSA is not risk free. There is a clear need for a comprehensive program to popularise CSA among coastal farming communities which would eventually contribute to their livelihood and food security. In this respect, a fresh program of capacity building for the adoption of climate smart agriculture that suits the requirements of coastal farming community could be launched in cooperation with research organizations, NGOs, local communities and local government agencies. However, to harness the full potential of CSA, it is really important to look at why households with certain characteristics are doing fine in one dimension of CSA while doing poorly in another. Although this study has successfully assessed the probability for households with certain characteristics to adopt CSA, it equally fails to unveil the underlying reasons of such contrasting behaviour. This requires a deeper insight for which fresh qualitative inquiry backed by sufficient grounded theory could be undertaken. Another important drawback of this study is that the CSA framework that this study has developed and tested assumes that all the 25 statements relating to natural agricultural practices equally contributes to achieving the goals of CSA. In practice, however, their contributions do vary. A final limitation of this research is that it tries to generalize the findings of a local area for a whole coastal setting although it is very common to increase the heterogeneity for the same agricultural practice across a large scale. Therefore, the essence of the findings of this research must be carefully reviewed before applying in an apparently similar context. Despite its limitations, this study's findings would help policy makers and planners to launch fresh new programs of intervention in order to encourage increased adoption of CSA, particularly through emission reduction from agriculture and sequestering carbon.

Acknowledgements The initial version of this chapter was presented in the Climate Smart Agriculture Conference held during 16-19 March 2015 in Montpellier, France. The Author gratefully acknowledge their support for participation in the conference.

References

Adger WN (2006) Vulnerability. Glob Environ Chang 16(3):268–281
Adger WN, Huq S, Brown K, Conway D, Hulme M (2003) Adaptation to climate change in the developing world. Prog Dev Stud 3(3):179–195
Adger W, Dessai S, Goulden M, Hulme M, Lorenzoni I, Nelson D, Naess L, Wolf J, Wreford A (2009) Are there social limits to adaptation to climate change? Clim Change 93:335–354
Agrawal A, Wollenberg E, Persha L (2014) Governing agriculture-forest landscapes to achieve climate change mitigation. Glob Environ Chang 28:270–280
Agrawala S, Ota T, Ahmed AU, Smith J, van Aalst M (2003) Development and climate change in Bangladesh: focus on coastal flooding and Sundarbans. OECD, Paris
Ahasan MN, Chowdhary MAM, Quadir DA (2010) Variability and trends of summer monsoon rainfall over Bangladesh. J Hydrol Meteorol 7(1):1–17
Ahmed N, Diana JS (2015) Coastal to inland: expansion of prawn farming for adaptation to climate change in Bangladesh. Aquac Rep 2:67–76
Ahmed N, Occhipinti-Ambrogi A, Muir JF (2013) The impact of climate change on prawn post larvae fishing in coastal Bangladesh: socioeconomic and ecological perspectives. Mar Policy 39:224–233
Alexandratos N, Bruinsma J (2012) World agriculture towards 2030/2050: the 2012 revision. ESA Working Paper No. 12-03. FAO, Rome. http://www.fao.org/docrep/016/ap106e/ap106e.pdf. Accessed 3 Oct 2014
Arbuckle JG Jr, Morton L, Hobbs J (2013) Farmer beliefs and concerns about climate change and attitudes toward adaptation and mitigation: evidence from Iowa. Clim Change 118:551–563
Banerjee L (2010) Effects of flood on agricultural productivity in Bangladesh. Oxf Dev Stud 38:339–356
Barnes A, Toma L (2012) A typology of dairy farmer perceptions towards climate change. Clim Change 112:507–522
Bennett EM, Carpenter SR, Gordon L, Ramankutty N, Balvanera P, Campbell BM, Cramer W, Foley J, Folke C, Karlberg L et al (2014) Resilient thinking for a more sustainable agriculture. Solutions (in press)
Berry PM, Rounsevell MDA, Harrison PA, Audsley E (2006) Assessing the vulnerability of agricultural land use and species to climate change and the role of policy in facilitating adaptation. Environ Sci Policy 9:189–204
Brooks N, Adger WN, Kelly PM (2005) The determinants of vulnerability and adaptive capacity at the national level and the implications for adaptation. Glob Environ Chang 15:151–163
Burney JA, Davis SJ, Lobell DB (2010) Greenhouse gas mitigation by agricultural intensification. Proc Natl Acad Sci USA 107:12052–12057
Calzadilla A, Zhu T, Rehdanz K, Tol RSJ, Ringler C (2014) Climate change and agriculture: impacts and adaptation options in South Africa. Water Resour Econ 5:24–48
Campbell BM, Thornton P, Zougmore R, van Asten P, Lipper L (2014) Sustainable intensification: what is its role in climate smart agriculture? Curr Opin Environ Sustain 8:39–43
Challinor AJ, Wheeler TR (2008) Crop yield reduction in the tropics under climate change: processes and uncertainties. Agric For Meteorol 148(3):343–356
Chartzoulakis K, Bertaki K (2015) Sustainable water management in agriculture under climate change. Agric Agric Sci Proc 4:88–98
Chowdhury SR, Hossain MS, Shamsuddoha M, Khan SMMH (2012) Coastal fishers' livelihood in Peril: sea surface temperature and tropical cyclones in Bangladesh. Center for Participatory Research and Development, Dhaka
Cisneros BEJ, Oki T, Arnell NW, Benito G, Cogley JG, Döll P, Jiang T, Mwakalila SS, Fischer T, Gerten D, Hock R, Kanae S, Lu X, Mata LJ, Pahl-Wostl C, Strzepek KM, Su B, van den Hurk B (2014) Climate change 2014: impacts, adaptation, and vulnerability: Contribution of working group II to the fifth assessment report of the intergovernmental panel on climate change. IPCC

Conway G, Waage J (2010) Science and innovation for development. UK Collaborative on Development Sciences, London
Dailidiene I, Baudler H, Chubarenko B, Navrotskaya S (2011) Long term water level and surface temperature changes in the lagoons of southern and eastern Baltic. Oceanologia 35(1–T1):293–308
Dasgupta S, Huq M, Khna ZH, Ahmed MMZ, Mukherjee N, Khan MF, Pandey K (2010) Vulnerability of Bangladesh to cyclones in a changing climate: potential damages and adaptation cost. Policy Research Working Paper 5280. The World Bank, Washington, DC
Dasgupta S, Huq M, Khan ZH, Ahmed MMZ, Mukherjee N, Khan MF, Pandey K (2011) Cyclones in a changing climate: the case of Bangladesh. Department for Environment, Food and Rural Affairs, London
Dastagir MR (2015) Modeling recent climate change induced extreme events in Bangladesh: a review. Weather Clim Extremes 7:49–60
FAO (2010) Climate-smart agriculture: policies, practices and financing for food security, adaptation and mitigation. FAO, Rome
FAO (2013) Food and agriculture organization of the United Nations: climate smart agriculture sourcebook. FAO, Rome
Field A (2005) Discovering statistics using SPSS, 2nd edn. Sage, London
German Watch (2013) Global climate risk index: who suffers most from extreme weather events? Weather related loss events in 2011 and 1992–2011
Government of Bangladesh (GOB) (2005) National Adaptation Programme of Action (NAPA), Final report: November. Ministry of Environment and Forests. Government of Bangladesh Press, Dhaka
Government of Bangladesh (GOB) (2009) Bangladesh climate change strategy and action plan 2008. Ministry of Environment and Forests. Government of Bangladesh Press, Dhaka
Haden VR, Niles MT, Lubell M, Perlman J, Jackson L (2012) Global and local concerns: what attitudes motivate farmers to mitigate and adapt to climate change. PLoS One 7:e52882
Hair JF Jr, Black WC, Babin BJ, Anderson RE, Tatham RL (2006) Multivariate data analysis, 6th edn. Prentice Hall International, Upper Saddle River, NJ
Harmeling S, Eckstein D (2012) Global climate risk index 2013. In: Who suffers most from extreme weather events? Weather-related Loss Events in 2011 and 1992–2011. German Watch, Bonn
Houghton JT, Ding Y, Griggs DJ, Noguer M, van der Linden PJ, Dai X, Maskell K, Johnson C (2001) Climate change 2001: the scientific basis, vol 881. Cambridge University Press, Cambridge
Howden SM, Soussana J-F, Tubiello FN, Chhetri N, Dunlop M, Meinke H (2007) Adapting agriculture to climate change. Proc Natl Acad Sci USA 104:19691–19696
Huntington TG (2006) Evidence for intensification of the global water cycle: review and synthesis. J Hydrol 319(1):83–95
Huq S, Karim Z, Asaduzzaman M, Mahtab F (eds) (1998) Vulnerability and adaptation to climate change for Bangladesh. Kluwer, Dordrecht
IFAD (International Fund for Agricultural Development) (2013) Smallholders, food security, and the environment. IFAD/UNEP, Rome
Iizumi T, Ramankutty N (2015) How do weather and climate influence cropping area and intensity? Glob Food Secur 4:46–50
Intergovernmental Panel on Climate Change (2007) Climate change 2007: synthesis report. IPCC, Valencia
Karim MF, Mimura N (2008) Impacts of climate change and sea-level rise on cyclonic storm surge floods in Bangladesh. Glob Environ Change 18:490–500
Khan TA, Mazid M, Mohammad F (2011) A review of ascorbic acid potentialities against oxidative stress induced in plants. J Agrobiol 28:97–111
Koohafkan P, Altieri AM, Gimenez HE (2012) Green agriculture: foundations for biodiverse, resilient and productive agricultural systems. Int J Agric Sustain 10:61–75

Kreft S, Eckstein D, Junghans L, Kerestan C, Hagen U (2014) Global climate risk index 2015. In: Who suffers most from extreme weather events? Weather-related loss events in 2013 and 1994–2013. German watch, Bonn

Kripalani R, Oh J, Kulkarni A, Sabade S, Chaudhari H (2007) South Asian summer monsoon precipitation variability: coupled climate model simulations and projections under IPCC AR4. Theor Appl Climatol 90(3–4):133–159

Kumar BM, Nair PKR (2012) Carbon sequestration potential of agroforestry systems. Opportunities and challenges. Springer, Berlin

Lasco RD, Habito MS, Delfino RJP, Pulhin FB, Concepcion RG (2011) Climate change adaptation guidebook for smallholder farmers in Southeast Asia. World Agroforestry Centre, Philippines

Mbow C, Smith P, Skole D, Duguma L, Bustamante M (2014) Achieving mitigation and adaptation to climate change through sustainable agroforestry practices in Africa. Curr Opin Environ Sustain 6:8–14

Morton JF (2007) The impact of climate change on smallholder and subsistence agriculture. Proc Natl Acad Sci USA 104:19680–19685

Nagayets O (2005) Small farms: current status and key trends. The future of small farms: proceedings of a research workshop, IFPRI, Wye, pp 355–367

Niles MT, Lubell M, Brown M (2015) How limiting factors drive agricultural adaptation to climate change. Agric Ecosyst Environ 200:178–185

Pervez MS, Henebry GM (2014) Projections of the Ganges–Brahmaputra precipitation—downscaled from GCM predictors. J Hydrol 517:120–134

Pervez MS, Henebry GM (2015) Assessing the impacts of climate and land use and land cover change on the freshwater availability in the Brahmaputra River basin. J Hydrol Reg Stud 3:285–311

Pethick J, Orford JD (2013) Rapid rise in effective sea-level in southwest Bangladesh: its causes and contemporary rates. Global Planet Change 111:237–245

Prokopy LS, Floress K, Klotthor-Weinkauf D, Baumgart-Getz A (2008) Determinants of agricultural best management practice adoption: evidence from the literature. J Soil Water Conserv 63:300–311

Rahman MA, Rahman S (2015) Natural and traditional defence mechanisms to reduce climate risks in coastal zones of Bangladesh. Weather Clim Extremes 7:84–95

Rahmstorf S (2007) A semi-empirical approach to projecting future sea level rise. Science 315:368–370

Reilly J, Tubiello F, McCarl B, Abler D, Darwin R, Fuglie K, Hollinger S, Izaurralde C, Jagtap S, Jones J, Learns L, Ojima D, Paul E, Paustian K, Riha S, Rosenberg N, Rosenzweig C (2003) U.S. agriculture and climate change: new results. Clim Change 57:43–69

Rosenzweig C, Elliott J, Deryng D, Ruane AC, Müller Ch, Arneth A, Boote KJ, Folberth C, Glotter M, Khabarov N, Neumann K, Piontek F, Pugh TAM, Schmid E, Stehfest E, Yang H, Jones JW (2013) Assessing agricultural risks of climate change in the 21st century in a global gridded crop model intercomparison. Proc Natl Acad Sci. doi:http://dx.doi.org/http://www.pnas.org/content/early/2013/12/13/1222463110.abstrac

Sabade S, Kulkarni A, Kripalani R (2011) Projected changes in South Asian summer monsoon by multi-model global warming experiments. Theor Appl Climatol 103(3–4):543–565

Salinger MJ, Sivakumar MVK, Motha R (2005) Reducing vulnerability of agriculture and forestry to climate variability and change: workshop summary and recommendations. Clim Change 70:341–362

Saroar MM, Routray JK (2010) In situ adaptation against sea level rise (SLR) in Bangladesh: does awareness matter? Int J Clim Change Strateg Manage 2(3):321–345

Schiermeier Q (2014) Holding back the tide. Nature 508:164–166

Schlenker W, Hanemann WM, Fisher A (2007) Water availability, degree days, and the potential impact of climate change on irrigated agriculture in California. Clim Change 81:19–38

Schuol J, Abbaspour KC, Srinivasan R, Yang H (2008) Estimation of freshwater availability in the West African sub-continent using the SWAT hydrologic model. J Hydrol 352(1–2):30–49

Shahid S (2010) Rainfall variability and the trends of wet and dry periods in Bangladesh. Int J Climatol 30:2299–2313
Sikder MAI (2013) Climate change and Bangladesh agriculture: country profile. In: Workshop on developing farming systems for climate change mitigation, 26–30 Aug 2013, Colombo
Singh H, Subash N, Gangwar B, Valdivia R, Antle J, Baigorria G (2015) Assessing economic impacts of climate change and adaptation in Indo-Gangetic Basin. Agriculture and climate change- adapting crops to increased uncertainty (AGRI 2015). Proc Environ Sci 29:229–230
Smit B, Wandel J (2006) Adaptation, adaptive capacity, and vulnerability. Glob Environ Chang 16 (3):282–292
Tanaka SK, Tingju Z, Lund JR, Howitt RE, Jenkins MW, Pulido MA, Tauber M, Ritzema RS, Ferreira IC (2006) Climate warming and water management adaptation for California. Clim Change 76:361–387
Thornton PK, Gerber P (2010) Climate change and the growth of the livestock sector in developing countries. Mitig Adapt Strat Glob Change 15:169–184
Thornton P, Lipper L (2013) How does climate change alter agricultural strategies to support food security? Background paper for the conference ‘Food Security Futures: Research Priorities for the 21st Century’, 11–12 Apr 2013, Dublin, 2013
Trenberth KE (2008) The impact of climate change and variability on heavy precipitation, floods, and droughts. In: Anderson MG (ed) The encyclopedia of hydrological sciences. Wiley, Chichester, UK. doi:10.1002/0470848944.hsa211
United Nations (1992) The United Nations framework convention on climate change. United Nations, New York
Vermeulen SJ (2014) Climate change, food security and small-scale producers. CCAFS Info Brief. CGIAR Research Program on Climate Change, Agriculture and Food Security (CCAFS)
Vermeulen SJ, Campbell BM, Ingram JSI (2012) Climate change and food systems. Annu Rev Environ Resour 37:195
Wollenberg E, Campbell BM, Holmgren P, Seymour F, Sibanda L, von Braun J (2011) Actions needed to halt deforestation and promote climate smart agriculture. CGIAR Research Program on Climate Change. Agriculture and Food Security
Wood SA, Jina AS, Jain M, Kristjanson P, DeFries RS (2014) Smallholder farmer cropping decisions related to climate variability across multiple regions. Glob Environ Change 25:163–172
World Bank (2012a) Thai flood 2011: rapid assessment for resilient recovery and reconstruction planning. Thailand, Bangkok
World Bank (2012b) Bangladesh and Maldives respond to climate change impacts. The World Bank, Washington DC, Press Release
World Bank (2008) World development report 2008: agriculture for development. World Bank, Washington, DC

Part II
Fostering Resilience and Handling Hazards

Chapter 11
Managing Natural Resources for Extreme Climate Events: Differences in Risk Perception Among Urban and Rural Communities in Sydney, Australia

Louise Boronyak-Vasco and Brent Jacobs

Abstract Lack of perception of the risks posed by climate change has been identified as a major constraint to social adaptation. Factors contributing to risk perception include experience of extreme weather events; socio-cultural factors (norms and values); knowledge of causes, impacts and responses, and socio-demographics. Qualitative data was collected from a series of participatory placed-based workshops conducted in the Greater Sydney and South East regions of New South Wales, Australia with participants drawn from a mix of 12 urban and rural communities. Workshop discussions were based on an Emergency Management Framework: Prepare, Prevent, Respond and Recover (PPRR) for the most important local climate hazards—bushfires, drought, storms, and flooding. Qualitative information from the workshops was examined for evidence of the role of risk perception in the management of natural resources for extreme climate events and the capacity of communities to adapt. Perception of risk differed among locations (urban vs. rural) and types of events, in particular bushfire and flood. Recent experience of an event, livelihood dependency on natural resources and the socio-demographic dynamics of communities were identified as factors contributing to adaptive responses to improve protection of natural resources (such as soils, water and biodiversity).

Keywords Climate change • Climate change adaptation • Natural resource management • Risk perception

L. Boronyak-Vasco (✉) • B. Jacobs
Institute for Sustainable Futures, University of Technology Sydney, Broadway, NSW 2007, Australia
e-mail: Louise.Boronyak@uts.edu.au

W.L. Filho et al. (eds.), *Climate Change Adaptation, Resilience and Hazards*, Climate Change Management, DOI 10.1007/978-3-319-39880-8_11

Introduction

Climate change has been identified as an evolutionarily novel risk (Griskevicius et al. 2012) because of the magnitude, complexity and scale of the problem. These characteristics coupled with the slow, cumulative and largely invisible nature of climate change (van der Linden 2015) make it difficult to experience directly (Weber 2010), and complicate the mental construction of the risk (Breakwell 2010). Projections of future climate indicate that changes to the frequency and or severity of extreme climate events are likely to occur (Richardson et al. 2013). Complex event-driven extremes, which include severe drought, storms and floods, do not necessarily occur every year at a given location (Easterling et al. 2000). However, they are increasingly viewed as drivers of rapid social and policy change (Johnson et al. 2005; Lujala et al. 2015; Marshall et al. 2013; Moser and Ekstrom 2010) and, therefore, offer opportunities to explore actions on adaptation with otherwise disengaged communities.

Risk perception is a complex and multi-faceted phenomenon that involves the interaction of multiple human and environmental factors over time. Paton and McClure (2013) argue that people decide either by necessity or by choice to live in locations that expose them to the risk of extreme natural events because they perceive that the physical, economic and aesthetic amenities outweigh the risks. However, problems for exposed communities arise when they fail to perceive a change in the frequency or intensity of events that exceeds their historical coping capacity (Bürgelt and Paton 2014). To support communities through disasters, integrated warning systems have been developed that generally consist of three components (Mayhorn and McLaughlin 2014): hazard detection, emergency management response and public communication. Although these systems are increasingly becoming sophisticated, it is the rational response of the community to the hazard that ultimately determines its level of engagement with emergency preparedness (Helsloot and Ruitenberg 2004). The relationship between perception and preparedness is complex and is influenced by a range of factors including experience of extreme weather events; socio-cultural factors (norms and values); knowledge of causes, impacts and responses, and socio-demographics (Scolobig et al. 2012; van der Linden 2015; Wachinger et al. 2013).

Many of the factors associated with community risk perception of natural hazards are influenced by the increasing urbanisation of the global population (MEA 2005). Unlike rural areas, cities are often dislocated from their supply of natural resources (in particular water) (Padowski and Gorelick 2014), food (Mason and Knowd 2010) and energy generation (Godschalk 2003). This means that urban populations may be disconnected from environmental signals that might otherwise stimulate behaviour change in communities dependent on natural resources to support their livelihoods (Pretty 2002). Differences among urban and rural communities in norms and values (Argent et al. 2010), knowledge (McGee and Russell 2003) and socio-demographics (Luck et al. 2010) are also likely to result in differing perceptions of risk to climate change and extreme events, and variations in adaptive capacity (Pretty 2003). For example, local ecological knowledge is an important component of capacity to manage natural systems. Local knowledge is considered to co-evolve with the ecosystem upon which it is based and is maintained through frequent interaction with the natural environment. Pilgrim et al. (2008) showed that ecological knowledge is inversely related to income levels among countries and suggested that the differences between countries were related to the level of urbanisation, reliance on services and the globalisation of trade and culture. In order to manage this disconnection from the environment in urbanized areas, emergency management institutions attempt to stimulate risk perception in exposed communities.

Over time governments have developed emergency service capability for dealing with natural disasters. In Australia, the State Government of New South Wales (NSW), in which the city of Sydney is located, has implemented a State Emergency Management Plan (NSW Government 2015). It is based on an adaptive management system and supported by local social capital (through community volunteer services such as the Rural Fire Service (RFS), State Emergency Service (SES) and local government). The Plan uses a four phase framework of prevent, prepare, respond and recover (PPRR, Fig. 11.1) to inform government and its emergency management agencies of the appropriate administrative and operational responses throughout the duration of an event (Bunker et al. 2015). It is becoming clear, however, that many communities, in particular those in urban settlements, are

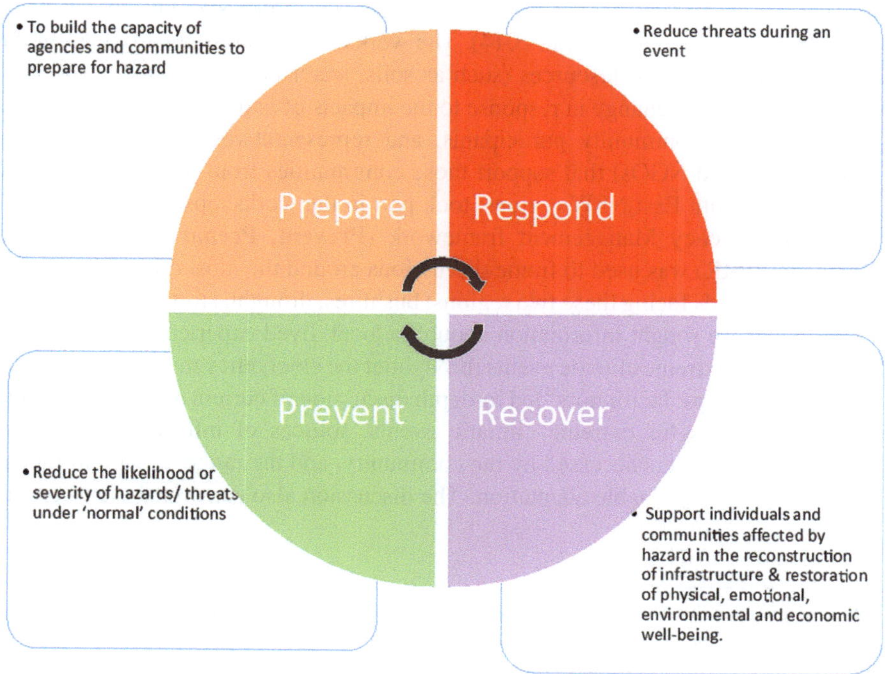

Fig. 11.1 Emergency management framework (PPRR) and definition of the phases. *Source*: authors

increasingly reliant on emergency services during the response phase of an event because of a failure to pro-actively undertake preparatory action. This has been in part attributed to the failure of risk communication to influence the at-risk individual's mental construction of risk and, in turn, how they act to mitigate consequences (Bostrom et al. 1994; Leiserowitz 2006; Reid and Beilin 2014).

This paper will examine how factors associated with perception of risk differed among locations (urban vs. rural) in relation to extreme climate events in Sydney and surrounding peri-urban areas. We anticipate that recent experience of an event, livelihood dependency of a community on natural resources and socio-demographic dynamics will be prominent among the components of risk perception that contribute to adaptive responses to extreme climate events.

Research Methodology

This methodology aimed to elicit information about community members' perceptions of risk of extreme climate events, the likely local impacts of these events and factors that increase local vulnerability to extreme climate events. We conducted a series of 12 placed-based, participatory workshops between May and November, 2014. Our approach was consistent with recent advocacy (e.g. Birkholz et al. 2014) for greater engagement with constructivist perspectives as an alternative to the more traditional reliance on rationalist approaches to understanding disaster risk perception (e.g. Reid and Beilin 2014). The workshops focused in particular on the management of natural resources (such as soils, water and biodiversity) and how management might change in response to the impacts of extreme climate events. In total 184 local community participants, and representatives from the agencies (government and NGOs) that support those communities from the Sydney Metropolitan and South East NSW regions, took part in the workshops.[1]

The Emergency Management framework (**Prevent, Prepare, Respond** and **Recover** (PPRR) was used to frame discussions around the most critical or disruptive local hazards facing these two regions: bushfires, drought, storms, and flooding. In particular we sought information about the local, lived experience (Lewis-Beck et al. 2013) of extreme climate events throughout the emergency management cycle.

In small groups, facilitators[2] led in-depth discussion of current actions to manage natural resources for extreme climatic events, sources of information and key information providers accessed by the community, and the range of support mechanisms available to enable adaptation. The discussion also canvassed views on the

[1]Representing landholders, emergency service volunteers, local and state government agencies, business owners, Indigenous peoples, financial institutions and a range of non-government community organisations such as Landcare.

[2]The workshop facilitators included the authors from the Institute for Sustainable Futures, UTS and staff from state government NRM agencies.

aspects of the natural environment and local livelihoods and lifestyles that may be lost to the community if extreme events become more frequent or intense. Finally, we asked the participants to identify management strategies that could either reduce exposure to extreme events or increase resilience to extreme events while simultaneously protecting the natural resource base throughout the PPRR cycle.

Participants were provided an overview of the climate drivers for their region and potential hazards faced by communities in their local landscape. Information was drawn from two sources:

- Historical climate analyses from the Australian Bureau of Meteorology including national and South East Australian temperature trends, rainfall trends, anomalies and seasonality, and the occurrence of East Coast low pressure systems.
- A climate summary prepared as part of the NSW Office of Environment and Heritage's Climate Impact Profile (Tables 11.1 and 11.2). It should be noted that these workshops pre-dated the release of regional climate data through AdaptNSW (OEH 2015).

The qualitative information collected about each stage of the emergency management cycle was coded for workshop location and extreme event type. This information was subjected to qualitative meta-synthesis (Sandelowski et al. 2007) to identify emergent themes and provide deeper insights than might be possible

Table 11.1 Major climate impacts for the Sydney Metropolitan and South East NSW regions

Climate attribute/ impact	Sydney Metropolitan	South East NSW
Temperature	Hotter with more heat waves	Hotter
Rainfall	Likely to increase with significant rainfall in autumn	Likely increase in summer and decrease in winter
Flooding	Increasing intensity, localised flooding of urban areas in the vicinity of rivers and tributaries	Run-off and stream flow likely increase during summer leading to a heightened flood risk
Sea level rise	Increased exposure to beach erosion and inundation	Increased risk to coastal property and infrastructure Increased inundation and acidification of agricultural soils
Snow fall	n/a	Likely decrease
Biodiversity	Large changes in areas of high biodiversity value especially in the Blue Mountains World Heritage Area Potential changes in extent and range of both native flora and fauna and invasive species	Changes to natural ecosystems (alpine, low-lying coastal and fire sensitive) Potential changes in extent and range of both native flora and fauna and invasive species
Soil erosion	Changes projected through increased annual surface run-off	Likely increase on erodible soil types

Source: OEH (2015)

Table 11.2 Commonly expressed perspectives on emergency events: differences between emergency services personnel and the community

Emergency management perspective	Community perspective
Most people unaware of the risk	Most people aware of the risk, but have other priorities that take precedence
Most people in the community rely on the emergency services to respond to bushfire, storms, and flooding	Many people do not want to rely on help from emergency services
High public expectation of service provision during emergencies	Individual householders see themselves as the most responsible for personal and home safety
Can advise but not direct residents to take action	Frustrated by lack of specific advice on what to do and limited help provided

Adapted from Cottrell and King (2008)

from studies of a single location or event type (Major and Savin-Baden 2011). Quotes from workshop transcripts are presented in the results to illustrate common aspects of risk perception related to the most important narratives emerging from the workshops.

The information collected during this process was rich and informative; however, we recognise that in utilising this methodology a compromise must be made between the scientific rigor of formal psychometric testing of risk perception (e.g. Lachlan and Spence 2007) and the need for policy relevance to stakeholder concerns. The latter particularly includes the ability to capture spatial and temporal context, and the need for a research process to resonate with the public in order to best understand society's concerns and aspirations (Stevenson and Lee 2001).

Results and Discussion

The Prevent-Prepare-Respond-Recover (PPRR) framework proved useful to engage both urban and rural communities in discussions about extreme climate events. The framing of extreme events into a logical sequence facilitated structured discussion and revealed differences in activities across a range of climatic events. The application of consistent, well-developed and tested frameworks in participatory assessments provides a common language that can be used to organise complex information and problems (Brown et al. 2010).

In the case of this series of workshops, using the PPRR framework as a unifying theme for discussion revealed differences in perception of risk to life, property and natural resources from extreme climate events that differed among locations (urban vs. rural) and types of events. Of the factors commonly associated with risk perception (Fig. 11.2) those most widely identified in our workshops as mediating adaptive responses included:

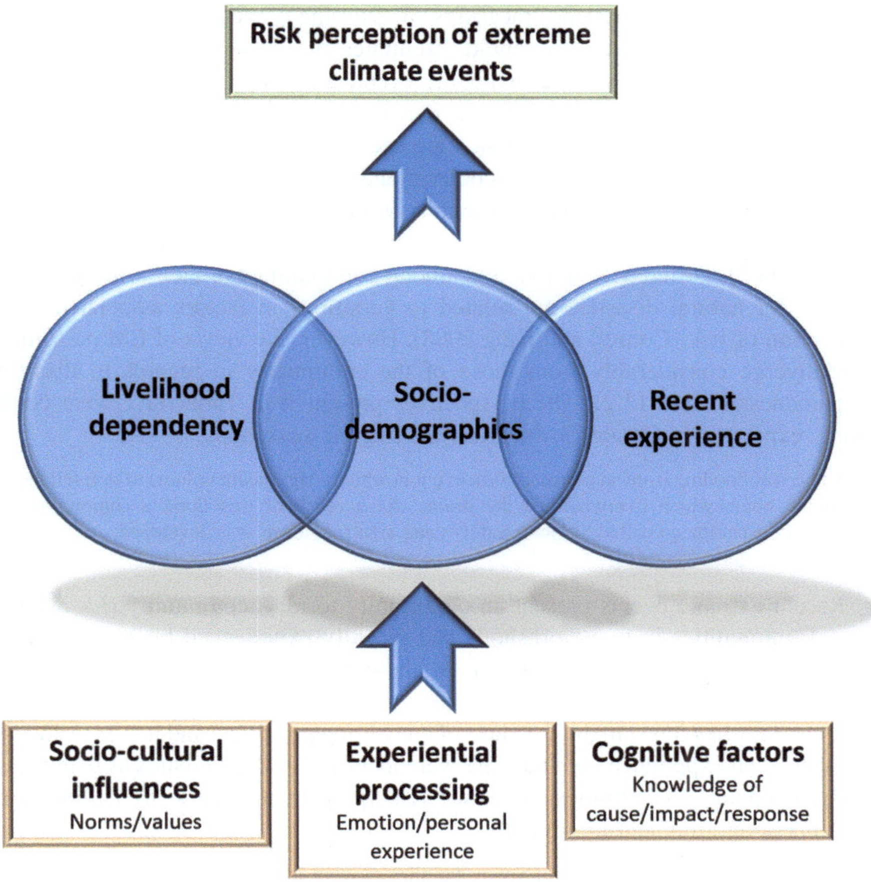

Fig. 11.2 Factors associated with risk perception among communities in South East NSW and the Greater Sydney regions. *Source*: authors

- Recent experience of an event,
- Livelihood dependency on natural resources, and
- The socio-demographic dynamics of communities.

Recent Experience of an Event

When it's cold, people don't think about the risks of bushfire.
 Peri-urban bushfire workshop

The focus is on preparing now because the experience of previous fires is fresh but it's hard to maintain that motivation five years down the track.
 Peri-urban bushfire workshop

Residents tend not to be pro-active in taking responsibility for protecting their own properties and expect agencies and others in the community to help.
 Urban flood workshop

The adoption of preparatory measures for extreme climate events reduces loss and injury within a given household, facilitates the capacity to cope with the temporary disruption associated with the event, and can minimise damage and insurance costs (Paton et al. 2006). For bushfires, Emergency Management (EM) agencies convey bushfire risk messages and strategies through multiple means to encourage preparation in cooler months of winter and spring. However, both EM service personnel and community workshop participants suggested that many people fail to act on these messages until the risk of fire risk is 'extreme' (AFAC 2005). Differences within and among communities in their willingness to prepare for natural disasters are related to variations in factors associated with perception of risk (Cottrell and King 2008). However, the views of EM personnel may diverge considerably from those of the community in regards to disaster preparedness (Table 11.2). The perspectives presented in Table 11.2 were commonly expressed in Greater Sydney and SE Region workshops.

> There was flooding from an unusual storm event in which a significant volume of rain fell in couple hours which overwhelmed the drains. As a result of this flood a community awareness campaign called Summer Safety campaign for storms was developed.
>
> Urban flood workshop

Recent extreme events present an opportunity to raise community risk awareness. Communities in two workshops had recently experienced bushfires, which had clearly raised awareness of risk in these locations, although this did not necessarily translate into greater preparedness of the community as a whole. Younger families (presumably because of competing priorities) and newer residents that lack local experience with bushfire were identified by participants as being less likely to engage in activities to reduce risk (prepare and prevent) without the stimulus of an imminent fire threat.

> This region is seeing an influx of young families moving in to take advantage of better housing affordability, they are ill-prepared and the motivation to prepare only comes over time.
>
> Bushfire workshop

For urban residents the preparation phase was commonly reported to be triggered by smoke in the air (often caused by hazard reduction burning) and media reports of actual fires (sometimes in other locations). This contrasts with communities in more rural areas where preparation is more anticipatory and driven by local knowledge of 'prolonged hot, dry and windy conditions' and 'signs in the surrounding landscape' such as water stressed vegetation, vegetation dieback or the build-up of fuel in natural areas. Connections to the local environment, observations of environmental change and personal weather experiences are factors known to be associated with climate risk perceptions (Higginbotham et al. 2014).

Lack of experience of extreme events is a barrier to effective engagement in preparedness (Weber 2010). Similarly, Higginbotham et al. (2014) found that "direct experience of extreme weather events appears to shape threat appraisal" (p. 701). In one location (the Blue Mountains located in Sydney's urban fringe), there was evidence that a recent major bushfire event had created novel social opportunities to foster communication about bushfire risk within the community, in particular with newer residents, through the production and screening of a film called *As the Smoke Clears* that illustrates local bushfire recovery (https://www.youtube.com/watch?v=5PjhyiYcMK0).

> The film night put on by the Blue Mountains World Heritage Institute is a novel way to engage community as well as getting people together to share experiences.
>
> Peri-urban bushfire workshop

> There is a sense of fear about being near the bush and this sometimes results in head in the sand attitudes.
>
> Urban flood workshop

In contrast, there was reportedly little community awareness of the risk of flooding in the Hawkesbury-Nepean floodplain located in western Sydney, which was attributed to a lack of "collective community experience' with flood. The flood plain contains expansive human settlements that have not experienced a major flood event since 1867 (Gillespie and Grech 2002). Water flows in the Hawkesbury-Nepean region are managed for water supply for Sydney rather than flood control (Gillespie and Grech ibid). However, the potential exposure to flood, particularly under the altered rainfall regimes expected with climate change is recognized as among the highest in Australia (Dowdy 2015; Brewsher et al. 2013).

> We [emergency management staff] did a letter drop to about 3000 households in the flood plain to invite participation in a flood awareness raising event. But only about 30 people came to the event.
>
> Urban flood workshop

The lack of response of Sydney's flood plain communities to traditional forms of emergency preparedness communication raises questions about its effectiveness. Paton et al. (2006) found that for communities exposed to bushfires the formation of "intention to prepare" and the formation of "intention to seek information" represented different cognitive pathways in relation to environmental hazards. Those who form "intentions to prepare" are more likely to prepare than those who form "intentions to seek information". For communication to be effective, engagement needs to be targeted specifically for each group, which may require EM institutions to embrace new forms of communication to increase the effectiveness of risk communication.

Livelihood Dependency

> The environment is a major tourist draw-card; consequently fire has huge socio-economic impacts especially on local businesses.
>
> Peri-urban bushfire workshop

> Houses, sheds, hay, equipment—all of these are required to cope with and recover from fire. If lost, this has a serious impact upon individuals, but also the community as a whole.
>
> Rural bushfire workshop

The extent to which communities understood and acted on the need to protect the environment from the impacts of extreme events appeared to be related to the dependency of community livelihoods on natural resources. Livelihood dependency would likely influence risk awareness through cognitive factors (rural communities have a greater knowledge of causes, impacts and responses) and differences in social norms and values between rural and urban communities (Fig. 11.2, van der Linden 2015). Pilgrim et al. (2008) demonstrated an association between economic growth and social capacity to manage the environment with wealthier, urban communities showing lower levels of ecological knowledge than rural communities. Increasing urbanisation of peri-urban areas adjacent to Sydney not only places these new communities at greater risk of extreme events such as bushfires, it also undermines their perception of the risk of these events.

Rural communities most often valued the natural resources they viewed as directly influencing agricultural production. In particular, they focused on protection of soils and surface water quality. They were also more aware of the impacts of events such as bushfire on animals both native and livestock, which often required euthanising during the recovery phase, a task rural communities reported as adding to the trauma of an extreme weather event.

> Native animals are likely to be hit by cars or attacked by feral animals after fire as they are disoriented from being forced out of their normal habitat.
>
> Rural bushfire workshop

In urban areas, the protection of natural resources was related primarily to their amenity value, such as surface water quality for recreational activities (fishing, boating). Urban communities living in close proximity to national parks (such as the World Heritage listed Blue Mountains area) also recognised the potential for local economic impacts through damage to environmental tourism and changes to their community's cultural identity. However, community concern for the environment is reduced when natural assets have the potential to impact on houses; this view is encouraged by government policy. For example, workshop participants were concerned that recent changes to vegetation regulations in the vicinity of housing will increase clearing of trees in lower parts of the Blue Mountains, greatly impacting biodiversity and giving 'a false sense of security'. Accordingly, many urban communities reportedly view trees as a source of risk rather than as a natural resource asset, eschewing tree maintenance in favour of complete removal. Despite this there was general agreement at storm and flood workshops that improved selection and maintenance of backyard trees, rather than wholesale clearing ('tree-hysteria'), could reduce damage to property and maintain the ecosystem service benefits that tree canopies provide (e.g. reduction of heat, enhancement of biodiversity, human health and well-being).

Socio-demographics

> There is a continuation to put people at risk by allowing housing developments on the flood plain and legislation to stop development on flood prone land—distorted by political process.
>
> Coastal storms workshop

> I've been working in the SES for 23 years and never, ever thought about NRM.
>
> Emergency service volunteer coastal storms workshop

Workshop narratives suggested that rapid population growth and urbanisation can erode local knowledge and community connection to local environment. There were quite marked differences between urban and rural-regional emergency management staff in their understanding of the importance of managing natural resources for extreme climate events. Emergency service personnel are generally drawn from the communities they serve and their perspectives on environmental protection might reasonably be expected to reflect those of the community. In rural areas it was common for volunteer EM staff to be drawn from the local farming community bringing to emergency management an understanding of natural resource management and resulting in informal changes to operating procedures where they do not conflict with the protection of life and property. In urban areas, emergency management staff frequently stated that they rarely considered the impacts of extreme climate events on natural resources. The exception to this was for bushfire management in the Blue Mountains, where the extensive areas of reserves in close proximity to residential areas require a deep understanding of the science of endangered ecosystems and collaboration between National Parks and Wildlife staff, the RFS and local government (e.g. Hammill and Tasker 2010).

Conclusions

This paper has demonstrated that while risk perception is critical to promoting preparatory action for extreme climate events, the relative importance of the components of risk perception varied among communities and with respect to extreme event types. We found recent experience of an event, natural resource livelihood dependency and socio-demographic change most closely associated with variation in risk perception among community members living in the Sydney metropolitan region and the South East regions of NSW. While this study was limited in scope to the socio-economic, biophysical and cultural context of two regions in South East Australia, the findings are consistent with research on responses to extreme climate events in other locations (e.g. Higginbotham et al. 2014; Scolobig et al. 2012; Van der Linden 2015). We identified two key findings. Firstly, the disconnection between urban communities and the natural resources on which they depend limits their level of understanding of the impact of extreme climatic events on the environment. Secondly, in contrast, rural communities with a greater reliance on and understanding of natural resource management are more likely to engage in preparedness for extreme events. The implication of these findings for EM institutions and policy-makers in seeking to improve the management of natural resources for extreme climatic events is that risk communication should be tailored to the ecological literacy of the community. Improvements to natural resource management may also require new governance partnerships between natural resource agencies, EM institutions and local communities to ensure that management actions are based on a shared understanding of the importance of environmental protection.

References

AFAC (2005) Position paper on bushfires and community safety. Australasian Fire Authorities Council Limited

Argent N, Tonts M, Jones R, Holmes J (2010) Amenity-led migration in rural Australia: a new driver of local demographic and environmental change? In: Luck GW, Black R, Race D (eds) Demographic change in Australia's rural landscapes, vol 12, Implications for society and the environment. CSIRO, Collingwood

Birkholz S, Muro M, Jeffrey P, Smith HM (2014) Rethinking the relationship between flood risk perception and flood management. Sci Total Environ 478(2014):12–20

Blue Mountains World Heritage Institute (2014) As the smoke clears. https://www.youtube.com/watch?v=5PjhyiYcMK0. Images published to highlight the recovery process in the Blue Mountains following the 2013 bushfires

Bostrom A, Atman CJ, Fischhoff B, Morgan MG (1994) Evaluating risk communications: completing and correcting mental models of hazardous processes, part II. Risk Anal 14(5): 789–798

Breakwell GM (2010) Models of risk construction: some applications to climate change. Wiley Interdiscip Rev Clim Change 1(6):857–870

Brewsher D, Grech P, Yeo S (2013) Hawkesbury's flood risk management plan: 15 years in the making. In: Proceedings of the floodplain management association national conference, 28–31 May 2013, Tweed Heads, NSW

Brown P, Nelson R, Jacobs B, Kokic P, Tracey J, Ahmed M, DeVoil P (2010) Enabling natural resource managers to self-assess their adaptive capacity. Agr Syst 103(8):562–568

Bunker D, Levine L, Woody C (2015) Repertoires of collaboration for common operating pictures of disasters and extreme events. Inf Syst Front 17(1):51–65

Bürgelt PT, Paton D (2014) An ecological risk management and capacity building model. Hum Ecol 42(4):591–603

Cottrell A, King D (2008) Understanding communities' needs for information and education. Risk Wise 2008:41–45

Dowdy A (2015) East Coast cluster report, Climate Change in Australia Projections for Australia's Natural Resource Management Regions. In: Ekström M et al (eds) Cluster reports. CSIRO and Bureau of Meteorology, Australia

Easterling D, Meehl G, Parmesan C, Changnon S, Karl T, O'Mearns L (2000) Climate extremes: observations, modeling, and impacts. Science 289(5487):2068–2074

Gillespie C, Grech P (2002) Reconciling development with flood risk: the Hawkesbury-Nepean dilemma. Aust J Emerg Manage 17(1):27–32

Godschalk DR (2003) Urban hazard mitigation: creating resilient cities. Nat Hazard Rev 4(3): 136–143

Griskevicius V, Cantú SM, Vugt MV (2012) The evolutionary bases for sustainable behavior: implications for marketing, policy, and social entrepreneurship. J Public Policy Market 31(1): 115–128

Hammill K, Tasker L (2010) Vegetation, fire and climate change in the greater Blue Mountains world heritage area. Department of Environment, Climate Change and Water, Sydney

Helsloot I, Ruitenberg A (2004) Citizen response to disasters: a survey of literature and some practical implications. J Conting Crisis Manage 12(3):98–111

Higginbotham N, Connor L, Baker F (2014) Subregional differences in Australian climate risk perceptions: coastal versus agricultural areas of the Hunter Valley, NSW. Reg Environ Change 14(2):699–712

Johnson CL, Tunstall SM, Penning-Rowsell EC (2005) Floods as catalysts for policy change: historical lessons from England and Wales. Water Resour Dev 21(4):561–575

Lachlan KA, Spence PR (2007) Hazard and outrage: developing a psychometric instrument in the aftermath of Katrina. J Appl Commun Res 35(1):109–123

Leiserowitz A (2006) Climate change risk perception and policy preferences: the role of affect, imagery, and values. Clim Change 77(1-2):45–72

Lewis-Beck MS, Bryman A, Liao TF (2013) Encyclopaedia of social science research methods: quantitative research. Sage, Thousand Oaks

Luck GW, Black R, Race D (eds) (2010) Demographic change in Australia's rural landscapes: implications for society and the environment, vol 12. CSIRO, Collingwood

Lujala P, Lein H, Rød JK (2015) Climate change, natural hazards, and risk perception: the role of proximity and personal experience. Local Environ 20(4):489–509

Major C, Savin-Baden M (2011) Integration of qualitative evidence: towards construction of academic knowledge on social science and professional fields. Qual Res 11(6):645–663

Marshall NA, Park S, Howden SM, Dowd AB, Jakku ES (2013) Climate change awareness is associated with enhanced adaptive capacity. Agr Syst 117(2013):30–34

Mason D, Knowd I (2010) The emergence of urban agriculture: Sydney, Australia. Int J Agric Sustain 8(1&2):62–71

Mayhorn CB, McLaughlin AC (2014) Warning the world of extreme events: a global perspective on risk communication for natural and technological disaster. Safety Sci 61(2014):43–50

McGee TK, Russell S (2003) "It's just a natural way of life..." an investigation of wildfire preparedness in rural Australia. Glob Environ Change B Environ Hazards 5(1):1–12

Millennium Ecosystem Assessment (MEA) (2005) Ecosystems and human well-being. Island Press, Washington, DC

Moser SC, Ekstrom JA (2010) A framework to diagnose barriers to climate change adaptation. Proc Natl Acad Sci 107(51):22026–22031

NSW Government (2015) https://www.emergency.nsw.gov.au/publications/guides-factsheets-brochures/emergency-management-arrangements/key-elements-of-em.html. Accessed 2 June 2015

Office of Environment and Heritage (OEH) (2015) http://climatechange.environment.nsw.gov.au/. Accessed 28 May 2015

Padowski JC, Gorelick SM (2014) Global analysis of urban surface water supply vulnerability. Environ Res Lett 9(10):1–6

Paton D, McClure J (2013) Preparing for disaster: building household and community capacity. Charles C. Thomas, Springfield

Paton D, Kelly G, Bürgelt P, Doherty M (2006) Preparing for bushfires: understanding intentions. Disaster Prev Manage Int J 15(4):566–575

Pilgrim SE, Cullen LC, Smith DJ, Pretty J (2008) Ecological knowledge is lost in wealthier communities and countries. Environ Sci Tech 42(4):1004–1009

Pretty JN (2002) Agriculture: reconnecting people, land, and nature. Routledge, NY

Pretty J (2003) Social capital and the collective management of resources. Science 302(5652): 1912–1914

Reid K, Beilin R (2014) Where's the Fire? Co-constructing bushfire in the everyday landscape. Soc Nat Resour 27(2):140–154

Richardson K, Steffen W, Liverman D (eds) (2013) Climate change: global risks, challenges and decisions. Cambridge University Press, Cambridge

Sandelowski M, Voils C, Barroso J (2007) Comparability work and the management of differences in synthesis studies. Soc Sci Med 64(1):236–247

Scolobig A, De Marchi B, Borga M (2012) The missing link between flood risk awareness and preparedness: findings from case studies in an Alpine region. Nat Hazards 63(2):499–520

Stevenson M, Lee H (2001) Indicators of sustainability as a tool in agricultural development: partitioning scientific and participatory processes. Int J Sustain Dev World Ecol 8(1):57–65

van der Linden S (2015) The social-psychological determinants of climate change risk perceptions: towards a comprehensive model. J Environ Psychol 41(2015):112–124

Wachinger G, Renn O, Begg C, Kuhlicke C (2013) The risk perception paradox—implications for governance and communication of natural hazards. Risk Anal 33(6):1049–1065

Weber EU (2010) What shapes perceptions of climate change? Wiley Interdiscip Rev Clim Change 1(3):332–342

Chapter 12
Integrating Microfinance, Climate Finance and Climate Change Adaptation: A Sub-Saharan Africa Perspective

Dumisani Chirambo

Abstract Climate change is arguably one of the World's gravest environmental and developmental challenge that has subsequently necessitated a re-contextualisation of many seemingly unrelated subjects such as poverty reduction, economic development, green growth, sustainability, equity and justice, trade, technology, investments and finance, and innovation.

In Africa, managing climate change is hampered by Africa's adaptation deficit as caused by a lack of institutional, financial or technological capacity to adapt effectively; and a lack of effective delivery mechanisms to channel climate finance resources at the sub-national level, particularly to target the poor who are also often the most vulnerable to the impacts of climate change.

In-order to encourage debate and discussion on the role to which microfinance may have in improving climate change mitigation and adaptation, this paper expounds upon the Microfinance-Climate Finance Framework that was shortlisted for the 2014 UNDP MDG Carbon Climate Finance Innovation Award. The paper shows that microfinance institutions in Africa may be sustainable mechanisms for financing climate change initiatives whilst promoting rural development and financial inclusion. Additionally, successfully adapting to climate change requires policy makers to focus on empowering the youth to transform them into (social) entrepreneurs capable of reducing social marginalisation and youth unemployment.

Keywords Clean Development Mechanism (CDM) • Entrepreneurship • Financial inclusion • Food security

---

D. Chirambo (✉)
Department of Civil and Public Law with references to Law of Europe and the Environment, Brandenburg University of Technology Cottbus–Senftenberg, Erich-Weinert-Str. 1, LG 10/334, 03046 Cottbus, Germany
e-mail: sofopportunity@gmail.com; chiradum@tu-cottbus.de

W.L. Filho et al. (eds.), *Climate Change Adaptation, Resilience and Hazards*, Climate Change Management, DOI 10.1007/978-3-319-39880-8_12

Introduction

Poverty is a contributing factor to environmental degradation and a loss of natural resources in Africa (Lesolle 2012) hence can be a contributing factor to the underdevelopment that is present in most African countries. In Sub-Saharan Africa (SSA) the average income per capita in real terms is currently lower than it was at the end of the 1960s, and life expectancy is lower now than 30 years ago as incomes, assets, and access to essential services are unequally distributed (World Bank 2013). Additionally, the greatest food security challenges overall remain in SSA, which has seen particularly slow progress in improving access to food, with sluggish income growth, high poverty rates and poor infrastructure, which hampers physical and distributional access (FAO, IFAD and WFP 2014a). The poverty of SSA has many dimensions and causes (both internal and external) which include bad luck, initial conditions, and an unfavourable international economic environment which are a manifestation of and an outcome of poor policy choice and bad governance (Luiz 2006). Climate change is anticipated to make efforts to attain the Millennium Development Goals (MDGs) challenging and/or reverse the achievements already made towards attaining the MDGs (GoK 2010). Consequently, there is therefore an urgent need for the international community and African governments to put in place effective poverty and development policies, and climate change policies that can simultaneously address poverty and climate change issues. Additionally, the proposed Sustainable Development Goals are urging stakeholders to reduce the vulnerability of poor people to climate-related extreme events and other economic, social and environmental shocks and disasters as a means to end poverty (UNSD 2014).

Climate change policy debates have largely focused upon estimating adaptation costs, ways to raise and to scale-up funding for adaptation, and the design of the international institutional architecture for adaptation financing consequently ignoring discussion on actual delivery mechanisms to channel these resources at the sub-national level, particularly to target the poor who are also often the most vulnerable to the impacts of climate change (Agrawala and Carraro 2010; Rong 2010; Hyder 2008; Bodansky and Diringer 2014). Within such a context, arguably the impact to which most climate change programmes could have on influencing inclusive climate resilient development could be limited due to a lack of emphasis on strengthening the participation of marginalised groups. In-order to ensure that the post 2015 Development Agenda effectively addresses inclusive growth and climate change issues, various policymakers in Africa developed The Common African Position (CAP) as Africa's official voice on the post-2015 Development Agenda. According to the CAP, Africa's post 2015 development priorities focus on six pillars namely: (i) structural economic transformation and inclusive growth; (ii) science, technology and innovation; (iii) people-centred development; (iv) environmental sustainability, natural resources management, and disaster risk management; (v) peace and security; and (vi) finance and partnerships (UNECA 2014a). However, how these pillars will be transposed into national development objectives, models and strategies, and how these pillars will be compatible with wider aspirations of the international community are issues that require further investigation as previous efforts on development and climate change agreements point out that uncertainties in predicting actual climate change and its impacts due to problems in estimating future global emissions of greenhouse gases (Hyder 2008) leads to disagreements on the roles and commitments to which developed and developing nations have to manage climate change.

Timilsina et al. (2010) consider the involvement of private sector stakeholders into socio-economic projects in Africa to be contingent on the removal of several barriers, such as market failures, lack of infrastructure and institutional capacity, lack of financial resources, and foreign investors' perception that investment in SSA is risky. These issues are particularly crucial as there is now a shared understanding that tackling climate change will not be possible without major mobilisation or a 're-channelling' of private finance hence tackling climate change requires unprecedented private investment and a transformation of common business practices (UNEP Finance Initiative 2014). Noting that climate change and poverty are two aspects that can seriously hamper the development of Africa, this paper aims to contribute to the existing body of knowledge on various policy responses and strategies that can be deployed in-order to enhance funding for climate change mitigation and adaptation initiatives.

Microfinance is a development tool that is considered to have positive impacts on poverty, income, savings, expenditure, and the accumulation of assets, as well as non-financial outcomes including health, nutrition, food security, education, women's empowerment, housing, job creation, and social cohesion (Van Rooyen et al. 2012). Arguably microfinance may be an ideal tool that can address the proximate causes of Africa's vulnerability to climate change. Consequently this chapter aims to encourage debate and discussion on the role to which microfinance may have firstly in supporting climate change adaptation in the agricultural sector and secondly in energy sector based climate change mitigation activities. Sections "Climate Change Impacts on Food Security and Agricultural Development" and "Climate Change Mitigation and Energy Access Policy Considerations" examine: SSA's food security and energy development challenges and opportunities in the light of current African development aspirations. Section "A Microfinance-Climate Finance Framework for Inclusive Growth" expounds upon the Microfinance-Climate Finance Framework that was shortlisted for the 2014 UNDP MDG Carbon Climate Finance Innovation Award. This framework is presented as a business model that microfinance institutions can use to enhance their mobilisation and disbursement of funds for climate change and poverty reduction activities in various contexts. A discussion then follows in section "Discussion" to highlight the enabling environment which could assist microfinance to improve climate change adaptation measures in Africa. A conclusion then follows in section "Conclusions". The paper shows that unlike market based climate finance instruments, microfinance institutions can utilise various business models hence presenting sustainable mechanisms for financing climate change initiatives whilst promoting rural development and financial inclusion.

Climate Change Impacts on Food Security and Agricultural Development

Climate change creates a double inequality through the inverse distribution of risk and responsibility. Developed states are responsible for climate change, but are forecast to confront only moderate adverse effects; least developed states are not culpable and yet experience significant threats to livelihoods, assets and security (Barrett 2013). About 70 % of Africa's population and roughly 80 % of the continent's poor live in rural areas (more than for any other region) hence improved agricultural performance in Africa can increase rural incomes and purchasing power for the continent's largely poor majority, but only if there are efficient, sustainable and widely accessible rural financial systems which can achieve pro-poor growth and poverty reduction goals (Mwenda and Muuka 2004).

Agriculture can be considered as the principal foundation for Africa's growth which could lead to the continent's structural transformation if the agricultural sector acts as the hub and a conduit that will influence the continent's thirst for a performing, productive, resilient, entrepreneurial and climate smart agricultural sector (UNECA 2014b). It has been pointed out that capital intensive industries and fast-growing sectors (e.g., mining, construction, communication, etc.) neither absorb a majority of rural and urban job-seekers nor provide adequate arrangements that are purposefully set to link up with and benefit slow-growing sectors by way of markets for their produce/products. This results in capital intensive industries and fast-growing sectors having a limited potential to promote growth that culminates into reductions in poverty, rural–urban inequalities and gender inequalities (UNDP-Zambia 2011; GoT 2011). However, a focus on agriculture sector based growth can lead to an improved development path that can lead to poverty reduction and inclusive growth because of the high labour requirements and employment opportunities from some agricultural systems. Moreover, UNECA (2014b) consider agricultural development that enhances climate change resilience to be feasible where there are effective climate policies and a presence of both public and private institutions that can efficiently implement those policies, for example, through improving coordination and mainstreaming of climate change policies in national development plans; and developing a low carbon, clean energy development pathway to rapidly scale up clean energy technologies. This therefore points out that the agriculture sector does not only have opportunities for creating employment but also has a role to play in improving the resilience of communities to climate change. Since a majority of African farmers are smallholder farmers, exploiting these opportunities arguably depends on how the capacity of the smallholder farmers is developed for them to embrace new technologies and strategies which can facilitate climate resilient agriculture and ensure food security.

Some of the factors influencing climate change adaptation on the continent include wealth, and access to extension services, credit and climate information in Ethiopia; and wealth, government farm support, and access to fertile land and credit in South Africa. Increasing access to information, credit and markets, and making a particular effort to reach small-scale subsistence farmers are therefore essential components of effective climate change adaptation (Bryan et al. 2009). Fankhauser and McDermott (2014) also point out that inclusive (and low-carbon) growth policies should be promoted as they can increase per capita income which leads to reduction in the impacts of extreme weather events and also increases the demand for substitutes to adaptation such as insurance cover. Microfinance programmes aimed at supporting inclusive growth by aiming at increasing the income levels of various communities to enable then to improve their agricultural output and develop other off-farm income generating activities could therefore enhance food security and climate change resilience.

Climate Change Mitigation and Energy Access Policy Considerations

Access to modern energy is a precondition for human development, and it has been stated that no country in modern times has substantially reduced poverty without a sizable increase in energy services (IGES 2014). Africa is the region with the least electrification rates and least per capita use of energy. Africa's electricity consumption of 571 kWh per capita is about five times less than the world average and when Northern African countries and South Africa are excluded, installed electricity generation capacity in the rest of Africa is about 31 GW suggesting the per capita electricity consumption in the sub-region is much lower than the African average (Gujba et al. 2012). SSA (if South Africa is excluded) is the only world region in which per capita consumption of electricity is falling even though the region and continent has the capacity to overcome the energy deficiencies (Eberhard and Shkaratan 2012). Consequently, by 2030 the number of people reliant on biomass for cooking in the region is expected to increase by some 200 million people and those living without electricity by 150 million (Leopold 2014). Improving energy access rates is desirable because access to energy increases incomes, improves healthcare, education, and security, and reduces labour-intensive practices of all kinds. However, as it stands, the global energy system is the single largest contributor to climate change since the energy sector represents the largest share of global GHG emissions (41 %) and global energy demand is estimated to grow by 33 % (IGES 2014). Reducing energy consumption and GHG emissions from the energy sector is therefore of paramount global importance to avoid catastrophic climate change, but it also means that increasing access to modern energy services for the over 620 million Sub-Saharan Africans living without access to any modern energy services should be done in a sustainable way (Leopold 2014).

Climate finance mechanisms and instruments such as the Clean Development Mechanism (CDM) were developed with the aim of promoting sustainable development and facilitating technology transfers that could promote the deployment of renewable energy. However, where such mechanisms are used, their success in achieving both roles of mitigating climate change and reducing poverty has been limited or had no effect in terms of poverty alleviation and sustainable development due to a lack of emphasis on strengthening the participation of marginalised groups (Cutter and Piguet 2014). Amatayakul and Berndes (2012) and Timilsina et al. (2010) further assert that carbon credits are there to provide an incentive for various stakeholders to implement projects that can contribute to reducing greenhouse gases; but fall short of solving the financing problem for the projects on their own as carbon revenues could be very small compared to the total investment required in many projects. These issues therefore highlight how community level socio-economic, political and environmental conditions differ from place to place hence make such mechanisms not to be suitable for some countries. In countries that are not suitable for carbon financing modalities, other measures that can facilitate the mobilisation of resources for renewable energy deployment should therefore be explored.

A variety of financial instruments and tools are available for supporting climate actions in developing countries (e.g., grants, carbon credits, concessional loans, green bonds and equity investments). While the effectiveness and suitability of different financial instruments and tools depends on specific climate activities, technologies, project types and the maturity of markets, problems arise in ensuring that vulnerable people are reached or become beneficiaries of such programmes (De Brauw et al. 2014; Agrawala and Carraro 2010). Some suggestions on how various financial incentives and financing schemes such as carbon credits can encourage project developers to reach out to vulnerable groups and improve the deployment of renewable energy particularly in rural areas include incorporating greater flexibility and simplifying the methodological and documentation procedures for the various schemes (Thomas et al. 2010). However, this might not address the other barriers such as the potential volatility of carbon prices which create uncertainties and risks for rural energy projects (Zerriffi 2011) and the preference of project developers to produce the cheapest credits but not the best environmental outcome (Pearson 2007). This therefore highlights how other financing modalities such as microfinance could play a significant role in improving the deployment of renewable energy in rural areas since microfinance models for renewable energy deployment can be designed to reduce the costs of accessing renewable energy technologies for rural people.

In addition to this, African farmers have the potential to implement various interventions which could facilitate climate change mitigation and adaptation in their local communities. For example, some biofuel crops have both climate change mitigation and adaptation benefits. Biofuel can provide a renewable energy source that can mitigate the use of fossil fuels for energy. Biofuel crops can also promote rural development and provide rural livelihoods with adaptation benefits when used as a primary or secondary crop. In Mozambique, depending on the production technology used and institutional frameworks in place (Fig. 12.1), biofuel investments could increase annual economic growth by 0.6 % and reduce the incidence of poverty by about 6 % over a 12-year period (Mwakaje 2012).

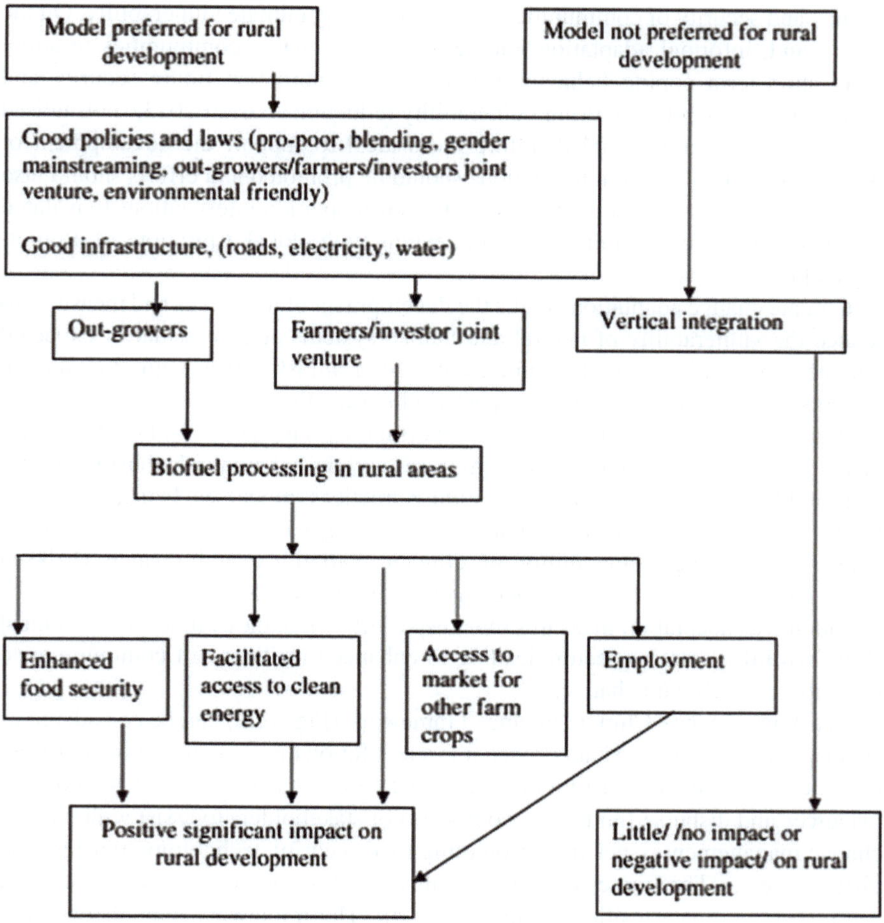

Fig. 12.1 Conceptual framework for biofuel plantations and rural development. *Source*: Mwakaje (2012)

A Microfinance-Climate Finance Framework for Inclusive Growth

Formalised adaptation includes the provision by external parties of additional funds and financing mechanisms to enable various stakeholders to undertake climate change related programmes, capacity-building and policy dialogue; while informal adaptations are on-going processes of human adaptation, occurring independently of external assistance, ranging from small adjustments in daily routines to significant changes in circumstance through particular disaster events (Birkmann et al. 2010). While both actions are shown to facilitate coping and adaptation to climate, however their effectiveness differs. Formalised adaptation interventions enable communities to address a greater number of climate risks; and enhance agency and security of communities thereby lessening climate vulnerability. On the other hand, informal adaptation interventions only enable communities to adopt more short-term coping behaviours that often compromise future security and agency and show little enduring vulnerability reduction (Barrett 2013). Formalised adaptation interventions are therefore more ideal to support the development of Africa and arguably such interventions should be promoted and efforts should also be made to improve the delivery of such formal adaptation interventions to isolated and marginalised communities to be in-keeping with global aspirations of promoting inclusive growth.

Climate change adaptation entails the development of initiatives and measures to reduce the vulnerability of natural and human systems against actual or expected climate change effects (de Oliveira 2009) and/or adjustment in natural or human systems in response to actual or expected climatic stimuli or their effects, which moderates harm or exploits beneficial opportunities (Glemarec 2011). Other commentators consider adaptation to encompasses various behavioural adjustments that households and institutions make (including practices, processes, legislation, regulations and incentives) to mandate or facilitate changes in socio-economic systems, aimed at reducing vulnerability to climatic variability and change (Eriksen et al. 2011). New products, services and business models can therefore also be considered as adaptation measures more especially when they influence a technical shift forward in various sectors leading to enhanced resilience of communities to the impacts of climate change.

The Microfinance-Climate Finance Framework (Fig. 12.2) can be considered as an adaptation to climate change since it is a revolutionalised or streamlined process to which microfinance institutions can use as their business model to mobilise resources and disburse funds to various kinds of stakeholders to assist with climate change management issues (i.e., promoting food security, technology transfers and diffusion, etc.). The framework can arguably also holistically address aspects that can promote climate resilient inclusive growth. The framework may also be considered as a new business model to enable microfinance institutions as well as entrepreneurs and businesses to explore socioeconomic opportunities that can come from enhancing climate risk management. Moreover, the sustainability of some microfinance institutions could be at risk from climate change induced stresses (i.e., where most of their customers are agriculture focused) (Agrawala and Carraro 2010) hence the framework highlights how microfinance institutions can mitigate these risks. Subsequently, the framework was shortlisted for the 2014 UNDP MDG Carbon/Mitsubishi UFJ Morgan Stanley Securities Co. Ltd Climate Finance Innovation Award.

The framework is based on the concept of a revolving loan fund that provides loans at concessional rates in-order to address the financial constraints that many organisations and individuals face when trying to implement inclusive growth focused initiatives. Revolving funds have been shown to be effective in easing credit constraints by reaching poorer communities, providing financial services to those households which rely on informal lending, and leveraging financial contributions from various sources (loan repayments from clients, loans from

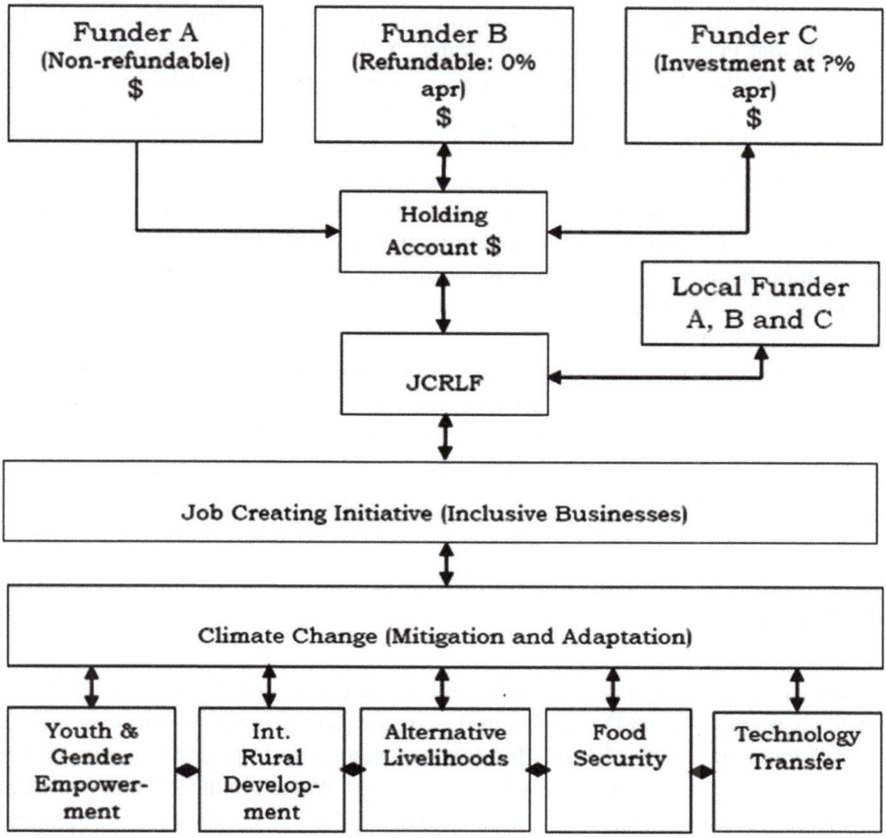

Fig. 12.2 Microfinance-Climate Finance Framework for inclusive growth. *Note*: JCRLF denotes a microfinance institution or revolving fund within a microfinance institution. *Source*: Author

development agencies, etc.) (Menkhoff and Rungruxsirivorn 2011). Revolving funds provide loans rather than grants thereby enabling microfinance institutions to become sustainable by recycling resources over and over again to deliver the 'holy trinity' of outreach, impact and sustainability (Kotir and Obeng-Odoom 2009). Funders in this framework include individuals, governments, multilateral and bilateral development banks, bilateral development cooperation agencies, the private sector, civil society, research and development institutions, and social investors. This framework therefore has a significant scope to tap into private finance more especially since public funding for environmental and socio-economic activities in many developing countries could be erratic.

Technological developments are making the sharing of information and transfer of money easier hence facilitating the growth of crowd-funding and peer-to-peer (P2P) lending/donating modalities. Technological developments are therefore enabling more individuals and institutions to be able to locally or internationally lend/donate money directly to other individuals and institutions. In this context, the microfinance institutions become mere intermediaries for the transfers or could also be principal beneficiaries of such funding. Additionally, there are many African migrants scattered globally who currently provide around $40 billion a year in remittances. These migrants have the potential to provide more than $100 billion a year to help develop Africa and there is also an estimated $50 billion in diaspora savings that could be leveraged for low-cost project finance (Arezki and Brückner 2012). The framework therefore assumes that these technological developments can facilitate the mobilisation of donations and investments from Africa's diaspora population hence further reinforcing an already robust and sustainable financing structure. More importantly, the framework can enable Africa's diaspora population to be able to receive commercial returns when they provide investments to microfinance institutions using such a framework. This can facilitate enhanced financial flows to Africa for socio-economic development through individuals and the private sector, and also enable private institutions and the civil society to be able to foster socio-economic development more especially where there are government institutional inadequacies.

Discussion

Due to the prevailing global uncertainties and fiscal consolidation in many developed countries, Official Development Assistance (ODA) should at best be seen as a complement and not a substitute for domestic resources, investment and trade. For example, tax revenues are already ten times larger than aid in the African continent and over the past decade, tax revenues have been rising across the developing world thereby making this a potential source for financing the post-2015 Development Agenda. Subsequently, Africa needs to improve domestic resource mobilisation by ensuring financial deepening and inclusion (e.g., domestic savings and micro-finance), and strengthening the tax structure, coverage and administration (UNECA 2014a). More effort should also go into developing innovative financing mechanisms and harnessing the private sector to conduct inclusive businesses (AfDB 2013). The financing mechanisms that can support Africa's development could focus on securitising and investing remittances; reducing remittance transfer costs and enhancing their effective management; and developing and strengthening long-term, non-traditional financing mechanisms (e.g., diaspora bonds) (AfDB 2013).

Global climate change mitigation and adaptation is incomplete in the absence of mechanisms for raising the capacity for effective climate change-related planning and management in least developed countries, including focusing on women, youth and local and marginalised communities (UNSD 2014). This follows that Africa's economic growth will not automatically reduce vulnerability to climate change unless growth policies incorporate investment in skills and access to finance to facilitate pro-poor inclusive growth, which will then reduce the vulnerability of communities to climate change (Bowen et al. 2012). Unemployment and particularly youth unemployment is already a significant challenge that can affect the social and economic development of the continent and poverty levels (UNECA 2012, 2013) hence needs to be addressed in-order to avoid incidences of crime, political violence, social backwardness and social unrest (Timilsina et al. 2010). Africa has the youngest population in the world whereby the continent has almost 200 million people aged between 15 and 24. If Africa's young population continues to grow rapidly, the number of young people in Africa will double by 2045 and the continent's labour force will be 1 billion by 2040, making it the largest in the world, surpassing both China and India (UNECA 2014a). Promoting private sector development and entrepreneurship among young people is consequently considered as part of the solution to address Africa's high unemployment and youth unemployment. Improving access to youth entrepreneurship focused finance, skills development, mentorships, social networks, and technology are considered as strategies that could enable young African entrepreneurs to drive economic growth and social progress in the years ahead (Brixiova´ et al. 2015). As elaborated earlier in this paper, the promotion of jobs in the agriculture and renewable sector have the potential to promote inclusive growth hence need to be integrated into wider aspirations for promoting entrepreneurship and reducing youth unemployment. However, the youth are noted to have limited access to capital and other assets, and they do not perceive agriculture employment (and living in rural areas) as attractive due to the slow modernization of the sector and dominance of traditional subsistence farming practices (FAO 2014b). Subsequently, it can be argued that there is a need to undertake studies on what policies and strategies the international community and African governments can pursue in-order to integrate the youth into local economies in-order to improve the resilience of rural communities.

Conclusions

Economic growth does not automatically lead to reductions in vulnerability to climate change. In SSA, average income per capita and life expectancy is lower now than in the past three–five decades, hence the vulnerability of many communities to climate change could be increasing. Without measures such as improving access to finance and climate information to marginalised communities, climate change will increase the rates of poverty and food insecurity in the sub-continent. Microfinance can potentially improve financing for both climate change mitigation and adaptation initiatives more especially if stakeholders create an enabling environment to utilise Africa's diaspora savings and remittances to stimulate climate resilient growth. Additionally, increases in Africa's population is leading to high youth unemployment rates, hence climate change management policies should also consider how to reduce the vulnerability of the youth to the impacts of climate change since youth unemployment is a proximate cause for poverty and environmental degradation.

References

AfDB (2013) At the centre of Africa's transformation: strategy for 2013–2022. African Development Bank Group, Abidjan

Agrawala S, Carraro M (2010). Economic aspects of adaptation to climate change assessing the role of microfinance in adaptation. Working party on global and structural policies, OECD Environment Policy Committee

Amatayakul W, Berndes (2012) Determining factor for the development of CDM biomass power projects. Energy Sustain Dev 16:197–203

Arezki R, Brückner M (2012) Rainfall, financial development, and remittances: evidence from Sub-Saharan Africa. J Int Econ 87:377–385

Barrett S (2013) Local level climate justice? Adaptation finance and vulnerability reduction. Glob Environ Chang 23:1819–1829

Birkmann J, Buckle P, Jaeger J, Pelling M, Setiadi N, Garschagen M, Fernando N, Kropp J (2010) Extreme events and disasters: a window of opportunity for change? Analysis of organizational, institutional and political changes, formal and informal responses after mega-disasters. Nat Hazards 55:637–655

Bodansky D, Diringer E (2014) Alternative models for the 2015 climate change agreement. FNI Climate Policy Perspectives 13. The Fridtjof Nansen Institute (FNI)

Bowen A, Cochrane S, Fankhauser S (2012) Climate change, adaptation and growth. Clim Change 113:95–106

Brixiova´ Z, Ncube M, Bicaba Z (2015) Skills and youth entrepreneurship in Africa: analysis with Evidence from Swaziland. World Dev 67:11–26

Bryan E, Deressa TT, Gbetibouo GA, Ringler C (2009) Adaptation to climate change in Ethiopia and South Africa: options and constraints. Environ Sci Policy 12:413–426

Cutter S, Piguet E (eds) (2014) Climate change 2014: impacts, adaptation, and vulnerability—chapter 13: livelihoods and poverty. In: Working group II contribution to the IPCC 5th assessment report—changes to the underlying scientific/technical assessment. Cambridge University Press, Cambridge

De Brauw A, Mueller V, Lee HL (2014) Role of rural–urban migration in the structural transformation of Sub-Saharan Africa. World Dev 63:33–42

de Oliveira JAP (2009) The implementation of climate change related policies at the subnational level: an analysis of three countries. Habitat Int 33:253–259

Eberhard A, Shkaratan M (2012) Powering Africa: meeting the financing and reform challenges. Energy Policy 42:9–18

Eriksen S, Aldunce P, Bahinipati CS, D'almeida R, Molefe JI, Nhemachena C, O'brien K, Olorunfemi F, Park J, Sygna L, Ulsrud K (2011) When not every response to climate change is a good one: identifying principles for sustainable adaptation. Climate Dev 3(1):7–20

Fankhauser S, McDermott TKJ (2014) Understanding the adaptation deficit: why are poor countries more vulnerable to climate events than rich countries? Glob Environ Chang 27:9–18

FAO (2014b) Case study series # 4—FAO, private and public partnership model for youth employment in agriculture (Experiences from Malawi, Tanzania Mainland and Zanzibar archipelago). Food and Agriculture Organization of the United Nations (FAO)

FAO, IFAD and WFP (2014) The state of food insecurity in the world 2014. Strengthening the enabling environment for food security and nutrition. Rome, FAO

Glemarec Y (2011) Catalysing climate finance: a guidebook on policy and financing options to support green, low-emission and climate-resilient development. United Nations Development Programme, New York, NY

GoK (2010) Kenya national climate change response strategy. Government of Kenya, Nairobi

GoT (Government of Tanzania) (2011) Tanzania: country report on the millennium development goals 2010. Government of Tanzania, Dar es Salaam

Gujba H, Thorne S, Mulugetta Y, Rai K, Sokona Y (2012) Financing low carbon energy access in Africa. Energy Policy 47:71–78

Hyder P (2008) Recycling revenue from an international carbon tax to fund an integrated investment programme in sustainable energy and poverty reduction. Glob Environ Chang 18:521–538

IGES (2014) Designing and implementing an energy goal: delivering multi-benefits for sustainable development. Policy brief number 30. Institute for Global Environmental Strategies

Kotir JH, Obeng-Odoom F (2009) Microfinance and rural household development: a Ghanaian perspective. J Dev Soc 25:85–105

Leopold A (2014) Making climate change mitigation more meaningful: the link to universal energy access. Practical Action Publishing, UK. doi:10.3362/9781780448732

Lesolle D (2012) SADC policy paper on climate change: assessing the policy options for SADC member states. SADC Secretariat, Gaborone

Luiz JM (2006) The wealth of some and the poverty of sub Saharan Africa. Int J Soc Econ 33(9): 625–648

Menkhoff L, Rungruxsirivorn O (2011) Do village funds improve access to finance? Evidence from Thailand. World Dev 39(1):110–122

Mwakaje AG (2012) Can Tanzania realise rural development through biofuel plantations? Insights from the study in Rufiji District. Energy Sustain Dev 16:320–327

Mwenda KK, Muuka GN (2004) Towards best practices for micro finance institutional engagement in African rural areas Selected cases and agenda for action. Int J Soc Econ 31(1/2): 143–158

Pearson B (2007) Market failure: why the clean development mechanism won't promote clean development. J Clean Prod 15:247–252

Rong F (2010) Understanding developing country stances on post-2012 climate change negotiations: comparative analysis of Brazil, China, India, Mexico, and South Africa. Energy Policy 38:4582–4591

Thomas S, Dargusch P, Harrison S, Herbohn J (2010) Why are there so few afforestation and reforestation clean development mechanism projects? Land Use Policy 27:880–887

Timilsina GR, de Gouvello C, Thioye M, Dayo FB (2010) Clean development mechanism potential and challenges in Sub-Saharan Africa. Mitig Adapt Strat Glob Chang 15:93–111

UNDP-Zambia (2011) Zambia: millennium development goals report (MDGR) for 2011

UNECA (2012) Assessing progress in Africa toward the millennium development goals: MDG report 2012—emerging perspectives from Africa on the post-2015 development agenda. Economic Commission for Africa, Addis Ababa

UNECA (2013) Assessing progress in Africa toward the millennium development goals MDG report 2013—food security in Africa: issues, challenges and lessons. Economic Commission for Africa, Addis Ababa

UNECA (2014a) Assessing progress in Africa toward the millennium development goals: MDG report 2014—analysis of the common African position on the post-2015 development agenda. United Nations Economic Commission for Africa, Addis Ababa

UNECA (2014b) Keeping climate impacts at bay: a 6-point strategy for climate-resilient economies in Africa. Economic Commission for Africa, Addis Ababa

UNEP Finance Initiative (2014) Demystifying private climate finance. UNEP, Nairobi

UNSD (2014) Open working group proposal for sustainable development goals. United Nations, New York

Van Rooyen C, Stewart R, De Wet T (2012) The impact of microfinance in Sub-Saharan Africa: a systematic review of the evidence. World Dev 40(11):2249–2262

World Bank (2013) Turn down the heat: climate extremes, regional impacts, and the case for resilience. A report for the World Bank by the Potsdam Institute for Climate Impact Research and Climate Analytics. World Bank, Washington, DC

Zerriffi H (2011) Innovative business models for the scale-up of energy access efforts for the poorest. Curr Opin Environ Sustain 3(4):272–278

Chapter 13
Climate Change Adaptation and Socio-Economic Resilience in Mexico's Grijalva-Usumacinta Watershed

Santiago Enríquez, Rodolfo Camacho, Michèle Olivier Laird, and David Wilk

Abstract The Grijalva-Usumacinta watershed in Mexico's Tabasco and Chiapas states is home to six million people and a rich biodiversity. It is also the major source of Mexico's hydropower, as well as in-land and coastal hydrocarbons. However, the area's close proximity to the Gulf of Mexico makes it highly vulnerable to climate change effects: rising sea levels, destructive hurricanes, heightened rainfall and floods. These climate change impacts could be devastating, particularly for the 31 % of the population that live in extreme poverty and face food insecurity.

This paper presents an inter-disciplinary assessment of future climate change scenarios and associated impacts in the region, particularly for vulnerable populations living in rural areas. It focuses on the role of institutions in mediating interactions between climate change impacts and livelihoods, as well as in shaping adaptation responses. The assessment used can be broadly applied in comparable settings in developing states and emerging economies, with increasing climate change risks and threats. In so doing, it provides a reliable methodology that can be used to assess regional vulnerability and design climate change adaptation initiatives in rural areas.

The paper draws on experiences gained from the preparation of the "Plan de Adaptación, Ordenamiento y Manejo Integral de las Cuencas de los Ríos Grijalva y Usumacinta", project ATN/OC-12432—ME, funded by the Inter-American Development Bank and developed by Abt Associates, Inc. The views expressed in the paper are those of the authors and do not represent those of the Inter-American Development Bank or Abt Associates, Inc. For further information see: http://blogs.iadb.org/cambioclimatico/2014/05/08/adaptacion-ordenamiento-y-manejo-integral-el-caso-del-sur-de-mexico/.

S. Enríquez (✉) • R. Camacho • M. Olivier Laird
International Economic Growth Division, Abt Associates, Inc., 4550 Montgomery Ave, Suite 800 North, Bethesda, MD 20814, USA
e-mail: Santiago_Enriquez@abtassoc.com

D. Wilk
Climate Change and Sustainability Division, Inter-American Development Bank, 1300 New York Avenue, N.W., Washington, DC 20577, USA

W.L. Filho et al. (eds.), *Climate Change Adaptation, Resilience and Hazards*, Climate Change Management, DOI 10.1007/978-3-319-39880-8_13

Keywords Climate change • Latin America • Mexico • Adaptation • Watershed management

Introduction

The Grijalva and Usumacinta river watershed is a complex hydrological system expanding over 13.2 million hectares in the Mexico-Guatemala border. This paper focuses on the Mexican portion, which comprises 64 % of the system.[1] These rivers are the source of 34 % of the Mexico's runoff and are mainly located in the Mexican states of Chiapas and Tabasco (CONAGUA 2011).

According to the Mexican government, the Grijalva-Usumacinta watershed is highly vulnerable to climate change because it is frequently affected by extreme weather events and is at risk from sea level rise (Nicholls et al. 2007; INE-SEMARNAT 2006). During the first decade of the twenty-first century, 34 events associated with extreme precipitations affected the states of Tabasco and Chiapas, impacting close to 3 million people and resulting in losses of close to US 5 billion (Abt Associates 2013). The 2007 flood of the Grijalva River affected 75 % of Tabasco's population and resulted in damages equivalent to almost 30 % of the state's GDP. Based on available data, Mexican agencies have indicated that the watershed's hydrological behavior is characterized by increasingly frequent extreme trends (SEMARNAT 2009).

Recognizing the region's vulnerability to climate variability and change, the Government of Mexico requested the assistance of the Inter-American Development Bank (IDB) to develop a long-term adaptation plan for the Grijalva-Usumacinta watershed. The plan was to integrate the development of water infrastructure with ecosystem management and sustainable land use, in order to maximize the hydrological regulation services provided by ecosystems, as means to reduce the exposure of people and infrastructure to severe hydrometeorological phenomena. In response to this request, the IDB supported the development of the Adaptation, Land Use, and Integrated Watershed Management Plan for the Grijalva and Usumacinta Watersheds ("the Plan").

This paper focuses on the components of the Plan that assessed the vulnerability of rural populations in the watershed. In particular, it draws on the analytical work underpinning the Plan to discuss the role of institutions in shaping climate change risks and in facilitating interventions to support adaptation and enhancing social resilience to climate change. This paper will be relevant to individuals and organizations interested in inter-disciplinary approaches to climate change adaptation at a regional level, as well as to those working on the preparation of climate change adaptation projects in watersheds and rural areas.

[1]All the data, analysis and other information presented in this paper refer only to the Mexican portion of the watershed of the Grijalva and Usumacinta rivers.

The rest of this paper is structured as follows: Section "Literature Review" summarizes the literature review. Section "Methodological Approach" describes the methodology used to develop the Plan, particularly those aspects that were relevant to assess the vulnerability of rural populations to climate change and the adaptation actions that could be adopted. Section "Results" presents the Plan's main results and we conclude in section "Conclusions".

Literature Review

Füssel and Klein (2006) note that climate change vulnerability assessments have evolved from approaches that basically superimposed climate change events on a passive system to newer approaches that recognize the ability of people and systems to adapt to climate change. The adaptive capacity of such people and systems is stronger when they are able to adapt by developing strategies that are robust against uncertain future developments and integrating them into policies.

However, the capacity of social groups to develop adaptation strategies is significantly shaped by institutions, which mediate the interactions between climate risks and social groups (Agrawal 2008; Tyler and Moench 2012). For this reason, research on climate change adaptation has increasingly paid attention to the role of institutions in reducing vulnerability. Part of the research has found that, in many cases, institutions exclude vulnerable groups from decision-making processes. Marginalization caused by institutional factors is an underlying cause of vulnerability and also limits the participation of vulnerable groups in the development of adaptive actions (Adager 2005). These findings are consistent with those of the broader body of social science research that has assessed the role of institutions and governance in development (Acemoglu et al. 2001; Rodrik et al. 2004; Ostrom 2005; Slunge and Loazya 2012).

Other researchers have stressed the important role of institutions in promoting adaptation in the face of uncertain climate change impacts. According to this approach, climate change impacts cannot be accurately predicted because of the limited use of historical data and available climate change projections. Thus, institutions that support adaptive management by incorporating learning and promoting good governance are more likely to prepare communities and other groups for climate change's dynamic and complex impacts (Armitage et al. 2007). Instead of focusing on specific perceived climate hazards, institutions can help to build resilience to unpredictable stresses and shocks (Walker et al. 2002).

The potential contributions of institutions to climate change adaptation will vary from one context to another. According to Agrawal (2008), in rural areas, institutions mediate between climate hazards and livelihoods in three ways. First, they structure environmental risks and variability, thereby influencing the nature of climate impacts and vulnerability. Second, they create the incentive structure that defines the adaptation strategies that individuals and groups can adopt. Finally, they shape the extent to which external interventions can contribute or undermine local adaptation practices.

Building on Agrawal's framework, this paper presents the findings on the institutional factors that limit climate change adaptation rural areas in the Grijalva-Usumacinta watershed, as well as on the opportunities to reform them so they can contribute to enhance resilience to climate change.

Methodological Approach

The methodological approach that was adopted for the preparation of the Plan can be described as consisting of three main steps. First, we characterized the watershed, with the aim of understanding priority development challenges. As a second step, we estimated the likely climate change impacts on the watershed and how they would affect vulnerable populations and sectors, including the extent to which these impacts would exacerbate development challenges. Finally, we identified potential interventions to enhance the resilience of vulnerable populations, focusing on the areas that could yield short term benefits while also contributing to building longer term adaptation capacity. In this section, we discuss only the methodological aspects that are relevant to understand the role of institutions in climate change adaptation in rural areas of the watershed.

In order to characterize the watershed, it was sub-divided into six regions, defined by hydrographic units consisting on the main primary basins. This sub-division helped to identify predominantly rural areas and to tailor the vulnerability assessment to them. For each of the watershed's sub-regions, data from government and academic sources was used to elaborate a bio-physical characterization, including the composition of the natural environment, the hydric dynamic, predominant land uses and vegetation cover (CentroGeo 2010). A socio-economic characterization was also developed, focusing on factors associated with vulnerability to climate change, particularly poverty; inequality; access to basic services, resources and information; quality of housing and infrastructure; and livelihoods (Cutter et al. 2003; Brooks 2003; Wisner et al. 2004).

The approach to understand likely climate change effects was based on the downscaling of global climate change scenarios to the watershed level. The resulting regional climate change scenarios projected future temperature and precipitation trends. These scenarios were developed based on data from 15 Global Circulation Models, combined in a weighted ensemble using the Reliability Ensemble Averaging (REA) method developed by Giorgi and Mearns (2002) and implemented in Mexico by Montero and Pérez (2008), which estimates the uncertainty of each model. The scenarios were fed with historical data for 1961–2000 and the Representative Concentrations Pathways (RCPs) 4.5, 6.0, and 8.5 from the Intergovernmental Panel on Climate Change's (IPCC) Assessment Report number 5 (AR5) for the near future (2015–2039) and the distant future (2075–2099).

Econometric analysis was used to estimate the impacts of climate change on key agricultural activities. The analysis focused on the linkages between the production functions of these goods and climatic factors (Seo et al. 2008; Gay et al. 2006). The analysis was modeled as an optimization problem showing how the producer maximizes yields through a combination of labor and inputs selected by him, for a determined level of temperature and precipitation, and soil characteristics (Olivera-Villarroel 2012). To run the model, we used geographic variables (coordinates and soil quality); economic variables (labor, capital, and agricultural inputs), and climate variables (historical and projected precipitation and temperature). Data on agricultural yields and socio-economic indicators were obtained from the official agricultural production database (SIAP) of Mexico's Ministry of Agriculture, Livestock, Rural Development, Fisheries, and Food (SAGARPA). Hydrological data were obtained from the National Meteorological System's hydrometeorological stations, and the regional climate change scenarios described above provided projected temperature and precipitation data.

The econometric analysis was conducted for corn and coffee. Corn was selected because it is the staple food and its production is the main economic activity in rural areas in the region. Thus, impacts on corn yields have implications on both livelihood and food security. Beans are the second most widely cultivate crop and a key ingredient of the regional diet, but lack of sufficient data at the municipal level precluded an analysis on the impacts of climate change on this crop.

Shade-grown coffee is an economically important product, particularly in the Chiapas' portion of the watershed. It was selected as part of the analysis because coffee growing is one of the more widely adopted economic diversification activities in the watershed. Farmers can grow coffee while continuing with their other agricultural activities and sell it to complement their income. Shade-grown coffee has the additional advantage of being an agricultural activity that is compatible with forest conservation (Moguel and Toledo 2004). Thus, reductions in coffee yields as a result of climate change would seem to threaten both local livelihoods and forest areas, which would face a higher probability of being converted into agricultural uses.

Based on the results of the impact analysis, we assessed a first set of adaptation activities that are based on current practices in the region and that would aim to compensate the falls in crop yields associated with climate change. However, some of adaptive strategies we initially considered proved limited for natural and institutional reasons, as discussed below. Consequently, we assessed adaptive strategies that included improvements in knowledge, technologies or investments that could help to adapt to the changing climate. An institutional component was integrated into each of these interventions. This component aimed to assess whether proposed adaptive actions were feasible in the context of existing intuitions, as well as to propose institutional reforms that would support better governance and learning. Cost–benefit analyses helped to assess the feasibility of these interventions.

A multi-level stakeholder engagement process was launched from the beginning of the Plan's preparation. More than 200 people participated in the process, through three workshops with representatives from federal agencies, three additional workshops with representatives of the state governments of Chiapas and Tabasco, two workshops with research centers in the region, and a workshop with Civil Society Organizations and international donors. More than two dozen in-depth interviews were conducted with experts from different fields with significant experience in the watersheds. The participatory process bolstered the preparation of the Plan by providing guidance in three specific moments: (1) at inception, in defining the scope of the analytical work to be conducted; (2) at the completion of the diagnostic assessment that identified priority climate change threats in key areas; and (3) towards the end of the preparation of the Plan, to validate proposed interventions and ensure their alignment with ongoing efforts in the Grijalva and Usumacinta watersheds.

Results

Watershed Characterization

The Grijalva-Usumacinta watershed is different from the rest of Mexico in many ways, including its higher economic dependence on natural resources. Extraction and use of natural resources contribute with close to 70 % of Tabasco's Gross Domestic Product (GDP), compared with 10 % at the national level. Most of this wealth stems from the energy sector: 17.4 % of the country's oil and 19.6 % of its natural gas are produced in the lower basin (PEMEX 2014), and over 40 % of Mexico's hydropower is generated by the Grijalva River (CFE 2012). Benefits from these activities are captured mainly by the state-owned productive enterprises PEMEX, and the Federal Electricity Commission. However, there are no institutional mechanisms to share the extraction of this wealth with local communities. For example, in spite of hosting the most important hydropower complex in the country, Chiapas is one of the three states with lowest electricity coverage in Mexico (INEGI 2010). Local communities have benefited from specific investments in the past. For example, the development of dams has resulted in construction jobs and other benefits for neighboring communities. However, these temporary benefits are the result of ad hoc negotiations.

The watershed is also more rural than the rest of Mexico. More than 50 % of Chiapas's population and 43 % of Tabasco's lives in settlements of <2500 inhabitants, compared with a national average of 23 % (INEGI 2010). As a result of this dispersion, local populations tend to have less access to basic services and infrastructure. Natural resources sustain the livelihoods of the population living in those localities. In Chiapas, the primary sector occupies 36 % of the economically active population, compared with 13 % at the national level (INEGI 2016).

The economic structure and geographic dispersion of communities in Chiapas and Tabasco is associated with high poverty and marginalization. According to official data, the watershed is home to more than six million people, out of which

31 % live in extreme poverty and 32 % face food insecurity (CONEVAL 2010). Out of the 116 municipalities in the watershed, 66 % are considered highly or very highly marginalized by Mexico's government (CONAPO 2010).

Agriculture and livestock constitute the main livelihoods of rural communities. Corn is by far the most important crop in the watershed, occupying 52 % of agricultural lands; beans are a distant second, covering <10 % of agricultural lands. Practically all the corn produced by small rural communities is rain-fed and used for self-consumption (CentroGeo 2012). In addition to its cultural and nutritional value, the large area devoted to corn growing is associated with governmental subsidies. Primary data collection in three rural communities in Chiapas (Nuevo San Juan, Tierra Nueva, and Veinte Casas, in the municipality of Ocozocuautla) found that subsidies provide about 47 % of the income that would be generated from the sale of the corn produced in each hectare[2]. In fact, if households considered the economic value of their own labor, corn growing would not be economically viable in most cases in the absence of the subsidy (Olivera-Villarroel 2011).

Regional Climate Change Scenarios

The developed regional scenarios indicate that climate change is expected to result in higher mean temperatures of 1 °C in the near future (2015–2039) and, under the RCP 6 scenario, up to 3.2 °C in the distant future (2075–2099).[3] Both maximum and minimum temperatures are projected to increase in the near and distant future. Temperature increases are expected throughout the year, with the highest increases for average and minimum temperature between March and May, and the highest maximum temperature between June and August.

Precipitation in the watersheds is likely to fall as a result of climate change. In the near future, under the RCP 6 scenario, rainfall decreases would be relatively small, between 0.04 and 0.4 %, while distant future modeling showed reductions of between 1.9 and 2.9 %. An analysis of the probability distribution functions of precipitation by season showed that seasonal rainfall increases and decreases would become more significant over time, resulting in higher probability of extreme events in the distant future, including both heavy rainfall and droughts.

[2]Estimate includes two different subsidies: "PROCAMPO" and "Maíz Criollo".

[3]In this paper, we only discuss the results of the RCP 6 scenario. The results of the other scenarios can be consulted in the reports available at http://blogs.iadb.org/cambioclimatico/2014/05/08/adaptacion-ordenamiento-y-manejo-integral-el-caso-del-sur-de-mexico

Climate Change Impacts on Agricultural Yields

The analysis found that, under the RCP 6 scenario, climate change is estimated to reduce average corn productivity by up to 2.9 % in the near future, resulting in more than 13,000 families facing food insecurity. In the distant future, average productivity would fall by 5.8 %, threatening more than 28,000 families (Figs. 13.1 and 13.2).

In the case of coffee, temperature is the variable that has a more significant effect on productivity. Low temperatures are needed for the plant to flower, while extreme high temperatures might stress the plant (Granados Solís and Zamora Castro 2012). Based on the regional climate change scenarios, coffee yields are expected to fall by around 2.4 % in the near future, and 7.3 % in the distant future (Figs. 13.3 and 13.4). The impacts might be even more severe, as coffee production might not be economically viable under such conditions.

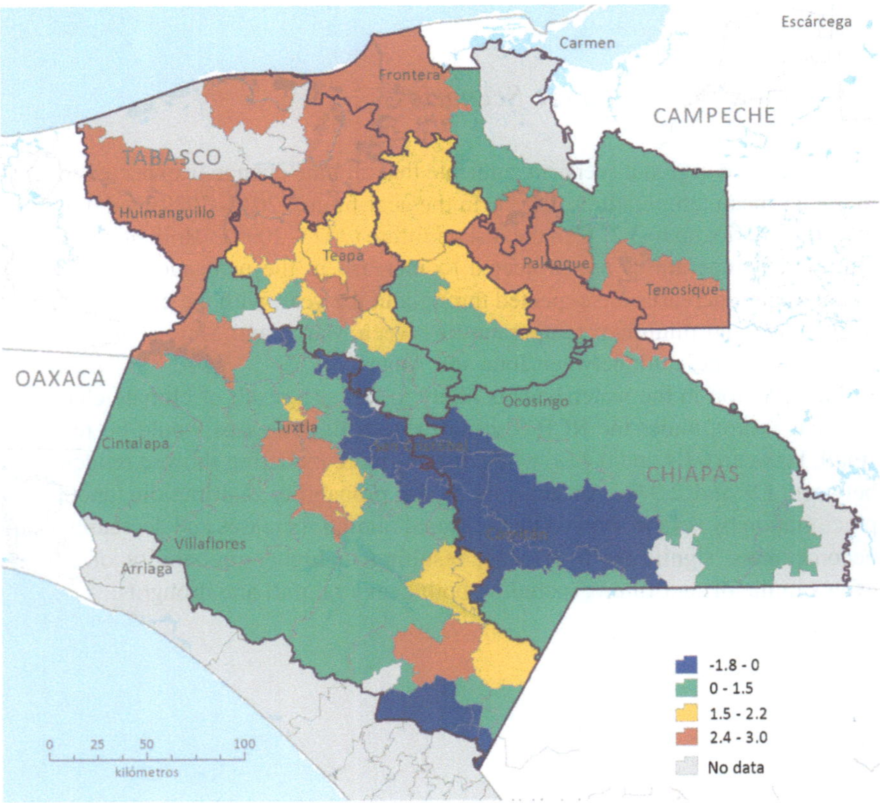

Fig. 13.1 Reductions in corn productivity (%), RCP 6 scenario, 2015–2039. *Source*: Abt Associates (2013)

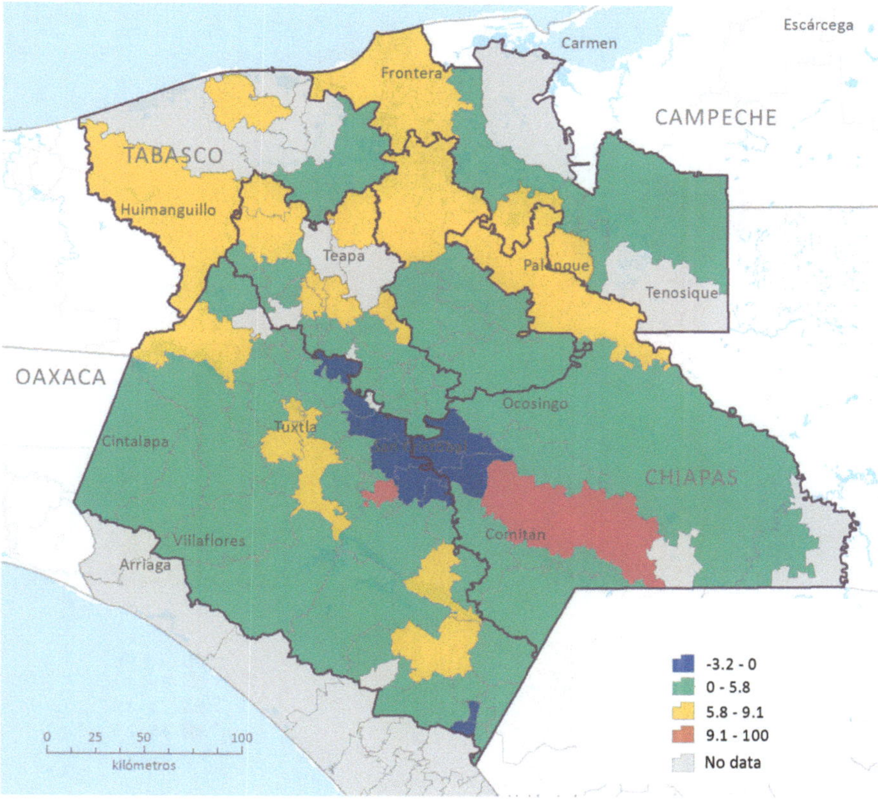

Fig. 13.2 Reductions in corn productivity (%), RCP 6 scenario, 2075–2099. *Source*: Abt Associates (2013)

Adaptation Options

A first type of adaptation response to the loss of corn productivity caused by climate change could consist of increasing inputs, labor, and other economic factors. However, given that corn is only economically viable because of government subsidies, most households would not be able to afford the additional inputs under current circumstances. Storage could be considered another adaptation alternative. However, the vast majority of rural farmers rarely have a production that is significantly above what they require to meet their self-consumption needs. Although market exchange is arguably one of the most effective adaptation strategies, it is constrained in the watershed's rural areas by institutional factors. These include: (1) the existence of monopolies and oligopolies that create gaps between seasonal sale and purchase prices; (2) lack of markets for other goods produced

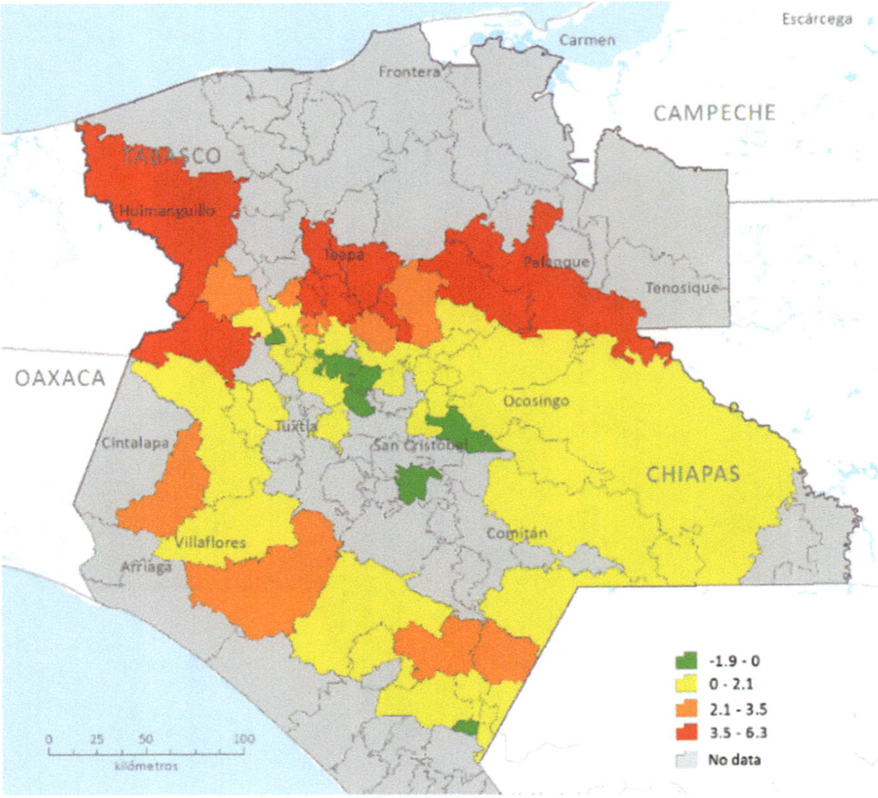

Fig. 13.3 Reductions in coffee productivity (%), RCP 6 scenario, 2015–2039. *Source*: Abt Associates (2013)

jointly with traditional corn; and (3) high transaction costs for participating in food markets (Olivera-Villarroel 2011).

In the case of coffee, mobility would seem the most likely adaptation action. Coffee plantations could be moved to higher altitude areas with colder temperatures. This alternative is constrained by geographic and institutional factors. In terms of geography, the region has only limited areas where such conditions are found. Institutionally, those areas have already been designated as national parks or are somebody else's property.

A different set of adaptation options would consist of developing the capacities of local communities to adopt climate resilient systems that integrate productive mosaics of forest, agriculture and livestock. The first of such options would consist of integrating corn and bean crops with fruit trees, such as peach and citrus. Experiences from the Mexican state of Oaxaca have shown that this type of activity can increase corn yields from between 0.5–3.9 and 21.2–24.9 t/ha because the trees provide organic matter and better nutrient recycling, and also reduce erosion. In Oaxaca, this intervention has also helped to increase and diversify households'

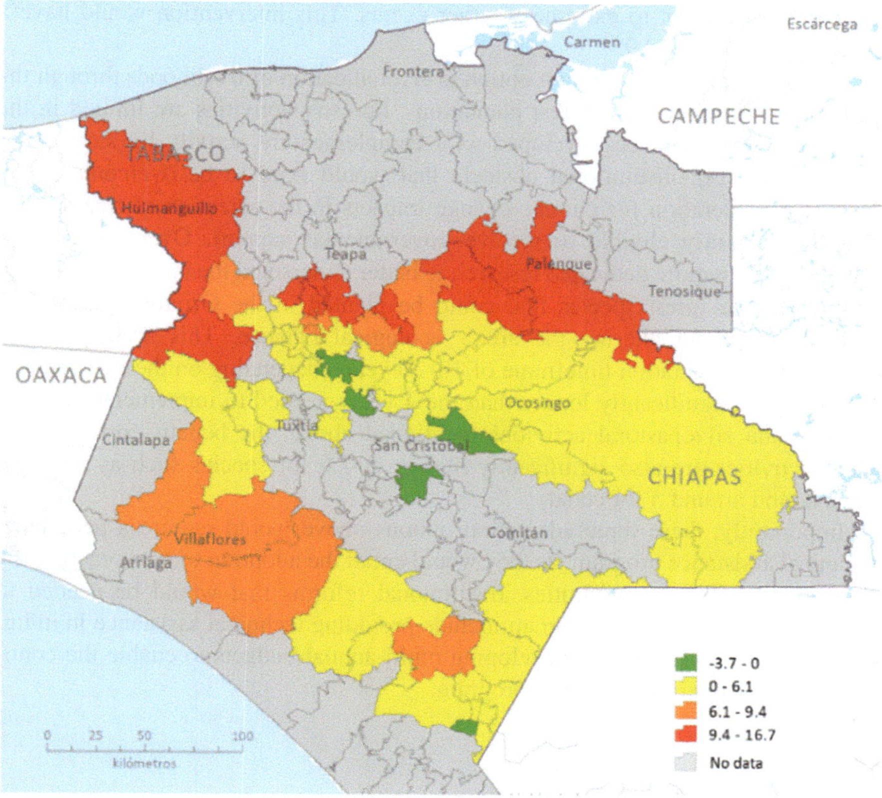

Fig. 13.4 Reductions in coffee productivity (%), RCP 6 scenario, 2075–2099. *Source*: Abt Associates (2013)

income, as well as to generate additional goods for self-consumption, such as fruits, wood and medicinal products (COLPOS 2008).

However, in order for this adaptation option to be feasible in the Grijalva-Usumacinta watershed, government programs would need to reduce their current focus on the intensive production of monocultures. In addition, technical guidelines would need to be developed, and continuously adjusted, to identify the tree species that could better adapt to climate change. We find that this program would have a benefit–cost ratio of 1.3.

A second adaptation option would be the promotion of silvopastoral systems that enable improved soil management and the diversification of livestock's feed. This intervention has been successfully implemented in Central and South America, where producers have been able to increase their incomes in US 70–1157 dollars/ha (Pagiola et al. 2009; Murgueitio 2009; Muhammad et al. 2009).

The main institutional barrier for the implementation of this intervention is the absence of programs in which silvicultural and agricultural activities can be integrated. The inclusion of fruit and fodder trees could also be used as a strategy to help livestock adapt to extreme weather events. This intervention would have a benefit–cost ratio of 2.9.

Another available adaptation option is diversification of livelihoods through the promotion of commercial forest plantations. Forestry activities are limited in the watershed and could be developed as a complementary activity by small rural communities. An institutional obstacle that would need to be overcome is the lack of consideration for climate change impacts in the selection of tree species that are currently eligible to receive governmental support. Overcoming such barrier would require dedicated research to better understand the impacts of climate change on the forest species that could be planted in the Grijalva-Usumacinta watershed without altering the area's ecological balance. This activity would need an implementation timeframe of 10–20 years, depending on the tree species. While this is significantly longer than the 4 years needed to implement the agro-forestry and silvopastoral activities mentioned above, the benefit–cost ratio for these activities are also significantly higher, at 7.6 for species such as teak and melina, and around 5 for cedar.

Importantly, these three adaptation options above would require a permanent technical assistance program, as they would entail the adoption of new practices by largely indigenous communities. Institutional reforms that would be needed to ensure the success of this program include providing technical assistance in indigenous languages, as well as developing multi-annual budgets to enable the continuity of the technical assistance program.

Conclusions

The findings of the analytical work underscore the Grijalva and Usumacinta watershed's vulnerability to climate change and advance a compelling argument to urgently initiate climate change adaptation actions targeting clearly defined geographic and sectoral climate change adaptation priorities. The case for supporting rural communities to adapt is particularly compelling.

Institutional factors have played an important role in shaping environmental risks and climate hazards in the Grijalva-Usumacinta watershed. The current institutional framework has favored the extraction of natural resources, particularly energy resources, without helping to translate natural resources wealth into other types of capital and sustained economic growth. As a result, an important portion of the region's population depends on low productivity agriculture to meet its needs. This dependence on primary activities, coupled with poverty and marginalization, results in high sensibility to climate change. As the results discussed in this paper show, projected climate change risks could result in more than 13,000 families facing food insecurity in the near future, and more than 28,000 families in the distant future.

Institutional factors also have an important role in enabling adaptation activities. As an example, the institutional factors that have constrained the development and integration of food markets in the watershed's rural areas have reduced the potential role of market exchanges as an adaptive strategy for rural households. This situation has also created incentives for farmers to focus almost exclusively on producing corn for self-consumption, which in turn limits the potential of storage as an adaptation activity. Coffee growing has worked as an economic diversification strategy, but it is highly vulnerable to climate change.

In this context, more effective adaptation activities will be those that can simultaneously help to improve agricultural yields and diversify households' incomes. The integration of forest, agricultural, and livestock productive mosaics has produced positive results in other parts of Mexico and other countries in the Latin American region. With dedicated research, these interventions could be continuously improved by incorporating scientific findings and community experiences. However, institutional reforms would be needed to enable such adaptation opportunities. In particular, technical assistance programs will be needed to create continuous learning opportunities and enhance the capacities of local communities to develop adaptation options.

References

Abt Associates (2013) Plan de Adaptación, Ordenamiento y Manejo Integral de las Cuencas de los Ríos Grijalva y Usumacinta. Programa de adaptación a las consecuencias de cambio climático en la provisión de servicios de la cuenca del Grijalva. Report commissioned by the Inter-American Development Bank, Washington, DC

Acemoglu D, Johnson S, Robinson JA (2001) The colonial origins of comparative development: an empirical investigation. Am Econ Rev 91(5):1369–1401

Adager WN (2005) Vulnerability. Glob Environ Chang 16:268–281

Agrawal A (2008) The role of local institutions in adaptation to climate change. Paper prepared for the social dimensions of climate change. Social Development Department, The World Bank, Washington, DC

Armitage D, Berkes F, Doubleday N (2007) Introduction: moving beyond co-management. In: Armitage D, Berkes F, Doubleday N (eds) Adaptive co-management: collaboration, learning and multi-level governance. UBC Press, Vancouver

Brooks N (2003) Vulnerability, risk and adaptation: a conceptual framework. Working Paper 38, Tyndall Centre for Climate Change Research

CentroGeo (2010) La Región del Usumacinta en México desde la perspectiva de la gestión del agua. En: Propuesta para una gestión territorial integrada de la cuenca Usumacinta en México. Report commissioned by the USAID Mexico Competitiveness Program, Mexico City, Mexico

CentroGeo (2012) Insumos para la Formulación del Plan de Adaptación, Ordenamiento y Manejo Integral de las Cuencas de los Ríos Grijalva y Usumacinta. Report commissioned by Abt Associates, Mexico City, Mexico

Colegio de Posgraduados (COLPOS) (2008) Investigación Agronómica y Transferencia de Tecnología en la Fase de Escalamiento del Proyecto Manejo Sustentable de Laderas en el Estado de Oaxaca Informe Anual 2007. Colegio de Posgraduados, Texcoco

Comisión Federal de Electricidad (CFE) (2012) Programa de Obras e Inversiones del Sector Eléctrico (POISE) 2012–2026. México, DF

Comisión Nacional del Agua (CONAGUA) (2011) Agenda del Agua, 2030. Programa Hídrico por Organismo de Cuenca, Visión 2030. Frontera Sur, México

Consejo Nacional de Evaluación de la Política de Desarrollo Social (CONEVAL) (2010) Informe de Pobreza en México. El país, los estados y sus municipios 2010, México, DF
Consejo Nacional de Población (CONAPO) (2010) Indicadores. http://www.conapo.gob.mx/en/CONAPO/2010
Cutter S, Boruff BJ, Shirley WL (2003) Social vulnerability to environmental hazards. Soc Sci Q 84(2):242–261
Füssel HM, Klein RJT (2006) Climate change vulnerability assessments: an evolution of conceptual thinking. Clim Change 75(3):301–329
Gay C, Estrada F et al (2006) Potential impacts of climate change on agriculture: a case of study of coffee production in Veracruz, Mexico. Climate Change 79(3):259–288
Giorgi F, Mearns LO (2002) Calculation of average, uncertain range and reliability of regional climate changes from AOGCM simulations via reliability ensemble averaging (REA) method. J Climate 15(10):1141–1158
Granados Solís A, Zamora Castro MV (2012) Impactos del Cambio Climático y la Gestación del Modelo Alternativo de Economía Verde en América Central. Fundación Friedrich Ebert, Costa Rica
INEGI (2010) Censo de Población y Vivienda 2010, Principales resultados por localidad (ITER). Instituto Nacional de Estadística y Geografía, Aguascalientes
INEGI (2016) Encuesta Nacional de Ocupación y Empleo. Instituto Nacional de Estadística y Geografía, Aguascalientes
Instituto Nacional de Ecología (INE)—Secretaría de Medio Ambiente y Recursos Naturales (SEMARNAT) (2006) México, Tercera Comunicación Nacional ante la Convención Marco de las Naciones Unidas sobre el Cambio Climático. Instituto Nacional de Ecología, Secretaría de Medio Ambiente y Recursos Naturales, Programa de las Naciones Unidas para el Desarrollo México, Environmental Protection Agency, Global Environment Facility, México DF
Moguel P, Toledo VM (2004) Conservar Produciendo: Biodiversidad, Café Orgánico y Jardines Productivos. Biodiversitas 55:1–7
Montero MJ, Pérez JL (2008) Regionalización de proyecciones climáticas en México de precipitación y temperatura en superficie usando el método REA para el siglo XXI. In: Martínez P, Aguilar A (eds) Efectos del cambio climático en los recursos hídricos de México. Instituto Mexicano de Tecnología del Agua, Jiutepec, pp 11–21
Muhammad I, Velásquez-Vélez R, Pezo D, Skarpe C, Mora J, Benjamín T (2009) Selectividad animal de forrajes herbáceos y leñosos en pasturas seminaturales en Muy Muy, Nicaragua. Agroforestería en las Américas 47:51–60
Murgueitio E (2009) Incentivos para los Sistemas Silvopastoriles en América Latina. Avances en Investigación Agropecuaria 13(1):3–19
Nicholls RJ, Wong PP, Burkett VR et al (2007) Coastal systems and low-lying areas. In: Canziani OF, Parry ML, Palutikof JP, van der Linden PJ, Hanson CE (eds) Climate change 2007: impacts, adaptation and vulnerability. Cambridge University Press, Cambridge
Olivera-Villarroel SM (2011) Aplicando programas de reducción de emisiones por Deforestación y Degradación: Costos de oportunidad en actividades productivas el Ocote, Chiapas. Consultoría desarrollada como parte de la propuesta Formulación de un proyecto piloto REDD (Reducción de Emisiones por Deforestación y Degradación Forestal) usando el Sistema Plan Vivo en la Reserva de la Biosfera Selva El Ocote, Fase 3. Cooperativa Ambio, Chiapas
Olivera-Villarroel SM (2012) La productividad del maíz de temporal en México: repercusiones del cambio climático. EUROCLIMA—CEPAL, Chile
Ostrom E (2005) Understanding institutional diversity. Princeton University Press, New Jersey
Pagiola S, Ríos AR, Arcenas A (2009) Poor household participation in payments for environmental services: lessons from the silvopastoral project in Quindío, Colombia. MPRA Paper No. 4794, World Bank, Washington, DC
Petro´leos Mexicanos (PEMEX) (2014) Anuario estad´ıstico de PEMEX. Petróleos Mexicanos, Mexico City

Rodrik D, Subramanian A, Trebbi F (2004) Institutions rule: the primacy of institutions over geography and integration in economic development. J Econ Growth 9(2):131–165

Secretaría de Medio Ambiente y Recursos Naturales (SEMARNAT) (2009) Programa Especial de Cambio Climático 2009–2012. Diario Oficial de la Federación, Viernes 28 de agosto de 2009, México, DF

Seo S, Niggol Dinar A, Kurukulasuriya Pradeep A (2008) Ricardian analysis of the distribution of climate change impacts on agriculture across agro-ecological zones in Africa. World Bank Policy Research Working Paper No. 4599, Washington, DC

Slunge D, Loazya F (2012) Greening growth through strategic environmental assessment of sector reforms. Public Adm Dev 32(3):245–261

Tyler S, Moench M (2012) A framework for urban climate resilience. Clim Dev 4(4):311–326

Walker B, Carpenter S, Anderies JM, Abel N, Cumming GS, Janssen M, Pritchard R (2002) Resilience management in social-ecological systems: a working hypothesis for a participatory approach. Ecol Soc 6(1), Art. 14

Wisner B, Blaikie P, Cannon T, Davis I (2004) At risk: natural hazards, people's vulnerability and disasters, 2nd edn. Routledge, London

Chapter 14
Adaptive Strategies Building Resilience to Climate Variability in Argentina, Canada and Colombia

Paula Mussetta, Sandra Turbay, and Amber J. Fletcher

Abstract Many regions of the world are experiencing the impacts of climate change, which include the increasing variability of weather as well as increased drought and flood. Although many areas have had a long history of this variability and have a strong historic practice of adaptation, increasing variability has had a significant impact on adaptive strategies of agricultural producers over the last several years.

Drawing on comparative vulnerability studies of agricultural producers in dry-land river basins in Argentina, Canada, and Colombia, this paper presents an analysis of the adaptive strategies employed by agricultural producers in responding to climate change impacts and an analysis of how these adaptive strategies have built resilience and improved producers' living conditions. Common exposures and sensitivities, linked with increasing variability, isolate important new adaptive strategies. Particular attention to the connection of these strategies to social, economic and institutional capital and the inter-relationship of these provides important insight for future adaptation. These research findings will be useful for governments, policymakers, and organizations assisting with adaptation.

This paper presents results of the research program Vulnerability to Climate Extremes in the Americas (VACEA) financed by the International Research Initiative on Adaptation to Climate Change [International Development Research Centre (IDRC), Canadian Institutes of Health Research, Natural Sciences and Engineering Research Council of Canada and social Sciences and Humanities Research Council of Canada].

P. Mussetta (✉)
Researcher at the Human, Social and Environmental Sciences Institute (INCIHUSA), Scientific and Technological National Research Institute (CONICET), Av Ruiz Leal s/n P., San Martín 5500, Mendoza, Ciudad, Argentina
e-mail: pmussetta@mendoza-conicet.gob.ar

S. Turbay
Department of Anthropology, University of Antioquia, Calle 34 A N. Ap. 203, 77-46 Medellín, Colombia

A.J. Fletcher
Department of Sociology and Social Studies, University of Regina, 3737 Wascana Parkway, S4S 0A2 Regina, SK, Canada

W.L. Filho et al. (eds.), *Climate Change Adaptation, Resilience and Hazards*, Climate Change Management, DOI 10.1007/978-3-319-39880-8_14

Keywords Adaptive strategies • Agricultural producers • Resilience

Introduction

This work addresses farmers' adaptations to climate variability in three regions of Argentina, Canada and Colombia. The purpose is to inform institutions and communities in planning a sustainable adaptation agenda. In drought- or flood-prone regions, adaptation to these natural phenomena is not a new issue. However, changing environmental conditions engender new challenges, not only to social actors inhabiting the territories but also to institutions that manage natural resources. Climate variability (understood as climate fluctuations above or below a long-term average value) involves problems of water quantity and quality to meet human, productive and ecological demands in arid and semi-arid ecosystems, as well as increasing flood probability for many regions. The occurrence of climate extremes, such as drought and flooding, are expected to increase as anthropogenic change affects natural climatic variability (IPCC 2015). Climate extremes that seriously affect crops undermine the conditions and possibilities in which producers can become more resilient to weather contingencies and to varying adversities.

Adaptation has been defined as adjustments made to human-environmental systems in response to expected or experienced climate impacts (Smit et al. 2000). Effective adaptation can reduce the vulnerability of a system (such as a household, community, or institution, for example) to climate extremes. Although mitigation of future climate change is important, the irreversibility—and thus inevitability—of existing anthropogenic climate change means that adaptation to future climate extremes is necessary (IPCC 2015).

The dominant approach to climate change adaptation has emphasized "technology, institutions and managerial systems" as key to successful adaptation (Eriksen et al. 2011). However, technology and infrastructure are inherently social, and more recent analysis has acknowledged the importance of social, economic, and political factors in determining adaptive capacity and adaptive strategies (Adger 2003; Below et al. 2012; Field et al. 2012). Adaptive strategies largely depend on economic and technological resources, information, knowledge, skills, social capital and institutional capital (Wisner et al. 2003; IPCC 2007; UNEP 2013). However, these strategies, whether institutional or autonomously initiated by farmers, are not always effective enough to become sustainable adaptations that increase the resilience of the stakeholders. Sometimes, this is due to "misuse" or inefficient use of resources. Other times, farmers' adaptive capacity is limited by social, political, or economic factors and resource constraints. "The underlying social, economic, institutional, and cultural conditions that contribute to a wider context for vulnerability" must be understood (Eriksen et al. 2011, p. 11).

The conditions that affect adaptation are context-specific (IPCC 2015). A 2006 review of the literature on adaptation (Smit and Wandel 2006) found a shortage of context-specific studies investigating adaptation in particular regions or communities, particularly studies where key vulnerabilities and adaptive strategies are identified through research with the communities themselves and not *a priori* (Smit and Wandel 2006). Understanding of context-specific adaptation needs is important, as are the local and traditional knowledge systems that also determine adaptation in a given location (IPCC 2015).

Adaptation is not a linear process where social systems simply adjust to natural stimuli. It is a struggle for access and control of material and social goods, and participation in political processes that determine the management, access, use, and distribution of natural goods. In other words, adaptation has also a sociopolitical dimension (Dietz 2013). Therefore, the possibility of developing adaptations that foster resilience should address the transformation of the root causes that make social systems vulnerable. Many of these are not even directly associated with climatic or environmental issues. From this framework, the aim of this paper is to assess farmers' adaptations and strategies for coping with climate events as well as with other relevant non-climatic stressors in three case studies.

The Wide Spectrum of Adaptations

Despite the widely recognized importance of adaptation, not all adaptation strategies are equally effective or beneficial. Moreover, what is often not questioned is the purpose of adaptations or their consequences. As not all adaptations lead to the same results, it is necessary to take into account the possible impacts of each. For example, many are focused strategies (such as the incorporation of technology, placement of hail nets, modification of irrigation shifts) that simply seek to adjust to the new conditions exclusively through cushioning the *impact* of climatic events. Some adaptive strategies may actually prove detrimental to adaptive capacity over the long term; for example, some strategies may increase social differentiation or environmental degradation. *Climate risk management* based on scientific information is a broader strategy than the impact focus. Climate risk management builds a standpoint that enables informed decision-making. In this way, it provides long-term solutions (Lozoya et al. 2011). The effect of climate risk management approaches on resilience will not be automatic; instead, it will depend on how that information is "translated" and distributed among producers.

A step towards *building resilience* and *sustainable adaptation* is to recognize that adaptive responses should go beyond climate and aim at troubleshooting and addressing systemic problems (Bassett and Fogelman 2013). Examples of such resilience-building efforts include the improvement of intersectoral coordination (Mearns and Norton 2010) or the enhancement of producers' capacities (Agrawal and Perrin 2009) through livelihood diversification or educational and health promotion. These practices are not exclusively related to environmental change policies, and fall within the spectrum of development and poverty reduction programs. Thus, adaptation to climate variability is an indirect effect of these other kinds of policies. Given the diversity of strategies and their impact on actors' resilience, the role of institutions that manage water, climate and production also becomes fundamental for adaptation planning.

Methodology

The analysis of adaptation discussed here is part of a broader international research project called "Vulnerability and Adaptation to Climate Extremes in the Americas" (VACEA). The project developed a model for assessing vulnerability to extreme events in rural contexts from an integrative and interdisciplinary approach.[1] The interdisciplinary project brought together natural and social scientists to examine vulnerability and adaptation to past and future climate events.

Our analysis here is based on participatory community vulnerability assessments (CVAs). The CVAs involved semi-structured interviews conducted with rural producers in the three case study areas; our analysis is, therefore, based on qualitative data. In Argentina, the sample consisted of 41 farmers and irrigation water managers along the districts of the basin and 25 institutional representatives from water, environmental and production institutional areas at national, provincial and local levels. The Canadian study included 170 participants: 100 were farmers, ranchers, and rural residents, and 70 were governance representatives working in the agricultural industry in the study area. In Colombia, 20 coffee farmers were interviewed in Chinchiná and Villamaría municipalities as well as 14 officials from institutions identified as key for governance of climate variability at the regional level.

Informants were selected through theoretical, non-probabilistic sampling based on their theoretical relevance to the research question and considering key variables for each study area: location of the farm in the basin, type of productive activity, farm size, use of labour force, etc.

The interviews were guided by a set of open topics that inquired about the different dimensions of vulnerability: exposures (climatic and non-climatic) that producers face; adaptive capacities for coping with exposures, and their perceptions of the future. Their responses highlighted the importance of underlying social, political, and economic factors in determining adaptive capacity to extreme events. Interviews were conducted until theoretical saturation was reached, i.e., until no

[1]The VACEA project (see footnote #1) involves three sets of interrelated activities: (1) Evaluation of past, present and future vulnerabilities related to climate, agriculture and natural resources using climatological scenarios and future climate projections for the areas; (2) Analysis of the vulnerabilities of communities using a combination of in-depth interviews—which inquired about risk exposures, sensitivities and adaptation strategies of farmers—and secondary socioeconomic data. This activity also included a study of the role of institutions in adaptability. (3) Finally, bringing together the insights produced by the first two sets of activities in order to assess future vulnerabilities based on how the current vulnerabilities will be affected by the expected future conditions. The analysis of adaptive capacity in this chapter corresponds to the second set of activities within the larger research project (For further information about the conceptual framework of VACEA project, see Harry Díaz in this volume).

new information was being produced (Glasser and Strauss 1967). Results from each case study were analyzed with NVivo 10 software for qualitative data.

Findings

Variability and Adaptations in Mendoza

Mendoza is an Argentinian province located in the west-central part of the country, near the Andes mountain range. As it is a semiarid region (it is part of the South American arid diagonal), agriculture and most human activities are only possible by the intentional handling of water coming from rivers originating in the Andes as a result of snowmelt and glaciers. Thus, the provincial territory is fragmented between the irrigated oasis (fed by a dense network of surface irrigation and where agribusiness has concentrated wine emblematic enclaves and all their material and symbolic representations) and drylands, that is, the part of the territory that has been excluded from irrigation system, so-called "deserts".

Hail, frost and the decrease in available water for irrigation are the main events of climate variability in Mendoza. The strategies producers deploy to adapt to these events mainly correspond to their economic capacity. To prevent the impact of frost, producers apply fairly precarious techniques such as moistening the soil and crops and setting fire to raise the temperature. Hail damage is absorbed with expensive protective nets partially subsidized by the state. In addition to these focused methods some producers obtain insurance; however, such insurance is insufficient because it fails to cover the complete loss. When producers receive their share they have already failed to recover profitability. However, insurance allows producers to *survive* from one cycle to another.

Only some large producers develop their own information systems to plan adaptations. In most cases, public institutions are the ones that develop preventive plans through the use of specific information. With the exception of these institutional strategies from the State, large producers are the ones who can be more resilient to hail and frost because they have easier access to investment, credits and allowances demanded by these adaptations.

Although producers in Mendoza have historically employed these mechanisms to adapt to these events, they have become increasingly more insufficient and ineffective. Hail and frost indirectly threaten the living conditions of families who fail to gather enough capital to meet production losses cycle after cycle. The inability to adapt is a bottleneck to the resilience of these actors.

Climate variability also affects *water availability*. In a basin irrigated by water coming down the mountain, the snow decrease seriously affects available flows downstream. This produces a direct impact on the quality and quantity of production. In a similar way to hail and frost, the range of adaptations to the decrease in water for irrigation is mediated by access to resources. In addition, not all producers get the same results, not only with respect to obtaining water but with respect to the situation of vulnerability and resilience of the producers.

Through specialized information, the water authority anticipates the annual availability of water. Based on those forecasts, it establishes how the quantity of allocated water will change. Based on this information at the beginning of a cycle, the producers would have the possibility of planning their practices and adapt to the announced volumes. The effect of this reduction of allocated water on farmers' livelihoods is different, since not all the producers depend, in the same way, on surface water. This is because access to underground water is the most important adaptation to diminishing flows. It can mean the difference between a good harvest and a bad one, so that those who have access to this alternative source are much more resilient than those who do not. This adaptation not only depends on the ownership of a license for using ground water, but on the economic resources to operate the well since the construction, maintenance and operation costs are very high.

Other adaptations to water shortage are the construction of private water reservoirs and irrigation technology. These two strategies make producers more resilient because by these mechanisms they are untied from the "variability of nature". Those farmers without the economic resources to develop these practices rely on precarious practices of water management on the farms. Although not as crucial as resource availability, individual aspects are also important. While some small producers receive drought relatively passively, non-technological practices do demand dedication and expertise. This initiative of change is relevant to increase resilience. Predisposition to adapt is also the basis of another important strategy for small producers. Agreements to share wells create possibilities that these producers could not engage in individually. Even though these practices are not abundant, they produce an outcome beyond the impact focus and build capacities to creatively address lack of water. One of the main limiting factors is the lack of confidence based on bad experiences. At the same time the State reveals an ambivalent attitude towards non-technological adaptive practices. On the one hand it promotes these practices giving financial aid and priority to these collective orders above the individual ones. But on the other, it restricts this access because in practice it does not facilitate the procedures.

Adaptations and Resilience Beyond the Climate

Associative organizations for common use of water, technology and machinery contribute to the improvement of producers' adaptive capacity. But cooperative networks in Mendoza are also a successful adaptation in another sense, not directly related to climate variability but through strengthening networks that ensure the commercialization of crops. Cooperative networks function together as an integrated, collective actor that is able to interact with the State and stands as a consolidated and valid interlocutor in the sector. However, on the other side, the system for purchasing commodities has acquired the modality of an oligopoly in which only a few companies[2] decide the market value in a manner that prevents farmers from obtaining favorable prices. Despite the inability to obtain desirable prices, the cooperative networks make a difference with respect to small producers of fruit and vegetables who otherwise would not have secured the ability to market their products.

Up to 60 % of farmers in Mendoza rely on off-farm income (CNA 2002*). Income diversification, creating alternatives to the agricultural economic activities (commercial, community, tourist services, etc.), is considered by mainstream adaptation literature as one of the adaptations that most contributes to building resilience. In the case of Mendoza, this correlation is not linear since these new revenues that producers get, usually come from precarious and not well-paid jobs. Therefore, it is probably an indicator of reduced climate exposures, but not necessarily an indicator of better life quality or less vulnerability (Montaña* 2008*). Whether diversification builds or not, resilience requires an assessment of the type of work these diversified incomes come from.*

While some producers are forced to diversify their incomes due to the failure of investment that the market demands, others opt for changing activity drastically. This means their defeat as producers. The sale of land for the real estate business is one of the most drastic options for producers in response to low profitability of their production. In contrast, income diversification acquires another meaning for large producers, for whom it is associated with the investment in new economic areas. This places large producers in a better position to cope with the consequences of climate change.

Variability and Adaptations in the Canadian Prairies

The South Saskatchewan River Basin (SSRB) runs for 998 km across the Canadian provinces of Alberta and Saskatchewan (PSRB 2009). Large portions of the basin are located within Palliser's Triangle, a region known for severe and prolonged drought. Although 30 % of the basin is classified as arid (Wandel et al. 2010), the SSRB is an important site for agricultural production in Canada. Farmers and ranchers in the area produce cattle, grain, pulses, and vegetable crops. The Canadian study included four communities across the SSRB: the communities of Taber and Pincher Creek in the Oldman River Watershed (ORW) of Alberta, and the communities of Rush Lake and Shaunavon in the Swift Current Creek Watershed (SCCW) of Saskatchewan.

Farmers and ranchers in the SSRB are generally well adapted to drought and dry conditions. Strong adaptive practices have resulted from generations of knowledge sharing between farmers, as well as from government investment in agriculture.

[2]Most small and medium farmers sell their grapes to four large corporations that produce common wines: Peñaflor, Baggio, FE.CO.VI.TA and Catena.

Recent years have brought new adaptive practices, such as the use of drought-resistant crop varieties and zero-till agriculture; however, recent years have also brought new political and economic challenges for producers in the region.

The majority of farmers in the SSRB have embraced minimum-till or zero-till practices to reduce soil erosion. These practices have been driven, in part, by memories of the 1930s, when a combination of land-degrading cultivation practices and severe drought caused suffering and starvation in the prairie region (Marchildon et al. 2008; McLeman et al. 2013). As a response to the historic drought, the federal government created the Prairie Farm Rehabilitation Administration (PFRA) to facilitate sustainable development in the prairies. Over its almost 80-year lifespan the PFRA implemented a number of programs to reduce erosion and manage natural prairie grasslands. The PFRA also established drought adaptation infrastructure, including a number of irrigation projects across the driest part of the prairies.

Although the majority of agricultural producers in the SSRB still lack access to irrigation, the communities of Taber and Rush Lake are an exception. Some Rush Lake producers have access to a limited number of flood-irrigated acres, which are used mostly for hay. The community of Taber stands out as a key site for irrigation in the basin. High-tech irrigation infrastructure allows production of high-value but water dependent crops such as sugar beets, corn, and potatoes. Several large facilities are located in the Taber area to process these products. Although irrigation increases adaptive capacity in general, producers of water-intensive crops face the threat of a severe or protracted drought that depletes irrigation reservoirs; this poses a risk not only for producers but also for those employed in the local processing sector.

Producers also engage in farm-level adaptation. For ranchers, a major challenge is ensuring a sufficient supply of hay, a drought-sensitive crop used to feed cattle. Many ranchers prepare for dry years by stockpiling hay. Crop producers make decisions based on the drought tolerance of crops; the introduction of new pulse crops to the region (e.g., lentils) has added new rotational possibilities to the historically dominant drought-resistant crops like wheat and barley.

Adaptations and Resilience Beyond the Climate

Despite the number of adaptive practices employed in the region, producers' adaptive capacity is affected by underlying economic, social, and political problems. Financial issues, especially the high cost of inputs, were the most commonly mentioned stressor for the producers in the study. Farmers now pay more for patented and certified seed varieties that promise high yields and increased resistance to drought and excess precipitation. However, these adaptive strategies may paradoxically increase vulnerability: a crop produced with more expensive seed is a more expensive crop lost to a climate disaster. Minimum and zero-till technology saves time and money by reducing cultivation; however, it also causes farmers to rely heavily on specialized equipment and chemical herbicides for weed control. This can increase input costs and agro-chemical use.

A key adaptive response to financial challenges, for many farmers, has been to increase the size of the farm. Over the past decade, the overall number of prairie farms has dropped dramatically (Statistics Canada 2011, 2006), and many farms have grown larger in a quest for financial sustainability and others leave the industry. Many study participants discussed the social changes this phenomenon has caused. Rural communities in the SSRB are now faced with declining populations, loss of services, and the associated impacts on social capital.

Similar to the case of Mendoza, farmers in the Canadian prairies rely increasingly on off-farm work to supplement farm income and to stabilize risk (Alasia and Bollman 2009; Jetté-Nantel et al. 2011). Approximately half of all farm men and women in Canada hold off-farm employment (Statistics Canada 2011). While off-farm work may increase adaptive capacity by providing a reliable source of income, this work can also add stress and additional hours to farmers' already lengthy workday. Although livelihood diversification is a commonly promoted adaptive strategy for agricultural producers, it is important to examine how such diversification can affect farmers' quality of life in both positive and negative ways.

Agricultural producers also face challenges related to policy and institutional programming. Over the past two decades a number of farm support policies and programs have been eliminated. In 2009, the federal government announced plans to divest of its PFRA irrigation projects. This development has been a significant source of stress for irrigators in the Rush Lake area, who now face the cost of privately operating the irrigation system. Community pastures—a significant source of protection for native prairie grasses—are also part of the divestiture. Some farmers are concerned about the impacts of these changes on rural communities, future adaptation, and long-term environmental sustainability.

Variability and Adaptations in Chinchiná River Basin

The Chinchiná River basin is located on the central mountain range of the Andes, in the department of Caldas, Colombia. Within the river basin there are receding tropical glaciers, *páramos*, temperate regions, as well as desert semi-arid landscapes. The rainfall in the Chinchiná river basin is bimodal, which means that during the year there are two periods with high precipitation and two periods with less precipitation (Poveda 2004). This annual distribution of precipitation defines the flowering and harvesting periods in the Colombian coffee producing areas (Jaramillo Robledo and Ramírez Builes 2013). At inter-annual timescales, El Niño-Southern Oscillation (ENSO) is the most important phenomenon related to climate variability in Colombia. During the warm phase of ENSO—what is called El Niño—the country experiences negative anomalies in rainfall, soil moisture, and river discharges. During the cold phase of ENSO—called La Niña—there are positive anomalies such as heavy rainfalls and increases in the average level of the rivers (Poveda 2004; Poveda et al. 2001, 2011).

La Niña periods have negative consequences on the growth and flowering of coffee trees, as well as increasing the incidence of coffee rust (*Hemilea vastatrix Berk and Br*). There is also an increase in the frequency of landslides, which destroy roads, houses and crops (Poveda et al. 2014; García Pineda 2013). El Niño, on the other side, extends the drier season, which damages the coffee fruit harvest and increases the incidence of pests such as the *coffee* berry borer (*Hypothenemus hampei*), the "minador" of the cafeto leaves (*Leucoptera coffellum*) and the "red spider" (*Oliganychus yotherse*) (Ramírez Builes et al. 2014).

Given this climate variability, there are several risk-reducing strategies sponsored by the Colombian Coffee Growers Federation in order to attain "climate intelligent coffee production". These strategies include: (1) Replacement of existing coffee plants with the Castillo variety, which is resistant to "coffee rust" (*Hemileia vastatrix).* This has been possible due to producers' better access to both financial credit and to a certain amount of free fertilizers provided by the state according to the number of new plants (García Pineda 2013; Silva Restrepo 2013). In total, 3250 million trees have been replaced between 2009 and 2014 (FNC 2015); (2) Sowing the coffee plants following the contour of the slopes, increasing plant density, and planting shrubs for both reducing rain erosion and the effect of the thermal amplitude (Ramírez Builes et al. 2014); (3) Cultural practices such as the integral management of weeds to maintain the coverage around each of the coffee plants, use of high quality seeds, timely and adequate fertilization, recollection of fruits left in the field after the harvest, gathering the fallen fruit to reduce the presence of "broca" (*Hypothenemus hampei*), monitoring of Lepidoptera, and use of miticide (Galindo et al. 2013; Ramírez Builes et al. 2014); (4) Ecologically beneficial coffee production processes, such as use of equipment that allows for the removal of pulp and the mucilage of the coffee seeds using less water, which reduce the amount of polluted water going into the local water sources (Oliveros et al. 2013); (5) Agro-forestry practices that allow for a control of shade in coffee plantations, diminish rain erosion, reduce the speed of the wind, and conserve soil humidity (Farfán Valencia 2013; Ramírez Builes et al. 2014); (6) Having fields with coffee plants of different ages to constantly maintain an area under production and avoid the same impact of extreme events over all the coffee plants (Turbay et al. 2014); (7) Creation of an agro-climatic website with data from the monitoring of climate in the coffee region during the last 65 years[3]; (8) Providing insurance against geological and climate risks.

The replacement of the coffee plants with the Castillo variety has allowed for an increase in coffee production from 7.8 million 60-kg bags in 2007 (Silva Restrepo 2013) to 12.3 million bags between April 2014 and March 2015 (FNC 2015). Small coffee producers have renovated their coffee plantations and increased both the

[3]http://www.federaciondecafeteros.org/particulares/es/buenas_noticias/nueva_plataforma_para_monitorear_el_clima (accessed April 6 2015).

density of plants in the fields and the area under technical management (Silva Restrepo 2013). The process has not been free of problems given the decline in production during several years due to the existence of young unproductive coffee plants. Fortunately, coffee prices were high during 2011, a situation that compensated the reduced incomes of the producers. Coffee exports have become normal during 2014 and coffee producers have reduced their use of fungicides, which were necessary to control the coffee rust.

It is important to emphasize that small producers have not always been able to keep their coffee plantations exposed to the sun due to the high costs of fertilizers and insecticides required by the monoculture. During periods of crisis, food insecurity has predominated among peasant households due to the need to eradicate all the food crops from the fields in order to ensure the supply of nutrients and sun exposure to the coffee plants. With low coffee prices in the market during 2012 and diminishing production, the only alternative left to many producers has been to protest for government subsides to maintain their families.

Different and more radical forms of adaptation among coffee growers are those that have converted traditional coffee farms to livestock or citric production, or to rural tourism. The results of these conversions have been the displacements of farm workers, who have no other alternative than to move into urban shantytowns.

The responses of coffee growers to climate variability do not automatically follow the technical recommendations of agronomists. Rather, they are related to the size of the farm, security of land tenure, the economic rationality that characterizes the unit of production (peasant or agri-industry), market coffee prices, the proportion of food crops produced by the farm, the availability of labor, the educational level of the grower, and his/her social networks. The producer balances all these dimensions in order to decide if he/she will ask for a loan to replace the old coffee plants, if it is possible to intensify the use of labor to establish new fields or to focus on the tasks required by the existing coffee fields, if he/she is in a position to increase the use of pesticides and fertilizers, if it is convenient to acquire a new water-efficient machine to remove the pulp, and other important decisions. In the meantime, government measures are limited to providing subsidies in the short-term and to strengthening the national system of risk management, which is still focused in prevention and mitigation of disasters rather than on the reduction of social and economic vulnerability and on the development of long-term adaptive strategies to climate variability.

Discussion and Comparison

Farm-level adaptation practices are very important in all cases. Some are technology-based, such as the use of new seed varieties to grow crops more resistant to extreme climates, use of zero-till, or the use of irrigation equipment. In both the Canadian Prairies and Mendoza, irrigation technology allows more efficient resource use and ensures better yields while causing an increase in water demand. However, because of the cost of these investments, mainly paid by producers—and, in the case of Canada, limitations on the number of water licenses available to producers—it is a restricted adaptation practice.

Other adaptation practices include strategic management of crops, land, and plantations, arising from the accumulation of traditional knowledge and generational inherited learning. This knowledge is expressed differently in each place. In Colombia, many farms still use coffee grown under shade combined with other food crops; in Mendoza and Canada, current practices make better use of water and withstand drought. These practices do not always coincide with the knowledge "handling" promoted by some agencies and government. Specialized information contributes to early management of climate risks, providing certainty and increasing resilience. However, we found that the mere existence of information is insufficient if its accessibility and usability is not ensured.

Beyond climatic events, high input prices and low commodity prices profoundly affect the profitability of producers' livelihoods, limiting the adaptability of producers and productive activity itself. A common process in the three case studies is to change productive activity within farms or more radically, the total abandonment of production. This enables, on one hand, a process of land concentration when small farmers are *forced* to sell their land to economically better positioned producers (Mendoza and Canada). On the other hand, it triggers a process of land use change, either from hand crop to livestock as in Colombia, or the urbanization of agricultural areas in Mendoza.

This analysis undermines the assumption, accepted by many in the scientific community on climate change, that income diversification is a form of sustainable adaptation. This is illustrated, for example, by the insecurity and instability of the off-farm income of small and medium farmers in Mendoza, and the increased work pressure on producers in Canada. This highlights the difficulties of making misleading theoretical generalizations without considering specific characteristics of contexualized cases.

Adaptation based on seed handling to get more resistant varieties deserves special attention. New seed varieties promising drought or flood resistance are often patented by corporations and this triggers an agricultural model that is technologically dependent and unsustainable (they demand many agrochemicals, usually petroleum-based), economically dependent (producers remain tied to royalties) and in broader terms this could lead to a food dependency model. In all three countries, the increasing dependence on patented seeds and agro-chemical inputs has important implications for future sustainable development. If adaptation and agricultural development practices are based on an environmentally unsustainable foundation, adaptation ultimately works against longer-term mitigation of future climate extremes, causing a cyclical effect that paradoxically increases vulnerability over the long term. Further, the high financial costs of such adaptation negatively affect farmers' economic capital, a key determinant of their adaptive capacity.

Government institutions acquire a similar character in Colombia and Mendoza, granting subsidies to producers, especially small ones. Such practices help, but are really insufficient and especially do not contribute to building producers' capacity or enhancing their access to critical resources necessary to generate genuine adaptations. In Canada, the divestment of government from agricultural support programs seriously affects the capacity to invest in irrigation and environmentally sustainable adaptive practices.

Conclusions

Farmers in the three basins have adopted a variety of adaptive strategies to reduce the negative impacts of climate variability. Many practices implemented by farmers help them to better cope with climatic events and do have a positive impact on crop yield. Access to improved seeds and irrigation infrastructure in the studied areas generate significant benefits for coping with extremes that affect each location.

But adaptation always has an economic cost. A number of factors associated with producers' economic and financial resources prevent (or facilitate) the implementation of adaptive strategies. Our comparative analysis illustrates that the impact of climate crisis (variability and extreme events) is strongly determined by economic conditions. One of the most important conditions in all three countries is the price of inputs. Commodity prices, which are determined primarily by international markets in all three countries, also have a powerful influence on adaptation. While high commodity prices may be able to save a poor harvest, as happened in Colombia in 2011, low prices can diminish or cancel out the success of certain adaptations.

Although the findings are specific have a limited and bounded extent to the studied areas, the analysis confirms the relevance of studying adaptation and resilience at the intersection of social, political and economic dimensions and physical factors. Adaptation assessment must be contextualized within broader socioeconomic processes that seriously condition adaptability. The paper reinforces a call to develop an approach that highlights certain dimensions that single discipline approaches fail to problematize, especially considering non-climatic aspects of the problem.

Further, another main finding of this study is the confirmation that not all adaptation strategies are equally beneficial. Our research also reveals that adaptations are not linear processes. Both the producers' own adaptations as well as adaptive policies have indirect impacts on actors, natural resources and ecosystems. Adaptation is a process of synergies and trade-offs: zero-till reduces soil erosion but may degrade it with agro-chemicals; the use of more resistant crop varieties generates technological dependence and increases input costs; efficient use of water or new water sources increases the demand for the resource.

Addressing this non-linearity of the adaptation process has critical implications for future studies and policies to foment adaptation: they must necessarily incorporate an assessment of those trade-offs. This indicates the need to seriously consider questions about *adapting to what, why, and with what consequences.*

While it is true that a measure of benefit to one group may disadvantage another, it should also be noted that the adaptive capacities of producers are often shaped by factors that, at the surface, appear to have little direct connection to the climate or environment but can dramatically affect both current and future vulnerability.

Our study supports the idea that there are no universal determinants generating sustainable adaptation in all cases. Technology, capital, institutional capital, and income diversification in general can strengthen adaptive capacities. However the actual impact of each of these determinants depends on the characteristics of local contexts. Therefore, it will not be possible to establish an all-purpose menu of practices and recipes that are successful in all cases.

Adaptation is a local and dynamic phenomenon. Therefore, practices that are successful in one place for coping with climate variability may not be feasible in other places; the measures that prove to be adaptive now may not be adaptive in the future. Hence, it may not be possible to replicate elsewhere the adaptation measures identified in this study. However, this work can alert other researchers and policy makers about going beyond the technical aspects of adaptation. Our study highlights the importance of context-specific social, economic and political factors that make some social groups more susceptible to climate variability than others and generate unequal access to adaptation measures planned by institutions. We must also anticipate the unexpected effects of adaptation as increasing production costs, the proportion of large holdings, the dependence on multinationals that provide seeds and agrochemicals and the loss of rural jobs. For this reason, it is necessary to explore further adaptation models based on agro-ecological principles that reduce the dependence on external inputs to the farm, enrich the biodiversity of the agricultural ecosystem, preserve natural ecosystems and ensure food sovereignty of communities.

Reduction of sensitivity to climate variability can only be achieved by reducing poverty and social inequalities. A rural population with high levels of resources, equality, and social infrastructure will be better prepared to cope with climate variability and may recover more quickly from the effects of extreme weather events. Adaptation, in these circumstances, is likely the result of a negotiation between the traditional knowledge that was successful to cope with climate variability in the past and the most innovative technical proposals of institutions dealing with issues relating to the environment, water and agriculture.

References

Adger WN (2003) Social aspects of adaptive capacity. In: Smith JB, Klein RJT, Huq S (eds) Climate change: adaptive capacity and development. Imperial College Press, London, pp 29–50

Agrawal A, Perrin N (2009) Mobilizing rural institutions: a comparative study of rural institutions for improving governance and development: Afghanistan, Ethiopia, India, Vietnam, and Yemen, Social development. World Bank, Washington, DC, p 80

Alasia A, Bollman RD (2009) Off-farm work by farmers: the importance of rural labour markets. Statistics Canada, Ottawa, ON, Retrieved from http://dsp-psd.pwgsc.gc.ca/collection_2009/statcan/21-006-X/21-006-x2008001-eng.pdf

Bassett TJ, Fogelman C (2013) Déjà vu or something new? The adaptation concept in the climate change literature. Geoforum 48:42–53

Below TB, Mutabazi KD, Kirschke D, Franke C, Sieber S, Siebert R, Tscherning K (2012) Can farmers' adaptation to climate change be explained by socio-economic household-level variables? Glob Environ Chang 22(1):223–235. doi:10.1016/j.gloenvcha.2011.11.012

CNA (2002) National agricultural census. National Statistics and Census Institute, Ministry of Economy. http://www.indec.mecon.gov.ar/agropecuario/cna_principal.asp

FNC (Federación Nacional de Cafeteros de Colombia) (2015) Colombia produjo 12,3 millones de sacos de café en los últimos doce meses. Retrieved from http://www.federaciondecafeteros.org/particulares/es/sala_de_prensa/detalle/colombia_produjo_123_millones_de_sacos_de_cafe_en_los_ultimos_12_meses1/. Accessed 6 Apr 2015

Dietz K (2013) Hacia una teoría crítica de vulnerabildad y adaptación. In: Ulloa A, Prieto AI (eds) Culturas, conocimientos, políticas y ciudadanías en torno al cambio climático. Universidad Nacional de Colombia, Bogotá, pp 19–46

Eriksen S, Aldunce P, Bahinipati CS, Martins RD, Molefe JI, Nhemachena C, O'brien K, Olorunfemi F, Park J, Sygna L, Ulsrud K (2011) When not every response to climate change is a good one: Identifying principles for sustainable adaptation. Climate Dev 3(1):7–20. doi:10.3763/cdev.2010.0060

Farfan Valencia F (2013) Guía para el establecimiento de barreras con árboles y sombrío del café. Avances Técnicos Cenicafé 428:8

Field CB, Barros V, Stocker TF, Qin D, Dokken DJ, Ebi KL, Mastrandrea MD, Mach KJ, Plattner G-K, Allen SK, Tignor M, Midgley PM (2012) Managing the risks of extreme events and disasters to advance climate change adaptation. Cambridge University Press, New York, NY

Galindo Leva LÁ, Villegas García C, Mantilla Afanado J (2013) Lepidópteros perforadores de los frutos del café. Avances Técnicos Cenicafé 434:8

Garcia Pineda AF (2013) Vulnerabilidad y adaptación al cambio climático en la caficultura colombiana. In: Ulloa A, Prieto AI (eds) Culturas, conocimientos, políticas y ciudadanías en torno al cambio climático. Universidad Nacional de Colombia, Bogotá, pp 107–142

Glasser B, Straus A (1967) The discovery of grounded theory: strategies for qualitative research. Aldine Publishing Company, Hawthorne, NY

IPCC (2007). In: Core Writing Team, Pachauri RK, Reisinger A (eds) Contribution of working groups I, II and III to the fourth assessment report of the Intergovernmental Panel on Climate Change. Geneva, p 104

IPCC (2015) Climate change 2014: synthesis report. In: Pachauri RK, Meyer LA (eds) Contribution of working groups 1, II, and III to the fifth assessment report of the Intergovernmental Panel on Climate Change. Intergovernmental Panel on Climate Change, Geneva

Jaramillo Robledo Á, Ramírez Builes VH (2013) Gestión del riesgo agroclimático. Fuentes de amenaza climática para el café en Colombia. In: Marín SM (ed) Manual del cafetero colombiano. Investigación y tecnología para la sostenibilidad de la caficultura. Federación Nacional de cafeteros de Colombia, Cenicafé, Manizales, pp 73–89

Jetté-Nantel S, Freshwater D, Beaulieu M, Katchova A (2011) Farm income variability and off-farm diversification in Canadian agriculture. Statistics Canada, Ottawa, ON

Lozoya JP, Sardá R, Gimenez J (2011) A methodological framework for multi-hazard risk assessment in beaches. Environ Sci Policy 14:685–696

Marchildon GP, Kulshreshtha S, Wheaton E, Sauchyn D (2008) Drought and institutional adaptation in the Great Plains of Alberta and Saskatchewan, 1914–1939. Nat Hazards 45(3):391–411. doi:10.1007/s11069-007-9175-5

McLeman RA, Dupre J, Berrang Ford L, Ford J, Gajewski K, Marchildon G (2013) What we learned from the Dust Bowl: lessons in science, policy, and adaptation. Popul Environ. doi:10.1007/s11111-013-0190-z

Mearns R, Norton A (2010) Social dimensions of climate change: equity and vulnerability in a warming world. World Bank, Washington, DC, p 319
Montaña E (2008) Central Andean foothill farmers facing global environmental change. IHDP 2: 32–40
Oliveros TCE, Sanz UJR, Ramírez GCA, Tibaduiza VCA (2013) ECOMILL. Tecnología de bajo impacto ambiental para el lavado del café. Avances Técnicos Cenicafé 432:8
Poveda G (2001) La hidroclimatología en Colombia. Una síntesis desde la escala interdecadal hasta la escala diurna. Revista Academia Colombiana de Ciencias Exactas, Físicas y Naturales XXVIII, 107:201–222
Poveda G, Jaramillo Á, Gil MM, Quiceno N, Mantilla R (2001) Seasonality in ENSO-related precipitation, river discharges, soil moisture, and vegetation index in Colombia. Water Resour Res 37(8):2169–2178
Poveda G, Álvarez DM, Rueda OA (2011) Hydro-climatic variability over the Andes of Colombia associated with Enso: a review of climatic processes and their impact on one of the Earth's most important biodiversity hotspots. Climate Dynam 36:2233–2249
Poveda G, Turbay S, Vélez JJ, Ocampo OL, Acevedo E, Bedoya M (2014) ¡No sé qué vamos a hacer con estos climas! Vulnerabilidad y adaptación a las variaciones climáticas extremas en la cuenca de la quebrada Los Cuervos, afluente del río Chinchiná. Universidad Nacional de Colombia, Universidad de Antioquia, IDRC-CIHR-NSERC, Medellín
PSRB Partners for the Saskatchewan River Basin (2009) From the mountains to the sea: the state of the Saskatchewan River Basin. Saskatchewan. Retrieved from http://www.saskriverbasin.ca/page.php?page_id=70. 12 Apr 2015
Ramírez Builes VH, Gaitán Bustamante Á, Benavides Machado P, Constantino Chaure L, Gil Palacio Z, Sadeghian Khalajabadi S, González Osorio H (2014) Recomendaciones para la reducción del riesgo en la caficultura en Colombia ante un riesgo climático de El Niño. Avances Técnicos. N. 445. National Coffee Research Center—Cenicafé Manizales, pp 1–12
Silva Restrepo S (2013) Evaluación de impacto de los programas de renovación de cafetales 2007–2011; efectos en la capacidad productive de los caficultores colombianos. Ensayos de Economía Cafetera 29:37–78
Smit B, Wandel J (2006) Adaptation, adaptive capacity and vulnerability. Glob Environ Chang 16(3):282–292. doi:10.1016/j.gloenvcha.2006.03.008
Smit B, Burton I, Klein RJT, Wandel J (2000) An anatomy of adaptation to climate change and variability. Clim Change 45:223–251
Statistics Canada (2006) 2006 census of agriculture. Government of Canada. Ottawa, ON. Retrieved from http://www.statcan.gc.ca/ca-ra2006/index-eng.htm. 12 Apr 2015
Statistics Canada (2011) 2011 census of agriculture. Government of Canada, Ottawa, ON. Retrieved from http://www.statcan.gc.ca/daily-quotidien/120510/dq120510a-eng.htm. 12 Apr 2015
Turbay S, Cruz BN, Vallejo FJ, Vélez JJ, Ocampo OL (2014) Adaptación a la variabilidad climática entre caficultores de las cuencas de los ríos Porce y Chinchiná, Colombia. Investigaciones Geográficas 85:95–112. doi:10.14350/rig.42298
UNEP (2013) PROVIA guidance on assessing vulnerability, impacts and adaptation to climate change. http://www.unep.org/provia/RESOURCES/Publications/PROVIAGuidancereport/tabid/130752/Default.aspx. Accessed 13 Apr 2015
Wandel J, Pittman J, Prado S (2010) Rural vulnerability to climate change in the South Saskatchewan River Basin. In: The new normal: the Canadian Prairies in a changing climate. University of Regina Press, Regina, SK, pp 245–258
Wisner B, Blaikie P, Cannon T, Davis I (2003) At risk: natural hazards, people's vulnerability and disasters, 2nd edn. Routledge, New York, NY

Chapter 15
Adaptation of the Bulgarian Water Sector to Climate Change Extremes

Mariyana Nikolova

Abstract Extreme weather events and the resulting hydro-climatic disasters have increased in frequency and intensity in recent decades, confirming trends outlined in the Fourth and Fifth IPCC reports on the increased susceptibility of Southeast Europe to drought, extreme temperatures, heat waves, and floods. Currently, climate change in the region is most apparent in the frequency and intensity of climate extremes, specifically in temperature and precipitation. These changes may significantly affect the water sector in Bulgaria by the end of this century.

The long-term strategic objective of the water sector in Bulgaria is to ensure sustainable use of water resources and to secure the future needs for water of the population, economy and aquatic ecosystems. The water sector operates in three main business areas: *plumbing* (supply, drainage and sanitation), *irrigation* (irrigation, drainage and protection from the harmful effects of water) and *hydropower systems and equipment* (dams and hydropower facilities). The operation of each one of these systems depends on the availability, quantity, and quality of water.

This paper analyzes the impact and sensitivity of the Bulgarian water sector to climate change and assesses the sector's vulnerability index.

The results show that the water sector in Bulgaria is characterised by higher sensitivity and vulnerability toward changes in water quantity and quality and to climate-related extremes such as drought and floods. The Climate Change Vulnerability Index describes the sector as moderately vulnerable to climate change over the 2016–2035 time horizon. Some adaptation measures are proposed in respect of these results. The need of better integration between Water Framework Directive, Sendai Framework for Disaster Risk Reduction 2015–2030 (SFDRR) and EU Strategy for adaptation to climate change (2013) are discussed.

Keywords Climate change • Water • Vulnerability index • Adaptation policy

---

M. Nikolova (✉)
Department of Geography, National Institute of Geophysics, Geodesy and Geography, Bulgarian Academy of Sciences, Acad. Georgi Bonchev str., bl. 3, 1113 Sofia, Bulgaria
e-mail: mknikolova@gmail.com

W.L. Filho et al. (eds.), *Climate Change Adaptation, Resilience and Hazards*, Climate Change Management, DOI 10.1007/978-3-319-39880-8_15

Introduction

The degree of vulnerability of a territory to natural disasters depends on many different factors (natural, human, social, management, etc.). In the context of climate change, these factors encompass the regional characteristics of hazardous phenomena as well as direct or indirect risk stemming from certain global processes. Therefore, risk management of disasters caused by climate change should take into account local factors of the environment together with global processes (climate change, demographic and socio-economic, changes, land use change, migration processes, etc.) and their possible regional projections. There is a lot of scientific evidence (supported by observations and models) linking climate change to increasing risk from water related extremes like heat waves, droughts, floods and forest fires (Parry et al. 2007; Trenberth 2012; Leal Filho 2012, 2013; Millan 2014; Denton et al. 2014). The impact of climate change on water resources and related risks and the need of adaptation are in the focus of many publications of IPCC, World Bank, UN and other leading institutions in the world (IPCC—SREX 2012; IPCC 2007, 2013, 2014a, b; UN 2009; UNISDR 2015; World Bank 2013).

There are many reasons to look for better integration between research and management agendas on climate change, water resources management, natural hazards, resilience and adaptation. The recent IPCC's publication "Climate resilient pathways: adaptation, mitigation, and sustainable development" (Denton et al. 2014) discuss climate change as a threat to sustainable development and frames climate-resilient pathways. In Sendai Framework for Disaster Risk Reduction (SFDRR) 2015–2030 and the EU Strategy for adaptation to climate change (2013) special attention is given to the need for better coordination between adaptation policies and the policies for disaster risk management at different levels of government. Such coherence is needed between national and sartorial strategies for adaptation and risk management plans (EU 2013).

Water management is a complex process whose aim is to ensure sustainable use of water resources and to meet the demand for water in various economic activities. Climate change and the uncertainty of its impact on water resources in different regions and countries necessitate the assessment of possible outcomes, the vulnerability of the water sector to climate change, and the corresponding policies for mitigation and adaptation to climate change.

The aim of this paper is to present implementation of DPSIR methodological approach for climate change vulnerability assessment of the water sector and to propose a method for evaluation of a Climate Change Vulnerability Index for water sector. It is estimated based on a set of indicators of the state of water resources, sensitivity to the regional climate change projections under different RCP scenarios, impact and the adaptive capacity (Nikolova 2014). The suggested index offers a comprehensive expert assessment of the level of vulnerability of the sector to climate change. It also proposes a framework that can incorporate more accurate data from regional models for change in climate and hydrology and vary the precision of the included indicators to meet a desired level of detail.

Data

The analysis is based on information for the current status of precipitation, air temperature, quantity and quality of water resources in the country and related hydro-climatic risks, climate models and climate scenarios according to the IPCC AR4 (2007) and AR5 (2013) and their regional projections for the territory of Bulgaria from KMNI (2013). Additional data were obtained from the National strategy for the management and development of the water sector in Bulgaria (MOEW 2012); the National Statistical Institute (NSI); the National Electric Company (NEC); data from the Ministry of Regional Development and Wellbeing (MRDW) and from the Ministry of Environment and Waters (MOEW) on the state of environment and water management, water supply, sanitation and water purification, water resources and water use over the period 2007–2013; publications for the state of the environment from MOEW, River Basin Directorates (RBD), National Institute of Meteorology and Hydrology, European Environmental Agency (EEA) and other sources of information.

Methodology

Driving-Pressure-State-Impact-Respond Research Framework (DPSIR)

Climate change is one of the driving forces of global change and the analysis and evaluation of the risks and the vulnerability of the water economy of climate change should be carried out in this more general context. Driving-Pressures-State-Impact-Response (DPSIR) approach, Fig. 15.1, has been successfully applied for an integrated multi-sectoral assessment of the regional impacts of climate change in the UK (Holman et al. 2005), where it is described briefly as follows:

- *Driving forces* are dealt with at the regional level and are analyzed in terms of climate change, socio-economic systems and the national and European policy;
- *Pressure* is also considered at the regional level, having analyzed variables that quantified the driving forces (temperature, rainfall, concentrations of carbon dioxide, extreme phenomena, GDP, regional development, etc.).
- *Status* is characterized by indicators for the variables, which relate to the sensitivity of the systems or the sectors to the pressure.
- *Impact* depends on what values are reached the system status indicators and how they are approaching critical levels.
- *Response* is expressed in terms of planned models for adaptation that aim to reduce the negative and enhance positive impacts of climate change. The results of these models are evaluated as possible future adaptation policies. They must also be supported by business and non-governmental.

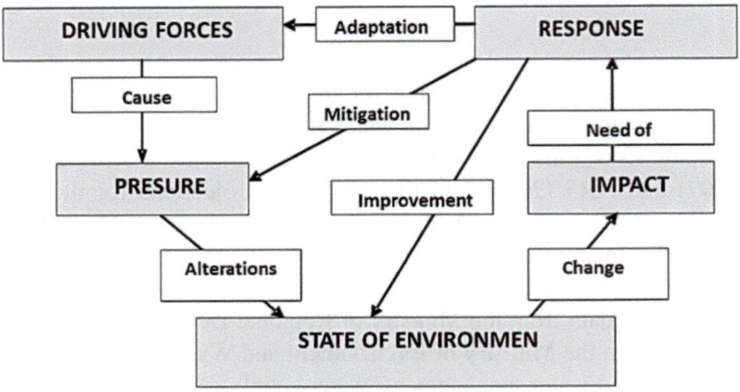

Fig. 15.1 DPSIR model (after Holman et al. 2005)

The DPSIR research approach provide a good platform for integration of both approaches, the general risk assessment and climate change risk assessment, and also for development of systematic responsive policy for adaptation, mitigation and improvement of the state of environment.

Climate Change Risk Assessment

The risk of climate change in the context of the general approach to risk assessment is a function of *probability, exposure and vulnerability*. Probability is determined by the degree of *uncertainty*, which can be realized under one or another scenario for climate change. *The hazard* of climate change stems from the uncertainty of the realization of one or another scenario of climate change and is defined as the likelihood that a particular outcome and impact against sensitive systems. This approach was adopted by IPCC in AR5 (AR5 2013), which distinguishes between 7 plus 3 additional measures of the likelihood that one or another scenario for climate change is realized. *Exposure* establishes which systems are exposed to the particular influence from climate change and how sensitive they are to it. *Sensitivity* depends on the combination of the likely outcome for realization of a scenario of climate change and the expected impact of climate change on the system. The impact of climate change depends on the exposure and sensitivity of socio-economic systems to it. Vulnerability is determined by many factors, both natural (which determine the strength of the impact), and social, economic and political (which give the adaptive capacity of the systems) Fig. 15.2.

While *vulnerability* in the context of the impact of climate change is "a function of the character, magnitude, and rate of climate change and variation to which a system is exposed, its sensitivity, and its adaptive capacity" (Parry et al. 2007), *resilience* deals with the change of uncertainty. "By "resilient" we mean a system's ability to anticipate, reduce, accommodate, and recover from disruptions in a

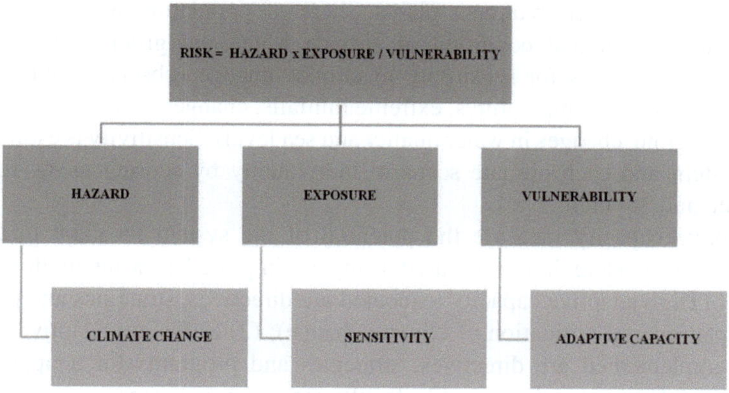

Fig. 15.2 Risk of climate change

timely, efficient, and fair manner" (IPCC—SREX 2012). The aim of adaptation policy relates very much to the need to strengthen a system's resilience. Adaptation is defined by IPCC as "adjustment in natural or human systems in response to actual or expected climatic stimuli or theirs effects, which moderates harm or exploits beneficial opportunities" (Parry et al. 2007). It is of crucial importance to assess the vulnerability of the system to climate change through this process.

Climate Change Vulnerability Index (CCVI)

To assess the CCVI for Bulgarian water sector we use an indicator based approach. There are three main groups of indicators: (1) *Indicators for climate change* (projected changes in the temperature (°C), precipitations (%) and extremes indexes like Maximum length of dry spell index and Extreme wet days index (IPCC 2013)); (2) *Water indicators* for observed changes in the quantity of surface water, quantity of underground water and water quality, drought and flood probability and observed and expected raising the sea level and (3) *Adaptive capacity indicators* for structure of operators providing services in the water sector, theirs status, the directives, programs, strategies and other available documents and tools for improvement of adaptation capacity.

The indicators used for assessment of the *systems sensitivity* refers to the following groups: (1) Indicators for the status of water resources (observed trends of change in fresh water resources, seized fresh leads on key economic activities per capita, seized fresh water per capita, underground water resources, seized groundwater on key economic activities, seized marine water in basic economic activities and Index of exploitation of water resources); (2) Indicators for the impact on water resources and water sector (observed change in water use by different sectors of the economy, treatment of wastewater, share of population with public sewer and municipal wastewater treatment plants, share of population with water supply mode, physicochemical condition of surface water and groundwater chemical status); (3) Indicators for sensitivity to climate change (observed and projected changes in extreme temperatures, extreme rainfalls, changes in the water discharge and flow regime, changes in water quality and sea level). Sensitivity is estimated for each system and each climate scenario individually by scoring as 1—low, 2—moderate and 3—high.

Adaptive capacity measure the potential of the system to adapt to changes (Brooks 2003). Here it is measured using a simplified scoring in three levels scale: (1) High adaptive capacity (executed are directives, strategies and programs for adaptation and mitigation of climate change); (2) Sufficient adaptive capacity (partly implemented are directives, strategies and programs for adaptation and mitigation of climate change); (3) Insufficient adaptive capacity (no action is taken to address the risk of climate change).

On the base of the observed and projected changes, a scoring of indicators was implemented and integrated in a climate change vulnerability assessment matrix, Table 15.3.

Climate Change Vulnerability Index is measured in this research by the ratio between sensitivity and adaptive capacity of the exposed systems (1) and defined in five categories according the vulnerability scale in Table 15.1.

Climate Change Vulnerability Index (VI):

$$\mathrm{VI} = \mathrm{S}/\mathrm{Ac} \tag{1}$$

Where:

S—Sensitivity.
Ac—Adaptive capacity.

Sensitivity is measured by the ratio between the sum of maximal scores for all indicators and sum of all scores, divided to the number of indicators (2):

$$\mathrm{S} = \sum \left(\sum \mathrm{Sn\ max\ scores} / \sum \mathrm{Sn\ scores} \right) / \mathrm{n} \tag{2}$$

Where:

n—Number of climate change indicators.
S—Sensitivity.

Table 15.1 Vulnerability scale (after Garcia et al. 2012)

Vulnerability index value	Vulnerability
0.80–1.00	Extremely vulnerable
0.50–0.79	Very vulnerable
0.20–0.49	Moderately vulnerable
0.01–0.19	Vulnerable

Source: Garcia et al. 2012

Results

Identification of Main Actors in Bulgarian Water Sector

The analysis is mainly based on data for water resources, water use and management and water related extremes in the period after the accession of the country in EU in 2007. The summarized results show that the Water sector in Bulgaria operates in three main business areas: plumbing, irrigation and hydropower systems and equipment.

In plumbing there are 66 "V&K" water operators, which serve 78.4 % of the settlements in Bulgaria. Their activity is regulated by the MRDW, MOEW, Ministry of Health (MH) and the State regulator. According to the form of ownership the water operators are divided into 4 types: 100 % municipal property (26 companies), 100 % owned (14 companies), 51 % state and 49 % municipal property (16 companies) and local private-owned (2). Eight private companies serve large industrial enterprises.

Management and control of companies and entities of irrigation is performed by the "Irrigation" Directorate in the MAF. Activity is carried out mainly by the company with 100 % state-owned "Irrigation Systems" EAD in Sofia and another 5 branches in major cities of the country and 86 irrigation associations of natural or legal persons. Agricultural irrigation and municipal public property serves 350,000 ha.

Management and control of Power Systems and Equipment is carried out by the Ministry of Economy through Trade Company "Dams and Cascades" at the NEC. It currently has 16 branches and 2 repair stations. TC "Dams and Cascades" managed 50.1 % of the total volume of the regulated reservoirs in Bulgaria and operates the 8 large cascades and 4 water supply systems, providing water resources for 35 hydropower plants with capacity of 2811 MW.

Three factors strongly influence the efficiency of the water operator "V&K": the form of ownership, company size and price of water. Utilization of irrigation and hydropower pool is determined by factors such as weather conditions, sown irrigated crops on irrigated land, availability of markets for production and financial conditions for the production and sale (MOEW 2012).

Assessment of Climate Change Sensitivity and Vulnerability of Bulgarian Water Sector

Projected changes in temperature and precipitation for the territory of Bulgaria show that, depending on the scenario, the average air temperature between 2081 and 2100 would be above the 1961–1990 average by 2 °C (RCP 2.6) to 7 °C (RCP 8.5), or with 3 °C (RCP 4.5) to 4 °C (RCP 6). Fluctuations in annual average rainfall are within 10 % (RCP 2.6; RCP 4.5; RCP 6) and 10–20 % (RCP 8.5). In the time horizon between 2065 and the end of the century, summer rainfall is projected to decrease by 10–20 % according to all models, or up to 30–40 % according to RCP 8.5 (2081–2100), AR5 (IPCC 2013).

Projections of extreme weather phenomena on the basis of temperature and precipitation indices in AR5 show probable increase the number and intensity of dry and hot periods in the summer. Droughts and floods will occur with greater frequency, as well as torrential rainfall and related dangerous natural phenomena and processes. The values of some of the indices predict less frequent, but more intense rainfall. This is confirmed by the values of the indexes for Maximum 1-day precipitation amount (Rx1day), for Annual maximum consecutive 5-day precipitation amount (Rx5day), as well as the values of the indexes for Very wet days (R95pTOT) and for Extreme wet days (R99pTOT) (IPCC 2013; MOEW 2014).

The sensitivity of water sector to climate change is evaluated in respect to the possible changes in time horizon 2016–2035, according to the Representative Concentration Pathways scenario (RCP 2.6) in AR5 (IPCC 2013). Because the time period is relatively short, the differences in the estimated average values of changes in temperatures and precipitation and projected changes according to the four RCP scenarios (RCP8.5, RCP6, RCP4.5 and RCP2.6) are very small and the likelihood of their realization is very high (MOEW 2014). For the evaluation of susceptibility to drought and flood, we refer to the trends of projected changes of indices for extreme climatic phenomena according to the same RCP scenarios in AR5 (IPCC 2013).

The results from analysis of indicators in Table 15.2 shows that in terms of fresh water resources and their allocation per capita there is no significant change and consumption is covered with water for domestic water supply and water for all economic sectors in the country under the current pace of development. The index of exploitation of water resources in Bulgaria is 0.6 % and confirms these conclusions. The Water Exploitation Index (WEI) shows the ratio of annual volume of water abstraction to the available fresh water resources. For Bulgaria the WEI was 5.4 % in 2012, and between 5.5 and 6.6 % for 2000–2010. These levels show that water use in Bulgaria does not stress the water ecosystem. The threshold for regions with scarce water resources is 20 % (NSI 2015). In terms of indicators for impact on water resources, the following positive developments are observed: increased proportion of purified waste water, increased share of the population with public sewer and municipal wastewater treatment plants, and reduced share of the population regularly subjected to the water supply rationing. Overall, the physic-chemical state of surface water and groundwater chemical status marks an improvement over the last few years, but the content of nitrates in groundwater is still significant. In addition, although a growing share of the population is served by municipal waste water treatment plants, it is still significantly lower than the average for the rest of the EU countries.

Table 15.3 presents the main activities carried out in the water sector and their relationship to specific indicators and trend of change. The Bulgarian water sector shows increased sensitivity to indicators for changes in quantity and quality of

Table 15.2 Indicators for assessment of sensitivity of water sector to climate change

Sensitivity indicators	Trend
Status of water resources	
Fresh water resources	→
Fresh water resources per capita	→
Seized fresh leads on key economic activities	↑↓
Seized fresh water per capita	↑
Underground water resources	↑↓
Seized groundwater on key economic activities	↑↓
Seized marine water in basic economic activities	↓
Water Exploitation Index	↓
Impact on water resources and water sector	
Water use in different sectors of the economy	↑↓
Treatment of wastewater	↑
Share of population with public sewer and municipal wastewater treatment plants	↑
Share of population with water supply mode	↓
Physicochemical condition of surface water	↑↓
Groundwater chemical status	↑↓
Sensitivity to climate change	
Extreme temperatures	↑
Extreme rainfall	↑
Changes in the quantity and flow regime	↑↓
Changes in water quality	↓
Changes in sea level	↑

water, to extreme temperatures and precipitation and, respectively, to the risk of floods and droughts.

The assessment matrix in Table 15.4 shows that the change in precipitations will have negative impact on water quality. The negative impact from changes in water-related extremes relates to water quality, drought and flood risk and sea level change. The water sector is most sensitive to the projected changes in related extremes (drought and floods) and in temperature. The index of the water sector's sensitivity to climate change in the time horizon 2016–2035, is estimated to be 1.25.

Adaptive capacity of the main actors in Bulgarian water sector is assessed as "insufficient" due to the luck of implementation of strategies for adaptation to climate change and is scored with 3 points.

As a result, the Climate Change Vulnerability Index (VI) for water sector in Bulgaria is estimated as "Moderately vulnerable" with a value of 0.42, Table 15.5. This result relates to the climate change in time horizon 2016–2035 and corresponds to the observed trends in most of the analyzed indicators over the last decade.

Table 15.3 Indicators that are sensitive to the activities of operators in the water sector

Water sector	Activity	Indicator	Trend
V&K	Delivery	Changes in the water amount and the flow regime Changes in water quality	↓↑ ↓
	Outlet	Extreme precipitations Changes in the water amount and the flow regime	↑ ↓↑
	Water purification	Changes in water quality	↓
Hydro-meliorations	Irrigation	Changes in water quality Extreme temperature Changes in the water amount and the flow regime	↓ ↑ ↓↑
	Drainage	Changes in the water amount and the flow regime Extreme precipitations	↓↑ ↑
	Protection from the harmful impact of water	Changes in water quality Extreme precipitations	↓ ↑
Hydro-power systems and equipment	Technical operation and maintenance of dams	Changes in the water amount and the flow regime Extreme temperature	↓↑ ↑
	Technical operation and maintenance of hydraulic systems and structures	Changes in the water amount and the flow regime Changes in sea level Extreme precipitations	↓↑ ↑ ↑

Conclusions

At present, the pressures (natural and human) on the quantity and quality of water resources is manageable by optimizing consumption; reducing water losses, which are unacceptably high due to old transmission facilities; coverage of all waste water with treatment facilities; quality control; and protection of ecosystems on which the resumption of water resources depends. Solving these problems would greatly increase the resilience of the water sector vis-a-vis possible climate change in the time horizon 2016–2035. In 2050–2100 time horizon, the challenges facing the water sector can be extremely high because “due to the inertia of the climate system” (UN 2009).

Implementation of the proposed CCVI provides a general notion about the level of vulnerability to climate change. It is very sensitive to the quality of indicator’s data and it have to taken in mind when it is used as a decision support tool from managers in water sector. More detailed assessment for each particular activity is necessary for better management of the process of adaptation in water sector.

Table 15.4 Assessment matrix for water sector sensitivity and vulnerability to climate change

Sector/Indicator	Climate scenario IPCC AR5	Probability of output in time horizon 2016–2035			Expected impact: positive (+) minor or no (0) and negative (−)			Sensitivity 1—Low 2—Moderate 3—High		
1	2	3	4	5	7	8	9	10	11	12
Water	Scenario	Δ T °C	Δ P %	Ex↓↑	ΔT °C	Δ P %	Δ Ex	ΔT °C	ΔP %	Δ Ex
Quantity of surface water	RCP2.6	1.5–2.0	0–10	↑↓	–	+	+/–	3	3	3
	RCP4.5	1.5–2.0	0–10	↑↓	–	+	+/–			
	RCP6	1.5–2.0	0–10	↑↓	–	+	+/–			
	RCP8.5	1.5–2.0	0–10	↑↓	–	+	+/–			
Quantity of underground water	RCP2.6	1.5–2.0	0–10	↑↓	–	+	+/–	3	2	3
	RCP4.5	1.5–2.0	0–10	↑↓	–	+	+/–			
	RCP6	1.5–2.0	0–10	↑↓	–	+	+/–			
	RCP8.5	1.5–2.0	0–10	↑↓	–	+	+/–			
Water quality	RCP2.6	1.5–2.0	0–10	↑↓	–	–	–	2	2	3
	RCP4.5	1.5–2.0	0–10	↑↓	–	–	–			
	RCP6	1.5–2.0	0–10	↑↓	–	–	–			
	RCP8.5	1.5–2.0	0–10	↑↓	–	–	–			
Drought	RCP2.6	1.5–2.0	0–10	↑	–	+	–	3	3	3
	RCP4.5	1.5–2.0	0–10	↑	–	+	–			
	RCP6	1.5–2.0	0–10	↑	–	+	–			
	RCP8.5	1.5–2.0	0–10	↑	–	+	–			
Flood	RCP2.6	1.5–2.0	0–10	↓↑	–	+	–	3	3	3
	RCP4.5	1.5–2.0	0–10	↓↑	–	+	–			
	RCP6	1.5–2.0	0–10	↓↑	–	+	–			
	RCP8.5	1.5–2.0	0–10	↓↑	–	+	–			
Raising sea level	RCP2.6	1.5–2.0	0–10	↑↓	–	+	–	3	2	3

(continued)

Table 15.4 (continued)

Sector/Indicator	Climate scenario IPCC AR5	Probability of output in time horizon 2016–2035			Expected impact: positive (+) minor or no (0) and negative (–)			Sensitivity 1—Low 2—Moderate 3—High		
	RCP4.5	1.5–2.0	0–10	↑↓	–	+	–			
	RCP6	1.5–2.0	0–10	↑↓	–	+	–			
	RCP8.5	1.5–2.0	0–10	↑↓	–	+	–			
Total scores sensitivity								17	15	18
Total maximum scores								15	9	16
Vulnerability								0.37	0.55	0.37

Table 15.5 Climate change impact and sensitivity assessment

	Degree of sensitivity per climate change indicators		
Scoring	Δ T °C	Δ P %	Δ Ex
Total scores sensitivity (Sn)	17	15	18
Total max scores for sensitivity (Sn max)	15	9	16
Sn max/Sn	0.37	0.55	0.37

The regional approach in this process will be essential due to the unequal distribution of water resources of the country, both in regions and seasons, and the expected increase of that inequality in the future. In the long-term perspective, to the end of this century and beyond, we'll need to develop more flexible and risk-responsive management of waters, build on the need for adaptation to climate change. It is of crucial importance to put in operation new construction standards and technological innovations, which will facilitate the synergy between mitigation and adaptation.

Public participation, conceived as a partnership between the institutions responsible for the implementation of adaptation policies and business entities (companies, associations, etc.), as well as citizen participation, are essential for achieving the goals of the adaptation strategy.

Acknowledgements This research was carried out in frames of the 7 FP-INCO-2011-7, SWAN Sustainable Water Action and partly implemented in „Analysis and assessment of the Risk and Vulnerability of the Sectors of Bulgarian Economy to the Climate Change" (MOEW 2014).

References

Brooks N (2003) Vulnerability, risk and adaptation: a conceptual framework. Working Paper 38. Tyndall Centre for Climate Change Research, University of East Anglia, Norwich, pp 6–15. Retrieved from http://www.tyndall.ac.uk/sites/default/files/wp38.pdf

Denton F, Wilbanks TJ, Abeysinghe AC, Burton I, Gao Q, Lemos MC, Masui T, O'Brien KL, Warner K (2014) Climate-resilient pathways: adaptation, mitigation, and sustainable development. In: Field CB, Barros VR, Dokken DJ, Mach KJ, Mastrandrea MD, Bilir TE, Chatterjee M, Ebi KL, Estrada YO, Genova RC, Girma B, Kissel ES, Levy AN, MacCracken S, Mastrandrea PR, White LL (eds) Climate change 2014: impacts, adaptation, and vulnerability. Part A: global and sectoral aspects. Contribution of working group II to the fifth assessment report of the Intergovernmental Panel on Climate Change. Cambridge University Press, Cambridge and New York, NY, pp 1101–1131

EU (2013) Strategy for adaptation to climate change. Available from the Internet http://ec.europa.eu/clima/publications/docs/eu_strategy_en.pdf

Garcia J, Wagan A, Medina S, Simplicio M (2012) VAST-Agro: community-based vulnerability and adaptive capacity assessment for agriculture (ppt). College of Agriculture, UP Los Baños. Retrieved from https://www.google.bg/#q=VAST-Agro:+Community-based+Vulnerability+and+Adaptive+Capacity++Assessment+for+Agriculture

Holman IP, Rounsevell MDA, Shackley S, Harrison PA, Nicholls RJ, Berry PM, Audsley E (2005) A regional, multi-sectoral and integrated assessment of the impacts of climate and socioeconomic change in the UK, Part I. Methodology. Clim Change 71(1):9–41, Springer

IPCC (2007) Climate change 2007: synthesis report. In: Core Writing Team, Pachauri RK, Reisinger A (eds) Contribution of working groups I, II and III to the fourth assessment report of the Intergovernmental Panel on Climate Change. IPCC, Geneva, p 104

IPCC (2013) Climate change 2013: the physical science basis. In: Stocker TF, Qin D, Plattner G-K, Tignor M, Allen SK, Boschung J, Nauels A, Xia Y, Bex V, Midgley PM (eds) Contribution of working group I to the fifth assessment report of the Intergovernmental Panel on Climate Change. Cambridge University Press, Cambridge and New York, NY, p 1535. Available from the Internet http://www.ipcc.ch/report/ar5/wg1/

IPCC (2014a) Climate change 2014: impacts, adaptation, and vulnerability. Part A: global and sectoral aspects. In: Field CB, Barros VR, Dokken DJ, Mach KJ, Mastrandrea MD, Bilir TE, Chatterjee M, Ebi KL, Estrada YO, Genova RC, Girma B, Kissel ES, Levy AN, MacCracken S, Mastrandrea PR, White LL (eds) Contribution of working group II to the fifth assessment report of the Intergovernmental Panel on Climate Change. Cambridge University Press, Cambridge and New York, NY, p 1132

IPCC (2014b) Climate change 2014: impacts, adaptation, and vulnerability. Part B: regional aspects. In: Barros VR, Field CB, Dokken DJ, Mastrandrea MD, Mach KJ, Bilir TE, Chatterjee M, Ebi KL, Estrada YO, Genova RC, Girma B, Kissel ES, Levy AN, MacCracken S, Mastrandrea PR, White LL (eds) Contribution of working group II to the fifth assessment report of the Intergovernmental Panel on Climate Change. Cambridge University Press, Cambridge and New York, NY, p 688

IPCC—SREX (2012) Managing the risk of extreme events and disasters to advance climate change adaptation. In: Field CB, Barros V, Stocker TF, Qin D, Dokken DJ, Ebi KL, Mastrandrea MD, Mach KJ, Plattner G-K, Allen SK, Tignor M, Midgley PM (eds) A special report of working group I and II of the Intergovernmental Panel on Climate Change. Cambridge University Press, Cambridge and New York, NY, p 582

KNMI (2013) Klimate explorer, climate change Atlas. In: IPCC WG1 AR5 Annex I "Atlas". Available from the Internet http://climexp.knmi.nl/plot_atlas_form.py

Leal Filho W (ed) (2012) Climate change and the sustainable use of water resources, climate change management. Springer, Berlin, Heidelberg. doi:10.1007/978-3-642-22266-5_2

Leal Filho W (ed) (2013) Climate change and disasters risk management. Springer, Berlin

MOEW (2012) National strategy for the management and development of the water sector in Bulgaria. Annex 1, analysis of water use and future needs for water (In Bulgarian). Available from the Internet http://www.moew.government.bg/?show=top&cid=569)

MOEW (2014) Analysis and assessment of the risk and vulnerability of the sectors of Bulgarian economy to the climate change. Report, General part, pp. 224, 458 (In Bulgarian)

Millan MM (2014) Extreme hydrometeorological events and climate change predictions. J Hydrol 518:206–224. doi:10.1016/j.jhydrol.2013.12.041, Elsevier B. V

Nikolova M (2014) Water sector. In: MOEW report (2014) analysis and assessment of the risk and vulnerability of the sectors of Bulgarian economy to the climate change. Special part, pp 25–59 (in Bulgarian)

NSI (2015) Sustainable development, WEI. www.nsi.bg

Parry ML, Canziani OF, Palutikof JP, van der Linden PJ, Hanson CE (eds) (2007) Climate change (2007): impacts, adaptation and vulnerability. Contribution of working group II to the fourth assessment report of the Intergovernmental Panel on Climate Change (IPCC). Cambridge University Press, Cambridge. http://www.ipcc.ch/publications_and_data/publications_ipcc_fourth_assessment_report_wg2_report_impacts_adaptation_and_vulnerability.htm

Trenberth KE (2012) Framing the way to relate climate extremes to climate change. Climate Change 115:283–290. doi:10.1007/s10584-012-0441-5, Springer

UN (2009) Guidance on water adaptation to climate change. In: Economic commission for Europe convention on the protection and use of Transboundary Watercourses and International Lakes. New York, NY, p 144

UNISDR (2015) Sendai framework for disaster risk reduction 2015–2030. ICLUX EN 5000 1st edition, pp 5–36

World Bank (2013) Climate change impacts on water resources and adaptation in the rural water supply and sanitation sector in Nicaragua. World Bank. https://openknowledge.worldbank.org/handle/10986/16586. License: CC BY 3.0 IGO

Chapter 16
Adapting to the Inevitable: The Case of Tanbi Wetland National Park, The Gambia

Adam Ceesay, Mathias Wolff, Ebrima Njie, Matty Kah, and Tidiani Koné

Abstract The role of human activities in degradation of estuarine resources has been well documented. Besides the effects of climate change, activities such as clearing of mangroves for tourism, use of inappropriate fishing gear and excessive use of pesticides for agricultural productivity are the most powerful ecological stressors. In the Sahelian climate zone, hydrological regimes are changing due to reduced river flow and increase in atmospheric temperatures leading to the formation of inverse estuaries. The evaluation and documentation of local adaptation practices is one way to prevent "conservation bottlenecks" and encourage sustainable use of estuarine resources. This study used a questionnaire-based approach to evaluate local adaptation strategies to climate-induced ecological changes in the Tanbi Wetland National Park (TWNP) over the past three decades, targeting the communities that are engaged in the four major socio-economic sectors in the wetland i.e. Fishing, Agriculture, Oyster collection and Tourism. The agricultural zone presented the best local adaptation techniques employed as a response to ecosystem changes in the TWNP (23.53 %), followed by tourism zones (7.35 %) and fishing (5.88 %). With the disappearance of many fish species within the same timeframe, this leaves much to be desired. Bearing in mind that fisheries and tourism are the second and third largest contributors to the Gambia's GDP, this paper provides useful recommendations for management of this important wetland.

A. Ceesay (✉)
WASCAL GRP on Climate Change and Biodiversity, Université Felix Houphouët Boigny, Cocody, BP V 34 Abidjan, Côte d'Ivoire
e-mail: ceesay.adam@gmail.com

M. Wolff
Leibniz Center for Tropical Marine Ecology, University of Bremen, Fahrenheitstr. 6, 28359 Bremen, Germany

E. Njie • M. Kah
Department of Natural Sciences, University of The Gambia, Brikama Campus, 3530, Serrekunda, The Gambia

T. Koné
UFR Environnement, Université Jean Lorougnon Guédé, BP 150 Daloa, Côte d'Ivoire

W.L. Filho et al. (eds.), *Climate Change Adaptation, Resilience and Hazards*, Climate Change Management, DOI 10.1007/978-3-319-39880-8_16

Keywords TWNP • Mangrove estuaries • Fisheries • Socio-economic • Climate change • Adaptation

Introduction

The effects of climate change on human systems have been reported by many, although the responses to recent changes are hard to identify due to adaptation and the presence of many non-climatic driving forces (IPCC 2007). In its fourth assessment report, the IPCC highlighted that Africa is one of the most vulnerable continents to climate variability and change because of multiple stresses and low adaptive capacity. The report goes further to predict a decline of up to 50 % in yields from rain-fed agriculture by 2020 due to a reduction in arable land and changes in length of growing season in the arid and semi-arid areas. Fisheries resources, it states, will also decline due to rising water temperatures, exacerbated by continued overfishing; meanwhile, up to 30 % of global coastal wetlands are already lost due to continued mangrove ecosystems degradation and sea level rise (Church and White 2006).

Adaptation strategies to the aforementioned climate-induced ecosystem changes are gaining momentum as confidence in climate projections is getting higher. As the achievement of the Millennium Development Goals (MDGs) and successful implementation of the Poverty Reduction Strategy Papers (PRSPs) in developing countries are being achieved, adaptation strategies are now viewed as important goals (Mertz et al. 2009). In the earlier years, and to the contrary, the focus was mostly on climate mitigation (IPCC 2007). Mertz et al. (2009) pointed out that most climate data studies are done in developed countries and as such more information is needed in developing countries with tropical and subtropical climates for more knowledge in vulnerability and adaptive responses.

Over the years, research into the drivers of climate change, the magnitude of its impacts on livelihoods and the adaptations strategies have also increased, especially for the tropics where the physical impacts are predicted to be more severe. Africa for instance is expected to be warmer than the global mean, with a general decline in annual rainfall (Mertz et al. 2009). This will have great consequences for countries in arid and semi-arid zones that depend on the exploitation of natural resources from fragile ecosystems such as mangrove estuaries/wetlands, for sustenance of their socio-economic sectors. Due to the decline in agricultural productivity in countries such as The Gambia, dependency on estuarine resources has risen tremendously over the past couple of decades and now the biggest contributors to The Gambia's GDP are Agriculture, Fisheries and Tourism i.e. 27 %, 12 % and 8 % respectively (Government of The Gambia 2010).

The ecosystem services provided by mangroves are valued at US$ 900,000 per year (Corcoran et al. 2007), one can safely assume that mangrove estuaries, while serving as favorable nursery grounds for diadromous fish species (Baran 2000) and as a hub for socio-economic activities in the tropics and sub-tropics, are thus at an ecologically precarious state; especially in arid zones where they grow slowly because of climate-induced hypersaline conditions, low humidity, high temperatures, and extreme light conditions (Alongi 2008). Hence, the superimposition of climate-induced changes such as prolonged tidal inundation on other impacts resulting from human activities such as over fishing, pollution and habitat loss put these coastal areas under great stress (Chen 2008).

Mangroves occupy sensitive intertidal zones that are more prone to the immediate effects of climate change. Drastic changes in hydrology for instance was reported to induce stunting of *Avicennia marina* stands and denaturing of terminal buds in *Rhizophora mangle* seedlings in the USA (Kathiresan 2002). Austin et al. (2010) also suggested that modest changes in rainfall and temperature caused significant reductions in mean annual run off and increased stream salt concentrations in Murray-Darlin Basin (Australia), resulting in loss of mangrove vegetation. The frequent fluctuations between climate events such as extreme floods and prolonged droughts have also been reported to cause massive mangrove diebacks in Sub-Saharan countries such as Senegal and The Gambia (Dia 2012).

With an 80 km long coastline and a continental shelf area of about 4000 km^2 (IUCN 2010), The Gambia boasts of 68,000 ha mangrove estuary that accounts for 2 % of the total coverage for Africa (Spalding et al. 1997). This important part of the Western Africa Marine Eco-Region (WAMER) shelters about 600 species of fish, 26 species of cetaceans, 6 species of Turtles and more than 200 species of birds (Lee et al. 2009). But like many ecosystems in Sahelian countries, climate change has taken a serious toll on the stability of coastal and estuarine services due to erratic rainfall, increase in atmospheric temperatures and persistent droughts for the past three decades following the great Sahel drought in the 1970s (Dai et al. 2004). Annual rainfall in The Gambia has decreased by 30 % between the year 1950 and 2000 alone (Fig. 16.1), now remaining at a range of 850–1200 mm, the bulk of

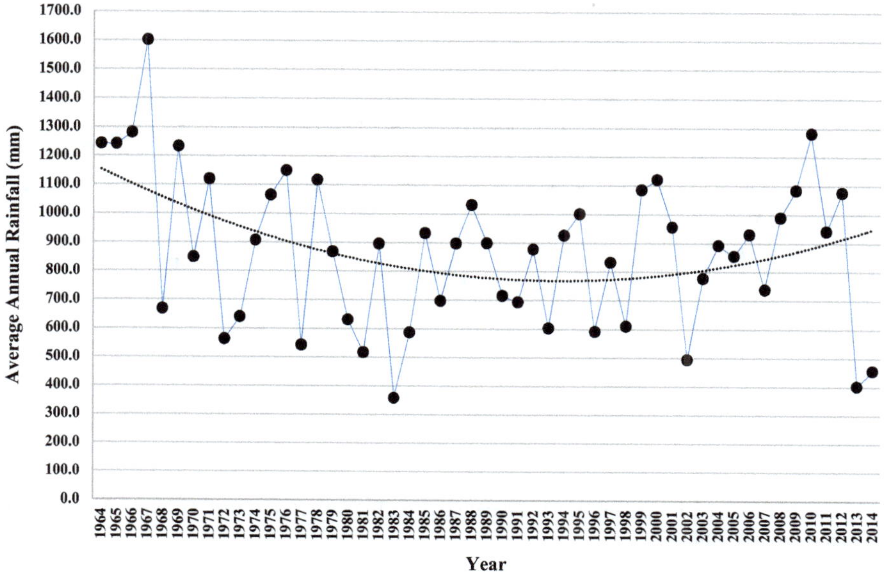

Fig. 16.1 Long-term rainfall pattern for Banjul, 1964–2014 (*Source*: The Gambia Meteorological Unit, 2015)

which occurs in August causing heavy floods in one-third of the country (Lee et al. 2009).

The progressive increase in atmospheric temperatures over the past five decades are not helping matters either. Long-term atmospheric temperatures for Banjul are shown in Fig. 16.2 below.

The Gambia's economy is heavily dependent on rain-fed agriculture, and as such the sector suffered the most decline (30 %) since the early 1970s (Government of The Gambia 2007). Like most countries in the Sahel, The Gambia responded by adapting and ratifying most (if not all) international climate-related accords including the UNFCC, Kyoto Protocol, Rio + 10 etc. (Lauer and Eguavoen 2016). In order to implement these accords at national level, The Gambia recently initiated programs such as the Program for Accelerated Growth and Employment (PAGE), the Gambia-Senegal Sustainable Fisheries Project, the Adaptation to Climate induced Coastal Changes Project etc., with the help of international bodies such as the UNDP, USAID and GEF (Lauer and Eguavoen 2016). In addition, positive changes were also effected in policies guiding the judicious use of natural resources. These include the formulation and enforcement of the Anti-littering Regulations to prevent indiscriminate waste dumping, and the Fisheries Regulations for safe-guarding the ecological integrity of fragile ecosystems such as the coast, which is already heavily influenced by the effects of Climate change (Government of The Gambia 2010).

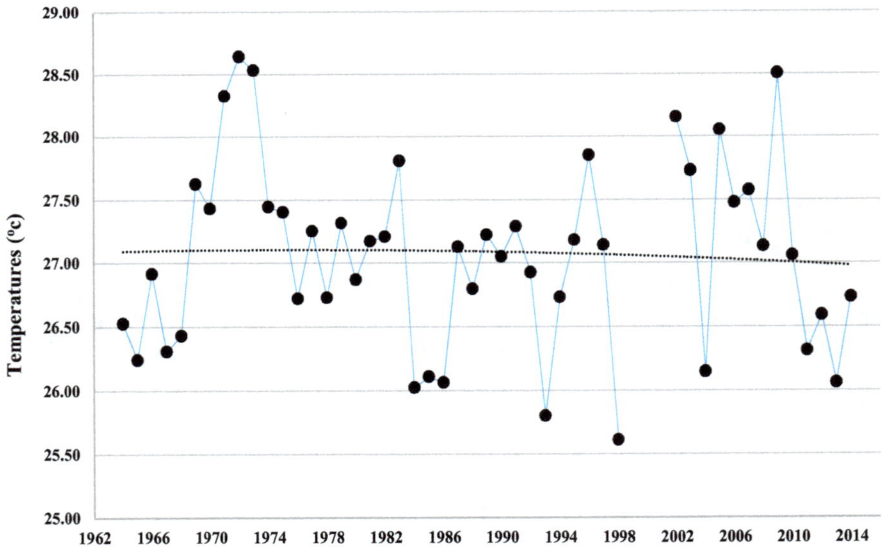

Fig. 16.2 Long-term atmospheric temperatures for Banjul, 1964–2014 (*Source*: The Gambia Meteorological Unit, 2015)

Making sure that the benefits of the aforementioned accords are felt by the local communities, farmer incentives such as provision of the drought resistant rice (NERICA) was introduced, to improve livelihoods of the poor communities involved in subsistence agriculture, as well as to create continued awareness of the farmers through agricultural extension workers and state-organized workshops (personal communication, Gambia Department of Agriculture). Successful implementation of projects such as the Gambia-Senegal Sustainable Fisheries project also gave birth to a bottom-top management approach to the fragile coastal resources such as the Tanbi Wetland National Park (TWNP), leading to the formulation of a co-management plan, which gave management rights to user groups such as the women oyster collectors. This, of course, is all based on the fact that there is an existent multi-sectoral team (comprised of all the relevant government institutions, CBOs and NGOs involved in conservation of natural resources, environmental sustainability and community development) (USAID-BANAFAA Project 2012).

Notwithstanding all the above, the socio-economic standing of vulnerable coastal communities in The Gambia leaves much to be desired. Socio-economic groups such as the fisher folks and tourism workers are not benefitting much from the institutional frameworks set up to reduce vulnerability of coastal communities within The Gambia. While abiding by all the rules set in conservation accords, these groups have little or no knowledge of the appropriate response strategies to apply/adopt at an individual level when faced with drastic ecosystem changes in their work environment. This leaves an information gap for the coastal communities, which needs to be urgently addressed for successful implementation of The Gambia's Climate Change adaptation strategies.

This paper aims to assess the socio-economic implications of climate-induced ecosystem changes in the mangrove estuaries of the River Gambia, where such information is almost non-existent. Using a questionnaire-based approach, this paper focused on evaluating relative vulnerability and adaptive capacity of the major socio-economic groups, an area that has not been well investigated in Sub-Saharan Africa. Filling this information gap will contribute toward successful implementation of National Adaptation Plans for mangrove-dependent countries such as The Gambia. Therefore, the main aim of this research was to gather and document local knowledge on climate-induced ecosystem changes, local adaptation practices adopted by the various socio-economic groups involved in Agriculture, Fisheries, Tourism, and Oyster collection, as well as the perceived changes in their economic gains in the TWNP.

Materials and Methods

Study Site: Tanbi Wetland National Park (TWNP)

The Tanbi Wetland National Park (TWNP) (Fig. 16.3) is a lowland area with a mean altitude of 1–1.6 m extending between 13°23–13°26 N and 16°34–16°38 W (The Ramsar Convention on Wetlands 2012). Located at the mouth of the River Gambia estuary, TWNP covers an area of about 6300 ha (Lee et al. 2009) and connects the three main urban settlements within the Greater Banjul Area (GBA). These are Banjul City (BC), Kanifing Municipality (KM) and Brikama (BA) (Government of The Gambia 2010). Due to its ecological richness, this wetland was designated a Ramsar wetland of importance in 2007 (Project 2012).

The TWNP falls within the Sahelo-Sudanian climate zone (Simier et al. 2006), having a long dry season (October–June) and a short rainy season (June–October) (Camara 2012). Mangrove vegetation in TWNP is comprised of *Rhizophora mangle* (red mangrove), *Avicennia germinans* (also known as *Avicennia africana*/black mangrove), *Laguncularia racemosa* (white mangrove) and *Rhizophora harisonni* (Maniatis 2005).

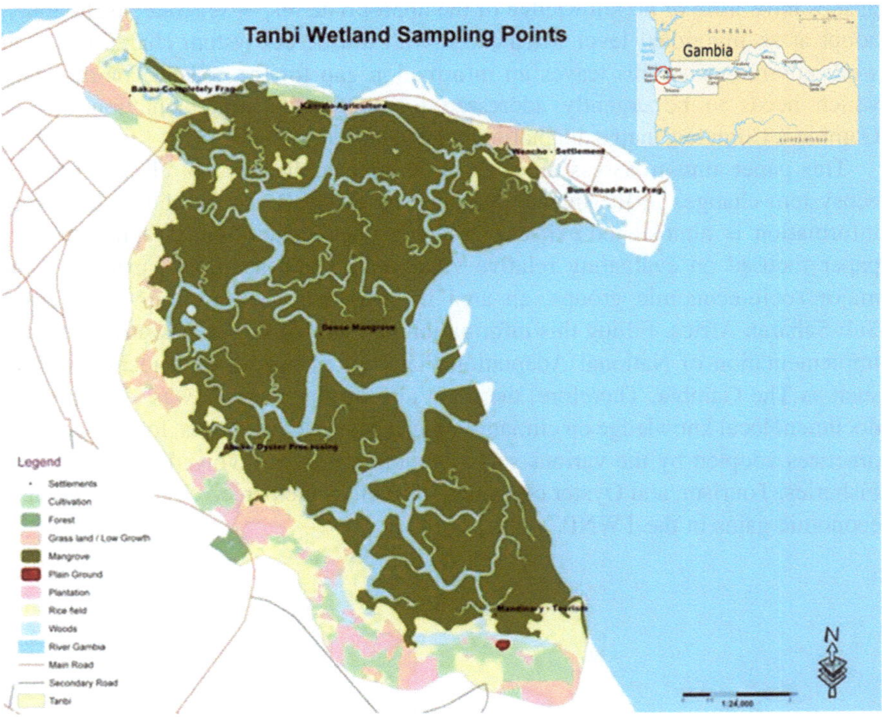

Fig. 16.3 Location of and socio-economic focus areas in Tanbi Wetland National Park on the Map of The Gambia

This mangrove habitat serves as an important nursery ground for fish species such as shad (*Ethmalosa fimbriata)* and sardine *(Sardinella maderensis)* (Baran 2000), African tilapias (*Sarotherodon* and *Oreochromis* species) (Albaret et al. 2004), as well as the pink shrimp (*Paneaus notialis)* and oysters (*Crassostrea gasar)* (Darboe 2002).

Economic activities in this area are dominated by Fisheries (including shell fishery) and Tourism (Satyanarayana et al. 2012). Agricultural activities (e.g. rice cultivation and gardening) are also common (USAID-BANAFAA Project 2012).

Methodology

Questionnaires

Questionnaires for this study were formulated to feature sustainability issues such as long-term (herein, three decades) ecosystem changes in TWNP. In addition, climate change matters were also included and their relationship with socio-economic setup based on the four categories below:

1. Ecosystem changes in TWNP (physical changes in terms of water quality, soil, vegetation and fisheries).
2. Climate change (as understood by the socio-economic groups in TWNP).
3. Economic status (as recorded in increase/decrease in daily earnings of the socio-economic groups in TWNP.
4. General observations (focusing on general understanding of the subject matter by the local people that are gainfully employed in TWNP).

Subject Groups (Interviewees) and Interviews

A group of 138 people belonging to four occupational groups were interviewed. These include: Farming, Fisheries, Tourism and Oyster collection. A subgroup of people residing within the wetland (in Wencho, locally known as Ndangane) were also interviewed. As this study was aimed at the local people's understanding of long-term changes in the ecosystem, the age group target was set at 18 years and above. For proportionate representation of the socio-economic groups, 15 tourist workers, 19 residents, 20 oyster collectors, 40 farmers and 44 fisher folks were interviewed.

Interviewers' group was comprised of five people (two from the National Environment Agency, one from the Water Resources Department, one from the University of The Gambia and the actual researcher). This team was put together based on their background in environmental monitoring, conservation and community development work. In addition, each of the team members is fluent in at least two local languages for translation purposes. All together the team is fluent in the four major local languages in The Gambia (Mandinka, Fulla, Wollof and Jola). To prevent any biases in the way questions were administered, the team prepared by "practice-asking" each other all the questions in the questionnaire and adopting a uniform introduction of the purpose and scope of the research in the four major local languages to ensure a uniform understanding of the questions.

Interviews were conducted under an informal roundtable chat with each socio-economic group at their place of work in TWNP. Where possible questionnaires were translated into the respondents' native language for better understanding. Interviews for each respondent lasted about 30 min. Questionnaire details are in appendix I.

Data Analysis

For statistical analysis, responses from the interviews were pooled in a similarity matrix and subjected to Principal Component Analyses (PCA) using Statistical analysis software R version 2.15.2 (2012). Descriptive statistical analyses were done using STATA 12 for windows. The methodology of this research was adapted from Satyanarayana et al. (2012).

Results

Based on the responses garnered during this study, 132 (95.65 %) out of the 138 respondents believe that the ecosystem of TWNP has changed over the past three decades, while the other 6 (4.35 %) disagreed/were undecided (Fig. 16.4).

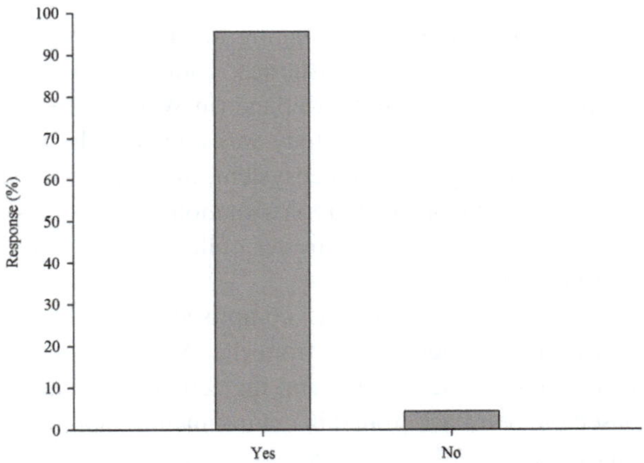

Fig. 16.4 Responses to the occurrence of ecosystem changes in TWNP

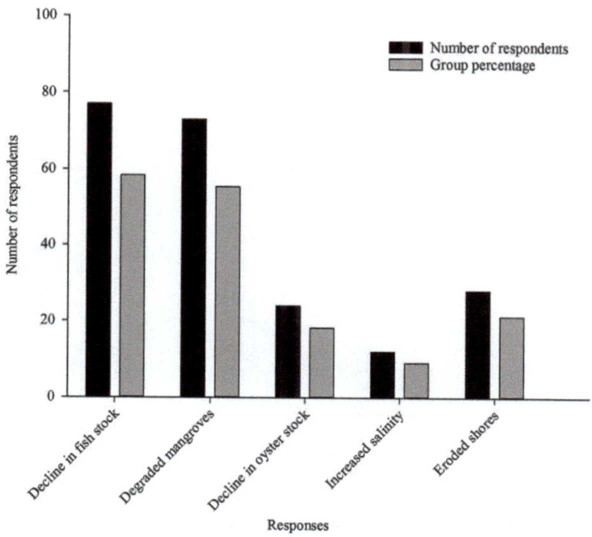

Fig. 16.5 Local perspective on specific ecosystem changes in TWNP

When asked about the specific changes within the TWNP, 58.33 % of the respondents mentioned a decline in fish stocks along with the disappearance of key species such as the giant African threadfin, which commands an attractive price compared to other native fish species. 55.30 % mentioned mangrove degradation, and 21.21 % of them mentioned soil erosion, while 18.18 % mentioned a decline in oyster stock (Fig. 16.5).

The local people's understanding of climate change is high despite the low literacy rates in The Gambia. 103 out of the 138 people (74.64 %) interviewed said they have heard of climate change before while 35 (25.36 %) said they have never heard of the concept before. When asked about the manifestations of climate change, 77.67 % of the respondents associated it with changes in rainfall pattern, 45.63 % associated it with changes in atmospheric temperatures, and 11.65 % to droughts (Fig. 16.6a and b).

Only one respondent was encountered (within the fisher folks group) who did not believe in the concept of climate change. Among those who have heard of the concept of climate change, 46.60 % claimed to have heard about it through the media, 26.21 % through workshops, 8.74 % from schools, 6.80 % from agricultural extension workers and 5.83 % from visiting tourists (Fig. 16.7).

Local understanding of climate-induced ecosystem changes such as seasonal variability and hyper-salinity was very high among the socio-economic groups in TWNP. On a scale of 0–10 (0 being poorest and 10 being excellent), the oyster collectors' group presented the highest level of awareness (8.25), followed by the fisher folks (7.52), residents (6.79), tourism group (6.53) and the least score was recorded with the farmers' group (6.23). In terms of awareness on seasonal hyper-salinity, responses followed a similar trend with the oyster collectors scoring 9.10,

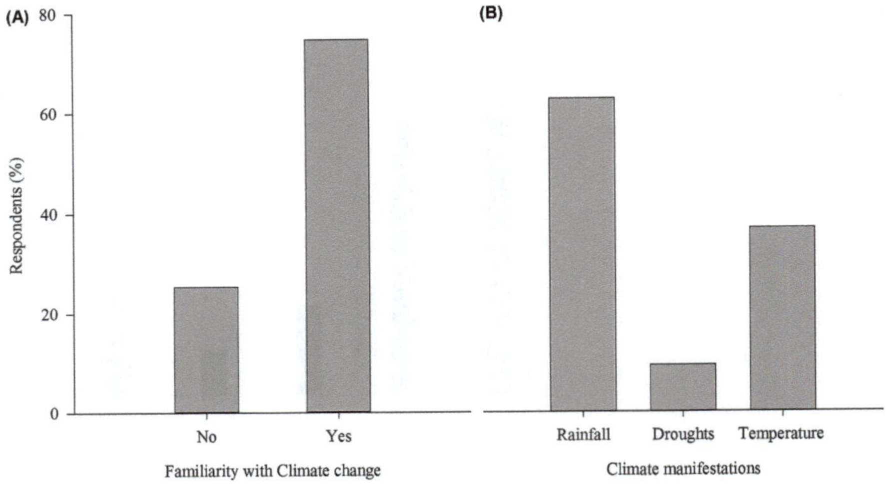

Fig. 16.6 Familiarity of the socio-economic groups with climate change concept (**a**) and local perception on manifestations of climate change (**b**)

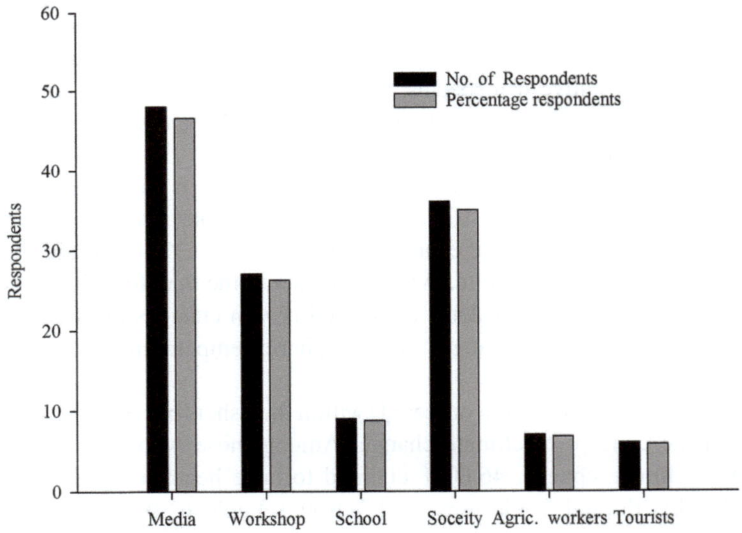

Fig. 16.7 Sources of Climate change information for the socio-economic groups in TWNP

fisher folks 8.18, farmers 7.30, residents 6.63 and the lowest scoring group being the tourism group (5.80) (Fig. 16.8).

In terms of the local adaptation measures/response strategies employed by the socio-economic as coping mechanism to the ecosystem changes in TWNP, a total of 32 different responses were given by the respondents. Most common among these was "nothing", 70 (50.72 %) out of the 138 respondents have no knowledge of

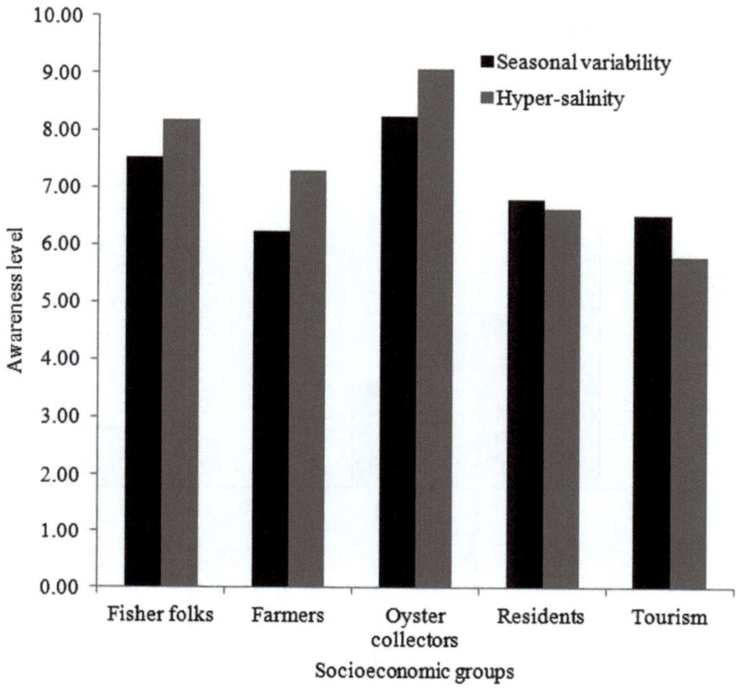

Fig. 16.8 Local understanding of climate-induced ecosystem changes by socio-economic groups in TWNP (score 0–10)

what measures they could take to maximize their daily output and prevent unnecessary economic losses in their place of work. 15 out of the remaining 68 respondents (22.06 %) seasonally shift their attention to exploitation of alternative resources allowing their main species of interest to "fallow" (case in point, oyster collectors who shift to crabbing and lobster fishing during the rainy season, when water quality is very poor in TWNP); 7(10.29 %) practice crop rotation based on observed changes in soil quality due to salt intrusion; while 9(13.24 %) change their usual sowing period in anticipation of late and insufficient rains (i.e. farmers' group). The "less fortunate" groups here 5 (7.35 %) deal with these changes by buying bigger boats (tourism group), while 4 (5.88 %) of all respondents deal with the ecosystem changes by expanding their fishing zones (this is the case of the fisher folks) (Fig. 16.9). Both of these measures in addition to being unsustainable are neither "pocket-friendly" nor "time-friendly", as they require huge financial inputs and longer fishing hours just for meagre catches.

In terms of economic growth, most of the respondents expressed a decline in income. About 65 % of the respondents from the oyster collectors expressed a decline in income over the past three decades, this was followed by the farmers group (61 % of the respondents), the residents (58 % of the respondents) and the

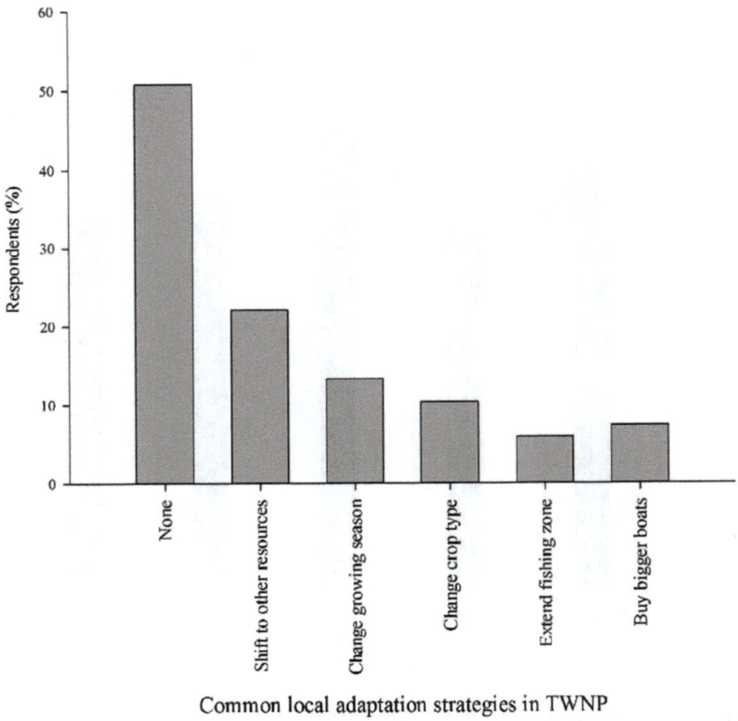

Fig. 16.9 Common local adaptation practices to climate-induced ecosystem changes in TWNP

Fisher folks (42 % of the respondents). About 57 % of the respondents from the tourism group expressed an increase in income (Fig. 16.10).

On average, the socio-economic groups considered here have registered a decline of 46 % in their earnings within the TWNP for the past three decades. The greatest average decline (72.73 %) was recorded among the fisher folks, followed by the oyster collectors (50 %) and the lowest (33.85 %) among the farmers groups' (Fig. 16.11).

Discussion

Mertz et al. (2009) advised for a careful formulation of policies and adaptation strategies to climate change for societies that are poor and vulnerable for a wide range of reasons. Hence, both physical and institutional assets need to be included when formulating adaptation strategies. In TWNP the former has been made available for the groups involved in shellfisheries through provision of canoes, sanitation facilities, oyster smoking pens and alternative livelihoods (setting up an oyster culture program), while the latter is available for groups involved in

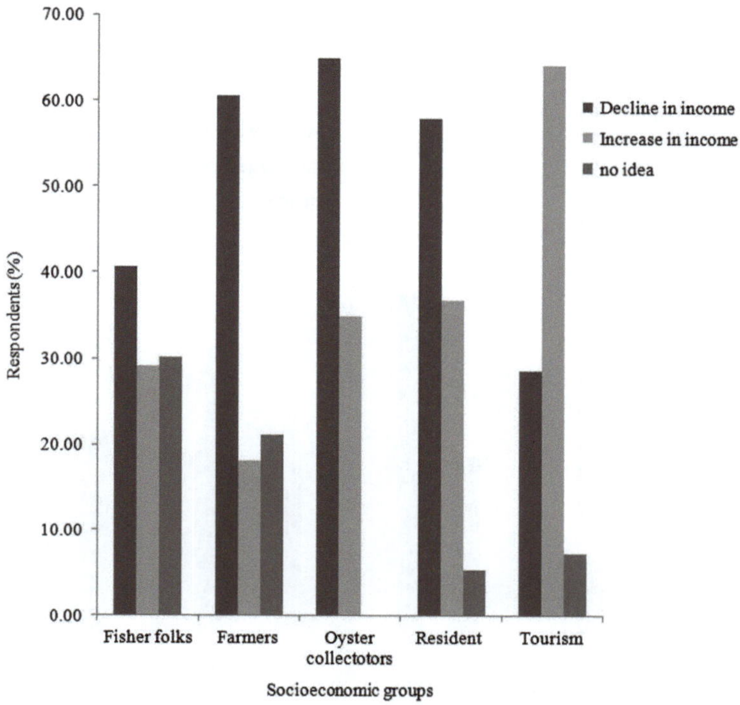

Fig. 16.10 Economic situation of the socio-economic groups in TWNP

agriculture through awareness creation by agricultural extension programs, credit schemes and subsidies for improved seed variety (National Environment Agency 2010).

The latter is also in part available for the fisher folks, through the coastal zone management plan, enforcement of fishing regulations, and guidelines on appropriate net usage within the wetland. While the aforementioned incentives are provided alongside numerous awareness creation campaigns for the socio-economic groups such as the farmers and oyster collectors, awareness levels are is still low for the fisher folks, as evidenced by their poor responses to adaptation mechanisms.

Kelly and Adger (2000) suggested that in cases concerning state-run adaptation programs, it might be necessary to "adapt to the adaptation," as some measures might solve one problem while creating another. As an example, (Barnett and Mahul 2007) suggested that credit schemes and new crops, accompanied by "weather insurance," have been tried experimentally in some developing countries. This approach was applied to a large extent in the agricultural sector, with provision of soft loan credit schemes with flexible payment plans that took account of the possibility of crop failure due to rainfall shortages (Government of The Gambia 2010). A similar approach was also applied for the shell fisheries sector, with the provision of alternative livelihoods through value-adding (packaging) of

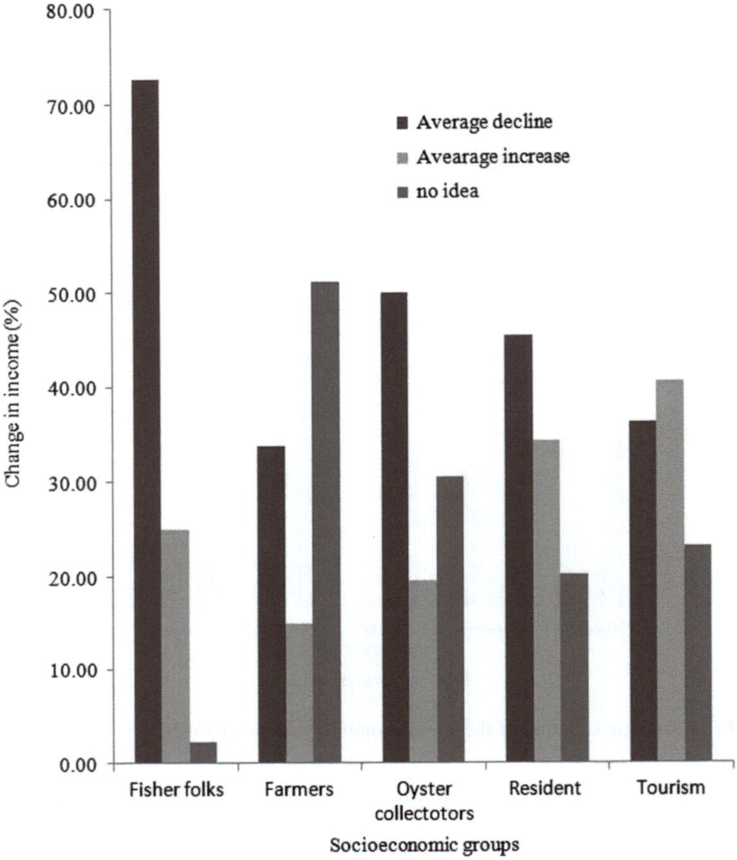

Fig. 16.11 Changes in average income of the socio-economic groups in TWNP

pre-processed oyster meat as well as setting up an oyster culture program (Crow and Carney 2013; USAID-BANAFAA Project 2012).

However, for the fisher folks only institutional assets are available in the form of the enactment of The Gambia's Fisheries Regulations (1995) and Act (2007) (Government of The Gambia 2007), as well as stricter netting regulations for sustainable fisheries in The Gambia allowing for no less than 20 mm mesh sized nets within the estuary of the River Gambia (personal communication, The Gambia Department of Fisheries). This has no doubt helped reduce overfishing, especially within the nursery grounds of the TWNP, but it still did not solve the problem of declining fish catches or the low economic returns as the physical assets needed by the fisher folks are not in place. This move might have also promoted complete colonization of estuary by hardy, but smaller sized fish species such as the native African tilapias (*Sarotherodon melanotheron*), which is known to respond to environmental stress by reducing its growth rate (Panfili et al. 2004).

In the words of Munang et al. (2010) it is necessary for policy makers to recognize the fundamental role of ecosystems as life-supporting systems first for successful implementation of adaptation strategies. In absence of a go-to activity for the fisher folks during times of low catches (as shown in this study), restricting fishing activity only lowers their earning potential without necessarily solving the problem of ecosystem change. This also does little or nothing to alleviate the declining fish catches, because fish species migrate in response to natural deterioration in their environment (Panfili et al. 2004).

Portner and Peck (2010) suggested that the implications of climate change for marine fish species are at four levels: organismal, individual, population and ecosystem level; individual level being the most relevant one when referring to the state of fisheries in TWNP. Over 23 % of the strict estuarine fish species for instance have been reported to seasonally migrate inland as a response to hypersalinity during the peak dry season and then to the creeks during the flood season (Ceesay et al. 2016—article in press). This is now believed to be the reason behind the drastic reduction in daily catches/earnings of the fisher folks interviewed during this research.

Similar findings were reported by Panfili et al. (2004) in a study of the impacts of salinity on the life traits of the native African tilapia (*Sarotheron melanotheron*) in the estuaries of Senegal and The Gambia, where he reported interruptions in fish migration patterns, reproduction timing and spawning success as well as stunting, leading to smaller market sized fish and thus earning lower prices. Such phenomenon was also reported to be the cause for the complete colonization of the inverse estuary of Sine-Saloum delta in Senegal by marine fish species; thus replacing the native estuarine species and causing a decline in fisheries output for over a decade (Dia 2012).

The aforementioned condition is worsened by the synergistic effects of human activities such as overfishing on climate-induced ecosystem changes, as evident in the responses from the fisher folks. In addition, the commonness of the phrase "in God's hands" among the fisher folks as a response to how they locally adapt to climate-induced ecosystem changes indicates their low levels of awareness. Thus leading to the deduction that the fisher folks are the most vulnerable socio-economic group when it comes to climate change adaptation in TWNP. This group has been relatively overlooked and uninvolved when it comes awareness creation campaigns by conservationists as well as in the provision of alternative livelihoods by state institutions.

The case is different for the farmers' group which apart from having its members frequently trained on alternative farming methods and provided with improved seed variety in order to adapt to climate-induced ecosystem changes in TWNP; also has the luxury of having an agricultural extension worker on site to guide them in responding to common farming problems. In comparison, the oyster collectors' group having been provided with suitable alternative livelihood programs such as the state-led oyster culture program. In addition, they were recently given management rights to the TWNP. Meanwhile, the tourism group finds it much easier to cope with the ecosystem changes, for instance by expanding sightseeing zones and proportionally increasing safari prices, which may not be a realistic approach for the other groups as one is catering to a local market, while the other targets an international market.

Of course, it is noteworthy that this current study also has its own set of limitations. For instance, difficulties in locally translating basic climate concepts to the mostly uneducated respondents might have downplayed their ability to accurately respond to questions about climate-ecosystem changes. The general mistrust between local resource users/end users and state-employed conservationists also played a dampening role in people's willingness to answer the questionnaires. Nonetheless, sufficient information was garnered to portray overall understanding and response mechanisms employed by the locals in order to deal with climate-induced ecosystem changes in the mangrove estuary under study (TWNP), as well as the socio-economic implications of these changes.

Conclusions

Interpretations of the responses gathered from the socio-economic groups during this research work in TWNP indicate the following:

- Even though slow, there has been a progressive increase in general awareness about climate change and how it affects socio-economic activities within the fragile mangrove ecosystems in The Gambia.
- Environmental awareness campaigns have been more successful with the groups involved in oyster collection and farming (who are mostly female), and least so with the groups involved in fishing and tourism (all male); as evident in their responses to observed changes and coping mechanisms. This calls for the need to make gender issues a principal component when designing awareness creation programs.
- Based on findings of this research, the reported decline in fisheries of the TWNP is mostly climate-induced. The reduction of fishing pressure and strict enforcement of Fisheries policies, legislations and net size regulations for sustainable fisheries has been going on for the past couple of decades. Yet, the positive effect of these measures are at best negligible; thus calling for the need to review the guiding fisheries policies, as well as provision of alternative livelihoods for the fisher folks.
- There is also a growing need for more research into the response mechanisms of aquatic species to environmental change and how this affects the lives of the common fisher folks. This will avail The Gambia a possibility of identifying the best modes of intervention needed in terms of sustainable alternative livelihoods for those gainfully employed within TWNP.

References

Albaret J-J, Simier M, Darboe FS, Ecoutin J-M, Raffray J, de Morais LT (2004) Fish diversity and distribution in the Gambia Estuary, West Africa, in relation to environmental variables. Aquat Living Resour 17:35–46. doi:10.1051/alr:2004001

Alongi DM (2008) Mangrove forests: resilience, protection from tsunamis, and responses to global climate change. Estuar Coast Shelf Sci 76(1):1–13. doi:10.1016/j.ecss.2007.08.024

Austin J, Zhang L, Jones RN, Durack P, Dawes W, Hairsine P (2010) Climate change impact on water and salt balances: an assessment of the impact of climate change on catchment salt and water balances in the Murray-Darling Basin, Australia. Clim Change 100:607–631. doi:10.1007/s10584-009-9714-z

Baran E (2000) Biodiversity of estuarine fish faunas in West Africa. Naga ICLARM Q 23:6

Barnett BJ, Mahul O (2007) Weather index insurance for agriculture and rural areas in lower income countries. Mississippi State University, Starkville, MS, 19p

Camara AS (2012) Protected Areas Resilient to Climate Change, PARCC West Africa. National Data Collection Report: UNEP-WCMC, Banjul, The Gambia, 42 p

Ceesay A, Wolff M, Koné T, Njie E, Matthias B (2016) Seasonal changes in water quality and fisheries of Tanbi Wetland National Park. Regional Studies in Marine Science. http://dx.doi.org/10.1016/j.rsma.2016.04.006

Chen CA (2008) Effects of climate change on marine ecosystems. Fisheries for Global Welfare and Environment, 5th World Fisheries Congress, pp 307–316

Church JA, White NJ (2006) A 20th century acceleration in global sea-level rise. Geophys Res Lett 33:94–97. doi:10.1029/2005GL024826

Corcoran E, Ravilious C, Skuja M (2007) Mangroves of western and central Africa. Regional Seas Programme/UNEP-WCMC, Cambridge, UK, 92p

Crow B, Carney J (2013) Commercializing nature: Mangrove conservation and female Oyster Collectors in The Gambia. Antipode 45:275–293. doi:10.1111/j.1467-8330.2012.01015.x

Dai A, Lamb PJ, Trenberth KE, Hulme M, Jones PD, Xie P (2004) The recent Sahel drought is real. Int J Climatol 24(11):1323–1331. doi:10.1002/joc.1083

Darboe FS (2002) Fish species abundance and distribution in the River Gambia Estuary. UNU–Fisheries Training Programme, pp 1–40. http://www.unuftp.is/static/fellows/document/famara2002prf.pdf

Dia MI (2012) Vulnerability assessment of Central Coast Senegal (Saloum) and The Gambia Marine Coast and estuary to climate change induced effects. Coastal Resources Center and WWF-WAMPO, Consolidate report. University of Rhode Island, United States of America, 40p

Government of The Gambia (2007) The Gambia fisheries Act. Supplement "C" to The Gambia Gazette No. 20. Banjul, The Gambia, 54p

IPCC (2007) Climate change: impacts, adaptation and vulnerability. Contribution of the Working group II to the Fourth assessment report of the Intergovernmental panel on climate change IV. In: Parry ML, Canziani OF, Palutikof JP, Van der Linden PJ, Hanson CE (eds) Cambridge University Press, Cambridge, 987p

IUCN (2010) National diagnostic report for The Gambia on climate change vulnerability. International Union for Conservation of Nature (IUCN), Banjul, The Gambia, 132p

Kathiresan K (2002) Why are mangroves degrading? Curr Sci 83(10):1246–1249

Kelly PM, Adger WN (2000) Theory and practice in assessing vulnerability to climate change and facilitating adaptation. Clim Change 47(4):325–352. doi:10.1023/A:1005627828199

Lauer H, Eguavoen I (2016) Mainstreaming climate change adaptation into development in The Gambia—a window of opportunity for transformative processes? In: Filho WL (ed) Handbook of innovation in climate change management. Springer, Switzerland, pp 87–98

Lee V, Tobey J, Castro K, Crawford B, Ibrahima Mat D, Ousman D, Tanvi V (2009) Gambia-Senegal sustainable fisheries project. In: Marine biodiversity assets and threats assessment for WAMER, p 50

Maniatis D (2005) Retrospective study of the Mangroves of the Tanbi Wetland Complex. Master's thesis submitted to the University of Vrije, Bussel, Belgium 129p

Mertz O, Halsnaes K, Olesen JE, Rasmussen K (2009) Adaptation to climate change in developing countries. Environ Manage 43:743–752. doi:10.1007/s00267-008-9259-3

Munang R, Rivington M, Takle ES, Mackey B, Thiaw I, Liu J (2010) Climate information and capacity needs for ecosystem management under a changing climate. Procedia Environ Sci 1(1):206–227. doi:10.1016/j.proenv.2010.09.014

Panfili J, Mbow A, Durand J-D, Diop K, Diouf K, Thior D et al (2004) Influence of salinity on the life-history traits of the West African black-chinned tilapia (Sarotherodon melanotheron): Comparison between the Gambia and Saloum estuaries. Aquat Living Resour 17:65–74. doi:10.1051/alr:2004002

Portner HO, Peck MA (2010) Climate change effects on fishes and fisheries': towards a cause-and-effect understanding, pp 1745–1779. doi:10.1111/j.1095-8649.2010.02783.x

Project U-B (2012) Report on Sanitary Shoreline survey within the Tanbi Wetlands National Park and other shellfish harvesting communities, The Gambia

Satyanarayana B, Bhanderi P, Debry M, Maniatis D, Fore F, Badgie D et al (2012) A socio-ecological assessment aiming at improved forest resource management and sustainable eco-tourism development in the mangroves of Tanbi Wetland National Park, the Gambia, West Africa. Ambio 41:513–526. doi:10.1007/s13280-012-0248-7

Simier M, Laurent C, Ecoutin JM, Albaret JJ (2006) The Gambia River Estuary: a reference point for estuarine fish assemblages studies in West Africa. Estuar Coast Shelf Sci 69:615–628. doi:10.1016/j.ecss.2006.05.028

Spalding M, Blasco F, Field C (eds) (1997) World Mangrove Atlas. The International Society for Mangrove Ecosystems, Okinawa, Japan, 178p

The Gambia Government (2007) Gambia National Plan of Action (NAPA) on Climate Change for The Gambia. GOTG, UNEP and GEF, Banjul, The Gambia, 97p

The Gambia National Environment Agency (2010) The Gambia State of the environment report, 2nd edn. National Environment Agency, Banjul, The Gambia, 263p

The Ramsar Convention on Wetlands (2012) Ramsar sites in order of their addition to the Ramsar list of wetlands of international importance, pp 1–37. http://www.ramsar.org/pdf/sitelist_order.pdf

Chapter 17
Climate Change Adaptation in Indian Agriculture- Assessing Farmers' Perception and Adaptive Choices

Chandan Kumar Jha and Vijaya Gupta

Abstract The impacts of climate change are expected to be the most devastating market failure in modern times. India's vulnerability to climate change is apparent with the frequent occurrence of flood, drought and cyclones in the recent past. Agriculture being one of the primary sources of livelihood of the country and the most climate centric activity, climate change is likely to significantly affect the key outcomes of agriculture systems and economic development. The most practical way to manage the undesirable climatic consequences is adaptation. Therefore, farm-level analysis of adaptive endeavors is prime requisite to understand the dynamics of adaptation to climate change.

This paper, tries to identify the major parameters which determine Indian farmers' awareness and expectation of climate change and the factors affecting their adaptive choices. The study also attempts to assess the key adaptive strategies which farmers intend to adapt depending upon agro-climatic conditions and constrained by their socio-economic situations. The observations of this paper will help in identification of micro-level barriers to adaptation and will facilitate appropriate policy formulation to ensure maximum returns out of the changing climatic conditions.

Keywords Climate Change • Vulnerability • Agriculture • Adaptation Strategies • Socio-economic condition

Introduction

Climate change is a continuous process and its reparation is becoming prominent with due course of time. Agriculture being the most climate centric economic activity, the sector is highly vulnerable to environmental shocks. Unpredicted

---

C.K. Jha (✉) • V. Gupta
Department of Economics and Strategy, National Institute of Industrial Engineering (NITIE), Mumbai 400087, India
e-mail: chandan1929@gmail.com; vijaya298@gmail.com

W.L. Filho et al. (eds.), *Climate Change Adaptation, Resilience and Hazards*, Climate Change Management, DOI 10.1007/978-3-319-39880-8_17

fluctuations in climatic factors like changes in temperature, the level and timing of precipitation, humidity etc. pose serious risks to global food production (Adams et al. 1998; Stern 2007); further raising complex challenges like food insecurity, malnutrition, rural poverty and environmental degradation especially for developing countries lying in the tropical and sub-tropical regions (Parry et al. 2007; Porter et al. 2014; Stern 2007; Mendelsohn and Dinar 1999; Adams et al. 1998; Mendelsohn 2008). Agriculture serves to be the primary source of livelihood in most of the developing countries (Mendelsohn 2008) like India and therefore assessment of climate change impacts is of utmost importance. Agriculture plays a key role in Indian economy, particularly in providing livelihood to about 60 % of the population. The sector contributes about 13 % to the economy (Ministry of Indian Agriculture 2014) however; continuous decline this trend is raising serious concerns for rural livelihood. Increasing population pressure, land-use change, shrinking natural resource base etc. is pressing the need for sustainable agricultural practices which requires that climate change is endogenously tackled within the production boundary.

Indian agriculture is highly vulnerable to climate change as agriculture output depends on vagaries of monsoon (Kumar and Parikh 2001). India's exposure to climate change is apparent with the frequent occurrence of flood, drought and cyclones in the recent past. The climate forecasts of Indian Meteorological Department (IMD) suggest a surge of 0.56 °C in annual mean temperature for the period of 1901–2009 for the country against the global warming of about 0.74 °C (IPCC 2007, referred as Parry et al. 2007). Spatially most part of the country experienced an increasing trend of mean annual temperature with maximum rise by 0.77 °C in post-monsoon season and in winter season by 0.70 °C. However, the annual and monsoon rainfall do not show any significant trend. According to Kavikumar (2009), India is set to experience an overall temperature increase of about 2–4 °C by 2100. Whereas, seasonal predictions suggest an increase in mean temperature by 0.4–2.0 °C during Kharif and 1.1–4.5 °C in Rabi season, with only a 10 % increase in mean rainfall during by 2070 (Khan et al. 2009).

Climate change impacts on agriculture mainly result in loss in farm net revenue or agricultural yield. A surge in temperature by 2 °C and 8 % increase in precipitation are likely to reduce agricultural net revenue by 12 % in India without carbon fertilization resulting into an annual damage of about 4–26 % (Sanghi and Mendelsohn 2008). Long-term projection suggests an estimated loss of about 30 % in agricultural productivity by 2080 for India (Pearce et al. 1996). Specifically, the most prominent impact of climate change is likely to be on rice and wheat yield which serves to be a staple food for the country (Jha and Tripathi 2011). The intensity of climate change impact however differs spatially within the country depending upon agro-ecological settings at regional scale and most importantly farmer's adaptation at farm level. The eastern wet regions of the country are expected to gain advantage from warming whereas the dry western regions are expected to witness huge losses (Mendelsohn 2008). Hijioka et al. (2014) suggest a decrease of about 50 % of wheat yield in the Indo-Gangetic plains due to surge in temperature.

The economic and agronomic impact of climate change mainly depends on variations in climate variables and the ability of agricultural systems to adapt to such changes. Adaptation entails local coping practices to restrain vulnerability of agro ecosystem to climate variations and extremes and ensure long-term resilience to future climatic turbulences. Guiteras (2009) find that crop yields in India are expected to reduce by 4.5–9 % over medium term (2010–2039) while in the long-run (2070–2099), no adaptation is likely to reduce yield by 25 %. Indian agriculture is primarily rainfed and therefore is more prone to climate risks as people dependent on rainfed agriculture lack adaptive capacity. Farmers being the key actors in agricultural system; their behavioral attitudes towards climate change are often complex and poorly understood (de Jalón et al. 2015) therefore, a clear understanding of farmers perception on climate change and their willingness to adapt is important (Arbuckle et al. 2013). Farmer's adaptation to climate change can ensure sustainable economic returns by taking benefits from changing climatic conditions. Farmers generally make rational choices from a set of adaptation strategies in the form of farm practices and technologies; available in their regions (Gbetibouo 2009). However, timely recognition of climate variations, incentive and most importantly, ability to adapt serves to be the three critical component of successful adaptation (Fankhauser et al. 1999).

Extensive research has been carried out to identify different adaptation strategies adopted by farmers. However, how socio-economic and other factors determine choices of adaptation strategies and adaptive capacity, especially for India has been seldom attended. Therefore, this study tries to understand the adaptation behavior of the Indian farmers and the key factors affecting their adaptive capacity and choice of adaptation strategies. This paper is arranged in four sections. The second section briefly discusses the methodology adopted for the study. The third section identifies the socio-economic factors determining farmers' adaptive capacity and evaluates how these factors play in Indian context. The last section concludes the critical findings of the study.

Methodology

This study has attempted to identify the key socio-economic variables and other factors determining adaptive capacity of farmers based on review of adaptation literature and how these factors affect Indian farmers. With a view to empirically support the selected farm and household characteristics specifically under Indian setting, the study has used secondary data Indian National Sample Survey (NSS) and Census (2011) for demographic details. This study has used the 70th Round of NSS conducted to assess the state of affairs of Indian farmers in the annual year of 2013 (January–December, 2013) covering rural India. The survey was found suitable for this study as the survey results gave detailed information on several facets of farming practices and socio-economic characteristics like income and expenditure of households, ownership of land and assets, access to agricultural resources including technology, agricultural awareness and extension services and credit availability and indebtedness of Indian farmers or agricultural households. The observations of the survey depicted an overall state of Indian farmers on parameters. Data from Census 2011 was used to get the demographic details of rural India.

Results and Discussions

Key Factors Affecting Adaptive Capacity of Farmers

Farm level adaptation strategies includes changes in farm inputs, managing crop sowing and tilling timings, alternate irrigation practices, strategic crop choices such as inclusion of warmer season crops, crop switching, livestock management, use of fertilizers and pesticides, improved weather forecasts, diversification to off-farm activities and soil and water conservation (Mendelsohn et al. 1996; Cline 2007; Adams et al. 1998; Darwin 1999; Mendelsohn and Dinar 2003; Gbetibouo 2009; Nhemachena and Hassan 2007; Kurukulasuriya and Mendelsohn 2008; Bryan et al. 2009; Below et al. 2010). Adaptation to climate change primarily requires that farmers perceive that climate is changing (Deressa et al. 2009; Maddison 2007; Below et al.; 2010). Education, access to extension services, external forces such as their peer's awareness, societal ethics, social capital, wealth, climate information and age of household head establish farmer's perception. However, farmers' perception on climate events does not guarantee taking adaptation measures (Bryan et al. 2009). In this respect, farmer's incentive and ability to adapt determines farmer's responses towards climate change. Farmers' decision making is an individual response often guided by intra household factors such as uncertain flow of income and environmental perceptions (Smit and Skinner 2002). The final response of adapting to climate change is determined by their adaptive capacity often determined by farmers' skill, education and personal ability (Tarleton and Ramsey 2008). Several studies have emphasized on farmer's household and socio economic characteristics as an important factor in determining farm-level adaptation capacities and decisions (Nhemachena and Hassan 2007; Deressa et al. 2009; Below et al. 2010; Falco and Veronesi 2013).

Socio-Economic Factors

- ***Age of the farmer:*** Age of the farmer or household head can determine farmer's perception, willingness to adapt and adaptive choices in two ways. Age of the farmer embodies farm experience which induces climate change perception and technological adoption (Maddison 2007; Nhemachena and Hassan 2007). Years of farm experience is expected to be related to the ability of farmer to attain;

process and use relevant information in better way and can therefore augment farmer's perception (Adesina and Zinnah 1993). More experience of farming system associated with age of the peasant increases the likeliness of perceiving soil erosion problem and its economic impact (Shiferaw and Holden 1998) and the possibility of crop diversification, changing planting dates and area under production (Gbetibouo 2009). However, aged farmers might often be reluctant to adopt conservation practices once they perceive the problem due to their attachment to traditional farm practices which they have been following for long. Younger farmers are more knowledgeable and aggressive about adoption of new technologies and risk taking (Adesina and Zinnah 1993) and are more prepared for long-term farm management such as irrigation and crop-livestock system.

- ***Gender:*** Usually there exists a positive relation between gender (male farmers or male head of the household) and farmer's decision to adapt. The head of the family is usually the main decision maker of the family and chiefly handles intra-household resource allocation and farm decisions. Gender of the farmer determines farmers' choice of adaptation strategies such as crops diversification strategies (Yegbemey et al. 2013) and agricultural technology as it influences access to forecasts and information (Bryan et al. 2009). The gender effect on adaptation however depends upon socio-cultural settings of the region, alternate sources of income for the females and type of family labour allocation. Ownership of assets is often gender biased as women usually lack ownership rights over land although they may have user rights thus limiting their decision making power. In rural Tanzania, women are restricted from access to land and credit which in turn limits their access to education (Below et al. 2010). Udry (1996) finds prominent gender division of labor in Africa as crop choice systematically differs by gender.
- ***Household size*:** The effect of household size has mixed impacts on farmers' adaptation responses (Nhemachena and Hassan 2007; Bryan et al. 2009; Gbetibouo 2009). Household size signifies intra-family labor supply which facilitates adoption of labor intensive adaptation measures in large households. Households with larger human capital invest more in conservation (Shiferaw and Holden 1998) and may also divert part of their labor to non-farm activities for income security (Gbetibouo 2009; Hisali et al. 2011) and cover up for weather uncertainty. As far as adaptation strategies are concerned, strategies like land allocation (Kokoye et al. 2013) double sowing instead of single sowing (Yegbemey et al. 2013), mixed farming systems, irrigation being more labor intensive; large family size facilitates adoption of such tactics.
- ***Education of the farmer*:** Several studies have established a positive relation between education and farmer's ability to perceive climate change and the likelihood of technological adoption (Norris and Batie 1987; Deressa et al. 2009; Gbetibouo 2009; Yegbemey 2013). Decision-making is a decentralized process and education can have intra-household spill-over effect on adaptation decision i.e. flow of knowledge from other family members to household head (Asfaw and Admassie 2004). Education enhances the ability of farmers to acquire, synthesize and respond to innovations such as chemical

fertilizer, farmer's perception on rainfall patterns (Gandure et al. 2013) and soil erosion.

- ***Off-farm income source:*** Occupational structure of the household; both primary and secondary also has major implication on adaptation decisions and choices. Off-farm income make farmers risk-averse as they can diversifying their livelihood source and helps in consumption smoothening in case of adverse weather and production outcomes, especially for households with large family in rural areas. Off-farm income has positive effects on adoption of fertilizer and pesticide technologies (Lamb 2003), ease liquidity constraint needed for soil conservation investments (Shiferaw and Holden 1998) and also helps in proper livestock management (Gbetibouo 2009). However, on the negative side, due to the assurance of income from non-farm activities, less knowledgeable farmers may be reluctant to adopt improved farm techniques as they might consider changes to be costly with uncertain returns and also contend for on-farm managerial time which may in turn increase reliance on crop insurance (Smit and Skinner 2002).
- ***Farm size:*** Farm-level adaptive response majorly depends on farm size as it determines the feasibility of adopting any particular strategy. Farm size is often considered as wealth indicator (Deressa et al. 2009) and may help ease liquidity constraint. Farm size can have both negative and positive consequence on adaptive decisions (Bradshaw et al. 2004). On positive side, advanced land management practices, farm mechanization and adoption of an innovation has proved to ensure more returns when applied to large farm size. A general perception prevails that farmers with large land holding are more willing and capable of adopting best suited farm strategies such as crop diversification, extensive irrigational arrangements, crop switching, adoption of pesticides and fertilizer technology. Although large farms can give lower yields at an initial stage but in the long run economies of scale is expected to lower the large fixed transaction costs of innovation. Small farm size may also often cause conflicts among household members with large family size and may affect individual decision making.

Institutional Factors

- **Extension services-** Farm extension services enunciate the process of 'social-learning' (Tazeze et al. 2012) among farmers and fasten ex-ante process of adaptation. Provision of free extension services have strong positive influence on the probability of choosing adaptation measures (Maddison 2007; Deressa et al. 2009). Extension services help overcome the problem of asymmetric information and generate distinct welfare effects through better flow of knowledge. Education and access to information can reduce costs of adaptation and risks enabling early responses (Wozniak 1987). Agriculture extension services can be in form of government extension services, farmer-to-farmer and information from radio, television or mobile phones (Falco and Veronesi 2013).

Farmers in villages usually observe farming activities of fellow farmers, including those experimenting with new technologies and accordingly update their own perceptions and decide on cultivation for the next season. Bandiera and Rasul (2006) find that in Mozambique, individual technology adoption decisions of farmers usually depend upon the others choices in the same social network and the network effect is stronger for farmers who engage in conversation with other farmers.

- **Access to Credit-** Financial access in form of credit and insurance from different agencies and individual cash holdings is an important catalyst for adaptation. Financial well being determines the adaptive capacity of farmers as it provides a sense of security to combat unpredicted impacts of climate extremes. Access to credit has been considered as a serious barrier to adaptation by several studies (Napier 1991; Deressa et al. 2009) especially for developing countries. Financial access induces farmers to change their management practices and increases the likelihood of adopting strategies like soil conservation, changing planting dates, irrigation, adoption of technology, use of high variety seeds, acquiring transportation, and hiring agricultural workers. In contrast, Hisali et al. (2011) find households without access to credit in Uganda are more open to technology adoption to adapt to livestock epidemic; probably due to availability of loan repayment options.

Socioeconomic Factors Affecting Farmers' Adaptation in Indian Context

Farm-level decisions in India are made over short-run and is by and large affected by inter annual or seasonal variations in climate elements. Indian farmer's adaptation to climate change is mainly dependent on their motive to minimize the risks associated with crop failure due to weather shocks. Farmer's usually adapt to maintain sustained flow of income throughout the year. Since farmers also utilize a certain part of their agricultural produce for household purpose; consumption smoothening is another objective of adaptation. However, it is often the case that even if farmers are willing to adapt they do not, due lack of adaptive capacity. Indian farmers usually are less capable of adapting to sudden shocks in climate and production system due to their dependence on natural inputs, the lack of technological know-how and limited access to institutional support systems. The basic adaptation rule followed by farmer's in India is maximization of net revenue constrained by their socio-economic situations (household size, age of the farmer, gender, education level, off-farm income and farm size), access extension services and access to credit.

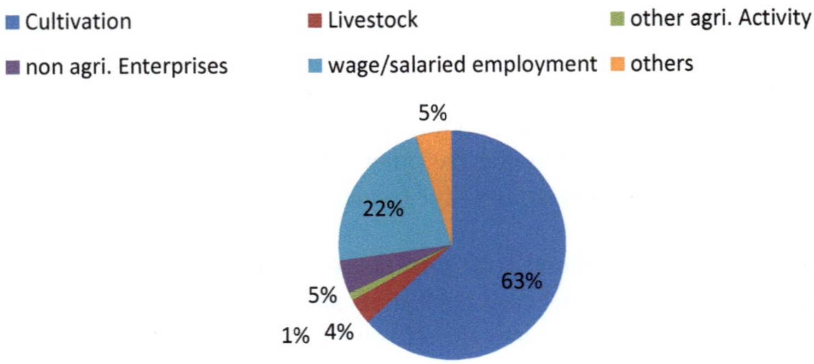

Fig. 17.1 Percentage distribution of agricultural household by principle source of Income. *Source*: National Sample Survey (NSS), 70th Round, 2014, NSSO

According to the 70th Round of NSS there were about 90.2 million agricultural households[1] in India comprising of about 57.8 % of total estimated rural households in the agricultural year of July, 2012 to June, 2013. Majority of agricultural households i.e. 63.5 % take up cultivation as their primary activity for livelihood followed by wage or salaried employment[2] i.e. 22 % (Fig. 17.1).

Agriculture being the primary source of income for the rural population; agricultural households are highly vulnerable to loss of livelihood due to climate shocks. On the other hand, population pressure is leading to frequent changes in land use arrangements. According to 2011 Census figures of India; the average household in rural area mainly lies in medium to large range. About 21 % of the rural households have four household members, 18.9 % five household members, 26.9 % have six to eight household members and 7.2 % have household members above nine. Households with large family size but small farm size often lead to clashes between household members due to property rights.

Principal source of income of agricultural households is largely determined by the extent of land possession. Table 17.1 shows that, among the agricultural households having less than 0.01 ha land (which included landless agricultural households also) about 56 % reported wage/salary employment as their principal source of income and another 23 % reported livestock as their principal source of income. Majority of the agricultural households which possessed more than 0.40 ha land reported cultivation as their principal source of income. The group of agricultural households which possessed little land (0.01–0.04 ha) earned their income both from cultivation (42 %) and wage/salary employment (35 %). Non-agricultural

[1]The 70th Round of NSS Survey India defines agricultural households as households receiving some value of their produce from agricultural activities including cultivation of crops, horticulture, fodder, plantation, livestock management including poultry and fishing.

[2]The different sources of income considered under 70th NSS survey are cultivation, livestock, other agricultural activity, wage/salaried employment, non-agricultural enterprises, pension, remittances, interest and dividends and others.

Table 17.1 Per 1000 distribution of agricultural households by principal source of income during last 365 days for each size class of land possessed

Size class of land possessed (ha)	Per 1000 distribution of households by principal source of income					
	Cultivation	Livestock	Other agricultural activity	Non agricultural enterprises	Wage/ salaried employment	Others[a]
<0.01	16	229	27	108	564	55
0.01–0.40	421	48	12	75	352	93
0.41–1.00	692	23	9	36	200	41
1.01–2.00	830	25	9	32	86	18
2.01–4.00	859	24	11	16	71	18
4.01–10.00	879	27	5	9	59	20
10.00 +	894	55	5	18	17	1
All sizes	635	37	11	47	220	51

[a]'others' includes pension and remittance also
Source: National Sample Survey (NSS) 70th Round, 2014, NSSO

enterprises were principal source of income for about 8 % and 11 % of the agricultural households, respectively, of bottom two size classes of land possessed.

Indian agriculture is majorly characterized by small and marginal operational holdings. As per Agricultural Census of 2010–11, average size of operational holding declined from 1.23 ha in 2005–06 to 1.15 ha in 2010–11 although the number of operational holdings in total increased from 129.22 million in 2005–06 to 138.35 million in 2010–11, an increase of about 7.06 %. Figures on share of operational holdings by size suggest 85.01 % of small and marginal holdings (below 2.00 ha.), 14.29 % of semi-medium and medium (2.00–10.00 ha.) and only 0.70 % large (10.00 ha. & above) operational holding in 2010–11 as compared to 83.29 %, 15.86 % and 11.82 % respectively in 2005–06 due to fragmentation of land holding after the distribution of land amongst siblings of the farmer. Another prominent outcome of Census 2011, suggests an increase from 11.70 % in 2005–06 to 12.78 % in 2010–11 in female owned operational holding. According to NSS survey estimates of India 93 % of the agricultural households possessed land other than just homestead land and only 7 % owned only homestead land while only 0.1 % of rural agricultural household were landless. In addition about 78.5 % of agricultural household owned land only in their residing village. Households holding small patches of land in India often opt for non-agricultural income source and are often reluctant in applying modern technologies, crop diversification and soil and water conservation as they consider it risky due to diseconomies of scale.

Agriculture in India is a traditional activity and therefore aged farmers upgrade their perceptions based on their past experience or follow their fellow farmers and accordingly adopt traditional farming practices. They follow risk-averse decisions and are often rigid in accepting new and advanced methods of climate forecasts and farming technology. According to Census 2011, in rural India 30.9 % population belongs to age group of 0–14 years, 61 % of 15–59 years and 8.1 % population are beyond 60 years of age. Although considering 15–59 years as middle age group for agricultural activity is misleading; these figures suggest that majority of rural population belongs to the middle age group. With the ongoing technological upgradation in the country; educated younger members of households are more open to risks and readily uptake modern agricultural practices and also show good entrepreneurial ability. Moreover, large households with educated members can develop information network and also contribute to family income by engaging in non-farm activities.

The rural female and male literacy rate for country is about 57.93 % and 77.15 % respectively according to Census 2011. Education builds decision making power and in India due to lower female literacy rate men dominate household decision-making. Farmers and head of the household are mostly men and women only assist their male counterparts. Small and marginal farmers are usually faced with increasing costs of production required for new modern agriculture. Under such cases, farmers often cut costs on labour and try to engage females of the household on farm. Moreover, men often switch from farm to non-farm jobs and because women cannot migrate as easily as men; they engage themselves in farm activities especially during sowing and harvesting seasons to support their family income.

Indian farm households usually belong to vulnerable social strata with limited access to institutional safeguards. The arrangements of agricultural extension in India have evolved since Green Revolution of 1966, in terms of activities, organizational structure and available human capital. The extension activities include farmer training, conducting exhibitions, capacity building aids and dissemination of information technology and are provided both at district and state level (Ministry of Agriculture, India; 2006–07). Although at the district level, Agricultural Technology Management Agency (ATMA) model which is a bottom-up approach is being followed in six districts of India, it has been criticized for its inability to reach the farmers. The outcome of NSS survey suggests that only 41 % of agricultural households had access to technical assistance from any agencies in period July, 2012- December, 2012 although they found radio, television, newspaper and internet were useful sources of technical information. Glendenning et al. (2010) observes that regardless of provision of various extension services, at the ground level Indian farmers have limited access to agricultural information as public extension services dominating the provisional arrangement largely focus on on-farm activities although farmers need information on entire food value chain.

Access to credit critically affects adaptive behavior of farmers in India as most of the agricultural households are poor. Absence of formal lending sources forces farmers towards informal credit sources at higher interest burden. Since majority of Indian farmers operate on small to medium scale; credit availability and crop insurance can encourage them to respond to sudden climate shocks, opting for new and advanced technology. In India, financial schemes like Kisan Credit Card, Agricultural Debt Waiver and Debt Relief Scheme are implemented to safeguard farmers from informal credit sources, to relieve them from previous agricultural debt and to bring small and marginal farmers, leaseholders, and share croppers under institutional credit coverage. However, the objectives of such scheme are yet far from achievement as around 52 % of agricultural households in India are still indebted and the average amount of outstanding loan is about Rs. 47,000 (USD 751.82) per agricultural household (70th Round, NSS India). Moreover, only a small proportion of farmers have their crops insured against crop losses mainly due to unawareness of such schemes.

Conclusion

The primary objective of this study was to identify the key socio-economic variables which determine farmers' adaptive capacity and how these factors affect farmers' decision making in India through country level secondary information. To conclude, farmers in India usually try to maximize their farm returns through adaptation and their decisions are often constrained by their socio-economic situations and regional institutional arrangements. The study identified farmer's age, gender, household size, education, off-farm income, farm size, extension services and credit as important factors affecting farmers' adaptive capacity in India. Factors like age, gender and household size are often beyond farmer's control and establish other factors like education, off-farm income, access to extension services and credit. These factors do not determine adaptation responses in isolation rather; decisions are outcome of how they interact with each other. The study finds that agricultural households in rural India usually have medium to large family size and operational holding. Farming serves to be the primary source of livelihood for the households and income from farm increases with the increase in farm size. Farmers mainly rely on farmer to farmer extension services and also find information through television and radio useful if effectively provided. However, the most serious barrier to adaptation in India as identified by this study is lack of credit facilities, less effective extension services and lack of awareness. In this respect, education can prove to be an efficient way to improve awareness among farmers about climate change and the effectiveness of agricultural extension services.

This study is however limited in several context. Firstly, the approach of the study is purely based on secondary data analysis and gives a general idea of Indian farmers' adaptive capacity and therefore, fails to capture the regional and agro-ecological differences within the country. Secondly, although other developing countries can draw similarities, there might be several economic, social and political differences and therefore the importance and relevance of factors might differ.

The study also finds that effective governance is crucial to foster sustainable agricultural practices. Policies should encourage farmers to take up adaptation and enhance farmers' individual adaptive capacities. Well targeted extension services spanning across different stakeholders engaged in agriculture value chain, appropriate credit and crop insurance arrangements, farmers' education and awareness can restrain vulnerability of farmers to climate variations and extremes and ensure long-term resilience to future climatic turbulences.

References

Adams RM, Hurd BH, Lenhart S, Leary N (1998) Effects of global climate change on agriculture: an interpretative review. Clim Res 11(1):19–30

Adesina AA, Zinnah MM (1993) Technology characteristics, farmers' perceptions and adoption decisions: a Tobit model application in Sierra Leone. Agric Econ 9(4):297–311

Annual Report (2013–14) Department of Agriculture and Cooperation, Ministry of Agriculture, Government of India, New Delhi, March 2014

Arbuckle JG Jr, Morton LW, Hobbs J (2013) Farmer beliefs and concerns about climate change and attitudes toward adaptation and mitigation: Evidence from Iowa. Clim Change 118(3–4): 551–563

Asfaw A, Admassie A (2004) The role of education on the adoption of chemical fertiliser under different socioeconomic environments in Ethiopia. Agric Econ 30(3):215–228

Bandiera O, Rasul I (2006) Social networks and technology adoption in northern Mozambique. Econ J 116(514):869–902

Below T, Artner A, Siebert R, Sieber S (2010) Micro-level practices to adapt to climate change for African small-scale farmers. A review of selected literature. Environment and Production Technology Division

Bradshaw B, Dolan H, Smit B (2004) Farm-level adaptation to climatic variability and change: crop diversification in the Canadian prairies. Clim Change 67(1):119–141

Bryan E, Deressa TT, Gbetibouo GA, Ringler C (2009) Adaptation to climate change in Ethiopia and South Africa: options and constraints. Environ Sci Policy 12(4):413–426

Census of India (2011) Ministry of Home Affairs, Government of India

Cline WR (2007) Global warming and agriculture: Impact estimates by country. Peterson Institute, Chicago

Darwin R (1999) The impact of global warming on agriculture: a Ricardian analysis: comment. Am Econ Rev 89:1049–1052

de Jalón SG, Silvestri S, Granados A, Iglesias A (2015) Behavioural barriers in response to climate change in agricultural communities: an example from Kenya. Reg Environ Chang 15 (5):851–865

Deressa TT, Hassan RM, Ringler C, Alemu T, Yesuf M (2009) Determinants of farmers' choice of adaptation methods to climate change in the Nile Basin of Ethiopia. Glob Environ Change 19(2):248–255

Di Falco S, Veronesi M (2013) How can African agriculture adapt to climate change? A counterfactual analysis from Ethiopia. Land Econ 89(4):743–766

Fankhauser S, Smith JB, Tol RS (1999) Weathering climate change: some simple rules to guide adaptation decisions. Ecol Econ 30(1):67–78

Gandure S, Walker S, Botha JJ (2013) Farmers' perceptions of adaptation to climate change and water stress in a South African rural community. Environ Dev 5:39–53

Gbetibouo GA (2009) Understanding farmers' perceptions and adaptations to climate change and variability: the case of the Limpopo Basin, South Africa. Intl Food Policy Res Inst

Glendenning CJ, Babu S, Asenso-Okyere K (2010) Review of Agricultural Extension in India. Are farmers information needs being met

Guiteras R (2009) The impact of climate change on Indian agriculture. Manuscript, Department of Economics, University of Maryland, College Park, Maryland

Hijioka Y, Lin E, Pereira JJ, Corlett RT, Cui X, Insarov GE, Lasco RD, Lindgren E, Surjan A, (2014) Asia. In: Barros VR, Field CB, Dokken DJ, Mastrandrea MD, Mach KJ, Bilir TE, Chatterjee M, Ebi KL, Estrada YO, Genova RC, Girma B, Kissel ES, Levy AN, MacCracken S, Mastrandrea PR, White LL (eds) Climate change 2014: impacts, adaptation, and vulnerability. Part B: regional aspects. Contribution of working group II to the fifth assessment report of the intergovernmental panel on climate change. Cambridge University Press, Cambridge, pp 1327–1370

Hisali E, Birungi P, Buyinza F (2011) Adaptation to climate change in Uganda: evidence from micro level data. Glob Environ Change 21(4):1245–1261

Jha B, Tripathi A (2011) Isn't climate change affecting wheat productivity in India? Indian J Agric Econ 66(3):353

Kavikumar KS (2009) Indian agriculture and climate sensitivity. Working paper 43/2009, Madras School of Economics, Chennai, India

Khan SA, Kumar S, Hussain MZ, Kalra N (2009) Climate change, climate variability and Indian agriculture: impacts vulnerability and adaptation strategies. In: Climate change and crops. Springer, Berlin, pp 19–38

Kokoye SEH, Tovignan SD, Yabi JA, Yegbemey RN (2013) Econometric modeling of farm household land allocation in the municipality of Banikoara in Northern Benin. Land Use Policy 34:72–79

Kumar KK, Parikh J (2001) Indian agriculture and climate sensitivity. Glob Environ Change 11(2):147–154

Kurukulasuriya P, Mendelsohn R (2008) Crop switching as an adaptation strategy to climate change. Afr J Agric Resour Econ 2:1–23

Lamb RL (2003) Fertilizer use, risk, and off-farm labor markets in the semi-arid tropics of India. Am J Agric Econ 85(2):359–371

Maddison DJ (2007) The perception of and adaptation to climate change in Africa. World Bank policy research working paper, 4308

Mendelsohn R (2008) The impact of climate change on agriculture in developing countries. J Nat Resour Policy Res 1(1):5–19

Mendelsohn R, Dinar A (1999) Climate change, agriculture, and developing countries: does adaptation matter? World Bank Res Obs 14(2):277–293

Mendelsohn R, Nordhaus W, Shaw D (1996) Climate impacts on aggregate farm value: accounting for adaptation. Agric For Meteorol 80(1):55–66

Mendelsohn R, Dinar A (2003) Climate, water, and agriculture. Land Econ 79(3):328–341

Napier TL (1991) Factors affecting acceptance and continued use of soil conservation practices in developing societies: a diffusion perspective. Agric Ecosyst Environ 36(3):127–140

Nhemachena C, Hassan R (2007) Micro-level analysis of farmers adaption to climate change in Southern Africa. Intl Food Policy Res Inst

Norris PE, Batie SS (1987) Virginia farmers' soil conservation decisions: an application of Tobit analysis. South J Agric Econ 19(1):79–90

Key Indicators of Situation of Agricultural Households in India, NSS 70th Round, January-December, 2013, Published by Ministry of Statistics and Programme Implementation, National Sample Survey Office (NSSO), Govenrnment of India, December 2014

Parry ML, Canziani OF, Palutikof JP, van der Linden PJ, Hanson CE (2007). IPCC, 2007: climate change 2007: impacts, adaptation and vulnerability. Contribution of working group II to the Fourth Assessment Report of the Intergovernmental Panel on Climate Change

Pearce DW, Cline WR, Achanta AN, Fankhauser S, Pachauri RK, Tol RSJ, Vellinga P (1996) The social costs of climate change: greenhouse damage and the benefits of control. In: Bruce JP, Lee H, Haites EF (eds) Climate change 1995: Economic and social dimensions—contribution of working group III to the second assessment report of the intergovernmental panel on climate change. Cambridge University Press, Cambridge, pp 179–224

Porter JR, Xie L, Challinor AJ, Cochrane K, Howden SM, Iqbal MM, Lobell DB, Travasso MI, (2014) Food security and food production systems. In: Climate change 2014: Impacts, adaptation, and vulnerability. Part A: global and sectoral aspects. Contribution of working group II to the fifth assessment report of the intergovernmental panel on climate change

Sanghi A, Mendelsohn R (2008) The impacts of global warming on farmers in Brazil and India. Glob Environ Change 18(4):655–665

Shiferaw B, Holden ST (1998) Resource degradation and adoption of land conservation technologies in the Ethiopian highlands: a case study in Andit Tid, North Shewa. Agric Econ 18(3): 233–247

Smit B, Skinner MW (2002) Adaptation options in agriculture to climate change: a typology. Mitig Adapt Strateg Glob Change 7(1):85–114

Stern NH (2007) The economics of climate change: the Stern review. Cambridge University Press, Cambridge

Tarleton M, Ramsey D (2008) Farm-level adaptation to multiple risks: climate change and other concerns. J Rural Community Dev 3(2):47–63

Tazeze A, Haji J, Ketema M (2012) Climate change adaptation strategies of smallholder farmers: the case of Babilie district, East Harerghe zone of Oromia regional state of Ethiopia. J Econ Sustain Dev 3(14):1–12

Udry C (1996) Gender, agricultural production, and the theory of the household. J Polit Econ 104: 1010–1046

Wozniak GD (1987) Human capital, information, and the early adoption of new technology. J Hum Resour 22:101–112

Yegbemey RN, Yabi JA, Tovignan SD, Gantoli G, Kokoye SEH (2013) Farmers' decisions to adapt to climate change under various property rights: a case study of maize farming in northern Benin (West Africa). Land Use Policy 34:168–175

Chapter 18
Towards Enhanced Resilience: Monthly Updated Seasonal Rainfall "Scenarios" as Climate Predictions for Farmers in Indonesia

C. (Kees) J. Stigter, Yunita T. Winarto, and Muki Wicaksono

Abstract This paper exemplifies how agricultural production suffers from climate change and how short term seasonal rainfall prediction improves its resilience under tropical lowland rice production in Indonesia. An introduction to and explanations on climate change and long term climate predictions in agriculture form the first parts of this paper.

Two problems haunt seasonal climate predictions for farmers to increase their resilience. These are the skill of predictions and the terminology chosen for monthly updated seasonal rainfall predictions. These "scenarios" are part of climate change adaptation attempts on the islands of Java and Lombok, Indonesia.

Originally, NOAA and subsequently NOAA/IRI monthly ENSO predictions for a period of 3 months were chosen to build planting season "scenarios", because they often explicitly mention Indonesia in their predictions, including recent higher atmosphere convection situations. More recently a check was added on these ENSO predictions, by reading IRI prediction maps for Asia, provided each third Thursday of the month. These maps are more detailed than the written predictions but not more accurate. However, these maps make it possible to separate Java from Nusa Tenggara (region of the islands east of Bali, like Lombok).

The wordings chosen for the monthly SMS messages on seasonal rainfall "scenarios" to farmers use terminology of probabilities as common in daily life. Replies to February 2015 questionnaires show how satisfied farmers are but also how they must get used to this wording.

C. (Kees) J. Stigter was deceased at the time of publication.

Y.T. Winarto • M. Wicaksono
Cluster Response Farming and Climate Change, Faculty of Social and Political Sciences, Department of Anthropology, University of Indonesia (UI), Fl.6/Building H, Selo Soemardjan Room, Kampus UI, 16424 Depok, Indonesia

W.L. Filho et al. (eds.), *Climate Change Adaptation, Resilience and Hazards*, Climate Change Management, DOI 10.1007/978-3-319-39880-8_18

Keywords Adaptation • Climate prediction • Rainfall scenarios • Resilience • Skill • Indonesia

Introduction

The planet earth has a unique but complicated climate that presently is changing due to the influence that mankind's activities appear to have on the composition of its atmosphere. It is called anthropogenic (man-made) climate change (e.g. Oreskes 2004; IPCC 2007; Rosenzweig et al. 2008). The world's agricultural systems face an uphill struggle in feeding a projected nine to ten billion people by 2050. Climate change introduces a significant hurdle in this struggle (e.g. Thornton 2012).

There is general and widely held scientific consensus that the observed trends in atmospheric and ocean temperature, sea ice, glaciers as well as climate extremes during the last century cannot be explained solely by natural climate processes and so reflect human influences (e.g. USGS 2014). The argument that what is measured could be natural climate change can also be refuted by the fact that present understanding of cyclic climatology of the past points to a cooling planet without the presence of mankind (Dr. Jim Salinger, private communication 2013; Easterbrook 2014).

On the simplest level, the weather is what is happening in the atmosphere at any given time. The climate, in a narrow sense, can be considered as the "average weather" (WMO 2015a). In a more scientifically accurate way, it can be defined as: "the statistical description in terms of the mean and variability of relevant quantities over a period of time" (WMO 2015a).

The issues are (e.g. Stigter and Ofori 2014a): (i) global warming, (ii) increasing climate variability, (iii) more (and possibly more severe) meteorological and climatological extreme events. Is global warming real? From worldwide observations WMO (World Meteorological Organization) concluded a long time ago that our planet is warming up. This has to be considered a fact (e.g. WMO 2015b).

The atmosphere gets its energy from two sources, both of course related to the basic source of solar energy:

(i) It is warmed from below by solar energy absorbed by the earth surface during the day. This heat gets distributed throughout the boundary layer.
    It should here already be indicated that there is a great difference between land and water surfaces. On land, in daytime, a tiny surface layer becomes much warmer, with the very surface becoming hottest, the rate depending mainly on water content. In water the absorption is over a certain depth, decreasing with depth. The water surface therefore does not become very warm from direct

absorption; ocean currents play a more important role here. But indeed most additional heat created is absorbed by the oceans. The large heat capacity of water prevents the oceans from overheating (WMO 2015c).

The second process of how the atmosphere gains heat is:

(ii) Its gases absorb the longwave radiation sent from the earth surface throughout day and night. This prevents the land surface from overheating, but only a part gets out to space.

This radiation loss from a cooling surface (and the cooling air due to this) may be recognized from nights without a cloud cover. When there are clouds, they send roughly as much longwave radiation back to the earth surface as they receive from that surface and no or appreciably less cooling occurs. So it must be concluded that our planet is actually heating up mainly because of this absorption of radiative heat by the greenhouse gases in the atmosphere. Increasing greenhouse gases mean additional heating. Among many others, IPCC (the Intergovernmental Panel on Climate Change) has indeed been stressing, with increasing confidence over the years, that the cause of this heat gain is an increase of greenhouse gases in our atmosphere (e.g. IPCC 2007).

The main source of this increase of carbon dioxide, methane and nitrous oxide appears to be human activities on this planet: e.g. generating electricity from coal, producing cement and driving cars are presently the main culprits (e.g. IPCC 2007). As to carbon dioxide, measurements show that it has increased from the start of the industrial revolution, but that changes in land use have also played an important role by large scale cutting of vegetation, including trees (e.g. Fearnside 2000). This is also why Indonesia has become a large contributor, by felling trees (sinks of carbon dioxide) in large scale (mostly illegal) logging, often planting oil palm trees instead, with appreciably less carbon dioxide absorption per hectare.

Climate Predictions

Recently a number of case studies have appeared that show that climate change is already affecting yields of various crops. In Box 1 it is illustrated how quantitative knowledge is helping to find the way to policies serving the purpose of adapting to the consequences of climate change. In the case of Arabica coffee, a solution could be to go to higher, still colder grounds, although this disrupts living conditions and biodiversity patterns (Craparo et al 2015; Stigter 2015a).

Box 1: Some Illustrative Data of Climate Related Predictions

IPCC (2007) shows that somewhere near 1800 the carbon dioxide concentration was something as 280 ppm, while it has recently reached 400 ppm. It also shows how steeply the curve is presently rising in comparison to a slow rise through time till roughly 1800.

As to temperature, from 1960 till 2010 the increase is estimated to have been less than a degree Celsius, but the projection for the next 50 years is in the order of 1 °C, with the emissions kept within the range of the IPCC (2007) scenarios. When all greenhouse gases and aerosols were kept constant at their 2000 levels, this heating would be half as much (WMO 2015b). It is generally accepted that, if for this century the temperature increase can be limited to 2 °C, the damages will remain much more limited than when the scenarios give a 4 °C increase at the end of this century (e.g. Pidcock 2014).

What do such figures mean in practice today? Here is an example for Arabica coffee grown on the slopes of the Kilimanjaro (Craparo et al. 2015). Coffee is the world's most valuable tropical export crop. Recent studies predict severe climate change impacts on *C. arabica* production. However, quantitative production figures are necessary to provide coffee stakeholders and policy makers with evidence to justify immediate action. Using data from the northern Tanzanian highlands, it was demonstrated that increasing night time (Tmin) temperature was the most significant climatic variable responsible for diminishing *C. arabica* yields between 1961 and 2012. The minimum temperature in that region of Tanzania rose in that half century by between 1 and 1.5 °C. The projection for the next 35 years for that region is 1.5 °C.

With the minimum temperature at 14 °C, the yields were about 500 kg beans per hectare. A non-linear (sigmoid) model constructed from data from local areas with different minimum temperatures gave the following results. With the night minimum rising to 15 °C, this would become about 450 kg ha $^{-1}$. With a night minimum temperature at 16 °C this decreases to about 300 kg ha^{-1}, while for 17 °C this becomes about 100 kg ha^{-1}. This means a prediction of average coffee production diminishing to 145 kg ha^{-1} by 2060 in those areas of Tanzania. Climate prediction was reduced here to prediction of minimum temperatures.

Consequently, without adequate adaptation strategies or substantial external inputs, coffee production will be severely reduced in the Tanzanian highlands in the near future (Craparo et al. 2015).

But for the lowland tropics, there is no way out apart from crop diversification and finding more heat (and drought) tolerant varieties (Stigter et al 2015). Box 2 shows how bad the situation is.

Box 2: Some Other Agricultural Upheavals

Here is another example of trouble: Maize (Thornton and Cramer 2012; Stigter and Ofori 2014a). The results are from something as 20,000 trials at 123 stations all over the world of CIMMYT (CGIAR, Columbia):

- Increased temperature significantly effects maize yield ($P < 0.01$).
- Possible gains in yield with warming at relatively cool sites.
- Significant yield losses at sites where temperatures commonly exceed 30 °C (corresponding to areas where the growing season average temperatures are >23 °C or average maximum temperatures are >28 °C).
- Daytime warming is more harmful to yield than night-time warming.
- Drought increases yield susceptibility to warming even at cooler sites.
- Under 'optimal' conditions yield losses occur over ca. 65 % of the harvested area of maize.
- Under 'drought stress' yield losses occur at all sites, with a 1 °C warming resulting in at least a 20 % loss of yield over more than 75 % of the harvested area.

And here is one more example: Rice. Data are obtained from a CGIAR umbrella study, the same as used for maize (Thornton and Cramer 2012; Stigter and Winarto 2013).

Temperatures beyond critical thresholds not only reduce the growth duration of the rice crop, they also increase spikelet sterility, reduce grain-filling duration, and enhance respiratory losses, resulting in lower yield and lower-quality rice grain. Rice is relatively more tolerant to high temperatures during the vegetative phase but highly susceptible during the reproductive phase, particularly at the flowering stage.

Unlike other abiotic stresses, heat stresses occurring either during the day or the night have differential impacts on rice growth and production. High night-time temperatures have been shown to have a greater negative effect on rice yields, with a 1 °C increase above critical temperature (>24 °C) leading to 10 % reduction in both grain yield and biomass. High day-time temperatures in some tropical and subtropical rice growing regions are already close to the optimum levels. An increase in intensity and frequency of heat waves coinciding with sensitive reproductive stages can result in serious damage to rice production.

The climate predictions that are exemplified in the Boxes 1 and 2 are long term predictions of which knowing the trends is an important indicator for adaptation to the consequences of climate change, food policies, crop planning, variety breeding and screening, farming system adaptations and modifications, extension policies and all other planning related to agriculture that has to be made to face climate change. For farmers these are important issues (e.g. Stigter 2015b) that can be discussed at "Science Field Shops" for their long term decision making (Stigter et al. 2015). But what about the shorter term?

Further Matters That Should Be Known

It is interesting to note that since the very end of the previous century, the rate of global warming has reduced by at least half of the rate in the last 50 years of that previous century. This has been baptized "the hiatus", a lack of continuity in the up going trend of global temperature. So climate change rates reduce. Is this going to change our thinking?

Stigter and Winarto (2014) stated that "Our lack of knowledge and understanding is best illustrated with the very recent discussion on the present global warming "hiatus" (e.g. Wikipedia 2014). Observations have shown that global warming presently takes place at a lower rate. Some deny its very existence (see for example Anonymous 2013a) but accurate worldwide measurements and comparisons show that this "hiatus", this break in continuity, is there, since the late 90s. Already four quantitative(!) reasonings in fully fledged or partly explanations may be found:

(i) more volcanic particles in the atmosphere (Stark 2014);
(ii) extremely strong large scale western winds in the Pacific (Milman 2014);
(iii) much warmer water being transported to deeper layers of the ocean (Anonymous 2013a, b);
(iv) indeed being in a down going phase of the Pacific Decadal Oscillation and/or another of such oscillations as atmospheric variations/imbalances (e.g. Anonymous 2013c, d; Wyatt and Curry 2013).

It is likely that all of these four explanations may actually be involved, if not more processes. But there is no clue about the ratios of their contributions, while in the last case the complexities are enormous".

There is a quarrel on the volcanic contributions. There was no one enormous contribution, but it was proposed recently that many smaller volcanic eruptions can cause such cooling (reduction of heating). But others think that this will not add up to more than 15 % of the cooling (e.g. Kalaugher 2015).

It is presently most likely that the cause of this hiatus is indeed more warmer water going to deeper layers, resulting in a (temporarily?) relatively cooler ocean. This also shows how important oceanic surface temperatures are for determination of our climate. It is one of the weakest rings in the chain towards climate predictions.

So much less is known about how the sea surface temperatures are determined by currents and deep waves than has been understood on the atmospheric resultants. Indeed, for decades radiosondes with balloons have been sent into the atmosphere, but only very recently have buoys been placed in the Pacific Ocean, particularly in those parts used for climate prediction purposes.

But looking at the early predictions of the 2014/2015 originally weak El-Niño (see below), it appears that the atmosphere sometimes does not want to behave the way it is known. That makes the little that is predictable suddenly also unpredictable.

Short Term (Seasonal) Climate Predictions: A Case Study in Indonesia

So, unpredictable ocean currents and deep waves that are not understood in sufficient detail, create the signals for El-Niño's to occur. They are very important in short term climate predictions (1–3 months). Now, it appears that the frequency of these phenomena, and how they follow each other, has changed in recent times! However, these actual changes cannot be simulated with the models that summarize our understanding, which at this moment is still very insufficient (e.g. Stigter and Ofori 2014b).

As a consequence of the above, simple growing season rainfall scenarios are very difficult to derive from existing raw or simplified (e.g. outlook fora, WMO 2015d) climate predictions. Two problems haunt seasonal climate predictions for farmers to increase their resilience, (1) the skill of predictions and (2) the terminology chosen for these monthly updated seasonal rainfall predictions.

Such "scenarios" have been made part of climate change adaptation attempts on the islands of Java and Lombok, Indonesia (UNIID-SEA 2014; Stigter et al. 2015). One approach is to follow the monthly El-Niño predictions NOAA (2015) is giving (these days together with IRI, see Appendix 1 for an example), because that influential phenomenon has been well unraveled (e.g. Gross 2014). These are monthly renewed predictions for a period of 3 months. They were chosen because they often explicitly mention Indonesia and/or the western equatorial pacific in their predictions, including recent higher atmosphere convection situations. They also express skills of the predictions where appropriate. These issues are exemplified in Appendix 1.

Every third Thursday of the month the International Research Institute for Climate and Society (IRI) at the Earth Institute of the University of Columbia (USA) issues updated maps on multi-model probability forecasts for precipitation (IRI 2015a) and updated quick look ENSO forecasts (IRI 2015b). These maps are more detailed than the written predictions but not more accurate. However, these maps make it possible to separate Java from Nusa Tenggara (region of the islands east of Bali, like Lombok) and other regions in Indonesia. As was exemplified earlier (Stigter et al. 2013), the way the NOAA monthly "ensemble" climate prediction knowledge, as presently brought to the participating farmers in Indramayu and Lombok, is given in a more client friendly wording that may be called an agrometeorological advisory. It is tried to derive the most likely start of the main rainy season and also indicate whether the dry season may be expected to be normal. However, this advisory, given as a scenario for the seasons concerned, remains a qualitative one, based on the raw climate prediction knowledge that is obtained from NOAA/IRI (Appendix 1).

As to these wordings in which the monthly SMS messages on seasonal rainfall "scenarios" are brought to farmers, terminology is used of probabilities as common in daily life. The tercile percentages system used in most Outlook Fora predictions are not used but the way in which for example doctors express chances of recovery of their patients is preferred. On 6 March 2015 Stigter wrote to Winarto: *The atmosphere is somewhat better adapting to ocean temperatures, which would mean that we will have somewhat more chance of actual but weak El-Nino effects, but the convection in the atmosphere is more conducive to rainfall above Indonesia. To this must be added that (northern hemisphere) spring predictions are usually low skill. This keeps the rainfall seasonal scenario conservative as*

"Most likely near normal rainfall for the coming months but more likely somewhat at the drier side of normal. Low skill should be admitted".

This part between quotation marks was then in translation sent by SMS to participating farmers. Replies to February 2015 questionnaires show how satisfied farmers are but also how they must get used to this kind of wording.

Farmer Questionnaire on the Monthly Updated Seasonal Rainfall Scenarios Provided

In February 2015, a questionnaire was used to interview 42 farmers in the project villages of Indramayu, NW coastal Java, that received the monthly seasonal scenario regularly for 6 months or more, and 42 farmers in the same villages that did not receive these scenarios as a control group. Reception of the seasonal scenarios is shown in Table 18.1. Of those receiving these scenarios, more than half received them for more than 2 years and 85 % for more than a year (Table 18.2).

Of the target group of farmers, more than 93 % received the seasonal scenarios via SMS on their mobile telephone while for more than 81 % this was the only way they received that knowledge.

Of the number of farmers receiving the seasonal scenarios, 55 % understood them regularly or better (see Fig. 18.1 for this terminology) but 42 % understood them only sometimes. This shows the necessity to improve the scenario messages as to the understanding required. It could be observed that those receiving the scenarios for at least 2 years had a much higher regular or better understanding than the others (Fig. 18.1).

Table 18.1 Reception of seasonal scenarios by interviewed farmers that participated in the project

Valid	Frequency	Percentage	Cumulative percentage
1. Every month	24	57.1	57.1
2. Often	9	21.4	78.6
3. Regularly	3	7.1	85.7
4. Sometimes	3	7.1	92.9
5. Never	3	7.1	100.0
Total	42	100.0	

Table 18.2 Length of period that interviewed farmers received the seasonal scenario

	Frequency	Percentage	Cumulative percentage
1. More than 2 years	20	47.6	47.6
2. More than 1 year	13	31.0	78.6
3. Less than 1 year	5	11.9	90.5
4. Less than half a year	1	2.4	92.9
5. Never received seasonal scenario	3	7.1	100.0
Total	42	100.0	

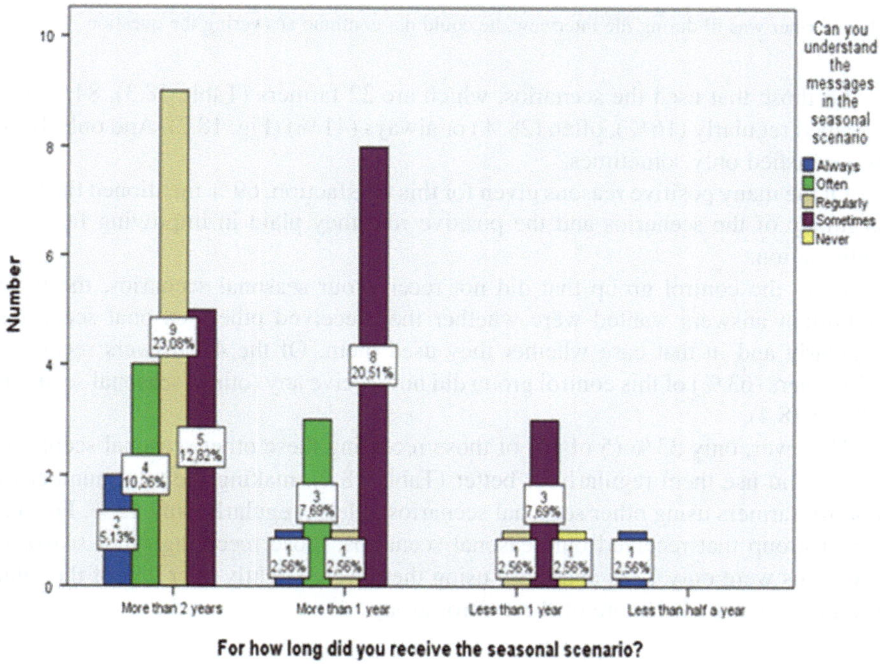

Fig. 18.1 Correlation of the period of receiving the seasonal rainfall scenarios with the understanding of the scenarios' contents. With three farmers not receiving the scenario, there were 39 replies

The difficulties were mainly of two kinds: (i) scientific terminology and (ii) the use of "below normal, normal and above normal" qualifications. Our farmer facilitators, selected from among the farmers by their peers, had the role of continuing to explain this, but that has apparently been insufficiently successful.

Of those farmers receiving the scenarios, 51 % used them regularly or better but 45 % only sometimes or never in their decision making (Table 18.3). The main reasons for not using the scenarios are that others make the farming decisions (40 % of those providing a reason) or that rain is not their main source of water (26 % of those providing a reason). For only 6 % the scenarios were not useful when followed.

Table 18.3 Use of the seasonal scenarios when received

	Frequency	Percentage	Cumulative percentage
Valid	39 interviewed farmers received the seasonal scenario[a]		
1. Always	9	21.1	21.1
2. Often	5	13.2	34.3
3. Regularly	8	21.1	55.4
4. Sometimes	10	26.3	81.7
5. Never	6	18.4	100.0
6. Never received seasonal scenarios	3		
Total	42	100.0	

[a]One farmer was ill during the interview, he could not continue answering the question

Of those that used the scenarios, which are 32 farmers (Table 18.3), 84 % was satisfied regularly (16 %), often (28 %) or always (41 %) (Fig. 18.2). And only 16 % was satisfied only sometimes.

Of the many positive reasons given for this satisfaction, 69 % mentioned the high accuracy of the scenarios and the positive role they plaid in improving farmers' anticipation.

From the control group that did not receive our seasonal scenarios, the most important answers wanted were whether they received other seasonal scenarios regularly and in that case whether they used them. Of the 41 answers received, 26 farmers (63 %) of this control group did not receive any (other) seasonal scenario (Table 18.4).

However, only 33 % (5 of 15) of those receiving these other seasonal scenarios (37 %) did use them regularly or better (Table 18.4), making the total number of control farmers using other seasonal scenarios at least regularly only 12 %. For our target group that received our seasonal scenarios, those receiving other seasonal scenarios were only 17 % and those using them were slightly over 9 % of the total group, similar to the figure of the control group.

Conclusions

Placing short-term climate predictions (1–3 months) in the context of climate change and long-term climate predictions, it could be appreciated how difficult shot-term climate prediction is. A case study on the Indonesian island of Java, on its NW coast, shows how far the progress was in the first 2 years of introducing such predictions. Monthly updated NOAA/IRI ENSO predictions were used to make our monthly seasonal rainfall scenarios. It was observed that those receiving the scenarios for at least 2 years had a much higher regular or better understanding than the others. The difficulties farmers had with the predictions were mainly of two kinds: (i) scientific terminology and (ii) the use of "below normal, normal and above normal" qualifications. Of those that used the scenarios, 84 % was satisfied

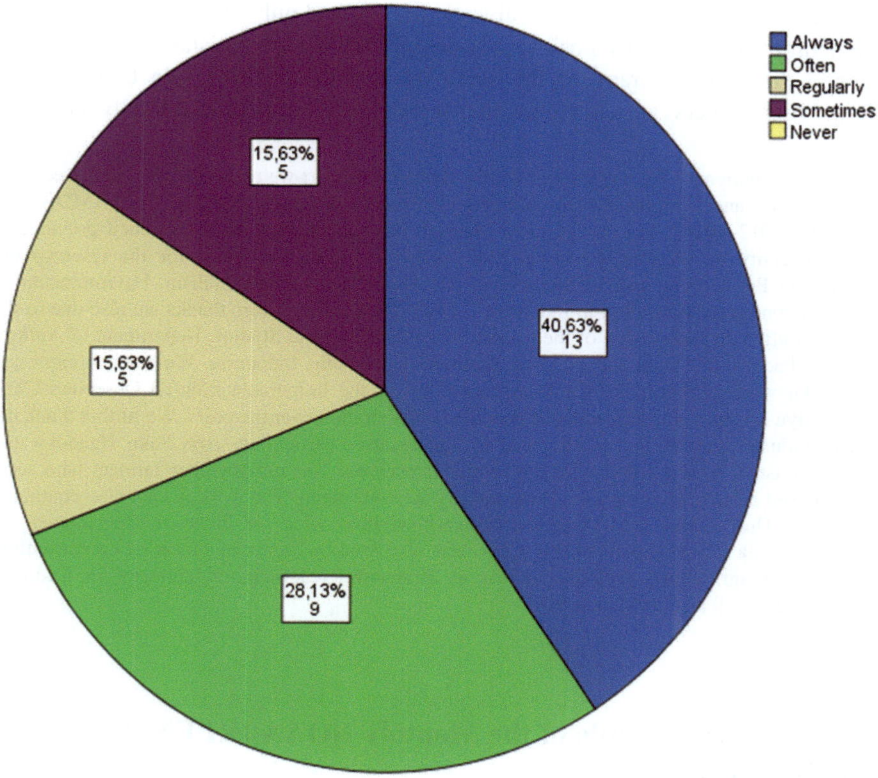

Fig. 18.2 Use of seasonal scenarios as a percentage of the total number of 32 farmers that received them. The "Nevers" have not been shown. One farmer of the target group could not be interviewed

Table 18.4 Use of (other) seasonal scenarios by the control group not receiving our seasonal scenarios

Valid	Frequency	Percentage	Cumulative percentage
1. Always	2	4.9	4.9
2. Regularly	3	7.3	12.2
3. Sometimes	3	7.3	19.5
4. Never	7	17.1	36.6
Sub-total	15	36.6	
Did not receive "Other Seasonal Scenarios"	26	63.4	
Total	41	100.0	

regularly (16 %), often (28 %) or always (41 %). And only 16 % was satisfied only sometimes. Of the many positive reasons given for their satisfaction, 69 % mentioned the high accuracy of the scenarios and the positive role they plaid in improving farmers' anticipation. Other seasonal scenarios played hardly any role.

Acknowledgements We are grateful for the financial support we received from the Directorate for Research and Community Development, Universitas Indonesia (2011); the SAMDHANA Institute (2012–2014); and via the Grant through the Australian Research Council provided to the University of Western Australia (Lyn Parker and Greg Acciaioli) for the research on: "Fostering Pro-Environment Consciousness and Practice: Environmentalism, Environmentality and Environmental Education in Indonesia" (2013–2015). Our sincere thanks are also due to the administration and leadership of the Center for Anthropological Studies, Department of Anthropology, Faculty of Social and Political Sciences, Universitas Indonesia. We want to express a particular word of thanks for the farmer members of the Indramayu Rainfall Observers Club, Indramayu Regency, for their much appreciated collaboration over the years. We further thank the undergraduate students that participated in the research, particularly Aria Sakti Handoko and Febry Sulistya, who assisted with the farmer interviews. The non-member farmers who were interviewed as control group are also gratefully acknowledged. The work is now also continued under the Universities (and Councils) Network on Innovation for Inclusive Development in Southeast Asia. The first parts of this paper were also used by Stigter in a meeting on curriculum revision for agricultural diploma courses at Zimbabwean agricultural colleges, in Kadoma, Zimbabwe, on 19 and 20 March 2015.

Appendix 1: Example of the Monthly NOAA/IRI ENSO Advisory/Prediction

EL NIÑO/SOUTHERN OSCILLATION (ENSO) DIAGNOSTIC DISCUSSION

issued by

CLIMATE PREDICTION CENTER/NCEP/NWS

and the International Research Institute for Climate and Society

5 March 2015

ENSO Alert System Status: El Niño Advisory

<u>Synopsis:</u> There is an approximately 50–60 % chance that El Niño conditions will continue through Northern Hemisphere summer 2015.

During February 2015, El Niño conditions were observed as the above-average sea surface temperatures (SST) across the western and central equatorial Pacific became weakly coupled to the tropical atmosphere. The latest weekly Niño indices were +0.6 °C in the Niño-3.4 region and +1.2 °C in the Niño-4 region, and near zero in the Niño-3 and Niño-1 + 2 regions. Subsurface temperature anomalies increased associated with a downwelling oceanic Kelvin wave, which was reflected in positive subsurface anomalies across most of the Pacific. Consistent with weak coupling, the frequency and strength of low-level westerly wind anomalies increased over the equatorial Pacific during the last month and a half. At upper-levels, anomalous easterly winds persisted across the east-central Pacific. Also, the equatorial Southern Oscillation Index (EQSOI) remained negative for 2 consecutive months. Convection was enhanced over the western equatorial Pacific and near average around the Date Line. Overall, these features are consistent with borderline, weak El Niño conditions.

Compared to last month, several more models indicate El Niño (3-month values of the Niño-3.4 index equal to or greater than 0.5 °C) will continue throughout 2015. This is supported by the recent increase in subsurface temperatures and near-term model predictions of the continuation of low-level westerly wind anomalies across parts of the equatorial Pacific. However, model forecast skill tends to be lower during the Northern Hemisphere spring, which contributes to progressively lower probabilities of El Niño through the year. In summary, there is an approximately 50–60 % chance that El Niño conditions will continue through Northern Hemisphere summer 2015 (click CPC/IRI consensus forecast for the chance of each outcome).

Due to the expected weak strength, widespread or significant global impacts are not anticipated. However, certain impacts often associated with El Niño may appear in some locations during the Northern Hemisphere spring 2015.

References

Anonymous (2013a) What ocean heating reveals about global warming. Real Climate. Climate Science from Climate Scientists. http://www.realclimate.org/index.php/archives/2013/09/what-ocean-heating-reveals-about-global-warming/comment-page-2/. Accessed 7 Dec 2013

Anonymous (2013b) Another piece of the global warming puzzle—more efficient ocean heat uptake. Skeptical Science.http://www.skepticalscience.com/watanabe-et-al-2013-another-piece-of-the-puzzle.html/. Accessed on 7 Dec 2013

Anonymous (2013c) The potential for successful climate predictions! Hindcast experiments capture long-term climate fluctuations. Helmholtz Center for Ocean Research, Kiel. Geomar. http://www.geomar.de/en/news/article/klimavorhersagen-ueber-mehrere-jahre-moeglich/. Accessed 7 Dec 2013

Anonymous (2013d) Solving the mysteries of hiatus in global warming. Scripps Institution of Oceanography, 28 August. https://scripps.ucsd.edu/news/13251/. Accessed 7 Dec 2013

Craparo ACW, Van Asten PJA, Läderach P, Jassogne LTP, Grab SW (2015) Coffea arabica yields decline in Tanzania due to climate change: global implications. Agric For Meteorol 207:1–10

Easterbrook DJ (2014) Global cooling is here: evidence for predicting global cooling for the next three decades. Global Research. http://www.globalresearch.ca/global-cooling-is-here/10783/. Accessed 15 Nov 2014

Fearnside PM (2000) Global warming and tropical land-use change: greenhouse gas emissions from biomass burning, decomposition and soils in forest conversion, shifting cultivation and secondary vegetation. Clim Change 46:115–158

Gross DA (2014) Fishing in pink waters: how scientists unraveled the El-Niño mystery. http://roadtoparis.info/2014/12/04/el-nino-mystery-pink-waters-scientists/. Accessed 25 Jan 2015.

IPCC (2007) Fourth assessment report, Working Group 1, The physical science basis. Section 9.7 Combining evidence of anthropogenic climate change. Intergovernmental Panel on Climate Change, WMO, Geneva, Switzerland and UNEP, Nairobi, Kenya

IRI (2015a) Seasonal climate forecasts. http://iri.columbia.edu/our-expertise/climate/forecasts/seasonal-climate-forecasts/. Accessed 19 Mar 2015

IRI (2015b) ENSO forecast. Quick Look. http://iri.columbia.edu/our-expertise/climate/forecasts/enso/current/. Accessed 19 Mar 2015

Kalaugher L (2015) Climate scientists tackle 'unfinished business'. http://environmentalresearchweb.org/cws/article/news/60005/. Accessed 5 Mar 2015

Milman O (2014) Global warming 'pause' due to unusual trade winds in Pacific Ocean, study finds. The Guardian, 10 February. http://www.theguardian.com/environment/2014/feb/09/global-warming-pause-trade-winds-pacific-ocean-study/. Accessed 25 Mar 2014

NOAA (2015) El Niño/Southern Oscillation (ENSO) Diagnostic discussion. http://www.cpc.ncep.noaa.gov/products/analysis_monitoring/enso_advisory/index.shtml/. Accessed 7 Mar 2015

Oreskes N (2004) Beyond the ivory tower: the scientific consensus on climate change. Science 306 (5702):1686

Pidcock R (2014) What happens if we overshoot the two degree target for limiting global warming? http://www.carbonbrief.org/blog/2014/12/what-happens-if-we-overshoot-the-two-degree-target-for-limiting-global-warming/. Accessed 2 Mar 2015

Rosenzweig C, Karoly D, Vicarelli M, Neofotis P, Wu Q, Casassa G, Menzel A, Root TL, Estrella N, Seguin B, Tryjanowski P, Chunzhen L, Rawlins S, Anton Imeson A (2008) Attributing physical and biological impacts to anthropogenic climate change. Nature 453:353–357

Stark AM (2014) Volcanoes contribute to recent warming 'hiatus'. Lawrence Livermore National Laboratory News Release NR-14-02-13. https://www.llnl.gov/news/newsreleases/2014/Feb/NR-14-02-13html#.U59GeU0U9yQ/. Accessed 5 Feb 2015

Stigter CJ (2015a) Peer review report 1 on "Coffea arabica yields decline in Tanzania due to climate change: global implications" by Craparo et al. Agric Forest Meteorol 201:15–16

Stigter CJ (2015b) A decade of capacity building through roving seminars on Agrometeorology/-climatology in Africa, Asia and Latin America: from agrometeorological services via climate change to agroforestry and other climate-smart practices. In: Leal Filho W, Dunk R, Illingworth S (eds) Information, communication, education and training on climate change. Springer, Heidelberg

Stigter CKJ, Ofori E (2014a) What climate change means for farmers in Africa. A triptych review. Middle Panel: introductional matters and consequences of global warming for African farmers. Afr J Food Agric Nutr Dev (AJFAND) 14(1):8420–8444

Stigter CKJ, Ofori E (2014b) What climate change means for farmers in Africa. A triptych review. Left panel: increasing climate variability and a response approach for African farmers. Afr J Food Agric Nutr Dev (AJFAND) 14(1):8445–8458

Stigter K, Winarto YT (2013) Rice and climate change: Adaptation or mitigation? Facts for policy designs. A choice from what recent summaries say and some critical additions for use with Indonesian farmers. In: Soeparno H, Pasandaran E, Syarwani M, Dariah A, Pasaribu SM, Saad NS (eds) Politik Pembangunan Pertanian Menghadapi Perubahan Iklim [Politics of agricultural development facing climate change], pp 474–485. http://www.agrometeorology.org/topics/climate-change/rice-and-climate-change-adaption-or-mitigation-facts-for-policy-designs/. Accessed 17 Jan 2014

Stigter CKJ, Winarto YT (2014) Climate prediction for farmers: should we laugh or should we cry? In: Memulihkan Ketangguhan Ekosistem dan Produksi Beras: Mengatasi "Ancaman Krisis Pangan" dan Menanggulangi "Keliru Pikir"? [Restoring ecosystem sustainability and rice production: managing "Threats of Food Crisis" and changing "Wrong Mindsets"?] AIPI (Indonesian Academy of Sciences), Jakarta, Indonesia

Stigter K, Winarto YT, Ofori E, Zuma-Netshiukhwi G, Nanja D, Walker S (2013) Extension agrometeorology as the answer to stakeholder realities: response farming and the consequences of climate change. Special issue on agrometeorology: from scientific analysis to operational application. Atmosphere 4(3):237–253

Stigter CKJ, Winarto YT, Wicaksono M (2015) Science field shops: an innovative agricultural extension approach for adaptation to climate change, applied with farmers in Indonesia. World Symposium on Climate change adaptation, Manchester, UK, 2–4 September 2015, Session 2:

Implementing climate change adaptation in communities, cities, countries and via outreach programmes. To appear in Leal Filho W, Adamson K, Dunk R (eds) Implementing climate change adaptation in communities, cities, countries, and via outreach programmes. Springer, Heidelberg

Thornton P (2012) Recalibrating food production in the developing world: global warming will change more than just the climate. CCAFS Policy Brief no. 6. CGIAR Research Program on Climate Change, Agriculture and Food Security (CCAFS). CCAFS, Frederiksberg, Copenhagen, Denmark. http://cgspace.cgiar.org/handle/10568/24696/. Accessed 7 Feb 2013

Thornton P, Cramer L (eds) (2012) Impacts of climate change on the agricultural and aquatic systems and natural resources within the CGIAR's Mandate. CCAFS Working paper 23. CGIAR Research Program on Climate Change, Agriculture and Food Security (CCAFS). Frederiksberg, Copenhagen, Denmark. http://hdl.handle.net/10568/21226/. Accessed 7 Feb 2013

UNIID-SEA (2014) Farmers, scientists, and students cultivate innovation in farmers' communities. http://uniid-sea.net/university-of-indonesia-students-cultivate-innovation-in-the-farmers-community/. Accessed 12 Mar 2015

USGS (2014) Distinguishing natural climate variability from anthropogenic climate change. United States Geological Survey, Reston, VA, USA. http://www.usgs.gov/climate_landuse/clu_rd/pt_nat_climate.asp/. Accessed 25 Feb 2015

Wikipedia (2014) Global warming hiatus. http://en.wikipedia.org/wiki/Global_warming_hiatus/. Accessed 12 Jan 2015

WMO (2015a) Understanding climate. World Meteorological Organization, Geneva, Switzerland. http://www.wmo.int/pages/themes/climate/understanding_climate.php/. Accessed 13 Feb 2015

WMO (2015b) Warmer trend continues in 2014. World Meteorological Organization, Geneva, Switzerland. http://www.wmo.int/media/?q=content/warming-trend-continues-2014/. Accessed 15 Feb 2015

WMO (2015c) Climate projections. World Meteorological Organization, Geneva, Switzerland. http://www.wmo.int/pages/themes/climate/climate_projections.php/. Accessed 16 Feb 2015

WMO (2015d) Regional climate outlook products. World Meteorological Organization, Geneva, Switzerland. http://www.wmo.int/pages/prog/wcp/wcasp/clips/outlooks/climate_forecasts.html/. Accessed 14 Mar 2015

Wyatt M, Curry J (2013) 'Stadium Waves' could explain lull in global warming. Georgia Tech Research News, 10 October. http://www.gtresearchnews.gatech.edu/stadium-waves-could-explain-lull-in-global-warming/. Accessed 3 Oct 2014

Chapter 19
Fostering Resilience Among Artisanal Fishers in Peniche (Portugal): An Exploratory Study

Vanda Viegas, Ulisses M. Azeiteiro, and Fátima Alves

Abstract This work seeks to comprehend how *covos*, *anzol* and *redes* (creels, hooks and net fishermen—artisanal fishing gear) artisanal fishermen in the Fishing Artisanal Local Community (small-scale fisheries for subsistence or local, small markets, generally using traditional fishing techniques and small boats −12 m) from Peniche, explain and deal with Climate Change in their daily lives. The study was based on a sociological theoretical and methodological approach with contributions of environment sciences. Qualitative methodological procedures included the application of a semi-structured interview to 12 artisanal fishermen between March and May 2013. Findings were analysed with a content analysis matrix built in the basis of the initial theoretical framework and, at the same time, integrating the new information resulting from and on the field, and reported the elements that characterize the relation of the fisherman with Climate Change. The findings in this study characterize the scenario where lay rationalities contextualize, highlighting the importance of considering the qualitative approaches to climate change by studying and interpreting lay rationalities.

V. Viegas (✉)
Centre for Functional Ecology, University of Coimbra, Coimbra, Portugal

Universidade Aberta, Lisbon, Portugal
e-mail: vnd.vgs@gmail.com

U.M. Azeiteiro
Department of Sciences and Technology, Universidade Aberta, Palácio Ceia, Rua da Escola Politécnica, 141-147, 1269-001 Lisbon, Portugal

Centre for Functional Ecology, University of Coimbra, Coimbra, Portugal

F. Alves
Centre for Functional Ecology, University of Coimbra, Coimbra, Portugal

Department of Social Sciences and Management, Universidade Aberta, Palácio Ceia, Rua da Escola Politécnica, 141-147, 1269-001 Lisbon, Portugal

Centre of Studies of Migration and Intercultural Relations, Universidade Aberta, Aberta, Portugal

W.L. Filho et al. (eds.), *Climate Change Adaptation, Resilience and Hazards*, Climate Change Management, DOI 10.1007/978-3-319-39880-8_19

Keywords Lay Knowledge • Climate Change • Artisanal Fisheries • Qualitative Methodologies • Interdisciplinary Studies

Introduction

A first exploratory study on the Impact of the Natural Reserve of the Berlenga Island (RNB) on the local life styles, particularly on the fishing community, highlighting their rationalities and lay practices was made by Santos et al. (2012). Revisiting this geographic, cultural, economic, social and environmental reality, and following the study of Viegas et al. (2014), that considers "Climate Change is a global problem which requires articulated local responses at the macro, meso and micro level and that its understanding involves the identification of mutually influent relations between Nature, Society and Culture", and bearing also in mind that any intervention—both to mitigate Climate Change and to adapt to Climate Change—involves necessarily that same society, in all the complexity of its social-cultural and environmental dimensions (Viegas et al. 2014), the work is carried on among the fishing communities in the Berlenga (Peniche) surroundings.

Lay knowledge, in its various denominations (as they highlight one or another characteristic), was subject of compilation in Delicado et al. (2012: 438) emerging as "traditional ecological knowledge" (Houde 2007); "indigenous knowledge" (Bohensky and Maru 2011; Bohensky et al. 2013); "local knowledge" (Paton and Fairbairn-Dunlop 2010), "stakeholders' knowledge" (Edelenbos et al. 2011); "lay knowledge" (Edelenbos et al. 2011; Brace and Geoghegan 2010) and "lay rationalities" (Alves et al. 2014). The dynamics and permanent reconstruction of lay knowledge, are rooted in interaction where makes sense and explains world and life phenomena, supporting the social action and interaction (Alves 2011).

The fishing communities due to their proximity and dependence on the coastal areas are particularly exposed to Climate Change and represent populations of great vulnerability and risk (Barbier 2014; Garai 2014; Nagy et al. 2014a, b; Verocai et al. 2014) justifying, thus, a focus of scientific, cultural, economic and socio-cultural interest (Fidalga et al. 2014; Seixas et al. 2014; Bitencourt and Rocha 2013), evidencing the understanding of its relation, adaptation and mitigation to Climate Change (Barbier 2014; Nagy et al. 2014a, b).

The main focus of this research is to understand how these fishing communities live with Climate Change and deal with Extreme Events. Therefore, the primary purpose of this analysis, focused on Lay Rationalities of Climate Change, is to get closer to understand how these people conceive and deal in *Hic et Nunc*[1] with Climate Change and which forms and sorts of knowledge these conceptions and actions stem from, bearing in mind the space and time contexts. Against this background the study seeks to: (i) observe and describe the way how Climate Change is perceived regarding its causes and effects; (ii) observe and describe the

[1]Authors' Note: Latin expression that means *Here and Now*

Fig. 19.1 Peniche Municipality. Lat. 39°21'N—Long. 9°23'O (*Source*: CMP, CEDRU (s/d) *Carta Educativa* (Education Charter) of Peniche Municipality. [Online]. [Consul. 12 Feb. 2015]. Available at: http://www.cmpeniche.pt/_uploads/educacao/carta_educativa_peniche.pdf. Pág. 26.)

Fishermen's sort of Climate Change related daily practices; (iii) understand which knowledge forms and sorts these conceptions and actions are based on, interweaved in socio-cultural dynamics.

Peniche

It is located on the Peniche peninsula, on the coast of the West sub-region, NUTS III, Peniche is defined by the municipality as the westernmost fishing city of the European continent (Fig. 19.1).

With traces of human activity stretching back to the Palaeolithic (Calado 1968), Peniche is believed to have been an island until the last decades of the fifteenth century. Only by late nineteenth century, beginning of the twentieth century XX, Peniche saw the tombolo, which links it to the continent fully consolidated and, "it was exactly that progressive transformation of the coast line which was at the genesis of all the social and cultural change process of the region" (Calado 1968: 93).

On June 30th, 2011, Berlengas archipelago was declared a Biosphere World Reserve by UNESCO.

The oceanic archipelago had its natural value, as island ecosystem, regarding the marine avifauna and their habitats, as well as the importance of its archaeological and geological heritage being officially recognised as of 1981, the year when it was declared Natural Reserve of the Berlenga Islands by Decree-law no. 264/81, of September 3rd and due to Resolution of the Council of Ministers (RCM) no. 142/97, of August 28th, the whole archipelago is included in the first phase of the national List of Sites, classified as: Berlenga Archipelago (PTCON0006). The Reserve is now called Natural Reserve of Berlengas (NRB) and its acknowledgement and protection status was progressive extended, by Official Gazette no. 0/98, of December 23rd, subsequently amended by Decree-law no. 32/99, of December 20th, and is included in National network of Protected Areas, pursuant to Decree-Law no. 142/2008, of July 24th and Natura 2000 Network, as provided for in RCM no. 115-A/2008, of July 21st.

However, given what was pointed out by Santos et al. (2012), the reactions to the application of the resolutions for this Protected Area do not seem to be consensual among the involved stakeholders, showing a top-down dynamics of legal provisions, scientific discourse and new tourism and sports activities which, when displaying local fishing knowledges and activities, must have boosted, in practice, the potentiation of competition/tension within various spaces traditionally used, among others, by the creels, hooks and net fishermen, which is under study.

Peniche's people has always been connected to the sea, and much of the local cultural heritage reveals this reality, regarding gastronomy handicrafts, where bobbin lace also reflect this deep connection (Guilherme 2010: 9).

Their history and stories are rich of elements to understand their identities and ways of living. And because the identities, when organizing meanings and experiences are also shelter and source for solidarity against disorder, the transformation or stigmatization appear to us (Castells 2007) as inevitable, trying to get closer to understanding how the creels, hooks and net fishermen in Peniche conceive, explain and deal with in their daily life with Climate Change and in which knowledge forms sorts they insert these conceptions and actions.

At the municipal scale and taking Mendes et al. (2011) study about *Social vulnerability to natural and technological hazards in Portugal* into consideration it can be seen that the social vulnerability level of Peniche municipality, weighing criticality and the support capacity is deemed to be medium.

The Sea, which in the case of Peniche is considered to be its development driver and a wide source of resources and opportunities, has been confirmed through Magna Carta, as the first general goal connected to the sea rank, its valuing "as structuring vector of the economic activities" (CMP 2009a, b: 272).

Fields as different as scientific research related to marine and energy resources, tourism, the production of electrical energy from the waves, or the search for the establishment of Peniche as a nautical centre, along with the wish to become a world centre for surfing, rely on the Sea as a main pillar.

Peniche, historically linked to fishing, maintains, following Magna Carta (CMP 2009a, b), a strong economic activity in the catching sectors, the canning industry, the freezing industry and the shipyards, despite the fleet and number of fishermen reduction, and the closing of some canning industries.

Peniche port is still considered to be one of the most important fishing ports in the country. However, and taking into consideration the reference as to the intention of valuing the fisherman profession (CMP 2009a, b: 147) or the promotion of initiatives which enable the youngsters to understand what it means to be a fisherman (CMP 2009a, b: 146), it is interesting to realize that on the 321 pages of the strategic diagnosis, *Peniche 2025* (2009), not just once comes across the expression "fishing community".

Although fishermen, together with the bobbin lace makers are figures of the collective imagery and are decisive factors for Peniche's economy, history and culture—the artisanal fishermen seem to be a community whose role and influence do not seem to be fundamental within any of the scenarios developed by the Strategy for Peniche until 2025. This situation makes, in our opinion, the issues pertaining to the identity and the subject relevant.

The identity concept is complex, dynamic and polysemous. For Castells (2007), whereas the identities organize meanings and experiences, the roles organize jobs, defending that the plurality of identities in actors develop tensions and paradoxes both for self-representation and social action, although he thinks that the more stable primary identity, structures the other ones.

Now, from a sociological perspective the identity is the result of a social construction, within a space and time context (historical, geographical, biological, religious, of the collective and the individual imagery) established by the power relations and shaped by the inter-influential dynamics of the (re)productive institutions, the collective memory and individual imagination is, however, from its processing—rationalities—by individuals, groups and societies, which are reorganized in their space and time meanings put into perspective, according to social structures and/or cultural shores (Castells 2007).

In turn, in the modern context of globalization and breakdown of social frameworks, Touraine (2005), advocating for the emergence of the subject in individuals based on the reflexivity, in an uneven struggle against the powers and standards which usurp the meaning of the existence. For the author, the meaning of existence is found by the collective social player—subject—in resistance acts and in the construction of singularity.

Bearing in mind that the subject of this study are the creels, hooks and net fishermen, also here, regarding the construction of identities and the subject's emergence, the lay rationalities, as an inclusive process, producer of holistic meanings and targeted responses, *i.e.*, which reach the meaning of the reflexive agent's action purpose based on his experience, appear, in our opinion, as a heuristic tool.

The characterization and the outline of the object of our study require the definition of 'artisanal fisherman', as it can't be find a formal definition of the concept "artisanal fisherman".

In Portugal, the legal framework of the fisherman activity, within the perspective of “a general framework of maritime safety requirements to protect the human life at sea and to preserve the marine environment” (Decree-Law (DL) no. 280/2001: 6731), is established by the legislation of inscribed seafarers. It could not be found, within the legal perspective, any reference to the ‘artisanal fisherman’.

But the research has enabled to get closer to the “artisanal fishing” concept: “traditional fisheries involving fishing households (as opposed to commercial companies), using relatively small amount of capital and energy, relatively small fishing vessels (if any), making short fishing trips, close to shore, mainly for local consumption. In practice, definition varies between countries, e.g. from gleaning or a one-man canoe in poor developing countries, to more than 20-m. trawlers, seiners, or long-liners in developed ones. Artisanal fisheries can be subsistence or commercial fisheries, providing for local consumption or export. They are sometimes referred to as small-scale fisheries” (after FAO’s website).

In the EU, reading with the User’s Guide of the Common Fisheries Policy (CFP) launched by the EC on 2009, not even once the expression “artisanal fisherman” occurs, although the expression ‘artisanal fishing’ is used for the African contexts.

Also Communication 417—final, of EC, of July (2011), on the Reform of FCP, does not mention ‘artisanal fisher’ and/or ‘artisanal fishing’, setting only a limit of 12 m for the small-scale fishing, excluding the towed gears.

On the other hand, Dawn and Gray (2005) in a critical analysis to the Common Fisheries Policy, in addition to underlining the huge failure of the EU with achieving a sustainable fisheries management, reflect on the limitations of the science: (i) the high levels of uncertainty of the fisheries sciences, (ii) the limited scope of analysis, which, traditionally, does not include multi-species analysis, nor environmental analysis and, finally, (iii) the gap between the scientific research, the fishing activity and the fishermen.

Relevant for this study is the absence of objective criteria for a precise outline of the ‘artisanal fishing’, as well as the absence of a legal framework for ‘artisanal fisherman’, by the dominant national and international players who establish the strategies for the sector. In addition to the possible semantic problems, this absence makes, in our opinion, its little economic, social, cultural and environmental valuing obvious, as these entities understand it, and the institutional void this way of life is subject to.

Against this background, some questions arise, among many others: Does it make any sense to be speaking of Artisanal Fishers in Portugal and, particularly in Peniche? After all, who are we talking about when we talk about Artisanal Fishers?

By the time of the Communication on *Artisanal Fishing Communities in Portugal* presented to the Naval Academy, Henrique Souto (2003) supports that artisanal fisheries is a fishing subsystem.

Rather diversified, artisanal fisheries, Souto advocates (2003), is defined by the specificity of the geographical, historical and economic context it is inserted and its traditional organization taking into consideration: (i) the size of the vessel—small or medium—and the vessels close to the coast; (ii) the ownership of the means of production—vessels and gears; (iii) the payment systems regarding the fishing income and with fish supplements; (iv) the great variety of usually passive fishing gears.

Based on the combination of these variables, Souto (2003) concludes that, there are four major types of Artisanal Fishing Communities: "1—Strictly family Artisanal Fisheries; 2—Family-based Artisanal Fisheries; 3—Transition Artisanal Fisheries; 4—Atypical Forms" (Souto 2003: 3).

With no legal existence but clearly present in the social and economic arena, unavoidable in social and environmental sustainability strategies and with indelible marks in the collective and individual imagery, it is before this framework full of diversities, semantic quarrels, imprecisions and paradoxes that the artisanal fishers emerge.

Always focused on the perceptions on Climate Change, trying to understand from which forms and sorts of knowledge stems the capacity to deal, here and now, with its possible effects, the Artisanal Fishers of Peniche under our study are the creels, hooks and net fishermen, while lay reflexive players.

These creels, hooks and net fishermen operate mainly off the coast of Peniche, and for the defence of their way of life have established the *Associação dos Armadores da Pesca Local, Costeira e Largo da Zona Oeste* (AAPLCLZO) (Association of Ship-owners of Local, Coastal and Off Coast Fishing of the West) and, for the organization/management of their activity *Cooperativa de Armadores da Pesca Artesanal* (CAPA) (Cooperative of Ship-owners of Artisanal Fisheries), under our study.

Methodology: Methods and Techniques

Sort of Study, the Methodological Procedures and the Participants

Location

This exploratory study is developed in the light of the comprehensive theories *i.e.*, it doesn't try to explain the causality of the players' social action, nor find regularity laws for living in society. The study seeks to comprehend the meanings of the social action, in specific individuals (Guerra 2012).

During the preparatory visits of the fieldwork, several talks took place and various contacts were established with potential informants, of which it's highlighted for the purposes of this study four preliminary exploratory interviews: (i) with historian Mariano Calado; (ii) with the President of the Fisherman Union; (iii) the Person in charge with the Peniche Port Authorities; (iv) with the President of AAPLCLZO.

In order to overcome the lack of a precise and consensual outline for the "Artisanal Fisherman" concept—both a scientific and a political one or even one defined by the fishermen themselves—for the present study having been selected to

Table 19.1 Sociographic characterization of the participants

	Age	Gender	Schooling	Residence
Fisherman 1	54	M	Former 2nd Grade	Peniche
Fisherman 2	56	M	Former 4th Class	Ribamar
Fisherman 3	54	M	Former 4th Class	Peniche
Fisherman 4	33	M	9th Grade	Peniche
Fisherman 5	56	M	Former 2nd Grade	Ribamar
Fisherman 6	40	M	7th Grade	Ribamar
Fisherman 7	41	M	8th Grade	Ribamar
Fisherman 8	52	M	Former 3rd Grade	Ribamar
Fisherman 9	35	M	12th Grade	Peniche
Fisherman 10	52	M	Former 4th Class	Peniche
Fisherman 11	39	M	8th Grade	Ribamar
Fisherman 12	54	M	Former 2nd Grade	Ribamar

adopt the expression of 'the creels, hooks and net fishermen'. Likewise it was selected CAPA, located inside Peniche Port, as the field for this research.

For the number of participants it was deemed to be convenient for the purpose and this sort of research, in addition to the availability of the fishermen, the availability of time and financial resources was decisive. In the case, it was chosen to use the word participants and not 'sample', as this is a qualitative research seeking a finer understanding from the respondents and according to their own terms. Therefore and to best meet the purpose and the sort of study, it was decided to select an externally homogeneous group (Guerra 2012). As it has been mentioned before, between March and May 2013, was interviewed a group of 12 'creels, hooks and net fishermen', which operate off the coast of Peniche in vessels up to 12 m and belong to CAPA (*Vide* Table 19.1).

Method for Data Collection

It was chosen the focused interview, according to the classical "Madelaine Grawitz Typology". This type of interview, such as the free interview is suitable for exploratory studies and has an intermediate level of informality and depth. The focused interview is different from the free interview only because it has a greater structuring level regarding the specific addressed topics (Grawitz 1993).

To carry out the interviews which were held in the facilities of CAPA, it was fundamental, on the one hand, the availability of the fishermen and, on the other hand, the inestimable support of CAPA, both providing a room for the interviews and promoting the scheduling of the meetings with the fishermen.

All interviews were recorded, with the previous verbal consent of the respondents and have an average duration of 60 min.

Taking into account the exploratory level of the work, a Pre-script was prepared based on the questioning and according to the purposes of the research, which was pre-tested.

According to Bardin (2013) "currently, and in general, the content analysis is known as: a set of analysis techniques of the communications aiming at getting by means of systematic and objective procedures of the content description of the messages indicators (quantitative or not) which enable the inference of knowledge regarding the production/reception conditions (inferred variables) of these messages" (Bardin 2013: 44).

Except for the tape recorder, the camera and the computer, used for the records during the field work, the analysis process to the content of the interviews did not turn to any dedicated technological tool.

And, although, the dialogue between deduction and induction has been permanent, this analysis may be classified as inductive, since it is from the discourse, and through its systematic deconstruction, shaped by categories and subcategories, from the analysis units, that it was attempted to conclude on the relations between the subjective meaning of the action, the social practices in their space and time context, considering, as Guerra (2012) states "the subject an active synthesis' of the social whole" (Guerra 2012: 31). This is even justified both by the sort of questions—semi-open– applied during the focussed interviews, and the welcome ease of the fishermen's responses and cooperation.

Taking into consideration the set indicators and questions, analysis categories and subcategories were then established, *a posteriori*, and the final grid of content analysis was prepared:

(i) ***Perception of the Climate Change***
This category seeks to analyse whether the respondents have a conception or not on the Climate Change and Global Warming. If they are aware or not of the Climate Change, on the one hand and if so, where and when had they become aware of them. Trying also to analyse where this knowledge stems from, their experience or another information source.

(ii) ***Perception of the environmental degradation***
This category addresses whether the respondents have a notion or not on the pollution in general and, particularly, on the coastal area and off the coast of Peniche, as well as the inter-relations between these effects. It was sought to understand how they perceived those impacts, both in the coastal habitats, surveillance in the quality of the fishing resources and where did they get this knowledge from. Finally, within this category, is searched for to assess which levels of surveillance/power could be, according to the respondents, related to the prevention of the perceived environmental degradation.

(iii) ***Causes of the Climate Change***
This category, assuming the existence of the phenomenon, sought to establish which causes the respondents ascribed the Climate Change to. Natural and/or anthropogenic? Others?

(iv) ***Risks***
With this category is tried to establish whether the respondents perceive the risks as a consequence of the Climate Change, and if so, which ones. Both in their family daily life, and within the scope of their fishing activity and/or at other levels.

(v) ***Behaviours***
Within the scope of this category, is tried to identify whether the respondents had/have had any change of behaviour in respect of Climate Change and, if so, which ones. On the other hand, it was sought to study how the respondents understand how to deal with the consequences of Climate Change and Global Warming and also, how the respondent, himself, dealt with the acknowledged phenomenon, identifying, or not, behaviours.

(vi) ***Responsibilities for finding solutions***
This category aims at understanding whether and/or how the respondents perceive or not, what should be done or how it should be done to deal with the phenomenon, on the one hand, and if they identify which agents they consider responsible for finding those solutions. Both at the global and national level, or local and/or/personal level.

(vii) ***History of the relation with the sea***
This category analyses the space and time contextualization of the respondents with the fishing activity.

(viii) ***Perception of the Fishing Resources***
This category seeks to understand the respondents' perception regarding the quantity of fish in the sea and the quantity of fish for consumption. Which species exist in the sea—if some have disappeared and/or others have appeared.

(ix) ***How to fish—Its effects on the sea, the waste and biodiversity***
This category seeks to identify the respondents' perception regarding the different fishing gears and their impacts.

(x) ***Perception about the Artisanal Fisherman***
This category analyses the notion of the respondents about the 'Artisanal Fisherman'. On the one hand his definition and on the other the perception the others have of the 'Artisanal Fisherman', as well as its own identification, or not with the 'Artisanal Fisherman'

(xi) ***Social Capital as members of the fishing community***
This category study's if the respondents identify a network of agents who support or hamper the activity. What is the role of these agents and which sort of relationship do they keep with the former. Trust? Fear? Other? Looking for to identify the existence or not of a social support network.

(xii) ***Reflexion on Climate Change***
This category aims at identifying if any of the respondents had already been asked about the Climate Change and Global Warming phenomena, by whom and when. As well tries to identify if they had reflected on the phenomenon.

(xiii) ***Identification of the Conception of Climate Change at the end of the interview***
In this category as tried to find out if the respondents, after having had this conversation (interview), could identify if there was any difference between their attitude before and after the interview. If so, which ones.

Characterization and Analysis of the Collected Information

In accordance with the design and the primary purpose of this study (to understand how this fisherman conceive, explain and deal with Climate Change), is presented the description and is sought to interpret their contributions.

The participants don't use the term or the concept of Climate Change, except one. They talk mainly about the changes in the weather and when they use the term Climate Change most of the times is with meanings that differ from the scientific jargon.

In most answers, the participating fishermen perceive the climate, its changes and effects, including Extreme Events, mainly from their life experience, although much of the information conveyed by the media, especially television, has been frequently mentioned:

> *We speak according to what we see. That's it! Some years we realize there are years with lots of rain, others with less and things are still practical the same. It was the same in past years too. It was always like this but people didn't pay much attention. An example, one year there was lots of rain and everyone would speak of that year: Man! It rained a lot and this and that ... Then, three or four years would go by when it didn't happen and people would forget ... Then again the same happened: Man! Everything was flooded and this and that ...* (Participating Fisherman No. 3)

> *Year after year, I think it seems to be ahead. As for the climate ... the season, come on! The weather is ahead or behind ... I guess it is, I don't know. The weather is not stable, it is not! It is not stable for the season we are living ... But okay, year after year, with our experience, it seems, there has always been a change!* (Participating Fisherman No. 6)
>
> *From what one hear and see I think so ... I think it's changing, but it was worse in the past! Winters were harder ...* (Participating Fisherman No. 8)

Being found in the speeches of the respondents the perception of a certain alarmism e and excessive emphasis, considered unnecessary, around Climate Change and Global Warming:

> *I don't mean it cannot happen, this is what I was telling you: People who are more aware of those things who are more into those things, it's possible there is a change, but I think, in my opinion, that there is no big change. There is no need to alarm people that much.* (Participating Fisherman No. 4)

The ignorance and indifference about the Climate Change topic, when addressed in the abstract, appears in the speeches. Nevertheless, when asked if they feel there are climate changes, the remarks become objective and changes are identified:

> *It's that, I am not sure what it is. But, sometimes I hear things on TV and that ... But I don't really get it. [...] Oh yes I do feel it! The seasons are ... They are not what they used to*

be. In the past they were more regular. The summer was summer, the winter was winter. Now, I think the seasons are a bit out of control. I don't know the reason why ... (Participating Fisherman No. 12)

It was also found that, although Climate Change has been reported, it was also mentioned that, the scientists are the people who are incumbent with identifying, understanding and explaining those phenomena. In other words, this downplaying of the knowledges resulting from the daily contact with these changes is highlighted, reflecting clearly the subordination of this knowledge to the knowledgeable people recognised as competent:

I can't understand that ... I ask myself. The extreme events ... They already happened when I was a boy, that's why I say this is very complex ... Everything can be. I don't reply too much and I don't reason too much about that, because I leave it to the scientists. Science is more competent to give answers. (Participating Fisherman No. 1)

Regarding the causes of Climate Change, the anthropogenic activity is predominantly considered. The action to fight against the pointed out causes is divided between prevention, promotion and the fatalistic view. In this sense either investment should be made in education or no way out is nowhere in sight:

We are the ones who make the Climate Change, we are the ones who pollute the environment, and we all do silly things! [...] *That starts at school ... There is no awareness regarding that. There isn't!* (Participating Fisherman No. 1)

The human being spoils everything! Because we do everything the wrong way. Okay, in the case of a volcano or something like that, that is different, it is of natural causes. Now... if the human being knows he is harming the climate, but keeps on doing the same ... there is no chance! (Participating Fisherman No. 11)

The complexity of the relations between climate, environmental and social systems is highlighted, mentioning in addition to the anthropogenic activities the interests at play. Furthermore, bioaccumulation and biomagnification processes by the trophic chain:

I think it is a bit of everything. Everything is connected. From my experience, I think this is it. Because nature is a phenomenon. [...] *But the human being is, in my opinion, a worse phenomenon. He pollutes very much. In the past there was more humanity, now there are more interests. There is more, a little more of everything. It is not even worth publishing it!* [...] *There is too much corruption! And we ... in the market, on TV, we see certain things, which...* [...] *For example: In the past in sewage everything was dropped in the sea. Are you following me? That was also very harmful, there were all those dirty things we all know about. And those things kill lots of things in the sea. It kills the coral it kills, if necessary, creates diseases among fishes. And we eat that. We don't know if we eat sick fishes. Outside we look nice but inside we may not be that way, right? It's the same with everything. Like the fish. And the veterinary turns a blind eye to it.* (Participating Fisherman No. 10)

There are certain water passages which are very contaminated because many sick fishes can already be found. Even with diseases such as those we call cancers. Those big lumps. Fish like that can already be found. Sometimes we see croakers with clots, spongy looking. The fish is sick, you know? Bibs, which are a fish from the edge, are seen all black... It is the disease that is appearing in the fishes: It can be because of contaminated water, or water currents... Water currents that move from one end to the other... It's based on that! Sometimes we see water spots of different colours... In my opinion this will impact the whole system. (Participating Fisherman No. 7)

When asked by the Climate Change's impacts, the emerging concern is remote. With regard to the *here and now*, which concern is interweaved with the human practices, particularly with anthropogenic activity and the fishing activity itself:

Very well, while the said ... the said hard periods of catastrophe do not come, isn't it? We don't feel them yet ... For us, things are going ... Everything's all right. Will it reach us? We don't know. (Participating Fisherman No. 5)

Of course I am worried with the changes, even at the level of fishing, in addition to the scarcity and the catch, the damage they make, but the climate has to do with it. If this cannot be controlled it will get worse and worse! Okay, they are trying to avoid certain things, but the main issue, I think they will never get there, I think ... But it's more dangerous for the future generations and the human race ... It doesn't affect me too much, anymore. In ten or twenty years I won't be here any longer ... (Participating Fisherman No. 9)

Fishermen's Practices Within Their Contexts

It is known from social research that the discourses on the practices may not coincide with the discourses on conceptions. In other words, when the analysis moves to the behaviours and attitudes included in the practices, other rationalities are summoned. Regarding the waste, one of the participants explains the reasons behind the new practices regarding the local waste handling by the seafarers, even if he is not sure about its final destination:

The waste we take to the sea web ring back to land. What we catch in the sea we bring to land. As a matter of fact I do this ... Be it old gears, or... the waste caught in the sea I bring to shore, I out it on top of the wall ... Afterwards that waste should go to ... As a principle I think it goes to recycling, isn't it? [...] *Because if we throw into the sea ... 'ah this is not ours' the other day we drop the gear and it resurfaces. It resurfaces and we bring it back again, we are breaking our art and we are contributing for further pollution also. Therefore, it resurfaces we bring it back to land, right?* (Participating Fisherman No. 11)

Although changes regarding the sea water temperature and atmosphere changes, more than directly it is the AG, the human practices and the anthropogenic activities which are perceived as the major cause of concern both about the depletion and the destruction of the resources and the marine and coastal habitats. However, in an inclusive narrative, in what can be thought of a search for an explanation and meaning for the present situation, the participating fishermen consider that the harmful practices are not limited, only, to their own fishing activity, and based on their life experiences, they interweave their professional practices and their life styles with moral, legal, political, safety, economic and financial issues, at the individual, local and global levels:

I have seen ships from all over the world! Chinese and Korean tankers! From all over the world! [...] *You have no idea of the oil spills made by these ships, the tanks washing in high sea, very year on our coast!* [...] *More pollution than that one? Beware! Beware! There are kilometres of that madam! Lack of surveillance, don't doubt it! There should be a sensor, a satellite to be able to detect from space the ships polluting the sea, the satellite would detect it immediately! This would be simple! Then... we were hoisting, from time to time, the little fishes even ... The hooked fish when going through those waters and would be all black of that ... Black? Yellowish!* (Participating Fisherman No. 1)

Defending their livelihood, another fisherman interlinks the pollution issue and that of the habitats and resources depletion issue:

We the fishermen are accused of putting stress, a huge stress on the animals. This is a lie! A lie! It is not us! Who are we?! If the problem was only the fishermen bringing to land then we would have no problems. There are other problems. The chemicals is something that has not been studied yet, because it doesn't need to be studied because it can be seen with the naked eye. If you look at a low tide what do we see? We see the rocks completely ... as if some acid has been thrown there and burned everything! Now if that can clean the rock ... imagine what it does to some being living near the coast! And we have many species living near the coast ... if in a very immature state ... some days after laying eggs ... That kills everything! Of course that after that species there is no recruiting and afterwards there are no more animals to catch! (Participating Fisherman No. 5)

The pollution resulting from human activities is also perceived as a major source of concern:

[...] *I started by saying that the sea was the largest waste bin in the world. Rivers flow into the sea.* [...] *Therefore, herbicides, pesticides, pig breeding, all that dirt into the rivers. Contaminated rivers, you know this is true. We cannot avoid this. And in the meantime the sea bears the blame. The sea is the largest waste bin in the world. Please note that it is I who says this. I say this!* (Participating Fisherman No. 2)

In addition to this the participant stresses also his relation with the sea, his perception of the practices which destroy the fishing activity and his concern about the future of the sea, warning for the confrontation of involved interests and the ethical and moral issues:

The sea is a dog. My grandfather died in it and I was about to die in it also! [...] *If everybody fished with hooks there would be lots of fish. Therefore, what happens is that each fisherman values his fishing gear. His sort of fishing. A sardine fisherman, pulls towards the sardine, the hooks fisherman pulls towards the hooks, the one who uses the nets pulls towards the nets, the one who fishes with trawl pulls towards the trawl and nobody yields. Envy, greed, all of that... So what happens? One of these days the sea has no future! Look here, the fishing has always been declining. And I will tell you why. Because we the fishermen are the murderers. Because there should be a close season and why isn't there one? Imagine this ... There is a pregnant woman there, and we go and cut the woman's neck. I don't know if this is a good comparison, I just used it so that you can understand what I am talking about. During these three months, January, February and March, it's a time of the year when the boats should not be at the sea. Nobody should be at the sea. Now what happens is that we have the trawl. In my opinion this is the worse murderer gear, that of the trawl. The trawl.* (Participating Fisherman No. 2)

In this context, one of the participants explains how trawling is operated nowadays and what its effects are:

And then, since the trawls have these iron balls ... to be able to trawl all the rocks. I tis very harmful ... The trawl trawled. They could only trawl in clean bottom. It only cleaned, it did not trawl rocks. Nowadays, wherever there is a bit of high rock: hillocks... They are called hillocks, everything goes. And now they trawl in the whole bottom. Do you understand? They go everywhere. And that harms the fishing a lot, kills the coral, where the little fishes survive, and all those things. They kill the little fishes, the smaller ones and destroy the coast. This is exactly what happens, isn't it?? (Participating Fisherman No. 10)

The diversity of the existing gears and the trawl as the most destructive sort of fishing, underlining the legality of this annihilating technology:

Now this is what happens, but there is much gear and that's it! And it gets more sophisticated each year. Like the trawls. The trawls didn't go to places where they go now ... Because they have other devices which detect the fish in a better way. And the trawl could not sail in places with very hard rock, because the device would break or the net would stay there ... And today they do ... with the rollers at the bow to break everything, to clean everything ... that's it...! Provided the mesh is legal that is legal! (Participating Fisherman No. 6)

In addition to destructing the habitats, in their discourses they highlight the waste of trawl fishing and the lack of attention given to these sort of practices, by Science and the security forces:

The artisanal fishing catches only the fish that can be consumed because of its size and, in fact, they are not juvenile. Unlike the trawl. Only a scientific observation to be able to see. When they lift the bag! To see what they do! The "restinga"[2] *of fish which is left ... All floating as they are no good. It would be good if someone wants to know how the fishing sector operates, that one day the State made available not too much money, and brought two scientists, rented a secret boat, made a deal with a ship-owner and that ship-owner would go on a route with those scientists and they, without being noticed, with no concerns for the people, without interfering in their lives—could film how the trawl works. How does it work!? And then they would raise their conclusions. Because what the trawls do... You have no idea about what they do. This is madness! 'Well, in fact, to estimate one day at the sea, it doesn't matter. We chuck it out!'... Everything is floating ... Such big mistakes, so big, so big ... And I say: 'Man! But where is the Navy to witness this?* (Participating Fisherman No. 1)

Gillnets, or equipment, is also legal, and considered a habitat destructing gear, in addition to 'marking' fish:

There are many people against trawl fishing, right? But I've been debating it with a few people and it goes like this: Trawl kills, damages, destroys the bottom of the sea, it kills from small to big fish, but trammel also does a lot of damage. Because fish create their own area. Fish has a home. But then it disappears from there! But, there it is, they've disappeared due to fishing net problems. Those that fall and end up there. The place where I used to catch a lot of fish about... thirty three and thirty four years ago, I used to catch a lot of fish there. Then we found out, the nets had been there, over and over... Now whenever I go there, I can't catch anything. Haven't caught a fish there anymore. Fishing nets destroyed the fishes' home. Hooks don't make any damage, they rotten up and disappear, while the net remains inside the fish, the fish grows around it... It stays there. [...] *There are many fish with disabilities due to that type of fishing.* (Participating Fisherman No. 3)

Speeches also note the impact on species and sin quality. Differences in fish types are pointed out:

[2]Authors' Note: Sandbank that may seem like a small island. It is the image that the fisherman has from the fish waste thrown away from the vessels.

"I sense the climate is different and we have fish we never had before. For example, hogfish, about twenty five years ago there was no hog fish in our coast. There was none. Now I also see a lot of meagre. On the other hand, we lost others. In other words, we lose some and we gain some. Many have disappeared. It's been years since I've last seen a "besugo trombudo (Lithognathus spp*)". I haven't seen it. I think it's almost extinct. "cação (Squalus acanthias or Mustelus asterias)", known as true "cação", we used to call it 'back run', we rarely see it. We rarely see it. Get it? "tramelgas (Torpedo* spp*)", a fish that gives electrical discharges. It gives electrical discharges. Do you get it? It's similar to skate but, attention, it's not skate. And if you touch it, it gives you an electrical discharge. I mean, it doesn't mean that there are none. You rarely see it. For example, "ruivo (Chelidonichthys cuculus, Trigloporus lastoviza, Chelidonichthys spp, Eutrigla gurnardus, Lepidotrigla, Trigla lyra* but most probable *Chelidonichthys lucerna)"... I remember when I was a kid, when I was about seven, eight years old... there were many."* (Participating Fisherman No. 2)

Normally, participating fishermen observe changes in seawater temperature and/or quality, from the type of fish and/or its quantity or from other plants and / or existing sea animals:

From what I hear, more on the basis of our sardine pub talk, they work more on temperatures... Because, well... A few years ago, around this season, the water was already fourteen, fifteen degrees, and now it's twelve. And, all of a sudden it rises over a month, and gets higher than before... I think we have a bigger gap between low and high temperatures. We have more fluctuation and previously the temperature was maintained for longer periods of time. (Participating Fisherman No. 9)

Cold waters, cold temperatures, more wind... This year, for example, it was a very bad year... These last six months, we couldn't stay a whole week at sea! (...) Due to wind and bad weather conditions, and the temperature, extremely cold waters. We would go out to sea and come empty handed! It's been changing quite a lot. And besides, at this point the water is much colder than in the previous years. The fish simply disappears, it doesn't come. This year there was no seabass. Not much sole fish... This has everything to do, in my opinion, with the water temperature. (Participating Fisherman No. 11)

Along our shorelines... water temperature is rising or decreasing? We are lead to believe... although there are no people studying it... We are lead to believe that the water temperature is tending to rise... We suppose. Why? Then what is being inserted? We have a new species... it has already being inserted... well... we've been catching some fish... meagre. About six, seven years ago... it wasn't normal... we never had it before in our country... we were invaded with that species, in immature condition... therefore, we are lead to believe that, in truth, that the temperature has been rising, because that animal, didn't use to live on these water temperatures. I relate it with... with what I've learned down south, along the shores of Africa... Because that species is abundant in those waters (Participating Fisherman No. 5)

One other participant questions about sea level rise and water pollution as the cause of sea weed disappearing:

A few days ago a North Pole report informed that those huge ice sheets, the icebergs, everything's breaking apart. Melting due to global warming... It's already happening. Of course! Maybe... sea level is rising because the water has to flow somewhere, right?!... Everything's related. It's probably happening now, because previously we could see seaweed level, which was one of best protections for fish... We were talking about that yesterday! It was a great protection for fish. Those "limbos-correia" (seaweeds) we had, huge, that used to float up... That was an amazing protection for fish! And it was good for people! They used to come all the way from farms, collect those "limbos (seaweeds)" to transform into manure... That has disappeared! We used to go through "malhadal" and you couldn't set one foot because it was all... Now you can't find a single "limbo". It has really disappeared. It might also be due to contaminated waters. We used to have a speedboat in Porto Novo, and there was an old man that used to go there every summer, to collect those few "limbos" still available and dry them off, to make medicinal products. Those older people... and he used to say the same: It's getting worse every year! (Participating Fisherman No. 7)

Sociocultural Dynamics and Rationalities

Seeking liaisons in memory construction, the perception and daily practices of this group of fishermen in creels, hooks and nets, it could be observed that every participant is a descendant of fishermen families not originally from Peniche, except one case, where the mother is a Peniche native. Better conditions offered by the safe haven for fishing activities was the main reason that helped them decide about coming to Peniche.

With regard to the term "artisanal fisherman" it could be observed that, besides the multiplicity of criteria in its definition, while for some participants the perception of their activity is defined, for others it's not:

I am an artisanal fisherman. What do you call artisanal? An artisanal fisherman is one that fishes on his own. That's artisanal. It becomes artisanal. Exactly. I mesh my own fishing nets, I make my own gear, you see here. Do you see it? Different categories. Do you see it? (Participating Fisherman No. 2)

We are like this, the life we live, we see ourselves as artisans. What's indeed artisanal, has no relation to creels, and has no relation to nets. We do everything manually, everything. Whatever you see after this, creels, nets, those are industrial gears. Everything we have, we pull our own gear... everything is done manually, we pull the gear manually, everything is manually, there are no machines... We release it manually, we have no machines, and we have zero machines. We have zero machines! (Participating Fisherman No. 4)

Another participant emphasizes fishermen plurality and some vulnerabilities:

When we speak about fishermen, we think about it in general... But that's not it. There are several types of fishermen... A lot! Let's see... There's the purse seine fisherman, which is definitely not related to artisanal fishing. There's the trawler fisherman, which is definitely not related to artisanal fishing. Then, we have distant-water fishing which is definitely not related to artisanal fishing. And then there's the artisanal fisherman... with than 10-meter boats an longer, which offer a certain stability, to bring a certain balance to his maintenance and sustainability, and there's the artisanal fishing fisherman, small boats which should have ended years ago.

He relates this situation to European policy guidelines to the sector and to technological, security issues and activity sustainability:

Let me tell you something: The European Union... You're aware our country makes many mistakes?! Many, many, many, many! And then some more! We all make mistakes and no one is perfect! Evidently, when the EU does this, it's 100 % wrong! We should practice artisanal fishing, which is an extremely sustainable fishing. But artisanal fishing with boats longer than X meters. A boat that is able to do some catches, not only along the coastline...

but with enough conditions to go 30, 40 miles away from the coast. The boat necessarily has to be in good conditions. Because a boat with 10 meters long, should never be less than 4 meters wide. That's the reason for stability. And there's another thing, The European Union is wrong, when it states it's doing it for artisanal fishing, it's a lie. If I own a small boat, and I have a 100hps engine, and that 100hps engine on my 10 meters long boat, runs at six miles. I need to use a 150hps engine. And the European Union doesn't give permission. The EU doesn't allow it. What's the policy? Where does stability and security stand? There's none! I never understood what's the reason behind the wrong policy adopted in Portugal.

Nevertheless, the same participant draws attention to the existence of a higher quantity of fish for human consumption:

There's more fish for human consumption due to the technological explorations carried out by trawlers. We have more fish here, but it's related to the technological explorations carried out nowadays. There are two types of fish for human consumption. Europe was invaded with fish for human consumption coming from non-partner countries... We are indeed the food source for, well not us, but Spain, and we get 50 % of fish coming from other countries. It shouldn't be like that. Coming from Chile, Morocco, down from Senegal, Mauritania, everywhere... (Participating Fisherman No. 1)

Another fisherman blames the markets, with their middle-men, in regard to the fishing effort he has to undergo and another participant refers the Government's liability for not introducing new technologies, less pollutant and more economic. Also the constraints created by the Government are mentioned, and pointed some reasons for maintaining greedy behaviours over money.

Regarding society's perception of fishermen, one of the participants shares his thoughts, his fears, his anger, and his strategy in social relations. His perception, in our opinion, indicates a low social capital:

We are not well attended, we are not treated respectfully. Why don't they speak to me just like you, madam, are doing now? When I have to go somewhere to take care of stuff, I'm put aside because I'm a fisherman. It's not what I sense, it's what I see! And I haven't been around during my father's years. You're really treated badly! And don't raise your voice, otherwise... you might be subjected to fines, fights and what else... And I can assure you it's because I'm a fisherman! Because if you're a fisherman you're the lowest in society... People live under the illusion that all fishermen are poor! And they are all dumb, they know nothing! Many people interact with me and aren't aware I'm a fisherman...

And, afterwards, his own perception filled with resistance:

I feel that because I'm a fisherman, I feel like they'll have to pay me whatever I ask for to go out to the sea. That's what will happen in fifteen or twenty years' time. They'll have to pay me whatever I ask for to go out to the sea! Because this will all be over! They find ways of finishing everything...Either they pay us whatever we ask for or we won't go out to sea. Either they comply with our rules... Because this... Things are changing. There are few young people, and they don't get it! Things are changing and in a few years' time, instead of complying with their rules, we will dictate the rules. They will ask us to go out and we'll say: No, no... (Participating Fisherman No. 3)

From a wider perspective, another fisherman, while comparing Climate Change and Global Warming, perceives poverty risks, social atomism and lack of security as his biggest fears for the future:

Either related to land or sea, in fishing. What we see and hear, face to face, is hundreds and hundreds of companies going bankrupt. And there's no help available. Nowadays you have the rich man and the poor. We used to have the middle class. Nowadays there's no middle class men, only rich and poor. Our country is in this situation and there are no improvements. If there's no assistance to fishing companies, and support, our country won't move forward. Why? Because there will be no companies paying taxes... Do you understand? To the Government and to the State, to say the least. There won't be. And then they will withdraw support policies. They will withdraw support for lower workers, with a miserable wage for his meals, surviving, sometimes with not enough money to pay his rent, not enough for his food, counting pennies. Loads of them... There has never been as much suicides as nowadays. The Company went bankrupt, people can't hold on, they can't stand it, their heads spinning the whole time, spinning, and spinning, and spinning until... leading to suicide. And it leads small couples that share... the same small wage, a minimum wage, a minimum wage, to the same situation. We can observe more people in the streets... Before there was no homeless people in Peniche. Before... we used to have more freedom. You could... walk the streets with no worries. Now you can't walk around in the streets. You can't walk around at night, it's terrible with the robberies. Robbing their own people! (Participating Fisherman No. 10)

Regarding ways of participating in behaviour's change in order to better deal with Climate Change mitigation, or adaptation, one of the participants comments, while expressing the complexity of the situation and the different levels of intervention:

If it's on my boat, I suppose I would have to initiate it. But if it starts coming from a superior hierarchy, at the highest position, and afterwards it starts flowing down, until everyone complies. Hardly, but... On my boat it would have to be me, or at my workplace, if I had a different workplace, I would try to do it, but if it's supposed to happen worldwide, only a wider organization like the UN, I suppose... That's my opinion. (Participating Fisherman No. 9)

It was observed that, sharing knowledge and responsibilities during their quest is always welcome, just as long communication forms and procedures are duly protected:

Look. Let me tell you something. I've always believed in science. Believe me. There's no doubt about it. I truly appreciate scientists worldwide. In every single field. Not only those related to Weather. In every single field, right? Every single field. And when I speak about Science, I'm referring to those whose ears stop to hear, and whose eyes stop to watch. I would say: knock at the right people's door and outline real values, so the boat reaches its destination... Because, you know there's a world of interests everywhere... If you'd knock another door instead of knocking the Municipality door first, they would be offended and it would not be possible to go through with it. Without the awareness and regard for the Municipality's interests, nothing can be done. Or some would do it, while others would ruin it.

And this participant shares an inclusive knowledge of an implementation proposal regarding a sustainable lifestyle, on the basis of goodwill, respect and work at every level:

Let's see. When children are born, parents should provide education, right? For the Municipalities to have this education to pass it on to its citizens, the order has to come from above and the governmental order has to come from the universe of all countries in the world... It begins with the universe of every country in the world, and then it's up to the governments... and then it reaches the municipalities, and afterwards the parish councils... and following parish councils, every single human being. But we should make a sum of all human beings' responsibility. That law should be easy to make, since we have a good relationship with people, we are at ease... Because it's not acceptable with a stranger... But if it's someone well integrated and capable of exposing... approaching the subject well, and carry it out with diplomacy, it shouldn't be difficult. But you have to listen, observe and plan everything very well. That's not difficult. There should only be respect. Respect for everyone. If a person is well accepted in the community, you can do anything. By force... there's no way! (Participating Fisherman No. 1)

On the other hand, every participant, including those who collaborated with the preparation interviews for field work, denied the existence of any contact related with Climate Change or Global Warming, on behalf of any political, scientific or technical entity.

Participating fishermen also referred this interview wouldn't change much their way of thinking or acting, since it's not 'their' field. Unless, of course, next time they hear someone speaking about Climate Change they will surely remember this conversation and pay more attention. And, although they don't consider themselves the right person, most of them were willing to participate whenever necessary, just as long it's not affecting their daily fisheries.

After verifying the non-existence of a clear incentive, by political impulse, for an active polyphonic dialogue, in order to promote an environmentally responsible behaviour, entrenched in cultural senses, the political action runs, in our opinion, into a discordant and indelibly uncomfortable situation, with the loss of human lives, particularly the weaker ones, non-human and habitats. i.e., in practice and as last resort, instead of an official speech, it is observed a strong contribution of political action in maintaining harmful practices, through the protection of power and knowledge which, when submitted to neoliberal hegemony and/or personal interests, do not promote society, nor inclusion, nor the pursuit of sustainable ecological resolutions.

Findings

In the course of the investigation process, and bearing in mind the study of a social reality is the study of an intrinsically complex and dynamic reality, it was particularly interesting to observe, throughout the work's structure and elaboration, the transmutation of the emotional relation associated to the object of study to a meta-position regarding the observed.

After a theoretical framework it was tried to build an investigation work that allowed, in its execution, to satisfy the curiosity raised by knowledge and lay rationalities and, with that, have access to a knowledge slice about a context sociability, centred on the professional activity of a small participating group, of creels, hooks and net fishermen, regarding Climate Change and Extreme Events, leading to a better comprehension of social action in this small group and, as a consequence, of the society where these people are integrated.

Thus, issues raised by the actual climate and environmental crisis and the way this crisis was built in its essence—on a diachronic perspective—is mainly related to human societies and their future. Either from the relations (de)constructed between them, or from relations (de)constructed with other beings and the environment, as a result of the effects of the inter-influences between those complex relations with the complexity of climate and environmental systems.

On the basis of what was observed during field work, and our thoughts, it's regarding a group of people trying to survive in a society where, besides not having a great social capital, isn't fully visible either for political power or scientific power. Nevertheless, these are people who pay their taxes, and work within a structured society bringing capital gain to that same society. Organized. Hierarchical.

From the work developed, it was observed in this group that this people are conscious of the social, environmental and climate crisis we are living and its complexity. They even have idea of how we, humanity, reached this situation and how we could track more sustainable paths. On the other hand, they are aware of games of interests—where they also play a role, but most of the times are victims—as well as they are aware of harmful social and environmental practices. Furthermore, they also use harmful practices, as a last resort, harmful for human beings, but on a daily basis that helps them survive in the society they are integrated.

Considering the scarcities in knowledge, power asymmetry and lifestyles, the uncertainty, space and time, ignorance, inter-influence and multi-dimension complexity, mostly conflicting or paradoxical, on the climate, environmental and sociocultural system, where the plural subject is a reflexive actor, lay knowledge is considered as a valid form of knowledge, and the observation of their practices as an heuristic capital gain.

In light of the viewed, amidst a climate, environmental and social crisis which, as a last resort, questions human survival, the question that arises, recalling Arendt (2013): If, just like a fungus, built over social structures and institutions, has spread itself and impregnated the gaps in social tissue, with the naturalization of social and environmental harmful practices amongst ordinary people?

References

Alves F (2011) A Doença Mental Nem Sempre é Doença. Racionalidades Leigas sobre a Saúde e Doença Mental—Um Estudo no Norte de Portugal. Afrontamento Ed., Oporto, Portugal, p 293. ISBN:978-9723611335

Alves F, Caeiro S, Azeiteiro UM (2014) Lay rationalities of climate change. Int J Clim Change Strategies Manage 6(1):1756–8692. ISSN:1756-8692, Emerald Group, Bingley, UK. http://www.emeraldinsight.com/journals.htm?issn=1756-8692&volume=6&issue=1

Arendt H (2013) Eichmann em Jerusalém—uma reportagem sobre a banalidade do mal, 3rd edn. Tenacitas Ed., Coimbra, Portugal, 385p. ISBN:978-972-8758-09-7

Bardin L (2013) Análise de Conteúdo, 5th edn. Edições 70, Lisbon, Portugal, p 281. ISBN:978-972-44-1506-2

Bitencourt NLR, Rocha IO (2013) Percepção das Populações Costeiras sobre os Efeitos dos Eventos Adversos no Extremo Sul de Santa Catarina—Brasil * Perception of coastal populations on the effects of adverse events in the extreme South of Santa Catarina—Brazil. Revista de Gestão Costeira Integrada/Int J Clim Change Strategies Manage 14(1):15–25. doi:10.5894/rgci408

Bohensky EL, Butler JRA, Davies J (2013) Integrating indigenous ecological knowledge and science in natural resource management: perspectives from Australia. Ecol Soc 18(3):20. doi:10.5751/ES-05846-180320

Bohensky EL, Maru Y (2011) Indigenous knowledge, science, and resilience: what have we learned from a decade of international literature on "Integration"? Ecol Soc 16(4):6. doi:10.5751/ES-04342-160406

Brace C, Geoghegan H (2010) Human geographies of climate change: landscape, temporality, and lay knowledges. Prog Hum Geogr 35(3):284–302. doi:10.1177/0309132510376259

Calado M (1968) Peniche na História e na Lenda. 2nd Author's Ed., Lisboa

Câmara Municipal de Peniche (2009a) PENICHE 2025—Diagnóstico Estratégico, Estratégia de Desenvolvimento e Programas de Actuação. Magna Carta, Peniche, 321 p. http://www.cm-peniche.pt/_uploads/PDF_MagnaCarta/Magna_Carta_Peniche2025_Completa_vFinal_imp_08_2011.pdf

Câmara Municipal de Peniche (2009b) Anexo—Metodologia de Prospectiva. In: Estratégia Peniche 2025, pp 287–317. http://www.cm-peniche.pt/_uploads/PDF_MagnaCarta/Magna_Carta_Metodologia_de_Prospectiva.pdf

Castells M (2007) O Poder da Identidade. In: A Era da Informação: Economia, Sociedade e Cultura. Vol. II., 2nd edn, Calouste Gulbenkian Foundation—Education and Scholarship Service, Lisbon, Portugal, 627 p. ISBN:978-972-31-1194-1

CE (2011) Reforma da política comum das pescas. In: Comunicação da Comissão ao Parlamento Europeu, ao Conselho, ao Comité Económico e Social Europeu e ao Comité das Regiões. COM (2011) Final. http://eur-lex.europa.eu/LexUriServ/LexUriServ.do?uri=COM:2011:0417:FIN:PT:PDF

Dawn T, Gray T (2005) Fisheries science and sustainability in international policy: a study of failure in the European Union's Common Fisheries Policy. In: Marine Policy. Elsevier, pp 189–197. doi: 10.1016/j.marpol.2004.03.003

Delicado A, Schmidt L, Guerreiro S, Gomes C (2012) Pescadores, conhecimento local e mudanças costeiras no litoral Português. Revista de Gestão Costeira Integrada/J Integr Coast Zone Manage 12(4):437–451. doi:10.5894/rgci349

Edelenbos J, Van Buuren A, Van Schie N (2011) Co-producing knowledge: joint knowledge production between experts, bureaucrats and stakeholders in Dutch water management projects. Environ Sci Policy 14(6):675–684. doi:10.1016/j.envsci.2011.04.004

Barbier EB (2014) A global strategy for protecting vulnerable coastal populations. Science 345 (6202):1250–1251. doi:10.1126/science.1254629

Fidalga ABP, Seixas S, Azeiteiro UM (2014) Estudo das Perceções da Comunidade da Palmeira (Ilha Do Sal, Cabo Verde) Sobre a Sustentabilidade das Pescas. Revista de Gestão Costeira Integrada/J Integr Coast Zone Manage 14(1):41–49. doi:10.5894/rgci446

Garai J (2014) The impacts of climate change on the livelihoods of coastal people in Bangladesh: a sociological study. In Leal Filho W, Alves F, Caeiro S, Azeiteiro UM (Eds) International perspectives on climate change: Latin America and beyond. Springer, Heidelberg, 316 p. 69 illus ISSN:1610-2010, 1610-2002 (electronic). ISBN:978-3-319-04488-0 ISBN 978-3-319-04489-7 (eBook). doi:10.1007/978-3-319-04489-7 Library of Congress Control Number: 2014934670, pp 151–163

Grawitz M (1993) Méthodes des sciences sociales, 9th edn. Dalloz, Paris, France, pp 569–631. ISBN:2247015581, 9782247015580

Guerra IC (2012) Pesquisa Qualitativa e Análise de Conteúdo—Sentidos e formas de uso. Ed. Principia, Cascais, Portugal. p 95. ISBN:978-972-8818-66-1

Guilherme I (2010) Amar Peniche. Author's Ed., Peniche, Portugal, 95p. Legal Deposit: PTI 312756/10

Houde N (2007) The six faces of traditional ecological knowledge: challenges and opportunities for Canadian co-management arrangements. Ecol Soc 12(2), 34. ISSN:1708-3087. Acadia University, Wolfville, Nova Scotia, Canada. http://www.ecologyandsociety.org/vol12/iss2/art34/

Mendes JM, Tavares AO, Cunha L, Freiria S (2011) Vulnerabilidade Social aos Riscos Naturais e Tecnológicos em Portugal. Revista Crítica de Ciências Sociais 93:95–128. Centro de Estudos Sociais, Coimbra, Portugal. ISBN:9789352541102

Nagy GJ, Muñoza N, Verocaia JE, Bidegaina M, Seijoe L (2014a) Adjusting to current climate threats and building alternative future scenarios for the Rio de la Plata coast and estuarine front, Uruguay. Revista de Gestão Costeira Integrada/J Integr Coast Zone Manage 14(4):553–568. doi:10.5894/rgci472

Nagy GJ, Seijoe L, Verocia JE, Bidegain M (2014b) Stakeholders climate perception and adaptation in coastal Uruguay. Revista de Gestão Costeira Integrada/Int J Clim Change Strategies Manage 6(1):63–84. doi:10.1108/IJCCSM-03-2013-0035

Paton K, Fairbairn-Dunlop P (2010) Listening to local voices: Tuvaluans respond to climate change. Local Environ 15(7):687–698. doi:10.1080/13549839.2010.498809

Santos AJFR, Azeiteiro UM, Sousa F, Alves F (2012) A importância dos conhecimentos e dos modos de vida locais no desenvolvimento sustentável: estudo exploratório sobre o impacto da Reserva Natural das Ilhas Berlengas na comunidade piscatória * The role of knowledge and the way of life of local inhabitants in sustainable development: an exploratory study on the impact of the natural reserve of the Berlengas Islands on the life of its local fishing community. Revista de Gestão Costeira Integrada/Int J Clim Change Strategies Manage 12(4):429–436. doi: 10.5894/rgci321

Seixas S, Hoeffel JLM, Renk M, Silva BN, Lima FB (2014) Percepção de pescadores e maricultores sobre mudanças ambientais globais, no litoral Norte Paulista, São Paulo, Brasil * Perception of fishermen and shellfish producers on global environmental changes in the Northern Coast of São Paulo State, Brazil. Revista da Gestão Costeira Integrada/J Integr Coast Zone Manage 14(1):51–64. doi:10.5894/rgci424

Souto H (2003) Comunidades de Pesca Artesanal em Portugal. Communication presented at the Marine Academy on March 23rd 2003. http://www.henrique-souto.net/resources/Comunidades%20de%20Pesca%20Artesanal%20em%20Portugal.pdf

Touraine A (2005) Um Novo Paradigma—Para Compreender o Mundo de Hoje. Col. Epistemologia e Sociedade. Piaget Institute, Lisbon, Portugal, 251 p. ISBN:972-771-822-1

Verocai JE, Gómez-Erachea M, Nagya GJ, Bidegain M (2014) Addressing climate extremes in coastal management: The case of the Uruguayan coast of the Rio de la Plata System. J Integr Coast Zone Manage 15(1):91–107. doi:10.5894/rgci555

Viegas VM, Azeiteiro UM, Dias JA, Alves F (2014) Alterações Climáticas, Perceções e Racionalidades * Climate change, perceptions and rationalities. Revista de Gestão Costeira Integrada/J Integr Coast Zone Manage 14(3):347–363. doi:10.5894/rgci456

Chapter 20
Analyzing Needs for Climate Change Adaptation in the Magdalena River Basin in Colombia

Ad Jeuken, Laurène Bouaziz, Gerald Corzo, and Leonardo Alfonso

Abstract In 2010 and 2011 Colombia was hit by severe floods. After this situation, Colombian government and river basin authorities started developing plans and preparing actions for adaptation to climate change. Together with Dutch institutes a demonstration study and capacity building program was executed in 2013–2014. A systematic analysis of future extreme discharges (as a proxy for the risks of flooding) for the upper and middle Magdalena river basin and water shortages for the Coello sub catchment was done using state of the art downscaling and hydrological modeling tools. In this analysis plausible future projections of climate (based on IPCC fifth assessment), land use and water demand (based on expert workshops and literature) were used to explore consequences of climate change. Recent maps, data and expertise of Colombian partners contributed for necessary input data and to validate the tools used. The study is presenting main results and is discussing its limitations and replicability. Climate scenarios show a persistent increase in the occurrence of extreme rainfall events and that, as a consequence, extreme discharges like during the recent floods in 2011 are likely to increase as well. The return period of the 2011 discharge is already quite high under current climate and might increase by a factor five under climate change. For water shortages the results are more ambiguous showing both the possibility of an increase and decrease of the unmet water demand depending on different future scenarios, the water use category and location within the catchment. The unmet demand is however also under the current climate substantial. The question whether this unmet demand is critical or not could not be answered. The modest length of the reference period in comparison with time period of the ENSO phenomenon, the relatively restricted modeling effort (only using hydrological and no hydrodynamic and damage-cost models) are limitations to keep in mind when interpreting these results. There are no principle barriers for replication the methodology in countries like Colombia that in addition is steadily increasing its data, modeling and

---

A. Jeuken (✉) • L. Bouaziz
Deltares, P.O. Box 177, 2600 MH, Delft, The Netherlands
e-mail: ad.jeuken@deltares.nl

G. Corzo • L. Alfonso
UNESCO-IHE, Westvest 7, 2611 Delft, AX, The Netherlands

W.L. Filho et al. (eds.), *Climate Change Adaptation, Resilience and Hazards*, Climate Change Management, DOI 10.1007/978-3-319-39880-8_20

knowledge capacities. For the adaptation tipping point analysis presented in this study the main challenge is to derive the necessary objectives, sufficiently supported by the relevant stakeholders and policies, as the starting point for a targeted adaptation process.

Keywords Climate change adaptation • Colombia • Adaptation tipping point analysis • Scenarios • Flood risks • Water shortages • Capacity building

Introduction

Likewise many countries do Colombia is preparing for climate change while improving its disaster risk preparedness, response and recovery capacity. The 2010–2011 la Niña (positive phase of El Niño) phenomenon affected four million Colombians, ~9 % of the total population, and caused economic losses of approximately US $7.8 billion, related to destruction of infrastructure, flooding of agricultural lands and payment of government subsidies (Hoyos et al. 2013). The government declared the state of emergency and during the flood situation released resources around 178 million us dollars. This was done while according to local media scientist estimated that the 9 % of the country was underwater.

Compared to an average flood in Colombia more than 70 % extra land was inundated (see Fig. 20.1). Poveda et al. (2001) indicate that there is a strong relationship between the ENSO phenomenon and river discharges in Colombia. Hoyos et al. (2013) in addition show relation between spatially varying flooding, social vulnerability patterns and damages that occurred due to the 2010–2011 floodings. It is only with low to medium confidence that these flooding events can be attributed to anthropogenic induced climate change (Harmeling and Eckstein 2013). Trenberth and Fasullo (2012) suggest that on a world wide scale the climate change induced increase in sea surface temperatures depend the extreme weather events that occurred in 2010 including the extreme precipitation in Colombia.

Within the Colombian territory the climate is however such extremely variable, being influenced by the extreme orography, the Caribbean and the Pacific weather phenomena, that both in observations and model projections trends are difficult to observe. Early climate and hydrological modeling studies (Nakaegawa and Vergara 2010) do not reveal clear trends in river discharges. Recent statistics and scenarios for Colombia give a geographically diverse picture and a large range of potential climate effects over the whole country. For the Andean region including the upper Magdalena basin the 16 model ensemble average however suggests a strong increase of average precipitation (IDEAM et al. 2015).

The National Plan for Adaptation to Climate Change -PNACC- supports the country's preparedness to cope with extreme weather events, and the gradual weather change. It directs the development of priority programs and projects and

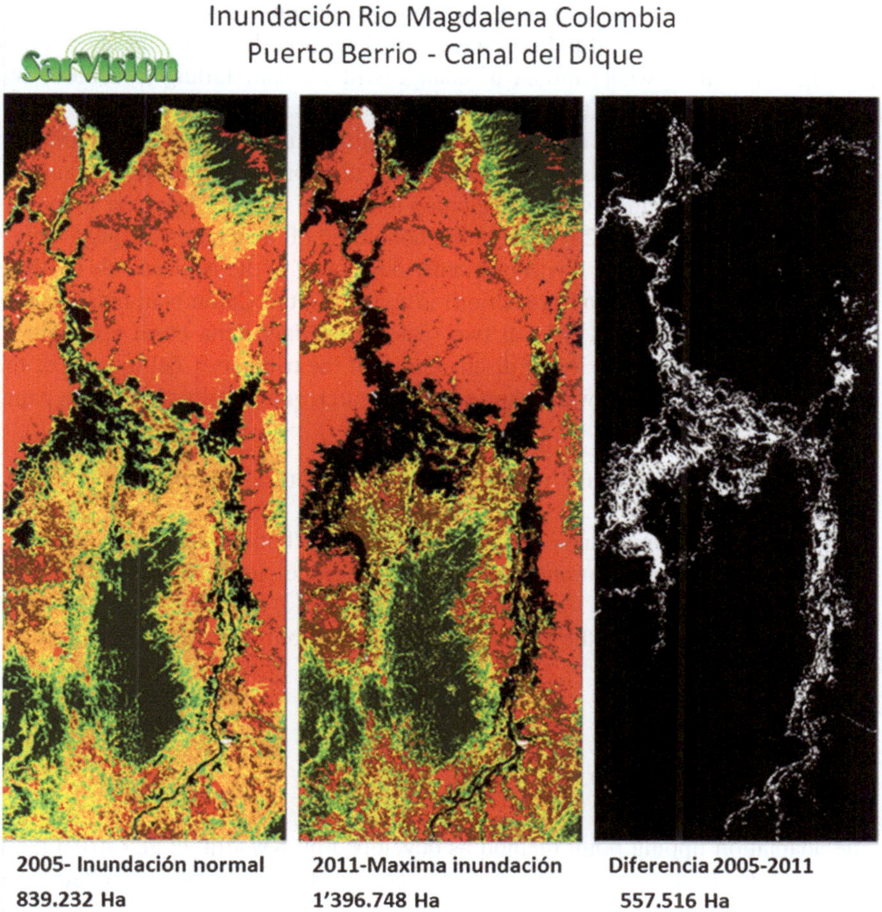

Fig. 20.1 Satellite (MODIS) derived maximum flood extent maps for the Magdalena basin from Puerto Berrio to canal del Dique for a regular year (2005 *left*), 2011 (*middle*) and the difference (*right*) showing how serious the floods in 2011 were (Quinones 2014)

strengthens the actions that are being already undertaken but that need to consider climatic variables in their planning and implementation, in order to reduce the negative consequences in the long term. The main objective of PNACC is to reduce socio-economic risks and impacts associated to the variability of climate. Specifically, the plan aims at generating a better understanding of the potential risks and opportunities, at incorporating climate risk management in planning of sectorial and territorial development, and at reducing the vulnerability of socio-economic and ecological systems to climatic events (Rojas-Laserna 2014).

In Climate Change Adaptation policy making, methods that are promoting a bottom up approach are gaining ground (Ward et al. 2014; Brown et al. 2012; Kwadijk et al. 2010; Wilby and Dessai 2010). Not the Climate Change itself is the leading argument for taking action but the key vulnerabilities of the area, sector, management practices and policies under consideration. Basically the question to formulate is: under what amount of change will we start failing to achieve our objectives or will we start to perform unacceptably?

Aim of this paper is to demonstrate the application of bottom up scenario and impact assessment methods in a case study in Colombia that could further support Colombian institutes involved in adaptation to climate change. The case study was executed from 2013 to 2014. The methodology of adaptation tipping points (Kwadijk et al. 2010) which was originally developed in the Netherlands (with its typical context of high flood protection) is applied in the Colombian management of flood risks and drought risks at two scales: the Magdalena basin upstream of Regidor, (for looking at extreme discharges) and the tributary Coello basin of the Magdalena (for looking at water shortages). Research questions treated in this paper are:

- What are the main results with respect implications of climate change for the areas under study.
- What are main limitations of the study referring both to technical (data and models) and organizational aspects (knowledge level, participation and cooperation)
- What can be said about the applicability and replicability of the methods applied.

Overall Methodology

The adaptation tipping point method as developed by Kwadijk et al. (2010) takes certain performance levels (derived from policy objectives) of the system as a starting point. Imagine for instance a river basin system with a basin authority providing water for consumption and irrigation and managing the flood risks by building and maintaining levees, reservoirs and flood pain management. To do this in a proper way, objectives are defined between authorities and stakeholders for the amount and quality of water that needs to be delivered or the safety level that needs to be maintained. These objectives need to be risk based since we have to deal with large natural variability within we won't be able to cope with all possible extremes. The main underlying question is: what is an acceptable risk? (Fig. 20.2).

Climate change will alter the river's hydrological regime which may lead to an increase of droughts and floods in the future causing at a certain moment that objectives can no longer be met and current management and policy has to be reconsidered, new measures might need to be taken. In scientific literature (Kwadijk et al. 2010; Haasnoot et al. 2012) these certain moments are called *adaptation tipping points*. The timing of occurrence depends heavily on the uncertainty in the speed and amount of climate change and on the definition of a critical level or acceptable risk level which may also change in time due to socio economic development or changing societal risk perception.

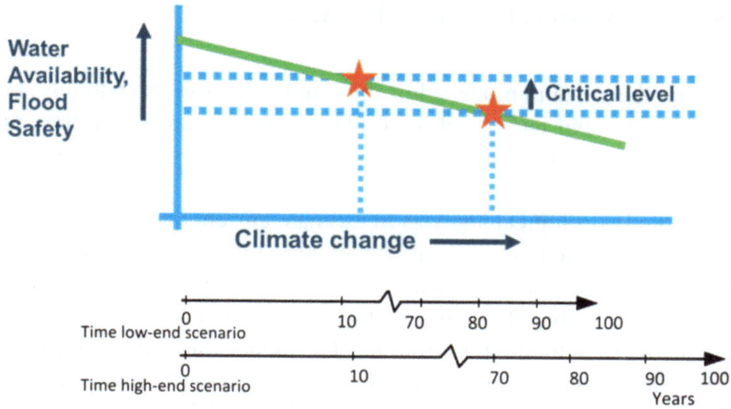

Fig. 20.2 Graphical depiction of adaptation tipping points. With increasing climate the critical or acceptable levels for flood safety or water availability may be threatened. The time range within this might happen is determined by the range of scenarios used. In a high-end scenario extremes are more likely to occur earlier than in a low-end scenario. Socio-economic changes can also alter the critical levels in different ways. For instance converting agricultural land to urban will increase the needed flood protection levels and population growth might increase the demand for fresh water

To determine the adaptation tipping points for the two pilot areas the following steps were taken:

(1) *Determine acceptable risk levels*. Relevant policy documents were analyzed and stakeholders consulted to find out what could be considered as accepted risk levels for floods in the Magdalena basin north of Puerto Berrio and for droughts within the Magdalena Basin. No clear flood protection standards neither clear service levels for water supply in the Coello basin could be found. Past floods however provide good information on hydrologic and hydrodynamic behavior of the flood and the casualties and damages that occurred to support such discussion. It is clear that the floods that occurred in the first half of 2011 are to be avoided in the future (see Fig. 20.1). Therefore we took this flood and it's a return period as a proxy for acceptable flood risks. For drought risks the drinking water supply to the city of Ibague was considered as a critical service.
(2) *Build system models* able to simulate risk variables for the two pilot areas (see Section "Build System Models for the Pilot Basins")
(3) *Derive climate and socio economic scenarios* to stress test the system (see Section "Development of Scenarios")
(4) *Stress test the system* with the scenarios and statistics developed in previous steps (Section "Deriving Adaptation Tipping Points for Flood Risks and Water Demand")

Build System Models for the Pilot Basins

In setting up the system models for this study the use of publicly available supported software and the use of local observational data was a prerequisite in order to really be able to improve and build upon existing capacities among the organizations involved.

To perform a hydrological analysis to estimate current and future discharges of the Magdalena river a hydrological model was built and calibrated using the WFLOW model platform (Bouaziz 2014; Schellekens 2014). The WFLOW model is a fully distributed version of the HBV model (Lindström et al. 1997) with a kinematic wave function for routing. The model was calibrated for the upper and middle Magdalena Basin at the stations of Payande (Coello basin), Puerto Berrio and Regidor. The NashSutcliffe (Nash and Sutcliffe 1970) scores obtained were respectively for each station 0.34, 0.80 (see Fig. 20.3) and 0.70. The input data used was inverse distance interpolated precipitation based on 490 stations measured by IDEAM (Instituto de Hidrología, Meteorología y Estudios Ambientales de Colombia). Minimum and maximum temperatures were also retrieved from IDEAM and interpolated using a lapse rate correction. They were then used in the Hargreaves formula (Hargreaves and Samani 1985) to derive potential evaporation. The point data was interpolated into a gridded product with a resolution of 0.02° (Fig. 20.4).

The unmet water demand under present conditions and under climate change was modeled using the WEAP model for five districts within the Coello basin (Droogers et al. 2014). The model used for the Coello basin is built using the WEAP framework. WEAP is selected as it is designed to work at basin scales and has the level of physical detail needed for this project. A detailed discussion on WEAP can be found in the WEAP manual which can be freely downloaded from the WEAP website (http://www.weap21.org/). An important component of the current project

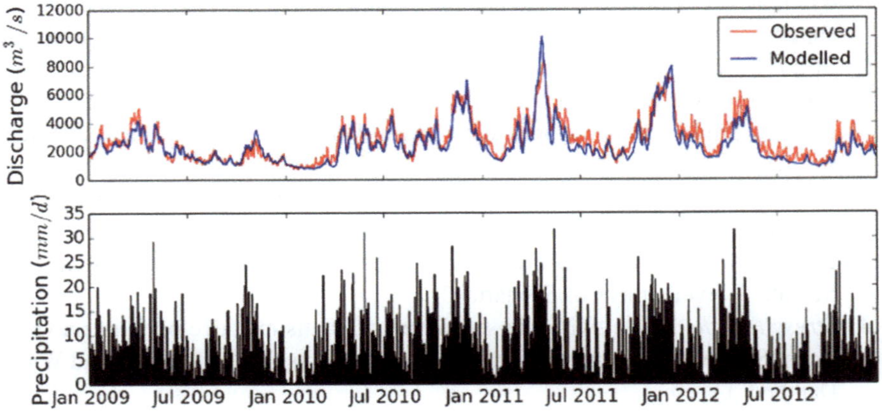

Fig. 20.3 Observed versus modelled discharge (*upper panel*) and observed precipitation (*lower panel*) at Puerto Berrio

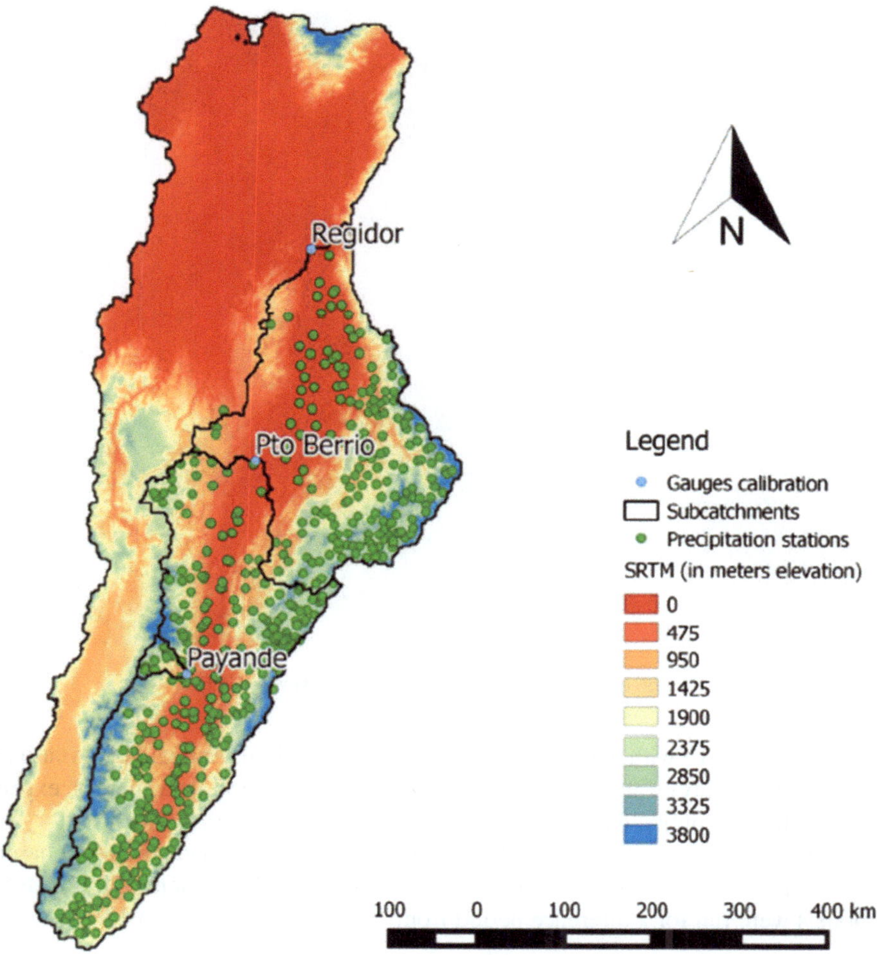

Fig. 20.4 *Upper* and *Middle* Magdalena basin with the location of the three gauging stations (Regidor, Puerto Berrio and Payande) used for calibration and precipitation stations

is the use of a two-tire approach in which data from one model feeds into another. Results from WFLOW model are used as available estimate for inflow in the WEAP model.

This has been done given the better capabilities of the WFLOW model to change the rainfall into runoff, which is determined by the highly-dynamic and complex hydrological processes. WEAP is not able to capture all these complex processes on a daily base and at fully spatial distributed extent. The Coello basin in the model is divided into five sub-basins according to major tributaries (see Fig. 20.5). Within each sub-basin the following water use categories are represented: Domestic (dominant in sub-basin 4 with the capital of Ibague), Industry (including mining dominant in the upper reaches of sub-basin 1–4) and Irrigation (dominant in

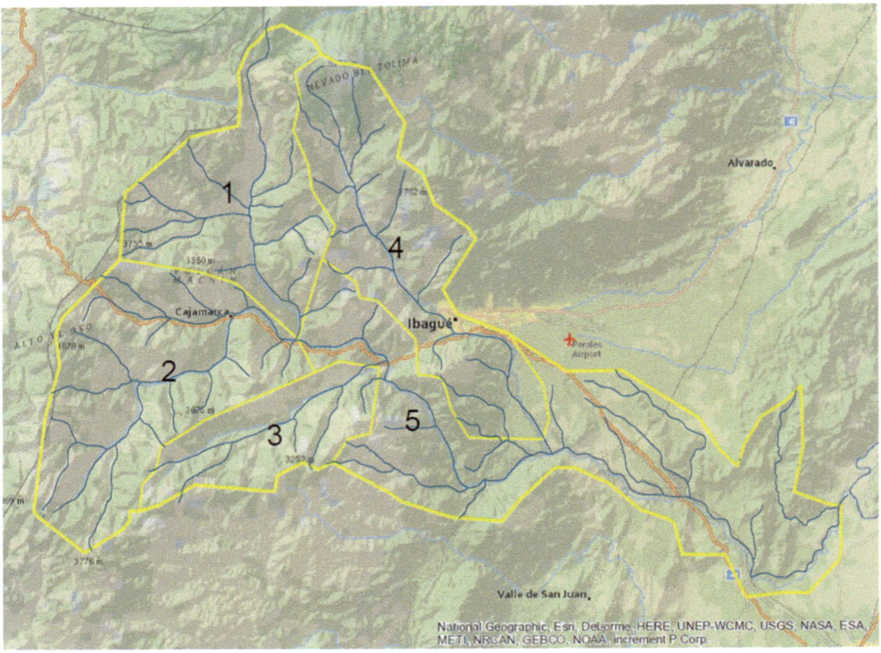

Fig. 20.5 Schematization of the Coello basin into five sub-basins for WEAP

sub-basins 4 and 5 with the extensive paddy rice cultivation). Annual (and in case of irrigation monthly) water demands were estimated using local data sources, expert knowledge and public data (see for more detail Droogers et al. 2014).

Input to both the WFLOW and WEAP model are detailed land cover data based on a map provided by IDEAM presenting the year 2005 (Quinones 2014). Both models were run for a reference period from 1980 until the end of 2013.

Development of Scenarios

To investigate possible impacts of climate and socio economic changes, targeted scenarios were developed. Three particular requirements were put to this scenario development: (1) they should cover a plausible range of uncertainty, (2) they should reflect the high variability key to the flood and drought problem and (3) they should be backed up by a stakeholder process.

To obtain a plausible range of climate scenarios we started with state of the art global available data from the Coupled Model Inter-comparison Project Phase 5 (CMIP5), which provides sets of future climate data by coordinating many institutions to perform climate model experiments (PCMDI 2014). The models in the experiment were driven by predetermined future paths of radiative forcing the representative concentration pathways (RCPs) (IIASA 2013). Model outputs for the RCP 4.5 (second lowest) and RCP 8.5 (highest) pathway were taken from two GCM's (out of 23): MPI-ESM-LR from Max-Planck Institute for Meteorology and HadGEM2-ES from the UK Met Office. These two models were chosen based upon their good reputation and also because their projected climate for Colombia was covering the range of all 23 available GCM's quite well.

Direct downscaling of the relatively course GCM output to a high resolution grid over Colombia would yield insufficient variability of and a bias in the variables under consideration. A solution is to combine GCM information with a reference meteorological time series. Statistics of a GCM control and future climate can be used to scale the values of an observational reference series so that they can be taken as representative for the future climate. This way the GCM signal is incorporated into the high spatial scale reference data that is more suitable for hydrological modeling. The observational data were obtained from IDEAM covering a 30 year reference period from 1980 until 2013. These point data were interpolated into a gridded product with a resolution of 0.02°. In order to assess climate change impacts on discharges of the Magdalena basin, the historical data (precipitation and potential evaporation) were transformed to the future climate as obtained from the two selected models using the advanced delta change method (ADC) (van Pelt et al. 2012) for mid (2036–2065) and end century (2071–2100). This method is especially suited to transform extreme precipitation to drive hydrological models and has been used successfully before for the Rhine Basin (van Pelt et al. 2012).

The results obtained show that an increase of the extreme precipitation values over the Magdalena basin can be expected of about 5–20 % compared to the reference data. This is a significant increase and it will likely affect flooding risks. This finding also appears to be quit consistent over the GCM and RCP projections used. Although these numbers are difficult to compare, since averaged over different geographical extend, the numbers are in line with projections given by IDEAM itself for some departments in the Magdalena valley. For instance values around 12–24 % (depending on timeframe and RCP) are reported for the increase in total precipitation for the departments Huila and Tolima until 2100 (IDEAM et al. 2015).

To develop scenarios for future land and water use in the Magdalena basin a participatory approach was chosen. Instead of extrapolating national and regional trend figures which are often difficult to obtain, experts from different sectors were asked to come up with estimates of water demand and land use in the basin in 2050.

First story lines (see textbox 1) were developed based on two future pathways, which were chosen by their selves as possible and plausible. One of these pathways was the shared socio-economic development path SSP2 and the other was SSP4 (O'Neill et al. 2012). Within each story line global estimates were given by the experts of percentage land use and water demand in the present and 2050 situation in the river basin (see for example Table 20.1). A group of around 12 experts developed the scenarios in a one day workshop. Good sectorial coverage was obtained with the relevant ministries of planning, environment, agriculture, mining and energy and some water management and municipal levels being represented.

Table 20.1 By experts estimated figures for water demand in the Magdalena basin

Sector	Demand in 2050		
	Actual use	SSP2	SSP4
Urban	10	10	20
Cattle	15	10	30
Agriculture	55	40	60
Mining and energy	13	15	18
Industry	7	15	20
Total	100	90	148

The national peace process, development of equality, the development of mining and some major cities were considered as key for the vulnerability to flood and drought risks (Jeuken et al. 2014).

Deriving Adaptation Tipping Points for Flood Risks and Water Demand

The hydrological model for the Magdalena was run for a reference period from 1980 until the end of 2013 and the in total 4 climate projections (see Section "Development of Scenarios") for both mid (2036–2065) and end of the century (2071–2100). An analysis of the effects of climate change on river discharges was done based on mean, minimum, maximum monthly flows and distributions of extreme values.

Extreme values based on a Gumbel distribution were analyzed for the different projections, as shown in Fig. 20.6. In this figure, as well as in Table 20.2, it can be seen that the central estimate of the return period of the discharge of the Magdalena river in the current climate is about 27 years. By the mid and end of the century this return period may be as much as once every 4 years on average.

The study on the impacts of climate change in the Magdalena basin shows an overall increase of the occurrence and the magnitude of extreme events. Additionally, overall mean discharges and especially during the first rainy period (April) are expected to increase. These results seem to be robust over the climate projections used in the study representing two different climate models and emission scenarios.

Based on this analysis of return periods there is good reason to act to improve the flood risks in this part of the basin. One may conclude that critical flood risk levels have already been exceeded in the current climate. The additional climate impact analysis indicates that if actions are taken now, the future climate projections should be taken into account as risks consistently increase over the multiple projections presented here. From Fig. 20.6 it could also be derived that the design discharge in 2050 for infrastructure to adapt to these increasing risks would be around 14.000 m^3/s if a 100 year return period is used as design standard for flood protection systems (like it often is in many countries) (Figs. 20.7 and 20.8).

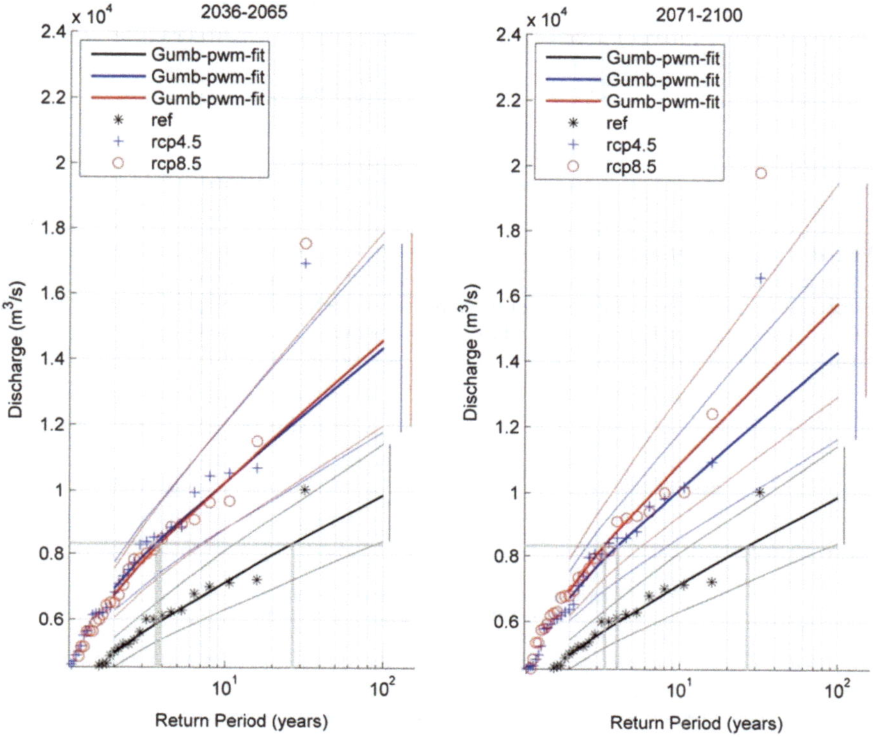

Fig. 20.6 Gumbel extreme value analysis for mid and end century for different RCP's of the HadGEM2-ES climate model. The points represent observations of annual maximum discharges and the *thicker lines* showing the different Gumbel fits for the different scenarios. The *finer lines* around the fits represent the 90 % uncertainty bounds. The *horizontal gray line* depicts the discharge that was measured at Puerto Berrio during the 2011 flooding

Table 20.2 Central estimates of the return period under climate change for extreme discharges at Puerto Berrio based on Fig. **20.6**

Discharge (m^3/s)	Current return period (years)	Return period in 2050 (years)	Return period in 2100 (years)
8300	27	5	4
10,000	100	10	10

There are however limitations to the above analysis that urge us to be careful to handle these results with care. There are quite large uncertainty bands associated to the central estimates given. For instance the bandwidth around the estimated return time of the 2011 maximum discharge ranges from 10 to 100 years. Discharges are used as a proxy for flooding using the recent example of the 2011 floods. Every flood situation is however different. Therefore additional hydrodynamic modeling is required to really be able to simulate flooding under different conditions.

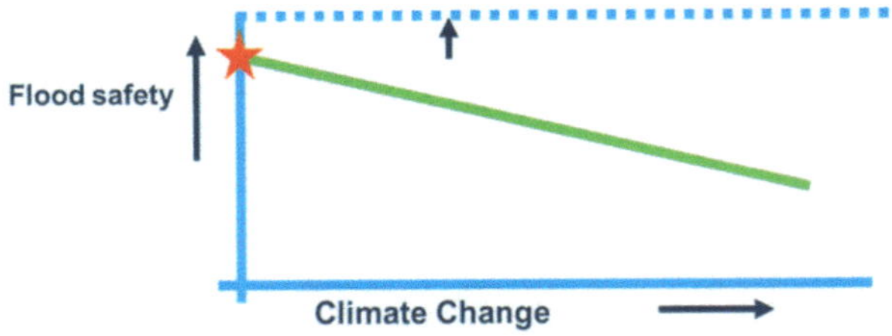

Fig. 20.7 In terms of adaptation tipping points it can be concluded that the current risk level (at which a discharge as occurred in 2011 has an estimated return period of 1/27 years (central estimate)) is already unacceptably high or graphically is below the accepted risk level (*dashed line*) Climate change will only increase the difference between actual flood safety and desired risk level

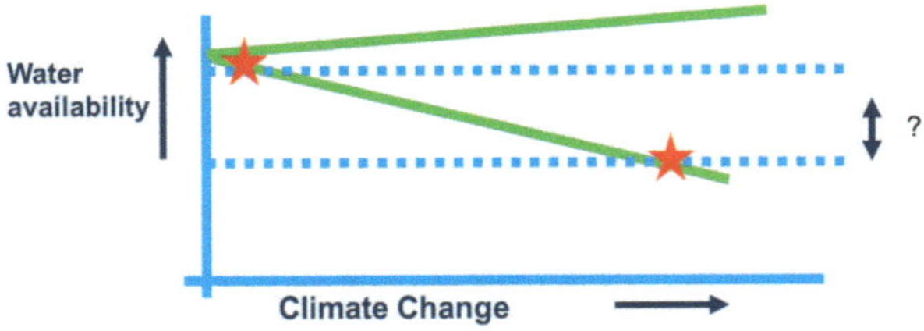

Fig. 20.8 In terms of tipping points for adaptation the situation in the Coello basin is unclear. There is clearly some ambiguity in the climate projections and no clear targets for water availability could be derived from the study due to a lack of clear objectives

The same climate projections were used to stress test the water supply and demand system of the Coello basin (as modeled described in Section "Development of Scenarios") under climate change. In addition changes in water demand (as suggested by the expert scenarios) were included in the water demand model.

For water availability the analysis gives ambiguous results as only one of the climate projections results in a decrease of water available and increase of unmet demand (see Table 20.3). This projection is the high emission RCP8.5 scenario simulated by the HADGEM2 model for the end of the century. Only under this scenario the total annual water demand for Ibague will increase from 57 MCM/year currently to 128 Million Cubic Meters year around 2100.

The average annual unmet demand will increase from 16 MCM/year to 42 MCM/year. However large variation in annual water shortages will occur. Equally important is to consider the number of months where water shortage occurs. The number of months without any water shortage does not change significantly (47 % currently and 43 % in 2100). However, the number of months with severe water shortages could increase significantly in the future considering this high end climate projection. The analysis by also shows that the influence of socio economic changes on the unmet demand is relatively low on the basin scale but locally could make a difference (Droogers et al. 2014).

Table 20.3 The unmet water demand in Million Cubic meters (MCM) per year for different projected combination of RCP, climate model and projection period (middle or end of century) and the reference period. Water demand scenarios are not included here

Scenario/GCM	Period	Unmet Demand (MCM/year)
Reference	1980–2013	878
RCP4.5/Hadgem2	1936–2065	509
RCP4.5/Hadgem2	1971–2100	609
RCP8.5/Hadgem2	1936–2065	618
RCP8.5/Hadgem2	1971–2100	1232
RCP4.5/MPI-ESM	1936–2065	608
RCP4.5/MPI-ESM	1971–2100	692
RCP8.5/MPI-ESM	1936–2065	668
RCP8.5/MPI-ESM	1971–2100	801

The analysis of future water shortages has yielded a lot of insight in the performance of the Coello water supply and demand system. However, in order to derive adaptation tipping points for water shortages or unmet demand in the Coello basin there are still too many unknowns and limitations to the analysis done. First of all total unmet demand is a too rough parameter to look at. The question if a water shortage is critical really depends on specific use at specific locations. Therefore a more detailed assessment is needed including locally acceptable service levels. A quick assessment of more local results indicate that for example for the drinking water supply to Ibague (sub-basin 4 a small tributary, with a growing population and upstream mining activities) a critical level will probably be reached earlier than for irrigation in sub-basin 5 (with a much larger base flow). Also to get a better grip on acceptability of water shortages it should be assessed first if the current climate variability already leads to critical conditions being surpassed.

Discussion and Conclusions

This demonstration study has provided a systematic analysis of extreme discharges (as a first proxy for flood risk) and water shortages for the upper and middle Magdalena and Coello river basin respectively. In this analysis plausible future projections of climate and water demand (based on expert workshops and literature) were used to explore extremes in rainfall and discharges, using state of the art downscaling and hydrological modeling tools, and to explore future water shortages using a state of the art water allocation model. Recent maps, data and expertise of Colombian partners were used to provide necessary inputs and to validate the tools used.

In summary the ***main results*** are that climate scenarios show a persistent increase in the occurrence of extreme rainfall events and that, as a consequence, extreme discharges like in 2011 are likely to increase as well. The return period of the 2011 discharge is already quite high under current climate and might increase by a factor 5 to once every 5 years under climate change. For water shortages the results are more ambiguous. One out of eight climate scenarios show decrease in precipitation in one of the growing seasons. Consequently only this one climate scenario leads to an increase of the unmet water demand in the Coello basin. This unmet demand is however also under the current climate substantial. All other scenarios lead to a decrease of the unmet demand. The question whether the unmet demand is critical or not could not be answered and needs additional analysis for specific water use categories at a local level both for the present as possible future situations. For the adaptation process in Colombia these results confirm the need for investments in improved flood risk management in the Magdalena basin that already have been started. In addition they show the need to take into account the potentially serious climate effects in the design of the flood risk management plans.

Being a demonstration project covering a wide range of topics by quick analysis, with the character of a sensitivity analysis more than of a thorough study, also implies that the above results should be handled with care. There are a number of key ***limitations*** to mention here:

Scenarios—In this study only 2 GCM's and 2 RCP's have been used to make downscaled projections of hydrological extremes using one novel statistical downscaling method. IPCC is promoting that a wide ensemble of models should be used. Although the models used in this study are selected to reflect a large range of inter-model variability this has not been tested thoroughly. Different downscaling methods take extremes in precipitation more or less into account. It is important to highlight that ENSO (El Nino Southern Oscillation) and the NAO (North Atlantic Oscillation) have shown to be important indices for climate variability in Colombia. There is a discussion at scientific level if and how these phenomena are affected by anthropogenic climate changes. According to Latif and Keenlyside (2009), Global circulation models do not show consistent indication of either amplification or reduction of ENSO under a warming climate. It is unclear what the implications are of downscaling climate model data using a 33 year data set of past observations. Therefore a critical review of different downscaling methods for the region (and its specific climatic characteristics) and looking at a wider model ensemble is therefore recommended.

From discharges to risks—In this study discharges in combination with past flooding data are used as a proxy for risks. Although this gives a first indication, for a proper risk analysis a much more extensive model and spatial analysis is needed. Such an analysis should include: an improved run off model taking into account the multiple reservoirs and its operations; a hydraulic model able to calculate water depths and two-dimensional flood patterns; flood vulnerability maps to be combined with flood hazards maps; and cost and damage indicator values.

To test the ***replicability*** of methods developed in the Netherlands in the Colombian context was one of the main research questions. The above presented limitations pose no principle barriers for replicability of the methods except that the high climatic variability as discussed provides an additional challenge compared to some other areas. At the moment Colombian authorities are investing in improving its capacity in climate adaptation. For example in the context of PNACC, the planning department (DNP) is aiming to improve its skills in its technical divisions. This includes the review of national and international experiences in adaptation, as well as a review of the relevant legislation framework. There already is a good data infrastructure present which together with modeling capacities are being strengthened at relevant institutions such as IDEAM and the Magdalena river basin authority (CorMagdalena).

The study further revealed that one of the main prerequisites, shared critical thresholds for adaptation needed to perform the adaptation tipping point analyses, were not easily to get. Clear objectives for managing climatic risks could not be derived from the policy documents or obtained during the stakeholder sessions neither at national nor at a more regional policy level. A more systematic participative process supported by proper knowledge and risk analysis studies as suggested above is needed to develop such objectives as a basis for an adaptation strategy.

Acknowledgement This paper is based on the study 'Adaptation to Climate Change in Colombia a tipping point analysis' which was supported by the Partners for Water program of the Dutch Government. The paper is further supported by EC FP7 funding grant agreement 308337 (BASE).

References

Bouaziz L (2014) Calibration of the daily openstreams wflow-HBV model for the Magdalena Basin. Report Deltares, Delft

Brown C, Ghile Y, Laverty M, Li K (2012) Decision scaling: linking bottom-up vulnerability analysis with climate projections in the water sector. Water Resour Res 48, W09537

Droogers P, Kraaijenbrink P, Lutz A (2014) Water allocation planning in Coello-Combeima Basin Colombia. Report FutureWater, Wageningen

Haasnoot M, Middelkoop H, Offermans A, van Beek E, Van Deursen WPA (2012) Exploring pathways for sustainable water management in river deltas in a changing environment. Clim Change 115(3–4):795–819. doi:10.1007/s10584-012-0444-2

Hargreaves GH, Samani Z (1985) Reference crop evapotranspiration from temperature. J Appl Eng Agric 1(2):96–99. doi:10.13031/2013.26773

Harmeling S, Eckstein D (2013) Global climate risk index. Briefing paper. www.germanwatch.org/en/cri

Hoyos N, Escobar J, Restrepo JC, Arango AM, Ortiz JC (2013) Impact of the 2010e2011 La Niña phenomenon in Colombia, South America: the human toll of an extreme weather event. Appl Geogr 39:16–25. doi:10.1016/j.apgeog.2012.11.018

IDEAM, PNUD, MADS, DNP, CANCILLERÍA (2015) Escenarios de Cambio Climático para Precipitación y Temperatura para Colombia 2011-2100. Herramientas Científicas para la Toma de Decisiones—Estudio Técnico Completo: Tercera Comunicación Nacional de Cambio Climático

IIASA (2013) RCP Database. http://tntcat.iiasa.ac.at/RcpDb/dsd?Action=htmlpage&page=welcome

Jeuken A, Corzo G, Quinones M, de Keizer O (2014) Adaptation to climate change in Colombia: a tipping point analysis: results of the workshop on socio-economic scenarios. Deltares, Delft

Kwadijk JCJ, Haasnoot M, Mulder JPM, Hoogvliet MMC, Jeuken ABM, van der Krogt RAA, van Oostrom NGC, Schelfhout HA, van Velzen EH, van Waveren H, de Wit MJM (2010) Using adaptation tipping points to prepare for climate change and sea level rise: a case study in the Netherlands. Wiley Interdiscip Rev Clim Change 1:729–740. doi:10.1002/wcc.64

Latif M, Keenlyside NS (2009) El Niño/Southern Oscillation response to global warming. PNAS 106(49):20578–20583. doi:10.1073/pnas.0710860105

Lindström G, Johansson B, Persson M, Gardelin M, Bergström S (1997) Development and test of the distributed HBV-96 model. J Hydrol 201:272–288

Nakaegawa T, Vergara W (2010) First projection of climatological mean river discharges in the Magdalena River Basin, Colombia, in a changing climate during the 21st century. Hydrol Res Lett 4:50–54. doi:10.3178/HRL.4.50

Nash JE, Sutcliffe JV (1970) River flow forecasting through conceptual models part I—a discussion of principles. J Hydrol 10(3):282–290

O'Neill BC, Carter TR, Ebi KL, Edmonds J, Hallegatte S, Kemp-Benedict E, Kriegler E, Mearns L, Moss R, Riahi K, van Ruijven B, van Vuuren D (2012) Meeting report of the workshop on the nature and use of new socioeconomic pathways for climate change research, Boulder, CO. http://www.isp.ucar.edu/socioeconomic-pathways. Accessed 2–4 Nov 2011

PCMDI (2014) Program for Climate Model Diagnosis and Intercomparison: coupled model intercomparison project phase 5. [online]. http://cmip-pcmdi.llnl.gov/cmip5/. Accessed 23 April 2014

Poveda G, Jaramillo A, Gil MM, Quiceno N, Mantilla RI (2001) Seasonality in ENSO-related precipitation, river discharges, soil moisture, and vegetation index in Colombia. Water Resour Res 37(8):2169–2178. doi:10.1029/2000WR900395

Quinones M (2014) Remote sensing products to support the project "Climate adaptation Colombia: Tipping Point Analysis". Report Sarvision, Wageningen

Rojas-Laserna M (2014) Adaptation to climate change in Colombia effective institutional arrangements for NAP formulation and implementation. Presentation at NAP Expo 2014, Bonn. http://unfccc.int/files/adaptation/application/pdf/colombia_mariana_rojas_laserna_session_3.pdf

Schellekens J (2014) Openstreams wflow documentation (Release 1.0-RC3-dev-31-45 M). http://schj.home.xs4all.nl/html/

Trenberth KE, Fasullo JT (2012) Climate extremes and climate change: the Russian heat wave and other climate extremes of 2010. J Geophys Res 117, D17103. doi:10.1029/2012JD018020

Van Pelt SC, Beersma JJ, Buishand TA, van den Hurk B, Kabat P (2012) Future changes in extreme precipitation in the Rhine basin based on global and regional climate model simulations. Hydrol Earth Syst Sci 16:4517–4530

Ward PJ, Jongman B, Kummu M, Dettinger MD, Sperna Weiland FC, Winsemius HC (2014) Strong influence of El Niño Southern Oscillation on flood risk around the world. PNAS 111(44):15659–15664. doi:10.1073/pnas.1409822111

Wilby RL, Dessai S (2010) Robust adaptation to climate change. Weather 65(7):180–185

Chapter 21
Livelihood Diversification as a Climate Change Coping Strategy Adopted by Small-Scale Fishers of Bangladesh

Apurba Krishna Deb and C. Emdad Haque

Abstract Coastal areas and wetlands are on the frontline of climate change and variability in Bangladesh. Small-scale coastal and floodplain fishers continually face a host of cross-scale stressors, some induced by climate change. This research is based on 21-month long field study carried out in two coastal and floodplain fishing villages represented by two distinct ethnic groups. Adopting nuanced people-centered ethnographic approach of field research, this study examines the ways small-scale fishers address the arrays of stressors in order to construct and reconstruct their livelihoods. Findings of our study highlight fishers' capability to plan and construct creative livelihood strategies and their adaptability in the face of stresses. We observed that fisher's coping strategies comprise a fluid combination of complex overlapping set of actions that they undertake based on their capabilities, socio-cultural embeddedness, and experiential learning under different adverse situations. Broadly, the coping strategies embody under economic, physiological, social, survival, institutional and religiosity-psychological factors. In this article, considering its predominant roles, only economic dimensions of coping actions that fishers undertake under unusual and abnormal stresses for survival and well-being are analyzed.

Keywords Bangladesh • Climate change • Coping • Small-scale fishers • Poverty • Livelihood diversification

---

A.K. Deb (✉)
Strategic Policy and Inter-Governmental Affairs, Environmental Stewardship Division, Department of Sustainable Development, Government of Manitoba, 123 Main Street, 2nd floor, Suite 160, Box 30, Winnipeg, Manitoba R3C 1A5, Canada
e-mail: apurba.deb@gov.mb.ca

C.E. Haque
Natural Resources Institute, University of Manitoba, 319 Sinnott Building, 70 Dysart Rd, Winnipeg R3T 3X8, MB, Canada

W.L. Filho et al. (eds.), *Climate Change Adaptation, Resilience and Hazards*, Climate Change Management, DOI 10.1007/978-3-319-39880-8_21

Introduction

'That you see I am still alive, means that I have gone through series of adversities, some happening on a day-to-day basis, and the others happening on a seasonal or annual basis. I have perfected my inner-self, my attitude and my body to the extent that I can survive any situation. Look, during calamities, the elephants die, little cockroach can survive somehow. We, the fishers, are like cockroaches.'- Anar Koli Jaladas, 45, fisherwoman and community leader, Cox'sbazar.

'I don't know about the climate change issues, but from my experience and observations in the last five decades in the sea, I can tell you that something goes wrong with the whole system. Things like species abundance, catch per trip, size and combination of species, etc. look extremely unusual. Probably the golden days of coastal and marine fisheries are gone. Who are to be blamed? We all are responsible- our lifestyle, deforestation, growth overfishing, population bloom, pollution, and so on. Who suffers most from this uncertainty and paucity in catch? It is us-the poor fishers.'- Ananda Jaladas, 70, Cox'sbazar.

Bangladesh and Climate Change

Climate change and variability is no more an issue of scientific thoughts, it is a reality. The low-lying deltaic physical characteristics, combined with the demographic and economic conditions make Bangladesh as one of the most vulnerable countries to climate change and variability (Government of Bangladesh 2009). It is one of the largest low-lying deltas on earth with a population of around 160 million living in an area of around 148,000 Km^2 with a GDP/capita of around US$ 2000 and life expectancy of around 65 years. Two-thirds of the country are below 5 m above the sea level, and get inundated by massive monsoonal floods at an interval of every 3–5 years- causing substantial damage to economy, crop production, lives and livelihoods.

The country is at the receiving end of about 40 % of the impact of total storm surges in the world. Between 1877 and 1995, Bangladesh was hit by 154 cyclones including 43 severe cyclonic storms and 68 tropical depressions, meaning an average of one horrendous cyclone every 3 years, at pre-or-post-monsoon periods creating storm surges that are sometimes >10 m. Even a median prediction from the General Circulation Models for Bangladesh would be 1.5 °C warmer and 4 % wetter by the 2050s, while a relatively warmer and wetter pattern of future climate that goes beyond historical variations will worsen existing climatic risks. These would cause increasing susceptibility by accelerating the severity and frequency of flooding, cyclones, storm surges, droughts, salinity intrusion, water-logging, and production failure. Notably, the impacts of existing climate variability and change are concentrated in geographical areas that are demographically characterised with higher concentrations of poor and socially vulnerable populations (World Bank Group 2010).

Climate Change, Small-Scale Fisheries and Fishers

Climate change can impact small-scale fisheries and fishers through multiple pathways. Its direct and indirect impacts on the lives and livelihoods of the majority of small-scale fishers in the tropical regions who comprise >90 % of worlds' fishers and fish traders are not well-researched and well-understood (Badjeck et al. 2010; Barbier 2015). The issue of climate change is alarming because of its direct and indirect linkages to food security, poverty, economies, well-being and sustainable fish production. Bangladesh is one of the most vulnerable countries to the impacts of climate variability and change on the fisheries, which is the source of economic livelihoods for millions of poor fishers, and around 80 % of all animal protein supply to the nation. Building fishers' resilience to adapt to short-term climate variability and long-term climatic changes through livelihood diversification and other development efforts are yet to be undertaken.

The real impacts of climate changes on fisheries are yet to be fully understood; it is assumed that climate change might amplify a boom-and-bust cycle of fish catches. But certainly, any change in production cycle is likely to negatively impact the marginalized fishers who are almost solely dependent on the profession. While a prolonged wet season flooding may be helpful in enhanced inland fisheries production, higher temperature regime might threaten the production, growth and distribution of both culture and capture fisheries. In the coastal areas, extreme weather events like cyclones usually hit the fishing communities, aquaculture operations, and fisheries infrastructures first. However, ascertaining precisely climate-induced changes from those of anthropogenic disturbances remain as a challenge to climate change researchers and development scholars (Brander 2010; Grafton 2010; Holbrook and Johnson 2014; Barbier 2015; Rita et al. 2015).

Poverty is a fundamental feature of the livelihoods of fishers- a harsh reality that thousands of small-scale fishers with so little economic endowments are forced to cope with, day-in and day-out. Such an unwanted situation continue to impede their capabilities, self-esteem, hope, happiness, dignity, level of tolerance, and other positive attributes for human well-being. Supported by the wide biodiversity and deltaic ecology, fishing is an age-old profession, especially for the birth-ascribed Hindu fishers. They are prone to multiple sets of natural and socio-culturally induced vulnerabilities that singly or synergistically impact the family well-being and livelihood resilience (Deb and Haque 2011). Given their dependence on fisheries resources, fishers live close to the shorelines, and hence, become first victims in the event of catastrophic hazards like cyclones and storm surges. We observed that alongside death and physical injuries, extreme climatic hazards like cyclones and floods damage their craft and gear- the basic economic capital they need for fishing and transportation on a daily basis, local fish landing facilities, houses, roads, and many other physical infrastructures.

By and large, fishers' economic livelihoods centre on the subsistence mode of income and employment. The corollary is that fishers falling below the minimum economic equilibrium with an extremely low level of income do critically risk, not only nutritional deprivation, but also their familial and social standing. In the face of shocks and stresses emanating from multiple sources, fishers as permitted by one's adoption of new skills and social networks, are more inclined towards adopting a multiplicity of livelihood strategies to sustain basic livelihood functions. Once used to be self-reliant, fishing villages have now turned to poverty catchment areas, especially in the last two decades. Struggling almost on a diurnal basis, the rural poor of the floodplain and coastal areas have proven their considerable capacities to adapt with layers of adversities including natural hazards and environmental degradation, primarily through mitigating impacts of stressors on fragile livelihoods.

Broadly speaking, fishers can reduce vulnerabilities in two ways. First, fishers as a whole can collectively resist the pressures. Second, they can define mechanisms at the individual or household level through actions like lessening consumption, employing more family labor, diversifying income sources, and seeking loans. While there are commonalities in nature and gravity, there are still remarkable variations in crises between families and among the fishing villages in different agro-ecological zones. Our knowledge is limited about how poor small-scale fishers cope with multiple stressing situations. Examining the ways how fishers try to cope with climate variability and change would help government policy makers and NGOs in designing and implementing development actions and strategies for ameliorating negative impacts.

The objective of this research is to examine the household level livelihood diversification efforts as coping actions of the small-scale fishers that they employ to sustain livelihood functions. This research excludes focus on long-term adaptation measures; rather we focused on the day-to-day coping actions. Our plea is simple: we simultaneously intend to reveal that although coping strategies and livelihood struggles of the poor fishers are primarily individual and familial issues, they are simultaneously telling of appalling social facts for this occupational class, coupled with irresponsiveness of the cross-scale political governance institutions. We explored these dimensions.

Conceptual Considerations: Climate Change, Coping and Livelihood of Fishers

The detailed effects of climate change on the physical and biological processes that singly or synergistically impact different biomes (freshwater, coastal and marine waters) and bio-agrological provinces are yet to be fully understood. Based on climate change vulnerability analysis of 132 countries including Bangladesh from the contexts of combined effects of predicted warming, the importance of fisheries in the national well-being and economies, and limited social adaptive capacity to these changes, Bangladesh has been categorized as the top 12th of the highly vulnerable countries (Allison et al. 2009). Historically, fishers coped with environmental challenges through developing, exercising and devolving their own socio-culturally appropriate strategies against the vagaries of nature. A comprehensive understanding and valuation of these adaptive strategies, and removing obstacles to further building adaptive capacity of fishers would obviously make them prepared to better cope with climate change vulnerabilities (FAO 2009). In simple interpretation, fisheries-based livelihoods of small-scale fishers, a climate-sensitive profession, become vulnerable when fishers fail to ensure livelihood functions, and cope with negative changes in catch availability supposedly induced by climate change variability and extreme events. Fisheries-resource based livelihoods become increasingly defenseless as the character and magnitude of climate-induced changes operate rapidly at a pace beyond the adaptive capacity of the fishers.

Livelihood at the household and community level as a concept is a good conceptual lens in understanding the likely impacts of climate variability and changes on the small-scale fisheries, and the fishers. Presumably, any climate-induced change in the production ecology and the productivity and distribution of fisheries resources can fundamentally affect the well-being of the small-scale fishers. A livelihood comprises the assets (natural, physical, human, financial and social capital), capabilities and activities for the means of living; it is considered sustainable when people 'can cope with and recover from shocks, maintain or enhance its capabilities and assets, and provide sustainable livelihood opportunities for the next generation' (Chambers and Conway 1992) (See Deb and Haque 2011 for details on *emic* views of livelihood and poverty in the fishing villages). A sustainable and vibrant livelihood system enables people to pursue robust livelihood strategies that provide, in effect, 'layers of resilience' to overcome 'waves of adversity', consequently enabling people to deal with and adapt to changes, and even transform multiple adversities into opportunities (Glavovic et al. 2002).

The notions of 'well-being' (Chambers 1995) and 'capabilities' (Sen 1981, 1987) provide a wider philosophical dimension along with other concurrent development issues. Simply, the more one manifests capabilities, the more secure is livelihood well-being, and individuals or households with higher levels of entitlements are anticipated to hold higher levels of livelihood resilience. Capability is viewed as *'what people can do or be with their entitlements'*, a concept surpassing material concerns of food intake or income that allows people to do things (Sen 1990). The concept of entitlement holds the fact that food insecurity and persistent hunger is an indicator of very low livelihood resilience of the poor who lack the requisite capacity either to produce sufficient food themselves or the financial ability needed to purchase food in a sustained manner, although food might be available in the market (Sen 1981, 1987; Devereux 2001).

From climate change point of view, vulnerability of fishers to climate variability and changes intricately relates to the extent of the distribution, productivity and availability of aquatic resources, and also the degree of the adaptive capabilities of aquatic resource itself to climate dynamics. From livelihood resilience and coping point of view, vulnerability also relates to the holding of assets; *'the more assets people have, the less vulnerable they are, and the greater the erosion of people's assets, the greater their insecurity'* (Moser 1998). Vulnerability is *'exposure to contingencies and stress, and difficulty in coping with them. Vulnerability has thus two sides: an external side of risks, shocks, and stress to which an individual is subject; and an internal side which is defencelessness, meaning a lack of means to cope without damaging loss'* (Chambers 1989). Adaptation as a concept is synonymous with 'downstream coping' (Downing et al. 1997), and usually refers to all those responses and actions of individuals and communities to climate change that may be used to reduce adverse effects on their health and well-being, and take advantage of opportunities that their climatic environment provides (Burton et al. 2002).

We consider the concept of coping both as process/strategies and sets of composite actions undertaken by people through experiential learnings in the face of adversities. Coping as a process or strategy comprises sets of active-behavioural actions or instrumental means that people decide to undertake based on their economic endowments and capabilities, and such actions gradually evolve, and the knowledge is transmitted. Under adverse situations, people undertake coping actions, deliberate or not. Coping is defined as *'the actions of ordinary people or disrupted remains of institutions, in contrast to official and planned response'* (Hewitt 1997). It is considered as *'the manner in which people act within the limits of existing resources and range of expectations to achieve various ends'* (Wisner et al. 2004). Some coping actions might be embedded to emotion, spirituality, and religiosity based on individuals' perceptions, human capital and the socio-cultural trend within the community.

Stability in livelihood functions can be considered as the *'coping dimension of response to climate change and variability and demonstrates the ability of people to retain basic livelihood functions while absorbing shocks, especially unpredictable climatic events'* (Osbahr et al. 2010). The concept of coping strategy is also connected to livelihood resilience; households with a higher level of livelihood resilience are expected to enjoy livelihood well-being and sustainability (Chambers 1995). Coping strategies involve people's cognitive or behavioral efforts, mostly situationally determined, or some long-term planned measures, that they apply when faced with stressors. At the family level, coping actions might be persuaded by experiential learning; such actions usually bring into play an image of collective behavior where members of households, as part of their parental or marital obligations and emotional attachments for providing necessary amenities, contribute differentially in their complex and multi-faceted areas of income, expenditure, power, influence and decision-making processes, as culturally appropriate for their age and gender. For different categories of people, numerous shocks in turn may result in different types of responses which include 'avoidance, repartitioning, resistance or tolerance mechanisms' (Payne et al. 1994).

Study Area and Methodology

This research was carried out in two fishing villages of Bangladesh, one coastal and the other floodplain, from January 2005 to September 2006 followed by another spell of group discussions with key informants during December 2009-February 2010 and November-December 2014. Small-scale fisheries in Bangladesh provide a poignant example of the dynamic setting of ecosystems, seasonality, geomorphologic process, productivity, poverty, and uncertainties to understand the impact of both anthropogenic and climate-induced impacts on fishers who understood and adapted to such changes for generations (Fig. 21.1a–d). The two villages selected for this study also represents 'climate change hotspots', each representing a primary hazard- the coastal village is most vulnerable to tropical cyclone, and the floodplain village is vulnerable to flooding. But never the less, both are again exposed to sets of stressors that reinforce the negative shocks of the aforementioned primary hazards.

The coastal fishing village *'Thakurtala'* is located in the Moheskhali Island of Cox'sbazar district along the Bay of Bengal of the Indian Ocean, and represents caste-based Hindu fishers *'Jaladas'* (literally, slaves of the water; one of the lowest in the caste-based social hierarchy). It has a population of 650 (male 300, female

Fig. 21.1 (**a**) Coastal erosion making fishing villages vulnerable to destruction, (**b**) water logging and tidal surge in the coast (**c**). house dilapidated by cyclone, and (**d**) women and children mostly become victims of cyclones

350, 78 households; approximately 45 % minor). The village is impacted by the semi-diurnal tidal pattern, especially during new moon and full moon phases; the periphery of the village is surrounded by a narrow strip of mangrove forests (a mix of *Avicennia* spp., *Sonneratia* spp., and *Rhizophora* spp.) and the terrain of the '*Adinath* temple hill', a sacred area for the Hindu pilgrims.

The floodplain fishing village *'Volarkandi'* is located in the Sujanagar union, Baralekha sub-district, Moulavibazar district of the north-eastern region of the country, and represents relatively new-entrant Muslim fishers (locally called '*Maimal*', bearing low-status in social hierarchy). It has a population of 1240 (male 640, female 600, total 184 households; approximately 40 % minor). The village is located within the '*Hakaluki haor'*- an ecologically critical area and the largest natural wetland system of Bangladesh. These low-lying fragile landscapes with interacting dynamic waterbodies render insightful historical and cultural geography of lower-Bengal delta, occupational niches of the inhabitants, their relentless struggle over territorial uses, and their access rights to the aquatic resources on which they primarily depend on for making simple livelihoods.

Both qualitative and quantitative methods were used with heavy emphasis on participatory techniques because they are more responsive to exploring complex phenomena that are situated and embedded locally. However, philosophically, field works in this study demanded a more nuanced ethnographic engagement with fishing families to get fine-grained understanding of their day-to-day livelihood constructions and coping actions for survival. Among the participatory tools, focus group discussions, key informant interviews, case studies, and participant observations were extensively used. Baseline survey was conducted among 78 coastal and 60 floodplain fishing households; 45 coastal and 27 floodplain key informants were interviewed, and 42 Focus Group Discussions (FGD) (coastal village: 23 events, floodplain village: 19 events) were carried out. Initially, prior to in-depth field research, around 3 months were spent with the respondents to develop a sense of intimacy and trust, and to learn more specifically about local cultures and fishers' sentiments.

The unit of analysis in this research spanned from individual to households to community. Direct observations and participation with the fishers on the fishing boats during day-and-night time fishing in the floodplains and sea were the most useful and straightforward way to learn about people's experience about livelihood dynamics, vulnerabilities, and coping strategies. To talk with women in a culturally appropriate manner, their time of cooking, weaving, chasing head-lice in a group, and other household activities in the open backyard areas of the homesteads was chosen. Given the fact that there is no authentic document on one's age in fishing villages, we used some historical markers (e.g., how old a respondent was during the 'liberation war' of 1971, 'big cyclone' of 1991, 'mega-flood' of 1998), followed by validation with parents or grandparents, as appropriate. This technique of 'back calculation' of age proved useful in building intimacy with the respondents.

Results and Discussions

Small-scale fishers in Bangladesh, specially the coastal/marine fishers, are prone to a host of cross-scale stressors or vulnerabilities (See Deb and Haque 2011 for details) including those emanating from climate changes, that impact differentially on different groups. We found it both flawed and difficult to precisely isolate the perceived impacts of climate change patterns from many other anthropogenic and natural resource management issues. However, based on their observations and experiential learnings, the older generation fishers agree that 'things' in mother nature are changing swiftly. Fishers from both the fishing villages agreed that climate change fetch poor catch, unpredictability in weather conditions, illness and other health-related problems (e.g., skin diseases), scarcity of potable water, threat to life, loss in income, and low livelihood status. Specifically, coastal fishers mentioned about the intrusion of salinity in both surface water and groundwater (tube well), increased frequency of cyclones and storm surges, change in species structure and size of fish, poor growth of mangrove forest, outbreak of disease in fish and shrimp culture farms, erosion along island edges, and accretion in shallow sea area. Floodplain fishers mentioned about increased flooding, drought, hot weather, worse infestation, poor fruit yields, and hence, increased dependence on insecticide, poor fruit yields, short retention of water in swamp areas, arsenic in ground water, prolonged fog, and deteriorating health status of livestock.

Common to both the fishing villages, some of the adaptation practices prior to ant hazard are: storage of essential items (like dry clothes, food, drinking water along with water purification tablets), building temporary platforms for livestock, raising the homestead land, making houses flood proof with bamboo as a base, protecting tubewell from contamination of floodwater, moving to safe areas like cyclone shelters, sending old people and children to upland areas, leasing livestock for disaster seasons to upland areas, protecting the tube well by sealing the openings, saving some cash for emergency uses, storage of essential medicine, etc. (Focus Group Discussion 2014) (Fig. 21.2a–d.). We exclude the importance of long-term adaptation hardware (like building embankments, cyclone shelters, green belting, developing salinity-resistant and floodwater resistant crop varieties, etc.) from discussion here. For most of the fishers, climate change issue is not a direct concern, rather they are heavily concerned with their daily needs.

A total of 73 % coastal and 45 % floodplain fishers have elicited their food insecure condition (*'pete bhat nai'*, literally no rice in the belly), followed by their worry about decline in natural resource base- the very basis of their livelihood making, and an increasing trend in climatic uncertainties and shocks (see Deb and Haque 2011 for details). Fishers' coping strategies against stressors may vary based on a host of factors: one's physical condition, age/experience, attitude, gender, literacy, frequency/repetition of stressors, social capital, and economic endowments (Table 21.1). For example, a food-insecure person with an acute disease will have to apply more coping strategies compared to an only food-insecure able-bodied person.

Fig. 21.2 (**a**) Making temporary shelter after cyclone, (**b**) raising the homestead land to combat flood, (**c**) tubewell base is raised high for protection from floodwater, (**d**) houses erected to make flood proof

Table 21.1 Categorization and prioritization of coping strategies by small-scale fishers

Coping strategy	What does it mean for individuals and households?	Fisher's priority
Reliogiosity-psychological	Making one-self and the collective behavior of the family members tuned to crises by signaling and informing in a culturally appropriate manner. Also includes some emotion-focused issues including extended prayer and more sacrificial actions to please the Almighty.	**
Physiological	Making one-self physically ready to cope with adverse situations through reduced food intake, taking little food or famine food, situation, and bear the pain out of persistent hunger.	***
Economic	Includes reduction in family expenses, employing more adult and minor's familial labors, intensification and extensification within natural resource sector, migration, use of assets and multi-tasking.	*****
Social	Communicating with the social networks involving neighbors, relatives, friends, well-wishers, and seeking beneficial or symbiotic actions (for example, exchange or lending of foods).	***
Institutional	Making best uses of the beneficial local level institutions and attempting to link with higher institutions.	*
Survival	Includes some immediate measures for survival of family members, community members, and domestic animals.	*****

Source: Based on FGDs with fishers, 2010, 2014. Number of stars reveals emphasis

Most of the climate change and livelihood related coping literature put emphasis on economic activities. It is difficult to isolate and judge coping actions using a single lens of analysis and often, such actions change within days or even hours. Based on empirical evidence, unlike academic and western constructions of coping, we follow a comprehensive course of coping following the socio-cultural landscape of the fishing villages. The broad groups mentioned here are dynamic, interactive, overlapping or intermingling in nature and some coping actions can be considered from more than one angle (See Table 21.1).

Each family has a different epic of struggle for addressing poverty correlates and traps, and socio-economic transitions and tensions. Each rural household has its own mental 'frame of living standard' or a benchmark for subsistence living, well-communicated within members of households. Responsible members singly or collectively take coping actions, deliberate or not, to address stress-provoking or dissatisfying conditions compared to that particular frame of living standard. How do the poor families figure out their coping actions? Salmi (2005) succinctly conceptualizes that *'for each life-mode, there is a corresponding specific conceptual universe which is quite distinct from that of other life-modes'*. Coping strategies of households may include defense mechanisms, active ways of solving problems, and techniques for overcoming stresses (Murphy and Moriarty 1976).

This study focuses on the aspects of livelihood diversification which fishers undertake as a mechanism to spread the risks and uncertainty inherent in small-scale fishery. First of all, we attempt to provide a theoretical discussion based on works of Shanin (1972), followed by a general picture of diversification, and then we look closely at what fishers do within and outside fisheries sector for livelihood diversification, one of the most important coping actions in addressing climate change. Small-scale fishing being a seasonal activity, livelihood diversification is extremely important for the fishers who are often considered not suitable for jobs other than manual and fishing-related activities. Diversification supports households to insulate themselves from multiple environmental and economic shocks, and access to the fisheries resources remains critical for such communities, sometimes even more so as a result of vulnerability (Baumann 2002). In the rural context, access to multiple sources of capitals not only helps people to make a living commendably, but also adds importance to the person's world, encompassing one's capability, moral strength, dignity, sustainability of living, empowerment process, and the ability to change institutional aspects.

Based on economic endowments, familial choices in actions, composition of families and overall condition of the food deficit situations, different households calculate their own sets of priorities with considerations to pros and cons of each action during a food deficit period. For all categories of families in the study areas, the most common economic strategy is to save and build up endowments, at whatever amount possible, to offset the sufferings of food deficit period. Typical of fishing families, there are both male economy and female economy, usually complementary and supportive to each other. When the male economy collapses, the female economy plays a distinct and cautious role. When both the economies collapse, families face serious crises. Female economy is usually based on rigorous savings through austerity and selling of products from home-based livestock, poultry, dry fish, smoked shrimps and vegetable products.

Theoretically, rural livelihood diversification is viewed as the 'process by which rural households construct an increasingly diverse portfolio of activities and assets in order to survive and improve their standard of living' (Ellis 2000). For ensuring food security, households in their livelihood pathways undertake multiple activities, comprised of 'complex *bricolage* or portfolio of activities' (Scoones 2009). Diversification of livelihood opportunities is extremely important for the 'persistence of subsistence', as familial incomes are seasonally affected by a host of factors. Livelihood diversification (What) involves activities undertaken by (Who) adult individuals and/or minor family members to (Why) augment individual or familial income from more sectors or activities than considered usual, (When) in the face of crises or as preparations for coping with adversities, (Where) centering one's own professions or involving some other sectors in which one can apply some acquired skills. It is the process by which 'rural households construct an increasingly diverse portfolio of activities and assets in order to survive and improve their standard of living'; such a process may comprise of 'reactive strategy' for family subsistence of small-scale farmers, and a relatively stable voluntary course of 'choice and adaptation' for 'big farmers' in the face of shocks and crises (Ellis 2000). The equilibrium thinking of 'fading away' and back to normalcy is often limited for a proper understanding of livelihood strategies (de Haan 2000), as household's coping strategies are now increasingly impacted by cross-scale issues like climate change, global market and politics.

Livelihood diversification, as a key to successful adaptation in the face of climate change vulnerability, is not only about acquiring new skills, but also about accumulating asset to capitalize the acquired skills. Diversification may be deliberately thought out by households as part of their survival strategies, or as a spontaneous reaction to offset crises during declining periods of livelihood well-being. The causes and consequences of diversification may be differentiated in practice by geographical locations, assets, family income, opportunities available and social networks (Davies 1996; Ellis 1998; Barrett et al. 2001), and thus, diversification may bridge over gaps or even accentuate economic and social inequality. The concept *'rural pluriactivity'*, an attempt to 'gaining an income from more than one economic activity' is popularized by Eikeland (cited from Salmi 2005). Such diversification activities are intended as mechanisms for coping, adaptation and accumulation, meaning there are differences between livelihood diversification of poor who are struggling to survive, and those of better-off households that diversify to mount up more capital for future (Salmi 2005).

Fishers' livelihood diversification can be hypothesized from a temporal and experiential point of view. For peasants, Shanin in 'Awkward Class' (Shanin 1972) put forward that *'a complex multi-dimensional mobility, involving both 'centripetal and centrifugal' tendencies simultaneously operating among peasant households is, therefore, at work and underlies the gross differentiation process in*

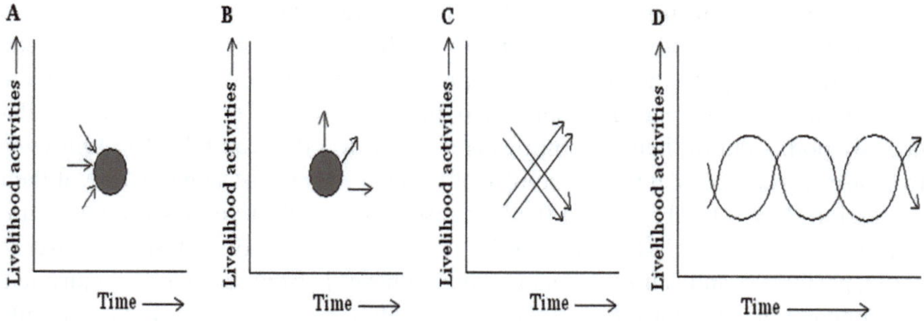

Fig. 21.3 Proposition (**a**): 'Birth-Ascribed profession and occupation of the last resort' (**b**) as opportunities surround, fishers diversify their incomes from multiple sources, (**c**) multi-directional professional mobility to maximize earnings, and (**d**) trade-off between fishing and non-fishing activities as cyclical mobility of livelihood activities (*modified from* Shanin 1972)

peasant society'. This hypothesis is used as a basis of our explanation of the livelihood diversity (Fig. 21.3) of fishers.

a. *Notion of centripetality:* Historically, caste-based Hindu fishers revealed *'centripetality'* towards their profession. Because of the lack of skill in other similar professions and a birth-ascribed occupational affinity, caste-based fishers tend to remain in fishing-related professions. This notion of 'centripetality' is also true for many fishers for whom small-scale fishing is literally a mouse-trap from which the exit is difficult though the entry is relatively easier (Fig. 21.3a).
b. *Notion of centrifugality:* The Muslim floodplain fishers revealed *'centrifugality'* in their professions (Fig. 21.3b). Being privileged by the availability of farming land in the vicinity of the fishing village (both upland and seasonally inundated drawdown areas), a relatively higher level of socio-political freedom and networks, subsistence fishers have the opportunity to earn from agriculture also.
c. *Notion of multi-tasking:* A section of the fishers in all the villages behaves smartly to pool more resources through multi-directional mobility adopting strategies like seasonal migration and migration for remittance earning. Up to a certain level of age and physical condition, multi-tasking is bearable. Such multi-tasking is physically demanding and stressful for some fishers (Fig. 21.3c).
d. *Notion of cyclical mobility:* Through the experience gained from multiple livelihood activities and based on the opportunities, fishers would make a trade-off between major livelihood activities in the form of cyclical mobility (Fig. 21.3d). The cyclical mobility model tends to conceal class differences that exists among small-scale fishers, but in real life conditions, some biases like the number of active earning members, number of women and old persons, inheritance of wealth, family partition, fisheries resource availability and price, political power, desire for social dignity and macro-economic issues would impact the proposed picture of professional diversity.

Efforts for livelihood diversification 'obeys a continuum of causes and motivations that vary across families at a particular point in time, and for the same families at different points in time' (Ellis 1998). It is more important for a marginalized section of the communities who are victims of the processes of deagrianization and defisherization. Diversification, in our opinion, under certain context is a rational for building economic endowments and minimization of risks, which simultaneously points out to their capabilities and access to other capitals. The following two cases reveal the adaptive capability of fishers given the seasonal nature of their profession. Interestingly, though the first case sounds reflective of seasonal migration or 'transhumance' (Keesing 1981), it is more a form of 'transcattle', where cattle, pastoralist and cattle-owners get mutual benefits from the change in physical environment. Cattle farming/numbers are central to the economy of not only agricultural families, but also for many rural households as a means of a safety net.

Case 1 His name is Nurul Islam; his life orients with seasonality; he is a fisher in the monsoons and a cow rancher (*bathanee*) in the dry seasons. "*During winter, there are hardly any grass covers left in the hilly area. There is a serious scarcity of fodder; the owners have no alternative left than to send the cattle at the downstream areas. As the uplands ('Kandis') wake up following the receding water, there is plenty of new grass. This new grass is very useful for the health of milking cows.'* says Nurul. Cows are from plain land, semi-hilly, and hilly areas. He charges Taka100 (US$ 1 = Taka 82)/month for oxen and pregnant cows for their maintenance. There is no charge for the milking cows as he earns money through selling milk. Animals are brought for 4 months- *Poush, Magh, Falgun and Chaitra* (mid-December to mid-April); number varies from 70 to 700 in the ranching areas of the *haor*. Nurul doesn't have any notebook or diary for keeping accounts of the cows and associated financial transactions. *'Everything is in head, there is no document'*- he adds proudly. There is a small shed for the milking cows and calves; Nurul lives in one corner of the cow-shed. Both Nurul and his cows drink polluted channel water. Around 20 % of the animals are milking cows (1–1.5 L/cow/day); milk is sold at US$ 0.25–0.30/L. His income ranges between US$ 350–850 (Taka 25,000 to 65,000 for the dry season) based on the number of cows. He notices unusual nature of floods in recent years. As water starts entering the *haor*, he returns cows and gets ready for subsistence fishing with a small lift net for rest of the year and deals with different sets of clients.

This case demonstrated that some households with enhanced social networks and different sets of knowledge can interplay with changing sets of ecological characteristics.[1] It shows that seasonal adaptations in the

[1]From literature, we get a similar picture of cow ranching known as *habbanae* in rural Cameroon which is based on social reciprocity. The *habbanae* is the loan of an animal (e.g., a heifer) from a herdsman to another. The receiver keeps the animal for a given period of time throughout which he can hold it in usufruct (e.g., a heifer's milk and calves). Afterward, the animal is returned to the provider, who may then enjoy a new *habbanae* in return. This system of gift giving—based on *habbanae* as a common norm of reciprocity—is a device for herdsmen to build and maintain social

haor shape the interrelations between social groups and their economic lives.

Case 2 'Sabbir is an influential young leader. With 235 ducks, he does duck ranching on a small piece of upland; he encloses the land with torn fishing nets and constructs a small house (L-7 m, W-7 m, H-2 m) with bamboo fence to protect the ducks against attacks of foxes and wild cats. He does not buy small ducks; he purchases juveniles to rear them quickly in a limited time. He provides some supplementary feed during *'Ashar-Srabhan-Bhadra'* (mid-June to mid-September). Around 20 kg of unhusked rice and 10 kg of rice bran are given everyday. Sabbir keeps a proportion of 5 % male (*haoa*) and 95 % female (*aai*). He gets 160–180 eggs each day. His wife and little sons take care of the duck when he goes outside. Sabbir earns around Taka 8000–15,000/month (US$ 100–190) for a period of 4 months. He knows that duck litter is very important in maintaining a higher productivity of the *beel* water. Once the business of ducks is over, Sabbir starts growing vegetables, thus making best use of his upland area.

Case 3 Her name is Urmila Jaladas. Her husband became injured during fishing at sea. He came back home, and could not do anything economically productive for months. Eventually Urmila had to come forward to sustain the family. She started vending fish and crabs from door to door. The cyclone of April 29, 1991 washed out whatever physical assets she had. This climatic hazard worsened her already precarious livelihood. She relied a lot on social support to sustain the family. She ended up working as a laborer in fish drying yard. To her, everyday is a different reality. She is desperate and employs host of manual works and *'dhaandha'* to feed the family.

Bengali word '*dhaandha*' indicates desperate efforts of poor for enhancing income from whatever sources possible. For fishers, sniffing around for earning from more than one source during the period of unemployment or disguised unemployment is critical for reducing stresses of food insecurity. There are scholars who believe that if income sources from the wider economy increase, individuals would leave fishing profession (Allison and Ellis 2001). We observed that such a generalization might not be applicable for all groups of fishers. There are caste-based fishers who adhere to fishing largely due to socio-religious obligations, typical psychological attitudes, a gambling nature with the seasonality of fishery, lack of skill for alternative activities, geographical remoteness, political powerlessness and lack of endowments that repeatedly tend to keep them dependent on a single profession. A fisher elicited such beliefs in the following words:

---

relationships and, incidentally, to protect themselves from exposure to natural disasters (dryness, epizootics, etc.) by dispersing part of their herds and asking members of their network for help as needed (synthesized from Ballet et al. 2007).

'I am a caste-based fisherman; fish curse on us before death; fish can reach heaven if/when touched by 'Jaladas' fishers. I have to wait for resurrection for obtaining purity for changing caste and profession'. Sunil Jaladas, 48, A Hindu fisherman, Thakurtala.

Sunil will wait for resurrection to change his birth-ascribed profession. While the caste-based affiliation is still strong in the Hindu fishing village, it is mainly the relatively higher income from low cost fishing operations that keeps Muslim fishers attached to fishing in the floodplain areas. For both these categories of fishers, centripetality of the fishing profession holds true.

Fishers employ their asset base, networks and experience accumulated over time to develop workable livelihood strategies. Results presented in Table 21.2 provide a perspective of the distribution of households' efforts for enhancing incomes from multiple sources. Considering the ancillary activities along with active fishing, around 84 % and 55 % of male-headed households in the coastal and floodplain villages respectively make their livelihoods directly from fisheries. Family members ranging from 2 to 4 individuals were engaged in earning activities whatever the amount is, and member earned from 1 to 3 activities with an average of two in most cases. However, the cases of multi-tasking and gear specialization[2] was found extremely limited in both fishing villages.

Access to and availability of natural resources from the ecosystem plays a significant role for the women-headed households, especially in the floodplains. Small groups of fishermen are found to take up some other jobs like terrestrial farming, daily labor, hair dressing, rural transportation, etc. on temporary or permanent bases. Women in both the villages did not exhibit any participation in some activities like rural transportation, carpentry, hair dressing and remittance earning. Begging was taken up as a means of survival by a few destitute fishers in the coastal village. For all these categories of people, our propositions of centrifugality and multi-tasking hold true.

Among the alternative sources of income other than fishing, there are professions which require substantial time for getting a full return (like terrestrial farming, cow and duck ranching). A small section of fishers remained engaged usually in a combination of two professions like- fishing and cow ranching, fishing and duck rearing, and fishing and farming. The nature of the secondary profession is such that engagement in the third category of activities is usually difficult. For these fishers, the notion of cyclical mobility holds true. They acquire substantial expertise in the alternative profession also. Only 56 % of the male-headed coastal households managed to gather secondary jobs while others are left with unemployment or unemployment eventually leading to the debt trap; the scope of earnings from secondary occupations is higher in floodplain areas (Table 21.2). The overall health of the ecosystem and catch directly or indirectly affects the earning of the female-

[2]Floodplain fishers adapted well to the banned mono-filament 'current nets' given their low price, poor regulatory measures and easy availability in the market. Similarly owners of larger marine set bag nets convert their nets to suit in the shallow coastal waters during bad weather periods through further reducing mesh sizes at the cod end.

Table 21.2 Combination of livelihood sources for male and female-headed households (expressed in %) as part of livelihood earnings *(multiple responses allowed)*

	Primary earning source				Secondary earning source			
	Coastal		Floodplain		Coastal		Floodplain	
Types of profession	M	F	M	F	M	F	M	F
Active fishing	68.44	–	49.08	–	6.77	–	9.43	–
Fish retailing/processing	5.27	29.42	1.88	–	3.76	7.53	–	–
Shrimp fry catching /selling	3.76	5.88	–	–	4.51	9.76	–	–
Lending boats and gear/ *beel* subleasing (floodplain)	3.76	–	3.77	–	–	–	–	–
Net weaving & repairing	5.27	29.42	–	14.28	9.54	5.41	3.77	2.50
Boat repairing/painting	1.5	–	–	–	4.50	–	3.77	–
Terrestrial agriculture	0.75	5.88	5.67	14.28	3.0	3.76	12.64	2.50
Daily labor	2.25	17.64	11.30	14.28	12.78	17.53	15.85	12.86
Carpenter	1.50	–	3.77	–	–	–	–	–
Rural transportation	1.50	–	5.67	–	3.76	–	3.77	–
Foreign remittance	–	–	1.88	–	–	–	–	–
Small business/tea stall	–	–	3.77	–	–	–	–	–
Tailoring	–	1.50	–	–	–	–	–	–
Barber	0.75	–	–	–	–	–	–	–
Seasonal ranching		–	3.77	–	–	–	–	–
Homestead poultry/livestock	0.75	5.88	–	14.28	2.83	9.76	7.55	37.14
Natural resource gathering	–	1.50	5.67	28.58	2.83	3.76	9.43	22.85
Service	0.75	–	1.88	–	–	–	–	–
Begging	1.50	3.00	–	–	–	–	–	–
Other	2.25	–	1.88	14.28	3.76	13.65	7.55	4.28

Source: Field survey

headed households; around 59 % of them depend on gathering natural resources as their main source of income.

Terrestrial agriculture ensures earnings for around 23 % and 29 % of the male- and female-headed households in the floodplain village which is more than double that of the coastal village (Table 21.2). This is reflective of the importance of the availability of arable land, which is extremely limited in the coast. Interestingly, female-headed households in the coast and floodplain village are in a better position of earnings through multi-tasking and compromising with a low wage rate compared to that offered to men for similar manual jobs. However, this situation also reveals an entry of unskilled labor force in the farm/non-farm activities. Recent trend of fishing in leased *beels* is a big concern both in terms of labor utility and sustainability. A 50 acre *beel* with multiple sets of *'katha'* (fish aggregation devices) and medium water level usually require a man-power size of 40 fishers for around 3 days; while the same *beel* can be dried up and totally harvested with one-fourth labor requirement using pump machines in the same duration. The former process is ecosystem-friendly while the latter is deadly for the wetlands.

A glimpse of subsistence-oriented diversification efforts within the fisheries sector can be seen in the results presented in Table 21.3. Small-scale fisheries, already overpopulated and overexploited, offer a limited scope for extensification and intensification. Only the wealthy owners of mechanized fishing vessels can take advantage of mechanization and intense fishing in the inshore and off-shore areas. For some fishers, foraging fish becomes a seasonal coping action. Serious food scarcity is a prime mover for children and women's active involvement in foraging for small discarded fish in fish landing areas. Socially accepted custom allows for small fish fallen from the baskets in course of handling and transportation of fish from mechanized vessels becoming the property of the poor foragers. Fish assembled from beach thus not only play a significant role in family nutrition and gourmandizing, but also in earning a small amount from sale of the surplus to low-income customers. Income earned through foraging (US$ 5–13/month/person), though insignificant, yet plays an important role in ameliorating grim situation of food scarcity. Also, such a scanty amount of income from foraging plays an invaluable role in those families where adult male fishers fail to cling to fishing due to disease or accidents. The immediate consequence of decline in catch level on the livelihood is that scope of foraging on beach for remnants becomes sharply limited.

Here, we would like to analyze the impacts of intensification on the overall fisheries resource (Fig. 21.4a–f). For the purpose, we consider shrimp fry (species: *Peneaus monodon*; harvested for semi-intensive shrimp farming) catching (coastal areas) and use of dense-meshed synthetic nets in the inland areas. *'Not bad; just 4-6*

Table 21.3 Intensification and extensification efforts for subsistence

	Actions undertaken for enhanced income	
Strategy	Coastal areas	Floodplain areas
Vertical integration or intensification	– Use of 'mosquito nets' for shrimp fry catching – Use of low-mesh 'tunnel' in E/MSBN – Catching of juvenile *Hilsha* fish – Forage for mud crabs – Catch unconventional foods like squids, eels, mollusks, etc. – Desperate inshore fishing ignoring bad weather signals – More fishing hours and child labor	– Use of synthetic 'current nets' – Use of illegal large encircling gears (*ber, jagat, dharma*) an lift nets – Fishing through obstruction – 'total fishing' by removing water from small ditches – More fishing hours and child labor – Use of multiple gears from same craft
Horizontal expansion or extensification	– Fishing beyond one's customary territory – Foraging fish – Targeting fish migration routes – Product diversification like fish drying, salting, fermentation (*Nappi*), smoking – Fish camping in the Sunderban mangroves areas in the south-west	– Poaching migratory birds – Gathering earthworm and frog for use as baits – Clandestine fishing in the leased waterbodies or sanctuaries – Night fishing – Angling by women – Product diversification through drying and fermentation

Fig. 21.4 (**a**) Ferrocement boats gradually replace traditional wooden boats, (**b**) front side of boat redesigned to face turbulent tide, (**c**). use of monofilament nets to catch small shrimp fry in the coast, (**d**) destructive low-mesh nets used in floodplain, (**e**). children seen engaged in shrimp fry segregation, and (**f**). duck farming in the backyard of house in the floodplain

hours/day at the advent of two high tides during full-moon and new-moon phases; if luck favors, we get good catch and good price': says Sazzad Islam, 35, a 'Rohinga refugee' from Myanmar. In catching one shrimp-fry, the catcher ultimately destroys around 100 other small creatures that include commercially valued shellfish and finfish larvae. In a similar vein, inland fishers evolved large-sized encircling gears, made of highly dense synthetic materials, which can filter everything from water including smallest plankton (See Deb and Haque 2011 for details). The former and later cases involved around US$ 3–5 and $2000–5000/gear respectively. These two cases demonstrate that intensification efforts within the unregulated small-scale fisheries sector would be unsustainable and negative for the resource base.

Members of about 2 % of floodplain fishing families managed to migrate as laborers to the Middle East earlier this decade. However, their remittances cover a much larger percentage of household income among the migrant families. There were demand of south-Asian fishers for work on fishing trawlers in Middle Eastern countries (Divakarannair 2007), but the process of emigration has become increasingly difficult for the poor and less educated members of fishing communities. Resultantly, none from the coastal village could afford to immigrate. We observed that remittance money did not succeed in uplifting the rural agricultural and fisheries enterprises. Rather it created a dependency syndrome among family members, started challenging the existing social and cultural institutions, and in some instances proved counterproductive to sustainable uses of the wetland resources. *'We struggle for earning little money from the haor, whereas he (Nurul Chairman, leaseholder of the highly productive 'Padma beel') fights with fishers for displaying power and earning more using his remittance money'*- comments Mamtaj Ali, 56, Volarkandi. Most of the leaseholders of productive waterbodies have remittance earners in their families, and once they manage lease order from the government through scrupulous process, their activities become subversive of the interests of common fishers and values for sustainable management of wetlands.

There is a difference in the trend of diversification among different wealth category groups. Our observation is that poor households rely more on income sources which are physically demanding, while the higher medium and rich classes use their networks for maximizing earnings. Remittance earning is almost impossible without strong financial backup and networks at home or abroad. Diversification of livelihood activities is closely connected to the growth of agriculture or non-agriculture-based labor-absorptive small and medium enterprises in the rural areas. Due to lack of macro-level support and inputs needed (like electricity, security, political stability, good governance, loan on easy terms, market outlets), labor-absorbing small and medium enterprises are yet to flourish in rural Bangladesh. Large-scale disasters impose a process of defisherization and subsistence crisis; fishers cannot go out for fishing if they lose boats and gear. Roles and efficient functions of macro-institutions are critical in such contexts.

Conclusions

This paper progressed with an implicit understanding that climate variability and change have begun to impact the small-scale fisheries, with the inherent recognition that fishers' livelihoods are at stake in the face of such effects. The Government of Bangladesh should be engaged to put more thoughts and actions towards making fisheries and fishers more resilient towards climate change vulnerabilities. It is critical that the Government seriously considers fisheries into its countrywide adaptation planning ensuring that macroeconomic drives in one sector do not become counterproductive for other important sectors. Creation of more income generation activities is a much-needed option. Policy interventions targeted at creating sustainable livelihoods for fishery-depended communities need to focus more on gender roles in the development process. Very nominal macro-meso-micro-scale adaptive strategies have been planned to build ecosystem resilience and social adaptive capacity of fishers under the national climate change adaptation program.

For the small-scale fishers at any point of time, livelihood functions are directly or indirectly impacted by the available opportunities, households' endowments and entitlements, macro-and-meso-institutional issues, and the stresses and hazards they face. The impact of climate change and variability is not homogenous across communities. Sustaining livelihood well-being requires classic skill to manage relationships and transactions in different spheres, making the best use of what can be achieved through one sphere, and then going together with more well-orchestrated actions in the other spheres (Bebbington 1999). This means that graduation from poverty requires not only enough income to move to a better economic status, but also the means to defend against negative forces of downward mobility so households can remain at that improved level (Rahman 2002).

Livelihood coping responses are strongly influenced by circumstances of individuals and households as there are strong inequalities and vulnerabilities accentuated at individual, household and community levels. Even in a fishing community where fishers are apparently equally exposed to climatic shocks, higher sensitivity and lower adaptive capacity combine to generate elevated level of vulnerability (Islam et al. 2014). One must not overemphasize that local level coping actions taken by small-scale fishers to deal with livelihood disturbances can consequentially sustain adaptive capacity in the face of climate change and variability. Research in Southern Africa revealed that *'local coping as a form of resilience to uncertain future climate change must not be overemphasized since the process at both the individual and household level is competitive, subtly differentiated by climate context, household adaptive capacity and individual perception of risk'* (Osbahr et al. 2010).

This research revealed that fishing households cope with adversities by adopting multiple livelihoods, in compliance to opportunities and capabilities they have, constraints they face, and the changing social relations dictated by external and internal forces. However, opportunity of multi-tasking is very limited in the study areas. Households' ability to ensure livelihood security over time is an outcome of a complex nexus of factors such as composition of the families, sex ratio and number of earners/dependents, endowment sets, socio-political linkages, biophysical settings, macro-level economic processes and political forces (Hesselberg and Yaro 2006). Based on Oshaug's classification of food security situation (Oshaug 1985), we conclude that majority of the fishers of Thakurtala village are fragile households, and there are limited numbers of enduring and resilient households. In both the villages, many households are forced to dwindle between their resilient status in the good fishing seasons and fragile status in the bad seasons. Around six and a half decade after Sen's (1981) observations of vulnerable groups like landless laborers, rural artisans and fishers on their entitlement failure over food and adequate alternative access, there has not been a significant change in the food security condition of these poor classes. Meanwhile, we notice added pressures on these vulnerable groups from climatic variability and change. Acute food deficits and a kind of hidden hunger or pseudo-famine is persistently prevalent for a sheer majority of small-scale fishers.

Poor people's livelihood systems are continuously buffeted by shocks and stresses, and hence, gravitate around some 'basins of attractions' as they fail to find equilibrium using a certain form of coping strategy, and they keep shifting or intermingling coping strategies until and unless their livelihood capabilities attain some resilience and transform to a new desirable stability landscape (Walker et al. 2004). Taking neoliberal precepts, some policy and management interventions suggested by the fishers are: (a) special recognition of the fishing villages as 'disaster prone' and 'seasonally food deficit' zones, (b) introducing food aid program for the most vulnerable section (widows, deserted women, physically weak persons, etc.), (c) providing microcredit support with zero interest, (d) financial support during no-fishing time, (e) providing government support in marketing fish products in overseas markets, and (f) taking special initiatives for accommodating unemployed and underemployed groups in labor-intensive industries (e.g., garments and shrimp processing).

Policy-makers of the country need to pay due attention to fishers' endogenous strategies for developing locally appropriate crisis management interventions. Each household, based on its capitals and capabilities, has an intangible line of coping threshold. As the coping threshold limits of households are crossed-over by multiple sets of stresses, households become extremely vulnerable and continue to experience serious livelihood struggles. When assets of a livelihood system are depleted and institutions are unable to adapt to change, available livelihood strategies become 'brittle'- resulting in reduced resilience- and vulnerability to disturbance increases (Glavovic et al. 2002). In the background of a worldwide signal of fisheries resource decline, crises of existence of poor people (FAO 2009), and recent trends in climatic and human-induced stresses, we consider that the situation of poverty in the small-scale fishing villages might be aggravated if appropriate types of resource governance are not adopted.

Disclaimer: The contents and views illustrated in this article are solely of the authors; these contents or policy issues do not reflect the position of the Department of Sustainable Development, Government of Manitoba or the University of Manitoba, Canada.

Acknowledgements We are grateful to the individuals of the fishing villages for their active participation in the research. We are thankful to the editor and reviewers for their valuable insights.

We are also grateful for the support received through SSHRC and CBRM project of Dr. Fikret Berkes, University of Manitoba, Canada.

References

Allison EH, Ellis F (2001) The livelihoods approach and management of small-scale fisheries. Mar Policy 25:377–388

Allison EH, Perry AL, Badjeck MC, Adger WN, Brown K, Conway D, Hall AS, Pilling GM, Reynolds JD, Andrew NL, Dulvy NK (2009) Vulnerability of national economies to the impacts of climate change on fisheries. Fish Fisheries 10(2):173–196

Badjeck M-C, Allison EH, Halls AS, Dulvy NK (2010) Impacts of climate variability and change on fishery-based livelihoods. Mar Policy 34:375–383

Ballet J, Sirven N, Requiers-Desjardins M (2007) Social capital and natural resource management: a critical perspective. J Environ Dev 16(4):355–374

Barbier EB (2015) Climate change impacts on rural poverty in low-elevation coastal zones. Estuar Coast Shelf Sci. http://dx.doi.org/10.1016/j.ecss.2015.05.035

Barrett CB, Reardon T, Webb P (2001) Nonfarm income diversification and household livelihood strategies in rural Africa: concepts, dynamics, and policy implications. Food Policy 26: 315–331

Baumann P (2002) Improving access to natural resources for the rural poor: a critical analysis of central concepts and emerging trends from a sustainable livelihoods perspective. LSP Working paper 1, Livelihood Support Programme (LSP), FAO

Bebbington A (1999) Capitals and capabilities: a framework for analyzing peasant viability, rural livelihoods and poverty. World Dev 27(12):2021–2044

Brander K (2010) Impacts of climate change on fisheries. J Mar Syst 79(3–4):389–402

Burton I, Huq S, Lim B, Pilifosova O, Schipper EL (2002) From impacts assessment to adaptation priorities: the shaping of adaptation policy. Clim Policy 2(2–3):145–159

Chambers R (1989) Editorial introduction: vulnerability, coping, and policy. IDS Bull 20(2):1–7

Chambers R (1995) Poverty and livelihoods: whose reality counts? IDS discussion paper 347. IDS, Brighton

Chambers R, Conway G (1992) Sustainable rural livelihoods: practical concepts for the 21st century. IDS Discussion paper 296. Institute of Development Studies, Brighton

Davies S (1996) Adaptable livelihoods. Coping with food insecurity in the Malian Sahel. St. Martin's Press, New York, NY

de Haan LJ (2000) Globalization, localization and sustainable livelihood. Sociol Rural 40(3): 339–365

Deb AK, Haque CE (2011) *'Sufferings Start from the Mothers' Womb'*: vulnerabilities and livelihood war of the small-scale fishers of Bangladesh. Sustainability 3:2500–2527

Devereux S (2001) Sen's entitlement approach: critiques and counter-critiques. Oxf Dev Stud 29(3):245–263

Divakarannair N (2007) Livelihood assets and survival strategies in coastal communities in Kerala, India. Unpublished PhD dissertation, University of Victoria, Canada

Downing T, Ringius L, Hulme M, Waughray D (1997) Adapting to climate change in Africa. Mitig Adapt Strateg Glob Change 2(1):19–44

Ellis F (1998) Household strategies and rural livelihood diversification. J Dev Stud 35:1–38

Ellis F (2000) Rural livelihoods and diversity in developing countries. Oxford University Press, London

FAO (Food and Agriculture Organisation) (2009) Fisheries and aquaculture in a changing world. Rome

Glavovic B, Scheyvens R, Overton J (2002) Waves of adversity, layers of resilience: exploring the sustainable livelihoods approach. www.devnet.org.nz/conf2002/papers. Accessed 15 May 2012

Government of Bangladesh (2009) Bangladesh climate change strategy and action plan. Ministry of Environment and Forest, Government of Bangladesh, Dhaka

Grafton RQ (2010) Adaptation to climate change in marine capture fisheries. Mar Policy 34(3): 606–615

Hesselberg J, Yaro JA (2006) An assessment of the extent and causes of food insecurity in Northern Ghana using a livelihood vulnerability framework. GeoJournal 67:41–55
Hewitt K (1997) Risks and damaging events. In: Hewitt K (ed) Regions of risk: a geographical introduction to disasters. Longman, Essex, pp 21–39
Holbrook NJ, Johnson JZ (2014) Climate change impacts and adaptation of commercial marine fisheries in Australia: a review of the science. Clim Change 124:703–715
Islam M, Sallu S, Hubacek K, Paavola J (2014) Vulnerability of fishery-based livelihoods to the impacts of climate variability and change: insights from coastal Bangladesh. Reg Environ Change 14:281–294
Keesing RM (1981) Cultural anthropology: a contemporary perspective, 2nd edn. Holt, Rinehart and Winston, New York, NY
Moser C (1998) The asset vulnerability framework: reassessing urban poverty reduction strategies. World Dev 26(1):1–19
Murphy LM, Moriarty AB (1976) Vulnerability, coping and growth from infancy to adolescence. Yale University Press, New Haven, CT
Osbahr H, Twyman C, Adger WN, Thomas DSG (2010) Evaluating successful livelihood adaptation to climate variability and change in Southern Africa. Ecol Soc 15(2):27
Oshaug A (1985) The composite concept of food security. In: Eide WB et al (eds) Introducing nutritional considerations into rural development programs with focus on agriculture: a theoretical contribution. Institute for Nutrition Research, University of Oslo, Oslo
Payne P, Lipton M, Longhurst R, North J, Treagust S (1994). How third world households adapt to dietary energy stress: the evidence and issues. Food Policy Review 2, IFRI, USA
Rahman RI (2002) Rural poverty: patterns, processes and policies. In: Toufique KA, Turton C (eds) Hands not land: how livelihoods are changing in rural Bangladesh. Bangladesh Institute of Development Studies, Dhaka, pp 79–87
Rita G, Teixeira CM, Costa MJ, Cabral HN (2015) Are regional fisheries' catches changing with climate? Fish Res 161:207–216
Salmi P (2005) Rural pluriactivity as a coping strategy in small-scale fisheries. Sociol Rural 45(1/2):22–36
Scoones I (2009) Livelihoods perspectives and rural development. J Peasant Stud 36(1):171–196
Sen A (1981) Poverty and famine: an essay of entitlement and deprivation. Clarendon, Oxford
Sen A (1987) The standard of living. Cambridge University Press, Cambridge
Sen A (1990) Development as capability expansion. In: Griffin K, Knight J (eds) Human development and international development strategies for 1990's. Macmillan, New York, NY
Shanin T (1972) The awkward class: political sociology of peasantry in a developing society, Russia 1910-25. Oxford University Press, London
Walker B, Holling CS, Carpenter SR, Kinzig A (2004) Resilience, adaptability and transformability in social–ecological systems. Ecol Soc 9(2):5
Wisner B, Blaikie P, Cannon T, Davis I (2004) At risk: natural hazards, people's vulnerability and disasters, 2nd edn. Routledge, London
World Bank Group (2010) Bangladesh: the economics of adaptation to climate change. 1818 H Street, Washington, DC 20433

Chapter 22
Towards Long-Term Resilience: The Challenge of Integrating Climate Change Related Risks into a Risk Analysis Framework

Christian Ploberger and Walter Leal Filho

Abstract This paper discusses the issue of how to integrate the impact of climate change into a risk analysis framework at both, business and country levels. It is argued that climate change related risk should be treated as a particular risk category. It is further emphasised that a lack of perception contributes to a systemic failure of integrating climate change related risk as a regular feature in a risk analysis framework, after all, there has to be acceptance of a particular risk before one can assess it. It is further argued that risk analysis and risk management is not just about identifying risk but equally in identifying opportunities as well, and the impact of climate change provides a good example for that.

Keywords Business risk • Climate change; • Country risk analysis • Risk management

Introduction

This paper addresses the various challenges which exist when embarking upon the task of integrating Climate Change within a risk management framework.

At one end, political risk analysis mainly focuses on issues such as market access, legal regulations, political systems, political stability, and quality of infrastructure, to name some examples. However, since the impact of Climate Change can undermine a country's development strategy and negatively impact the living conditions of a population and by extension carry implications for political

C. Ploberger
Laternenring 4, 4061 Pasching, Austria

W.L. Filho (✉)
Research and Transfer Centre "Applications of Life Sciences", Faculty of Life Sciences, Hamburg University of Applied Sciences, Lohbrugger Kirchstraße 65, 21033 Hamburg, Germany
e-mail: walter.leal@haw-hamburg.de

W.L. Filho et al. (eds.), *Climate Change Adaptation, Resilience and Hazards*, Climate Change Management, DOI 10.1007/978-3-319-39880-8_22

stability, climate change represents a fundamental political risk for the stable development of countries, and should therefore be integrated as a regular feature when undertaking a political risk evaluation.

What's more, climate change related risks can also generate serious economic risks for business, by undermining a company's business strategy, consequently reducing the prospect of arriving at the strategic goals identified within an organisation, or indeed, represent a danger to the very physical existence of a company's assets. Therefore, the impact of climate change should be considered as an inherent part of any company risk assessment as well.

In identifying the challenges of integrating Climate Change as a particular risk category in risk management it is necessary to discuss the definition and roots of global environmental change, since this is fundamental in understanding how risk frameworks should be contextualised. This is followed by addressing the question what constitutes political risk analysis, before continuing with the subject of how to integrate climate change related risk into a risk analysis framework.

Issues Involved in Global Environmental Change

Global environmental change is defined by the Intergovernmental Panel on Global Environmental Change (IPEC) as a complex and changing system, where life evolved with ups and downs while transforming water, air, soil, flora and fauna. Natural systems do not easily respond to change and when they do, it takes long time scales of thousand to a million years.

The environment is considered as part of our lives and is constantly changing (Hinchliffe, et al. 2003). It includes the ecosystems and their constituent parts; all natural and physical resources; the social, economic, aesthetic and cultural conditions which affect the environment or which are affected by changes to the environment. Due to the accelerating technological progress, a rapidly expanding population and increasing consumption the pressure on the environment intensifies continuously.

Human activities and associated pressures have caused a more rapid global change now than the Earth has experienced over the past 12,000 years (Barnosky and Hadly 2012; Ehrlich et al. 2012). The damaging act of human activities is not only changing the ecosystems and the world's climate (Hinchliffe et al. 2003), but they have induced impacts on atmospheric composition, waters, land and biodiversity. All these occurring so rapidly that natural systems no longer have time to adapt.

The elements responsible for environmental change are interlinked through physical, chemical and biological processes. An example of this trend is global warming, which among other things has impacts on ozone depletion, land cover and biodiversity changes, impact negatively on the climate which in turn causes water stress and affecting agriculture. Due to this rapid change, the rules that govern society's relationship with the biosphere must be radically redefined in order to promote a healthy and sustainable human-earth relationship (Chapin III et al. 2011).

The major issues related to global environmental change are both political and scientific. Politically oriented issues includes the sustainable management of natural resource and sustainable land use planning for agriculture, urbanization and habitat conservation. Secondly, the management of water resource, water pollution, wetland loss, watershed degradation, coastal ecosystem degradation, over-fishing, marine pollution and damage, ozone protection, air pollution and health, biodiversity and habitat loss.

Scientific-oriented issues includes technology transfer (renewable energy technology, pollution control and clean production technologies), biomass energy, monitoring global land cover/land use change, sound management of chemicals and hazardous waste, the scientific assessment of the global marine environment, energy efficiency, control of greenhouse gases, air pollution control, monitoring impacts of climate change and air pollution, control ozone depleting substances.

According to Kullenberg on a 1999 review of the approaches to addressing the problems of pollution of the marine environment, coastal zone degradation, physical disturbances, marine pollution from land-based sources and over-fishing are a result of the multiple and conflicting demands of rapidly increasing populations.

There is a particular concern about such impacts on the oceans, which play a key role on the balance of emissions and absorption of greenhouse gases, especially carbon dioxide. The combination of sea-level rise, and changes of the North Atlantic circulation due to global warming is known to have far reaching consequences to the global environment.

Around the world, urbanization will lead to greater exposures to flood and drought hazards (Burak et al. 2015), with hotspots in Asia and Africa. Urban areas in low-elevation coastal zones are predicted to increase worldwide to 234,000 km2 by 2030 (i.e., from 11 % of global urban extent to 13 %). This is higher than the prediction made in 2012. Burak et al. on a 2015 study estimate that by 2030, India, Southern Asia, and Southeastern Asia will have almost three-quarters of the urban land under high frequency flood risk while the largest percentage of urban lands facing high-frequency flood risk is expected in Mid-Latitudinal Africa.

Globally, water scarcity continues to be a threat (Veldkamp et al. 2015). According to Veldkamp et al. (2015), between 1960 and 2000, the population affected by water shortage rose from 473 million to 2.55 billion, whilst the population affected by water stress increased from 326 million to 1.9 billion. Relative to the total population, this represents an increase from 17 % to 45 % for water shortage, and about 11.7 % to 33.6 % for water stress. Over this period (1960–2000), 8.9–28.6 % of the global population lived under both water shortage and stress conditions. The study also showed that hydro-climatic variability accounts for more than 79 % of the yearly change in water scarcity conditions, that it is the largest driver of change within the short-term.

Challenges in Global Environmental Change, Including Climate Risks

The above outlined challenges of global environmental change are to devise and implement a sustainable balance in alleviation of poverty and promotion of fair trade, good health, food security, and access to clean energy, to reduce the loss of biodiversity; land and water degradation, depletion of stratospheric ozone; and accumulation of waste and persistent organic pollutants in the environment while minimizing the impacts of environmental changes. Analysing environmental changes should always include the impact of such changes to human wellbeing. In doing so it will be easier to engage the general society to protect the environment.

A study released by IPEC in 2003 outlined the challenges posed by data collection that will enable new technologies to monitor global environmental change. It stated that, different countries used different methods in collecting data, making it incomparable, data gaps in developing countries, inadequate resources to process and interprets data so most assessments are based on existing data. Making it difficult to determine the actual state of the changing environment since the past cannot determine the present and the future. Unfortunately, we still see the same problem today.

The IPCC-AR4 report, released in 2007, stated the main key knowledge and research gaps, which include: the consequences of abrupt change, impacts of multiple drivers, costs of impact and adaptation, communicating risk to stakeholders, adaptive capacities and resilience of natural and human systems. AR5, released in 2014, has indicated that such gaps still exist and need to be addressed.

The wide range of research needs entail an integrated monitoring system for natural and social aspects; studies on integrated modelling, and on better regional scenarios. Including climate change adaptation into development agendas has been a challenge to most governments and communities (Murdiyarso 2010). Therefore, climate change adaptation strategies need to integrate with reduced socioeconomic inequality and poverty linking to development (Murdiyarso 2010). Climate change and economic development have conflicting fundamental issues, that is a challenge to global environmental change and addressing it can only be successful if there is a transition to low carbon economies (Lahsen, et al. 2010). Without a commitment to the transition to low carbon economies, economic development and addressing climate change will continue to be in fundamental conflict.

Technology transfer is also a challenge in addressing global environmental change. There has been insufficient cooperation between North–south around green technology development (Burton et al. 2002). Park et al. (2008), argue that, the issue of technical knowledge as a challenge to global environmental change is to integrate it into the existing institutional government arrangements.

Global Environmental Change also faces international, regions and national challenges as implementing international relevant treaties have been very difficult in some part of the world. Different countries either lack the scientific inputs to support effective implementation, policies not put in place due to political reasons or the general public is not educated. Talking about the impacts and vulnerability of countries and region exposed to global environmental change, the middle and low income countries faced the highest challenges as adaptation efforts, and vulnerability reduction always tend to deferred in the local and national level (Lahsen et al. 2010).

According to Lahsen et al., the governance of middle-income and low-income countries needs to become central objectives of empirically based, detailed, multiscale and action-oriented research. This will help reduce global inequities in power and resource distribution. Addressing these issues will help reduce the challenges of global environmental change and its impacts. There is also the fear that, the implementation of international treaties on climate change adaptation tends to deferred at the local level because they failed to pay sufficient attention to local realities, failing to adopt bottom-up approaches to decision making.

Dominant institutions and international organizations do not currently place equity and environmental issues as top priority of policy agendas (Brulle 2010). According to Park et al. (2008),

> the 1972 United Nations Conference on the Human Environment in Stockholm has arguably failed because they have integrated flawed ways of understanding the problem and its politics, reflecting inclinations toward the status quo by privileging states and market solutions and Misrecognizing the underlying dynamics of development and economic globalization.

At the national level, while the market forces are more interested in making profit, the States interest is to protect national security (Schlosberg and Rinfret 2008; Brulle 2010). Less concern is shown at the research level and by major sponsors in defining and reducing current global environmental threats, including their interactions with other human driven dynamics (Lahsen et al. 2010) and given the fact that data collection or compilation is a challenge internationally; therefore global environmental change depends on projections. This indirectly affects the behaviour of national and local decision makers (Lahsen et al. 2010).

The following section will define challenge of addressing climate change related threats within the context of development which is defined in a wider sense as jut economic growth by arguing the climate change related threats need to be integrated as a regular feature in country risk analysis.

Defining the Context and the Challenges

A first step in approaching the task of identifying the challenges of integrating climate change within a risk analysis framework is to develop an understanding of what risk refers to and the context of risk analysis.

When characterising risk, we are following the definition of risk as advocated by the Management-of-Risk (MoR) framework,[1] where risk is identified as 'an

[1] See official web pager: http://www.mor-officialsite.com/AboutM_o_R/WhatIsM_o_Raspx

uncertain event or set of events that, should it occur, will have an effect on the achievement of objectives' (Management of Risk c2010, 4). Hence, risk management represents the task of applying a systematic framework and process for identifying and assessing risks before implementing a particular risk response (Management of Risk c 2010, 4). At the outset, we can assess that risk management focus on particular political, economic and social risk and the potential impact these risks can have on planning and realisation of national development goals as well as company targets and objectives.[2]

Traditionally, political risk analysis was associated with risks like the threat of assassinations and kidnapping, the impact of civil war, political instability, failing state institutions, the existence and extent of corruption, market access and market regulations, to name just a few. However, it should in no way be denied that such threats analyses continue to be as relevant today as they have been in the past. Even so, it is now widely acknowledged that what actually constitute security and security issues underwent a critical revaluation and a number of potential new security subjects, transcending a mere traditional military focus and a solely state focus, are already identified.

Security threats like environmental degradation or challenges to economic development are now added to the security agenda (Terrif et al. 1999) and Buzan offered a significant contribution by articulating a broadening of the security agenda by adding five sectors to the security agenda, namely: political, economic, societal, and ecological security sector (Buzan 1991). It is also worth considering an assertion made within the Critical Security Studies approach, which emphasises that one can identify a complexity of security challenges when confronted by the multidirectional cases of locally and global security challenges, adding that the starting point for conceptualising security lies in the real conditions of insecurity suffered by people and collectivities (Smith 2005).

In a similar effort the UN also emphasized a re-interpretation of security by putting forward the concept of Human Security, by bringing together the various aspects of human insecurity as well as highlighting the complexity and interdependence of security. It is further emphasized, that an inter-disciplinary approach is required when addressing this security interdependence, consequently, the concept of Human Security is people-centered, multi-sectoral, comprehensive, context-specific and prevention-oriented (Human Security in Theory and Practice 2009). As such it is comparable to the above identified process of re-interpreting security.

The UN conception of Human Security also moves away from a traditional state focused interpretation of security, a process which can also be identified within the framework of Security Studies. Indeed, the protection of society and especially of individuals is broth into perspective and with it the multitude of threats that cut

[2]Throughout this paper the authors refrain from using the familiar term business environment, as it may conflict with the use of the term environment in the context of climate change and would like to ask the reader to keep this in mind.

across different aspects of human life, like highlighting the relationship between security and development, after all, insecurity is context-specific. Consequently, if the impact of climate changes signifies an actual or potential threat to various communities, it certainly should be recognised as a security risk and therefore assigned a specific risk category.

Yet, climate change and the various negative impacts it can generate, or already has generated on business operations and national development goals, are largely missing from a traditional conception of political risk. Therefore, when professional risk analysts are undertaking a country specific risk evaluation, identifying climate change related risks should form an integrated feature of their risk management evaluation process and climate change related risks should be integrated in every political risk management strategy, at both the company and the country level. After all reducing and managing the challenges of climate change, signifies a fundamental political issue.

In overcoming this deficit, one has to acknowledge that risk perception, or more accurately, the lack of it, represents an essential issue, though it is worth pointing out that a fundamental requirement is that an issue is identified as a potential risk in the first place. The failure to include climate change related impacts as a standard feature in political risk analysis may be related to a continuing existing lack of risk perception as to the potential impact climate change can have on business operations and a country development process. To highlight, that it is not as straightforward as it may seem to identify a specific issues as a risk, we shall take another short excursion into the discipline of security studies, as this will demonstrate that awareness alone is only one part of the process of identifying a particular risk issue.

We are quite familiar with different topics which are labelled as security issues and rather willing to accept these as given (like different issues of national security issues). Yet even within national security issues, we should be rather sensitive when labelling an issue as a national security issue as this does not constitute an ordinary process as one may assume. In the words of Katzenstein (1996, 2): 'State interests do not exist to be "discovered" by self-interested, rational actors. Interests are constructed through a process of social interaction.' Buzan et al. (1998) and Waever (1998) re-mind us that it is a political choice to securitize an issue and such an undertaking should be understood as an inter-subjective process, adding that securitization implies that a specific issue is presented as an existential threat, requiring emergency measures and thus justifying actions outside the normal bounds of political procedure. Hence the implications are, that security is not something objective which just has to be uncovered, indeed security issues can take many forms.

This position is also advocated by Smith (2005), stating that the conceptualization of security is a product of different understanding of what politics is and should be about, emphasising that there is no politic-free definition of security, adding that security is something we choose. Thus advocating that critical explorations of the realities of security should take in our head, before labelling something as security issue.

Consequently, when we highlight the need to increase the awareness that climate change represents a particular risk category we should be aware of the above mentioned process in identified and labelling s specific issue as a particular security risk. Considering that we are en-route to miss critical climate change related targets, it seems that climate change and the related risk it represents to are still not recognised.

Failing to Address Climate Change Related Risks

The target agreed in the UN Framework on Climate Change (Climate Convention in 2009), was to keep the temperature increase below 2° relative to pre-industrial levels. However, the recent UNEP Emission Gap Report alerts us that it is doubtful that we will manage to keep global average temperature below 2° after 2020 (The Emission Gap Report 2013). Even as we are reminded that limiting global warming to 2°, at least by 2100, remains technically and economically feasible, if the required measures are to be introduced now, though the window for reversing the current emission trajectory narrows rapidly (Vieweg et al. 2012, 2).

Yet a report from the World Bank indicates a much more challenging development trajectory, by stating that contemporary and future greenhouse gas emissions combined implies higher emission levels by the 21st century, with the consequence that it is likely that a warming of 4° will be reached. With the implication that the risk of breaching vital thresholds for human life such as crop yields, dry season irrigation systems, coral-sea reefs, grassland—this may in turn lead to abrupt system change with considerable negative implications for human development and life (Turn down the Heat: Climate Extremes, Regional Impact, and the Case for Resilience 2013).

Another good example for climate change related risks is the threat those risk pose to dense urban areas. Many cities and mega-cities along coastlines or in close proximity of it are under threat from sea-level rise and storm surge in addition to other climate related issues like heat waves. As stated in the UN report 'Cities and Climate Change', urban areas with their high population density, their concentration of industries and infrastructure, are highly likely to experience the most severe impact of climate change (Cities and Climate Change 2011).

Consequently, developed and developing countries will need to formulate and implement strategies to address their current and future emissions trajectory. As stated in the UNEP Emission Gap Report (2013) the relative contribution to global emissions from developing and developed countries changed significantly between 2000 and 2010, seeing the fall of developed countries share from 51.8 % to 40.9 % whereas the contribution of developing countries' emissions increased from 48.2 % to 59.1 %. Hence, without the participation of major contemporary emitters, including key developing counties like China, India, Brazil, and Russia, regional and global mitigation efforts will be undermined considerably.

However, this is not to deny that developed countries have to take responsibility for the current emission levels and that they must be at the forefront of emission reduction activities, but it would be wrong to ignore the changing dynamic in global emissions trajectory and the context of ongoing emission growth within developing countries. Especially since an early change of the development model, from an emission intensive model towards a low carbon economy, would lessen the lock-in effect of future emissions trajectory of developing countries. Indeed, an early shift towards a low carbon economy will be instrumental for developing countries to avoid, what Zhengzhong (2011) describes as 'development emissions', emissions which are generated during the process of industrialization, modernization and urbanisation.

It should also be remembered that climate change related risks representing a potential fundamental challenge for developing countries as they have the potential of undermining gains made with regard to poverty reduction or advances already made in providing health care and managing diseases. Yet, synchronizing development strategies with adaptation and mitigation measures forms an elementary change of the development framework, given that new risk and uncertainties with fundamental political, economic and social implications are added.

Another feature of risk analysis and risk management is linked with opportunity maximisation. This forms a critical aspect for our discussion, as climate change and related shifts in economic activities also offer potential business opportunities for companies around the globe and opportunities to governments for generating employment and income and to reach various aspects of national development strategy. However, it will be the task for risk analysis and risk management to identify the opportunities climate change may offer for businesses within a particular country by either referring to a county's development strategy with regard to the impact of climate change in addition to the implications climate change may have for parts or the whole country in question.

Just to take a prominent example: energy efficiency and energy savings. Both representing critical issues when addressing current and projected future emission levels and consequently in addressing some of the underlining dynamics which drives climate change. A vital point is that these targets are based on existing technologies, a point not to be missed, as this implies that resistance to change is rather politically motivated than based on technological challenges. Even as increasing energy efficiency is a critical example of how to manage rising emission levels, other topics like better urban infrastructure management and planning are further significant issues, as is behavioural change within society.

Integrating climate related risk as a standard risk category in country and business risk evaluations, will highlight the related risk as climate change related risk will became more visible and thus should contribute to the integration of emission reducing and emission avoiding strategies within national development strategies.

Conclusions: Climate Change—A Fundamental Risk Category

It is essential to recognize that climate change is a multifaceted issue and that it is closely linked with economic, technical, social and political factors. Equally, the implications are multi-layered as well, spanning from specific local impacts, to regional or global issues manifested in various forms, like sea-level rises, changing global weather patterns and changes in the local climate.

However, it should not be denied that the negative impacts of climate change vary in intensity and consequences have to be differentiated between particular levels (global, regional, national, local) when identifying climate change related risks. After all, the specific impact within a defined geographical area climate change generates is of critical relevance when assessing its influence and the degree of political response it may generates.

Even when treating climate change as a distinguished risk category, related risk should not be analysed independently of wider development goals as the potential implications are far reaching with regard to human welfare, economic costs and for political stability. Critical, when addressing and identifying both climate change related risks and opportunity maximisation, political risk management should be essentially pro-active rather than reactive.

It may be helpful to remember that country analysis, despite that it has its merits, is not solely a statistical data exercise. Indeed country risk analysis does not represent a mechanistic undertaking. After all, the selection of a particular development strategy is a foremost political decision, and one which is formulated within a specific political, economic and social context. Alike sharing and dissemination knowledge of climate change and climate change related threats represents an essential undertaking, not only be technology transfer mechanism but in communication international and local experience at local, national and international and interdisciplinary forums like the recent World Symposium Conference for Climate Change Adaption (Manchester UK, 2015).

References

Barnosky AD, Hadly EA (2012) Approaching a state shift in Earth's biosphere. Nature 486:52–58

Brulle R (2010) From environmental campaigns to advancing the public dialog: environmental communication for civic engagement. Environ Commun 4(1):82–98

Burak G, Inci G, Ying L (2015) Changing global patterns of urban exposure to flood and drought hazards. Glob Environ Chang 31:217–225

Burton I, Huq S, Lim B, Pilifosova O, Schipper EL (2002) From impacts assessment to adaptation priorities: the shaping of adaptation policy. Clim Policy 2(2/3):145–159

Buzan B (1991) People, States & Fear. Harvester Wheatsheaf, University of Michigan

Buzan B, Waever O, Wilde J d (1998) Security: a new framework for analysis. Lynne Rienner Publisher, Boulder

Chapin FS III, Power ME, Pickett ST, Freitag A, Reynolds JA, Jackson RB et al (2011) Earth stewardship: science for action to sustain the human-earth system. Ecosphere 2

Ehrlich PR, Kareiva PM, Daily GC (2012) Securing natural capital and expanding equity to rascal civilazation. Nature 486:68–73

Hinchliffe S, Blowers A, Freeland RJ (2003) Understanding Environmental Issues. OU-Wiley Environment Series

Human Security Unit United Nation (2009) Human Security in Theory and practice. United Nation Trust Fund for Human Security, 2009, Human Security Unit Office for the Coordination of Humanitarian Affairs, United Nations, New York, USA. http://www.tr.undp.org/content/dam/turkey/docs/news-from-new-horizons/issue-41/UNDP-TR-HSHandbook_2009.pdf

Katzenstein PJ (1996) Introduction: alternative perspectives on national security. In: Katzenstein PJ (ed) The Culture of national security: norms and identity in world politics. Columbia University Press, New York, pp 1–32

Kullenberg G (1999) Approaches to addressing the problems of pollution of the marine environment: an overview. Ocean Coast Manag 42:999–1018

Lahsen M, Sanchez-Rodriguez R, Lankao,RP, Dube P, Leemans R, Gaffney O et al (2010) Impacts, adaptation and vulnerability to global environmental change: challenges and pathways for an action-oriented research agenda for middle-income and low-income countries. Science Direct:364–374.

Management of Risk: Guidance for Practitioners C (2010) 3rd ed. Office of Government Commerce, London: TSO (The Stationery Office). = no author name exist

Murdiyarso D (2010) Climate and development-the challenges in delivering the promises: an editorial essay. WIRES Clim Chang 1(6):765–769

Park J, Conca K, Finger M (2008) The crisis of Global Environmental Governance. Towards a New Political Economy of Sustainability

Schlosberg D, Rinfret S (2008) Ecological modernization, American style. Environl Politics 17 (2):254–275

Smith S (2005) The Contested Concept of Security. In: Security C (ed) Studies and World Politics, edited by Ken Booth. Boulder, Lynne Rienner, pp 27–62

Terrif T, Croft S, James L, Morgan PM (1999) Security studies today. Polity Press Cambridge, Cambridge

UNEP (2013) The Emissions Gap Report 2013. United Nations Environnent Programme (UNEP), Nairobi, Accessed 29 Nov 2013 http://www.unep.org/pdf/UNEPEmissionsGapReport2013.pdf

UN-Habitat (2011) Cities and climate change: global report on human settlements. United Nations Human Settlements Programme. Earthscan Ltd, London

Veldkamp IT, Wada Y, Moel H d, Kummu M, Eisner S, Aerts CJ et al (2015) Changing mechanism of global water scarcity events: impacts of socioeconomic changes and inter-annual hydro-climatic variability. Glob Environ Chang 32(May):18–29

Vieweg M et al (2012) 2° be or not 2° be—Climate Action Tracker Update. Accessed 05 Nov 2013. http://www.ecofys.com/files/files/cat-doha-update-2012-final.pdf.

Waever O (1998) Securitization and desecuritization. In: Lipschutz RD (ed) On security. Columbia University Press, New York, pp 46–86

World Bank (2013) Turn down the heat: climate extremes, regional impacts, and the case for resilience. A report for the World Bank by the Potsdam Institute for climate impact research and climate analytics. World Bank, Washington, DC, file:///C:/Dokumente%20und%20Einstellungen/ploberger/Eigene%20Dateien/Downloads/9781464800559s.pdf

Zhengzhong. X (January 2011) Models and policy system for low-carbon economic development in China, online paper. Chinese Academy of Governance Secretary General. Accessed 31 Jan 2011. http://www2.warwick.ac.uk/fac/soc/economics/research/centres/cage/events/conferences/climate11/xu_zhengzhong-models_and_policy_system_for_low-carbon_economic_developmentin_china2.pdf.

Chapter 23
Resilient Architectural Design: Considerations in the Design of Airports to Withstand Climate Change Effects

Paolina Ferrulli

Abstract Airports ask for environmental, social and economic sustainability instances as the core of the project development. The project design must focus on the role of sustainability taking into account new performance's classes to face higher levels of building resilience to the changing climate effects. Airports need to adopt long-term strategic approach to address the need of adaptation to the climate change and mitigation of impacts on the environment. This will require new designs and materials for future airport infrastructure and retrofitting of the existing ones.

The paper reports ongoing research analysis of a broad study focused on the evaluation of the project compliance with the green building requirements and the definition of technological strategies for climate change adaptation. The correct identification and evaluation of climate sensitivities enable to prioritize adaptive responses. It also creates new opportunities for airport authorities and operators to anticipate the consequences of project changing conditions during operation using sustainable strategies to enhance building resilience. European and international airports examples and case studies are presented in this paper in order to illustrate the good practices and methods already adopted for assessing and minimising the risk of climate change and to formulate a proposal for the design of airports for the future.

Keywords Climate change • Adaptation • Resilience • Airport design • Good/best practices

Introduction

Airports are critical nodes in the transport system and can have a vital role in supporting the socio-economic development of city regions. The structure and organisation of the transport systems have determined the evolution and changes

P. Ferrulli (✉)
Dipartimento di Architettura, Università degli Studi di Firenze, Via di San Niccolò, 93, Florence 50125, Italy
e-mail: paolina.ferrulli@unifi.it

W.L. Filho et al. (eds.), *Climate Change Adaptation, Resilience and Hazards*, Climate Change Management, DOI 10.1007/978-3-319-39880-8_23

of human settlements in each age, influencing the creation of public spaces designed to accommodate nodes and connections (Button et al. 1995; Trinder 2001; Woudsma and Jensen 2003). Therefore strategies for the development of air transport must be considered a priority, integrating them in the context of broader strategies for economic development and the infrastructure of the country.

Europe faces a particular challenge in respect of its airport infrastructure because of limited capacity that prevents aviation responding to demand when and where it arises and the difficulty of securing planning approval for new airport infrastructure development (Advisory Council for Aviation Research in Europe (ACARE) 2008). This is due to the dense urbanisation of the continent, the complex system of rules and planning regulations that have arisen as a result and opposition from local residents and their politicians to airport growth.

Airport operations are not only limited by their infrastructure. Airports can be constrained by environmental issues which restrict current operations and limit future growth potential. It is evident that the debate on the subject is not only focused on the noise and air quality impacts on the areas surrounding the airport, but has expanded the focus on the effect that airport and aviation activities have on climate change through carbon emissions (European Environment Agency (EEA) 2007, 2012; Airport Council International (ACI) 2009; Thomas et al. 2010; Eurocontrol 2013a) and vice versa on how the climate changing conditions ask for adaptation and affect the airport design as it needs to comply with resilience requirements (National Air Traffic Services (NATS) 2011; Department for Environment, Food and Rural Affairs (DEFRA) 2012; Airport Cooperative Research Program (ACRP) 2012; Eurocontrol 2013b; Burbidge 2014).

Defining the Issue: The Challenge of Developing a Sustainable Airport

ACI defines airport sustainability as a "holistic approach to managing an airport so as to ensure the integrity of the economic viability, operational efficiency, natural resource conservation, and social responsibility of the airport" (www.aci-na.org). The design of the airport—as infrastructure consisting of multiple functional spaces and facilities and integrated with the surrounding territory—requires many levels of analysis and assessment to evaluate the development constraints and the impacts on the environment at different scales, in function of traffic capacity.

A wide range of impacts on local communities and the natural environment can constrain the operation of airports and restrict their ability to secure planning approval for future growth (Upham et al. 2003). Airport infrastructure growth depends on the assessment of those issues and the opportunity to strategically manage them during an integrated design process. Environmental impacts are associated with the operations of the airport and the specific conditions and characteristics that pertain the area in which the airport is located—proximity to the houses, other polluting sources and industries, water supplies, energy resources and materials availability, sensitive habitats, climate changing conditions, and others (Thomas et al. 2004).

The European Commission (EC) explicitly notes that "the development of transport systems must not be at the expense of the quality of life of citizens or the destruction of the environment. The indefinite continuation of current trends in transport in certain modes (road, air) would be unsustainable in relation to its environmental impact, in particular as regard climate change" (1998). Today, environmental constraints affect European airports (Eurocontrol 2013a) and these constraints can be predicted to grow, as they are related to the pressure of traffic growth, competition for resources (e.g. water, energy, fuel) with other sectors, increasing democratisation and changing public attitudes, and the consequences of climate change. Adverse environmental and community impacts can result in failure of legislative compliance and planning approval for new infrastructure development.

The development of sustainable airport infrastructure depends on achieving correct balance between social and economic objectives within the limits imposed by the environment (Upham et al. 2003). The integration of these concepts implies the definition of what are the environmental constraints to airport development and how this is affected by the design of infrastructure and its configuration, and technological, operational and business features. Therefore, environmental and operational capacity and infrastructure resilience can be maximised through a long term planning ensuring an effective environmental management that compensates for growth through the introduction of eco-efficient design, technological, and operating strategies (Thomas et al. 2001).

Methodology

This paper aims at developing a framework of considerations in the design of airports to withstand climate change effects. The study described in this paper forms part of an ongoing doctoral research project—titled *Green Airport Evaluation Design* (*GrADE*)—focused on the process of evaluation of project compliance with green building requirements during preliminary stages of the design process. The primary aim of the research is to develop method and tools to check and evaluate the sustainability design performances during the whole project development. The research project is organized around eight categories of analysis that represents the main environmental threats for airport projects, namely noise, local air quality, carbon and greenhouse gas emissions, energy and water use, waste management, water pollution, biodiversity, and resilience and climate change adaptation.

In order to achieve the proposed aims, analysis and study have been carried out concerning the changing in the climate conditions and how this affects air travel, airport operations and environmental capacity. The analysis has been carried out through the scientific literature review and the study of international research results and European aviation authorities' regulations and reports. Literature and web review focused on two main topics:

1. the carbon emissions from airport operational activities which increasingly affect climate change and require the use of effective management and **mitigation** actions;
2. the implications of climate change which imply infrastructure **adaptation** strategies and new design specifications in order to comply with the change in the environmental conditions.

Furthermore, the analysis of international airports examples and case studies has been carried out in order to investigate the scene of the good/best practices and methods already adopted for assessing and minimising the risk of climate change.

Greenhouse Gas Emissions Mitigation Strategies

The current contribution of aviation sector to the global greenhouse gas emissions is estimated at about 10.62 % (Intergovernmental Panel on Climate Change (IPCC) 2014). In the short and medium term emissions from the aviation sector will continue to increase; this implies that the main focus for the future development policies is to adopt methods for assessing and monitoring quantity of carbon emission minimising their impact.

Airport operators and international bodies involved in the control and regulation of the activity and development of airport infrastructure have an important role in identifying and prioritizing the best practices for assessing and minimising greenhouse gas emissions effects. Voluntary programmes, national policies and international regulatory regimes (such as Airport Carbon Accreditation, www.airportcarbonaccreditation.org) provide a set of requirements and define a comprehensive approach which involves strategies for the identification of sources and pollutants and emissions calculation and quantification.

In order to meet the emissions reduction targets aviation stakeholders have set strategies considering all sources of emissions based on four main *pillars* (International Air Transport Association (IATA) 2013):

- Improvement in aircraft technology (including the deployment of sustainable low-carbon fuels).
- Efficient aircraft operations.
- Positive economic instruments.
- Improved airport infrastructure.

As regard the infrastructure improvements, Table 23.1 includes three categories of design and technological strategies for minimising carbon emissions.

These improvements are also closely related to the infrastructure resilience strategies, as shown in the following paragraphs.

Table 23.1 Infrastructure improvements for minimizing carbon and greenhouse gas emissions

Landside and ground transport system
Airports can develop themselves as inter-modal transport hubs by including local and region bus and coach facilities, and train stations for local trains, light rail, subway/metro systems, and regional/international trains. Other possibilities include dedicated fast train services between an airport and city centre, and facilities for off-airport and city centre check-in.
Renewable energy
Programmes to improve building energy efficiency and reduce vehicle fuel use can provide meaningful cost savings, while at the same time contributing to reductions in greenhouse gas emissions.
Airport design and resources management
Infrastructure design can be the single most significant factor affecting the greenhouse gas emissions associated with operating an airport. Engineering and architectural features of new terminal buildings can greatly enhance energy efficiency (Leadership in Energy and Environmental Design (LEED) and Building Research Establishment Environmental Assessment Methodology (BREEAM) building certification programmes can provide guidance). For example, the new Midfield Terminal complex under construction at Abu Dhabi Airport is one of the first in the region to achieve the highest level of green building design and operation with the award of LEED platinum status (www.kpf.com).

Source: Adaptation of the author from ACI (2009)

Climate Change Adaptation and Airport Resilience

Airport infrastructure operation and development affect the environment but are also affected by the changing climate. Recent extreme natural events—the spread of Icelandic volcanic ash over European airspace in 2010 and the Hurricane Sandy in 2012, to mention the most recent—have shown the need to make airport infrastructure more resilient to the new climate conditions. This also is a *green building* requirement class which asks for more secure and resistant buildings and civil works (Larsen et al. 2011). Extreme weather events, rising temperatures and sea-level, changes in precipitation, put airport infrastructure and operations at risk (ACRP 2012). Although aviation deals with disruptive weather on a regular basis, such events are likely to become more extreme and more frequent (www.eurocontrol.int/Resilience). The main objective of the aviation industry is to guarantee infrastructure resiliency, safety, efficient operations and service quality in a changing climate condition.

Resilience is the measure of a system to buffer negative climate effects while maintaining its structure and function (IPCC 2007). A resilient system is the converse of a vulnerable system. It is not sensitive to climate variability and change and has the capacity to adapt (Blanco et al. 2009). In the context of future climate change, a resilient system would be able to operate at its normal capacity given more extreme climate effects such as higher or lower temperatures, greater wind speeds, and increased or decreased precipitation levels. Due to the interconnections between airports and the surrounding region, a resilient infrastructure could contribute to building overall network resilience encouraging the spread and development of climate change adaptation measure on the whole territory (Burbidge 2014).

Key factors when determining the appropriate responses to this challenge are based on a climate change risk-assessment methodology and are aimed to:

- identify the met consequences of climate change;
- indicate time scales over which they are likely to occur;
- assess implications for airport infrastructure, operations, and capacity;
- evaluate the magnitude of the impact;
- specify which are the most cost-effective and sustainable design, operational and business adaptation practices;
- determine the time frames limits for the proposed practices to be effective (short, medium, long term planning);
- develop an adaptation action plan.

An example is represented by Birmingham International Airport which produced a climate change adaptation action plan based on a tool called BUCCANEER (Birmingham Urban Climate Change and Neighbourhood Estimates of Environmental Risk), included as a future method of assessing the airport's risk to climate change (www.birminghamairport.co.uk). Climate modeling was used to demonstrate how climate change could affect the airport and the services it provides and provide information on climate variables. The use of probabilistic projections of climate change for different scenarios provide a set of information (e.g. rainfall, temperature, cloud, sea level rise, storm surge, etc.) that directs project design decision process. The definition of the high priority climate related risks represent the basis for the definition of the adaptation action plan. Many airports in the UK are adopting this approach to define their own climate adaptation strategies (e.g. Heathrow, Stansted, Edinburgh, Glasgow).

The development and delivery of the adaptation strategy is a continuous and ongoing activity that responds and adapts to new and improved information on the science and understanding of critical thresholds.

Airport Design Climate Adaptation Strategies

In this paragraph the main climate threats to airport are identified based on the literature (NATS 2011; Larsen et al. 2011; ACRP 2012; DEFRA 2012; Eurocontrol 2013b). For each impacts, constraints to the infrastructure development and operation are described and technological and design solutions are suggested in order to comply with the resilience requirements, as shown in Table 23.2. European and international case studies are presented in order to illustrate some of the good practices already adopted for minimising the risks of climate change.

Table 23.2 Potential climate change effects and illustrative responses for airports

Implications of Climate Change	Constraint on infrastructure and operations	Design solutions
Temperature change		
• Temperature rise • More hot days • Increasing in extreme temperature days • Fewer cold days • Temperature swing above and below freezing	Extreme heat can produce **pavement buckling** (e.g. concrete expansion while remaining rigid) and **loss of pavement integrity** (e.g. tarmac melt) Increase in temperature will **increase demand in energy** for cool air conditioning systems **Permafrost thaw** causing subsidence of runways and other airport infrastructure and damages on airfields, airstrips, access roads Changes to **freeze/thaw cycle** of road subsurface and building envelope materials Increasing **urban heat island effect**	• Higher temperatures will mean a decrease in aircraft lift, requiring longer runways at some airports • Utilize heat-resistant paving materials • At 40–100 years in the future, consider possible significant impact on pavement and structural design (e.g. select materials and equipment with durability in high temperatures; design concrete structures for subsidence) • New design specifications will be required for future airport terminals and retrofitting of existing terminals to improve thermal efficiency and reduce energy requirements for passenger comfort • Include energy efficiency and renewable energy measures and design for redundancy • Existing infrastructure will need to be refurbished with new construction materials • Design building and envelope considering: – Interior and exterior shading devices – High performance glazing – Proper insulation systems and materials – High albedo roofing and paving – Green roofs – Ice dam resistant construction – Cross ventilation **Case study #1**: **New Chitose Airport**, **Hokkaido**, **Japan** The terminal building has adopted a heat-insulated storage system that collect snow during the winter. The collected snow is used in summertime to chill the liquid of the building's cooling system, providing 30 % of the building's cooling needs (www.new-chitose-airport.jp/en).

(continued)

Table 23.2 (continued)

Implications of Climate Change	Constraint on infrastructure and operations	Design solutions
Precipitation change		
• Increase in heavy precipitation events • Increased extreme rainfall events • Higher intensity /longer duration of precipitation	Airfield, roads, bridges, stormwater **decreased capacity of drainage systems** Increased **runoff** from paved areas	• Increased overload and backup of stormwater drainage systems and combined sewer system have to be considered • Increase capacity of stormwater conveyance and storage • Install or improve stormwater treatment systems systems • Include rainwater harvesting and reuse systems • Increase envelope capacity collecting rainwater • Oversized roof drainage • Minimise impervious surfaces **Case study #2: O'Hare International Airport, Chicago, Illinois, USA** The Chicago Department of Aviation promotes the installation of vegetated green roofs in its airports: around 340 square feet have been installed at O'Hare International Airport on 12 different facilities, and more than 20 square feet at Midway International Airport on 4 different buildings (www.flychicago.com). Green roofs are pervious areas that can minimise stormwater runoff and improve roof performances related to noise and heat insulation, durability and air quality.
• Seasonal changes in fog	High levels of **humidity**	• Consider damages to the envelope materials and components • Evaluate decreasing in the envelope capacity to ensure indoor air quality (temperature, mould, etc.)
• Drought	Changes in the ability to meet demand due **water shortages**	• Maximise opportunity for water harvesting • Improve water management practices

Sea-level rise		
• Sea-level rise • River-level rise	**Rising water levels** in coastal areas and rivers	• Airport relocation because of the long-term threat from increasing sea levels—which will lead to an increasing land use • Include protective dikes, containment walls and levees • Elevate facilities and runways • Salt damage to materials and components • Design building and envelope considering: – Materials resistant to brackish and saline waters **Case study #3**: **Brisbane Airport**, **Australia** The Antarctic Climate & Ecosystems Collaborative Research Centre (ACE CRC) developed a sea-level calculator tool called "Canute" (http://canute2.sealevelrise.info) which has been used to support Brisbane Airport's operator to analyse different configurations proposed for the new runway and choose the optimal runway height, maximising the resilience of the infrastructure to sea-level rise and minimising the cost of the runway (www.bne.com.au).
Extreme events		
• Increase hurricane intensity • More intense aspects of storms: precipitations, winds, wind-induced storm surge, greater wave height	Increased **pressure** on structures, components and materials Increase in **winter storms**, with increases in winds, waves Potential changes in **storm track location and strength** **Snow**, **sleet**, **blizzard**, **ice**	• New structural and materials specifications will be required in order to resist to pressure and erosion • Use pervious pavements for roadways, shoulders, non-traffic pavements, maintenance roads, utility yards and airside and landside parking facilities • Use vegetated green-roof systems to reduce runoff from buildings • Design for curb breaks, drainage ditches, basins and/or bioswales **Case study #4**: **Pulkovo Airport**, **St. Petersburg**, **Russia** The large flat roof of the new terminal is designed to cope with heavy snowfall and distribute the weight to different parts of the structural system (www.pulkovoairport.ru; http://grimshaw-architects.com).

(continued)

Table 23.2 (continued)

Implications of Climate Change	Constraint on infrastructure and operations	Design solutions
Wind load		
• Increases and decreases in wind speed and loading • Change in prevailing wind	**More turbulence**, which has a primary effect on runway utilization through reductions in take-off and landing rates causing backlog, delays etc.	• Consider decreased ability in natural ventilation • New structural and materials specifications will be required in order to resist to pressure, wind load and erosion **Case study #5**: **McCarran International Airport**, **Las Vegas, Nevada (USA)** Preliminary wind analysis have been carried out during the structural design of the new control tower. Wind effects on the building are minimized by aerodynamic properties of its geometry, dynamic characteristics of mass and stiffness, application of damping devices (www.mccarran.com; http://peer.berkeley.edu).

Sources: Adaptation of the author from ACRP (2012), NATS (2011), DEFRA (2012), Eurocontrol (2013b) and Larsen et al. (2011)

Discussion and Conclusions

Infrastructure design can be the most significant factor in determining many of the impacts that act as environmental capacity constraints to current operational capacity and growth potential of an airport. Very often this occurs because the design of infrastructure can determine the design of operations and therefore the effectiveness of the airports environmental management system. Airports should address some of the uncertainties of climate change outcomes. This includes planning for new infrastructure with climate change impacts in mind such as considering the criteria for drainage, erosion protection, wind loads, and so on, also integrating them with those environmental issues that are currently on the agenda of sustainable airports development (i.e. noise, local air quality, use of resources, water pollution, waste management, biodiversity).

Sustainable development at an airport concerns developing an infrastructure that facilitates the long term growth of the site so that the airport can continue to respond to demand when it arises. Architects have played a significant role, historically, in the development and design of airports that deliver additional capacity whilst meeting the essential requirements of the air transport industry (e.g. operational efficiency, costs, safety). While increasingly they are having to address environmental issues, their focus has been primarily related to terminal design and not environmental capacity issues. It is becoming apparent, however, that the sustainable development of airports will require that a much wider variety of environmental issues have to be addressed not at the level of an individual building but site wide including an holistic approach to those impacts arising from climate changing conditions.

Climate change, regulations and resource constraint, environmental impacts and constraints and the improving of passengers flows and changing safety requirements, increasingly demand systemic changes in the definition of the design process and of the green building requirements to comply. Strategies to make airports environmentally sustainable and climate resilient should include design methods and tools, allowing a proper design process focused on the analysis, evaluation and management of all the airport infrastructure environmental constraints. Designers have also to deal with impacts on urban planning, business costs and opportunities, new financial opportunities, increased security challenges. The balance between environmental, social and economic assessment criteria represent the core of a sustainable development.

Acknowledgments This paper is part of a research carried out during an internship at the Centre for Aviation Transport and the Environment of the Manchester Metropolitan University under the supervision of professor Callum Thomas. The author wish to thank professor Thomas for his patient mentoring, enthusiastic encouragement and constructive suggestions.

References

ACARE (2008) Addendum to the strategic research Agenda. ACARE, Brussels

ACI (2009) Guidance manual: Airport Greenhouse gas emissions management. ACI, Washington, DC

ACRP (2012) Airport climate adaptation and resilience. A synthesis of airport practice, Airport Cooperative Research Programme. Transportation Research Board of the National Academies, Washington, DC

McCarran international Airport (2015) http://www.mccarran.com/. Accessed 21 Mar 2015

Airport Carbon Accreditation (2015) http://www.airportcarbonaccreditation.org/. Accessed 13 Feb 2015

Birmingham Airport (2015) http://www.birminghamairport.co.uk/. Accessed 13 Feb 2015

Blanco H, Alberti M, Forsyth A, Krizek K, Rodriguez D, Talen E, Ellis C (2009) Hot, congested, crowded and diverse—Emerging research agendas in planning. Progr Plann 71:153–205

Brisbane Airport Corporation (2014) http://www.bne.com.au/. Accessed 13 Feb 2015

Burbidge R (2014) Airports and climate change: identifying the risks and building resilience, Airports in urban networks. In: Proceedings of the Airports in the Urban Networks 2014, Paris

Button K, Leitham S, McQuaid RW, Nelson JD (1995) Transport and industrial and commercial location. Ann Reg Sci 29:189–206

CANUTE 2.0 (2015) Sea-level calculator. http://canute2.sealevelrise.info/. Accessed 21 Mar 2015

DEFRA (2012) Adapting to climate change: helping key sectors to adapt to climate change Government Report for the Adaptation Reporting Power. DEFRA, London

EEA (2007) Climate for transport change. TERM 2007: indicators tracking transport and environment in the European Union. EEA, Copenhagen

EEA (2012) Climate change impacts and vulnerability in Europe 2012. EEA, Copenhagen

Eurocontrol (2013a) Challenges of Growth 2013 summary report. Eurocontrol, Brussels

Eurocontrol (2013b) Task 8: Climate change risk and resilience. Eurocontrol, Brussels

Eurocontrol (2014) Adapting aviation to a changing climate. http://www.eurocontrol.int/ Resilience. Accessed 21 Mar 2015

European Commission (1998) The common transport policy. Sustainable mobility: perspectives for the future, Commission Communication to the Council, European Parliament, Economic and Social Committee and Committee of the Regions. European Commission, Brussels

Grimshaw Architects (2015) http://grimshaw-architects.com/. Accessed 21 Mar 2015

IATA (2013) Responsibly addressing climate change. IATA, Montreal

IPCC (2007) Climate Change 2007: Synthesis Report. Contribution of Working Groups I, II and III to the Fourth Assessment Report of the Intergovernmental Panel on Climate Change. IPCC, Geneva

IPCC (2014) Climate Change 2014 Mitigation of Climate Change. Working Group III Contribution to the Fifth Assessment Report of the Intergovernmental Panel on Climate Change. Cambridge University Press, Cambridge

Kohn Pedersen Fox Associates (2015) http://www.kpf.com/. Accessed 21 Mar 2015

Larsen L, Rajkovich N, Leighton C, McCoy K, Calhoun K, Mallen E, Bush K, Enriquez J, Pyke C, McMahon S, Kwok A (2011) Green building and climate resilience Understanding impacts and preparing for changing conditions. University of Michigan and U.S., Green Building Council, Washington, DC

O'Hare and Midway International Airports (2015) http://www.flychicago.com/pages/landingpage.aspx. Accessed 13 Feb 2015

NATS (2011) Climate Change Adaptation Report, Issue 1.0. Available at http://webarchive.nationalarchives.gov.uk/20130402151656/http://archive.defra.gov.uk/environment/climate/documents/adapt-reports/08aviation/nats-climate-change-report.pdf. Accessed 13 Feb 2015

New Chitose Airport Terminal (2015) http://www.new-chitose-airport.jp/en/. Accessed 21 Mar 2015

ACI North America (2015) http://www.aci-na.org/. Accessed 13 Feb 2015

Pacific Earthquake Engineering Research Center (2015) http://peer.berkeley.edu/. Accessed 21 Mar 2015

Pulkovo Airport (2015) http://www.pulkovoairport.ru/. Accessed 21 Mar 2015

Thomas C, Upham P, Raper D (2001) Environmental capacity of aviation: theoretical issues and basic research directions. J Environ Plan Manag 44(5):721–734

Thomas C, Hume KI, Hooper PD (2004) Aircraft noise, airport growth and regional development. In: Proceedings of the Royal Aeronautical Society/American Institute of Aviation Acoustics Conference, Manchester

Thomas C, Hooper P, Raper D (2010) Air transport in an environmentally constrained world. J Airport Manag 5(1):4–6

Trinder D (2001) Transport infrastructure and economic growth. Structural Issues Development Group Working Paper 01(06)

Upham P, Thomas C, Raper D, Gillingwater D (2003) Environmental capacity and airport operations: current issues and future prospects. J Air Transp Manag 9:145–151

Woudsma C, Jensen JF (2003) Transportation's influence on land use development: a historical spatial-temporal approach. Louisiana Transportation Research Center, Baton Rouge

Chapter 24
Temporal Relationship Between Milk Production and Meteorological Variables in Southern Brazil

Alexandre Mossate Gabbi, Jaime Araújo Cobucci, Vivian Fischer, and Concepta Margaret McManus

Abstract Much of the Brazilian territory is located in tropical regions, where year-long high temperatures are an obstacle to the best expression of milk potential of high performance dairy cows from Continental breeds. In southern Brazil, some areas have subtropical characteristics, with mild summers and cold winters where average temperatures are within the thermal comfort range for dairy cattle breeds of European origin. To assess the relationship between meteorological variables and milk production from Continental breeds in southern Brazil was the goal of this study.

A multivariate analysis was carried out using meteorological variables, including minimum, maximum and average compensated temperature, relative humidity of air, wind speed, solar radiation, temperature and humidity and temperature-humidity corrected by wind speed and solar radiation index obtained from official weather station the National Institute of Meteorology located in Castro, Paraná, Brazil along with the milk production of Holstein dairy herds controlled in this municipality during the years 1996 to 2010. Although the minimum and maximum temperature variables showed higher relationships with milk production when compared with other variables, they were not responsible for a marked change in the milk production. The mesoregion of Castro, Parana, Brazil has adequate weather for milk production from high performance European dairy breeds.

Keywords Dairy cattle • Environmental temperature • Subtropical climate • Historical data series

A.M. Gabbi (✉) • J.A. Cobucci • V. Fischer • C.M. McManus
Departamento de Zootecnia, Universidade Federal do Rio Grande do Sul, Av. Bento Gonçalves, 7712, Porto Alegre, RS, Brazil
e-mail: nuplac.ufrgs@gmail.com

W.L. Filho et al. (eds.), *Climate Change Adaptation, Resilience and Hazards*, Climate Change Management, DOI 10.1007/978-3-319-39880-8_24

Introduction

In Brazil, Portuguese colonization led to the introduction of cattle and sheep from Iberic breeds, as well as horses for transport and traction. These animals were precursors to today's locally adapted breeds such as Pantaneiro, Pé-duro, Lageano or Franqueiro, as well as the native Crioula sheep breed, characterized by a poor to medium performance but with at least four centuries of adaptation to subtropical and tropical Brazilian climates (McManus et al. 2011a, b).

The continuous flux of immigration, as well an increase in economic security of Brazilian farms, led to the introduction of specialized meat or milk cattle breeds, such those from the British Islands and Continental Europe, including Holstein, Jersey and Brown Swiss, where the use of the Holstein breed is more disseminated due high capacity for milk production (Costa et al. 2014).

Almost 66 % of Brazilian territory is localized in the tropical region, where environmental temperatures are high year round and animals from the above breeds have comfort zones similar to those in temperate regions. Heat stress is therefore an important trait in the assessing of dairy systems. Management tools to mitigate these effects include crossbreeding with Zebu breeds, the use of equipment and installations to mitigate heat stress, as well as adaptations of the diet and animal handling in Central Brazil.

On the other hand, there are areas in Southern Brazil that can be classified as subtropical, due to its high latitude and altitude, leading to mild summer temperatures. The highest milk production in Brazil is found in these areas (Costa et al. 2014), due to the use of technology and probably by a positive influence of climate on the animal welfare. The Castro region (State of Paraná) is an important region for benchmarking milk production based on use of the Holstein breed in Brazil.

Research on the climate influence over milk production in Brazil is based on the evaluation of extreme environmental temperature during short experimental periods and examining how these influence milk production. But, long time studies, with 15 to 30 years, or more, of observations, are lacking and these data will determinate true climatic influence on milk production and which technological measures are necessary to the maximize performance capacity of dairy herds.

The goal of this study was to assess the relationship between meteorological variables and milk production over a 15-year period in controlled Holstein herds in Castro, Paraná, Brazil.

Material and Methods

Milk controls were obtained from the Holstein Cattle Association of Paraná collected in Castro, Paraná, Brazil, with localization 24°47′ 27″ S, 50° 0′ 43″ W, average altitude of 988 m and Köppen climatic classification Cfb (humid temperate), with mild temperatures in spring and summer and occasionally rigorous winters (Peel et al. 2007), in the Center-west of Parana State in the Southern Region of Brazil region. Data include animal identification, herd, milk control date, calving date, lactation order, days in milk, animal age, daily milk production and milk production per lactation.

Meteorological data utilized were provided by the National Institute of Meteorology (INMET) from Meteorological Station N° WMO 83813 in Castro municipality, with geographical coordinates 24°48′S, 50°0′W, altitude 1008.80 m, with farms with a maximum distance within a radius of 40 km of weather station. Data included precipitation, environmental temperature (minimum, maximum and average), relative humidity of air, wind speed, solar radiation and Piché evaporation index, gathered daily at 0:00 GMT and 12:00 GMT. From these data, the temperature and humidity index (THI) and temperature and humidity index corrected for solar radiation and wind speed (THI corrected) were calculated using equations described by Mader et al. (2004), which are these below:

$$\begin{aligned} \text{THI} = {} & 0.8 \times \text{ambient temperature} \\ & + (\%\ \text{relative humidity}/100)*(\text{ambient temperature} - 14.3) \\ & + 46.4\text{THI corrected} = 6.81 + \text{THI} - (3.075 * \text{wind speed}) \\ & + (0.00055 * \text{solar radiation}) \end{aligned}$$

Milk production and meteorological variables were used from January 1, 1996 to December 31, 2010, as milk production and meteorological variables were paired only in this period. There were 161,351 milk control data measured as milk/cow/day from 6155 herds and 10,162 data generated by the Castro Meteorological Station.

Data was submitted to a multivariate statistical analysis. In addition, a general analysis of data per period (1996–2000, 2001–2005 and 2006–2010) and season (Summer, Autumn, Winter and Spring) was carried out. All statistical analysis were carried out using SAS® for Windows version 9.3 (Statistical Analysis Institute, Cary, North Carolina).

Statistical Analysis

Data were standardized (PROC STANDARD), with mean zero and standard deviation 1. Variance inflation factors were evaluated (PROC REG option VIF), where variables with VIF value over 10 were deleted from the analysis. Meteorological variables were correlated with milk production by Pearson's correlation analysis (PROC CORR). Analyses included principal component analysis (PROC PRIMCOMP), to observe relations between variables. The importance of each year and five-year periods were determined. In the principal component analysis, input variables included cow age, days in milking and number of lactations. The importance of each climatic variable was studied in relation to milk production using a step-by-step discriminant analysis (PROC STEPDISC). For all multivariate analysis, were used default values determined by statistical package SAS for Windows version 9.3, that is 0.15.

Milk Production and Meteorological Characterization of Castro, Paraná, Brazil

Castro's region have most developed dairy system of Parana State, based in farms with area between 60 to 80 ha. Milk production is intensive, with use of confinement or free-stall, using alfalfa, annual and perennial forages to harvest and supply, as well use of concentrate during year. Artificial insemination is utilized in 100 % of females, and calving are distributed in whole year. Milking is 100 % mechanized, in some cases robotized, and milk cooling is realized in own farm. Much of labor is hired.

During 15 years of survey, means observed for some performance and management variables were: average daily milk production was 30.57 L/cow, where 10.69 % of cows produced until 20 L/day, 74.74 % of cows produced between 20 to 40 L/day and 14.57 % of cows produced up to 40 L/day; also 67.18 % of cows were milked twice a day and 32.82 % milked three times a day.

Regarding the number of lactations, 29.58 % of data collected corresponded to cows of first lactation, 30.19 % of cows in second lactation, 28.11 % of cows in third lactation and 12.12 % corresponded to cows in fourth lactation; and cow's age were observed than 36.41 % of animals had between 21 to 36 months, 46.66 % between 37 to 60 months and 16.93 % of cows had age between 61 to 96 months. Frequency of days in milking verified was: 1–100 days in milking, 28.28 %; 101–200 days in milking, 31.17 %; 201–305 days in milking, 30.82 % and observations up to 305 days in milking, 9.73 %.

Meteorological variables during 1996 to 2010 and seasons are presented in Figs. 24.1 and 24.2. No significant alterations in meteorological variables were seen over the 15 years analyzed (Fig. 24.1), although maximum temperature presented low variability in the period. Positive variation in milk production in the period from 2004 to 2009 seems to be linked to technological or market changes than to be associated with meteorological variables.

In Fig. 24.2, no high monthly variation in meteorological variables was seen, except for maximum temperature, characteristic in regions classified as humid temperate Cfb (Peel et al. 2007). It is important to mention that the greatest average maximum temperature during 15 years of observations did not ultrapass 30 °C at Castro's Meteorological Station. Milk production was maintained constant (approximately 30 L/cow/day) year round, with a high increment during the winter due to increased milk prices. As average THI and THI corrected for solar radiation and wind speed were maintained within acceptable values for animal thermal comfort (THI between 64 and 68) (Mader et al. 2004; Bohmanova et al. 2007; Hill and Wall 2015), meteorological variables did not influence milk production,

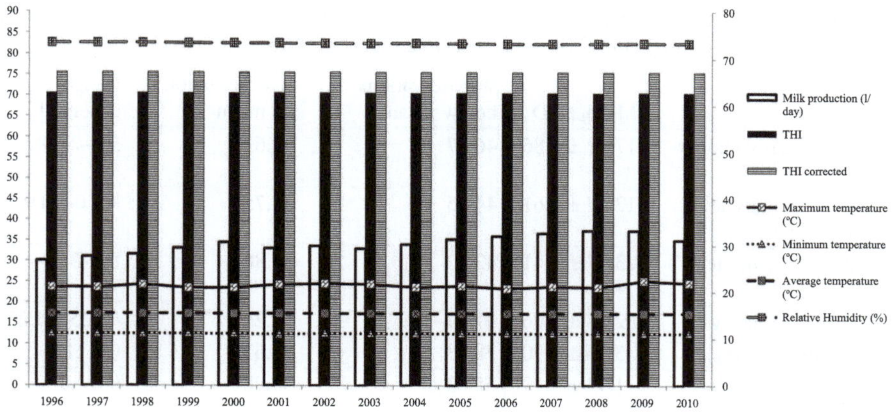

Fig. 24.1 Milk production and meteorological variable in period from 1996 to 2010 in Castro, Paraná, Brazil

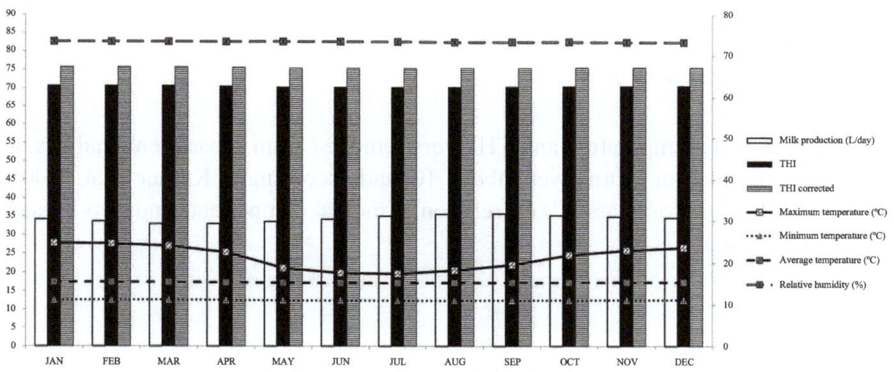

Fig. 24.2 Monthly milk production and observation of meteorological variable in period from 1996 to 2010 in Castro, Paraná, Brazil

due an ideal climatic condition or by the use of technology to mitigate the problem of heat stress.

Average daily temperatures (Table 24.1) are in accordance with those described by West (2003) and Pires and Campos (2011) as ideal environmental conditions for commercial exploitation of dairy herds. In this study, maximum daily temperature up to 27 °C were found in 30 % of observations and temperatures up to 30 °C were registered in only 275 observations of a total of 4934 observations.

THI and THI corrected for solar radiation and wind speed were within the range of thermal comfort for Holstein cows (Bohmanova et al. 2007). Extreme values of THI and THI corrected found in this study, based on the Livestock Weather Safety Index (LCI 1970) corresponded to approximately 2 % and 14 % of alert level observations and only 0.31 % of observations for THI corrected were in the danger level.

Table 24.1 Means, standard deviation and distribution of frequencies of meteorological variables observed in Weather Station in Castro, Paraná, Brazil, in period from 1996 to 2010

Variable	Mean ± SD	% observations below mean	% observation up mean	Range observed
Average daily temperature (°C)	17.31 ± 3.86	46.37	53.63	3.2–24.98
Minimum daily temperature (°C)	12.47 ± 4.61	44.25	55.75	−5.0–20.9
Maximum daily temperature (°C)	23.97 ± 4.41	42.12	57.88	8.6–34.4
Relative humidity (%)	82.48 ± 8.06	47.79	52.21	48.25–99.25
Rainfall (mm/day)	3.96 ± 9.90	78.83	21.67	0–111.8
Piché's evaporation (mm/day)	1.94 ± 1.19	55.02	44.98	0–21.9
Solar radiation (kWh/($m^2 \cdot$ day))	4.50 ± 3.27	50.12	49.86	0–11.4
Wind speed (m/s)	0.80 ± 0.61	59.90	40.10	0–4.67
THI[a]	62.72 ± 6.21	46.37	53.63	35.85–74.08
THI[a] corrected	67.16 ± 6.41	46.59	53.41	42.89–80.42

[a]Temperature and humidity index

Average daily temperature and THI were removed from subsequent analysis, as the variance inflation factors were above 10, and according to Kutner et al. (2005) should not be used for Pearson's correlation, principal components and step-by-step discriminant analysis.

Correlations Between Milk Production and Meteorological Variables

Tables 24.2 and 24.3 show correlations between milk production and meteorological variables for climatic seasons and five-years periods. In all evaluations, correlations were low, and at times significant, indicating that these variables are important in milk production, but are not determinants of milk production. Minimum daily temperature had the most consistent correlation with milk production, probably by climatic classification, with mild temperatures in warm seasons and severe minimum temperature in cool seasons. According results found by Liang et al. (2013), in an environment similar to this study and using reticulorumen temperature as indicator of welfare in lactating cows, in Holstein breed, reticulorumen temperature is constant until 20 °C in ambient temperature and increase 0, 0118 °C each 1 °C of increment in ambient temperature.

In the different seasons (Table 24.2), among the four key meteorological variables used to determinate the welfare of dairy cows (temperature, relative humidity, solar radiation and wind speed), only maximum and minimum temperatures,

Table 24.2 Pearson's correlation between milk production and meteorological variables year round during 1996-2010 in Castro, Paraná, Brazil

Milk production						
Rainfall (r) (P>|t|)	Maximum temperature (r) (P>|t|)	Minimum temperature (r) (P>|t|)	Relative humidity (r) (P>|t|)	Solar radiation (r) (P>|t|)	Wind speed (r) (P>|t|)	THI[a] corrected (r) (P>|t|)
Summer						
−0.00804 NS	−0.01879 **	−0.02476 ***	−0.00425 NS	0.00314 NS	−0.00088 NS	−0.01238 *
Fall						
0.00364 NS	0.00240 NS	−0.04308 ***	−0.00555 NS	−0.00520 NS	0.00608 NS	0.00093 NS
Winter						
−0.01525 **	0.02186 ***	0.01487 **	−0.00448 NS	0.00004 NS	0.00782 NS	0.01923 **
Spring						
−0.00367 NS	−0.00950 NS	−0.00489 NS	−0.00487 NS	0.00526 NS	−0.00357 NS	−0.00876 NS

NS no significant
[a]Temperature and humidity index
*(P < 0.05), **(P < 0.01), ***(P < 0.001)

showed statistically significant correlations in summer and winter, which represent that extremes of temperature, characteristic of these seasons, thereby affecting milk production. Dairy farms in the region use technologies to mitigate deleterious effects of these variables, such as mist & fan housing and cloth shades (Martello et al. 2004).

For the evaluation of five-year periods (Table 24.3), as well as temperature, relative humidity was also showed significant correlations with milk production for all five-year periods. The importance of relative humidity in dairy production is supported by a significant correlation of rainfall in 2001–2005 and 2006–2010. In the 2006–2010 period, maximum temperature presented a significant correlation, but minimum temperature did not. The low non-significant correlation with THI corrected means we cannot confirm that environmental warming will affect milk production in the region.

Principal Components and Step-by-Step Discriminate Analysis Between Milk Production and Meteorological Variables

In the analysis of whole period (1996–2010), the first three principal components explained to 48.23 % of total variance (Fig. 24.3). Maximum temperature, solar radiation and Piché's evaporation showed larger eigenvectors values in principal component 1, cow age and number of lactations larger eigenvectors values in

Table 24.3 Pearson's correlation between milk production and meteorological variables in three different five-years periods of evaluation in Castro, Paraná, Brazil

Milk production						
Rainfall (r) (P>|t|)	Maximum temperature (r) (P>|t|)	Minimum temperature (r) (P>|t|)	Relative humidity (r) (P>|t|)	Solar radiation (r) (P>|t|)	Wind speed (r) (P>|t|)	THI[a] corrected (r) (P>|t|)
1996–2000						
−0.00815 NS	0.00041 NS	−0.07731***	−0.00914*	0.00584 NS	0.00786 NS	−0.00005 NS
2001–2005						
−0.01452**	−0.00186 NS	−0.05430***	−0.00955**	0.00780 NS	−0.00086 NS	−0.00102 NS
2006–2010						
0.01440**	−0.01270**	−0.00576 NS	0.01463**	−0.00531 NS	0.00261 NS	−0.00651 NS

NS no significant
[a]Temperature and humidity index
*(P < 0.05), **(P < 0.01) ***(P < 0.001)

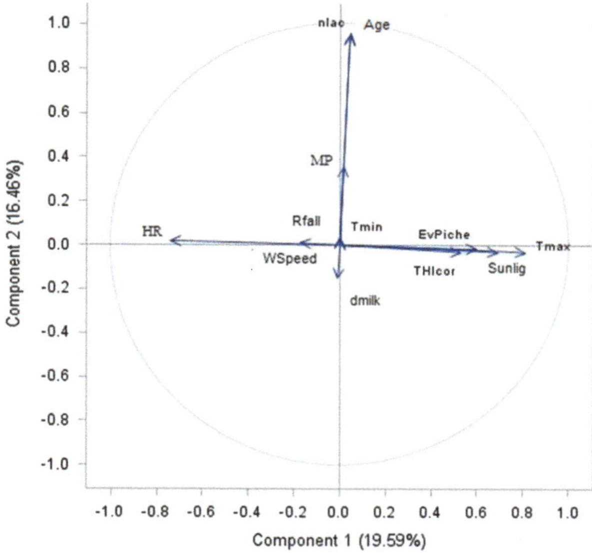

Fig. 24.3 Vector analysis of principal components in relation with milk production, animal performance and meteorological variable during 1996–2010 in Castro, Paraná, Brazil. *MP* daily milk production, *nlac* number of lactation, *Age* cow's age, *dmilk* days in milking, *T min* minimum temperature, *T max* maximum temperature, *Rfall* rainfall, *WSpeed* wind speed, *RH* relative humidity, *Sunlig* solar radiation, *EvPiche* Piché's evaporation, *THIcor* temperature and humidity index corrected for solar radiation and wind speed

principal component 2 and THI corrected had a large eigenvector values in principal component 3. When studying the vectors of the orthogonal plan of principal components, we observed an association or direct relation between milk production with minimum temperature and with rainfall, but at the same time no relationship was observed between milk production with maximum temperature or with THI corrected. This demonstrates that in a medium temporal series (15 years) analysis, for Castro, (Paraná, Brazil), minimum temperature and rainfall are important meteorological variables to consider as part of a welfare program of Holstein cows and for the viability of milk production in a Cfb climate.

Among the five-year periods analyzed (Fig. 24.4), as in the general analysis, the first three principal components explained 45 to 50 % of data variance, and larger eigenvectors values were the same as in the whole period analysis. A difference was seen in the first five-year period (1996–2000), where days in milking showed major eigenvector values for principal component 3. This leads us to presume that animals were well adapted to the Castro climate, as major changes in the relation between milk production and meteorological variables over the five-year periods suggesting the absence of significant change in meteorological patterns that could modify the milk production profile in the fifteen years of observations (Silva et al., 2009; Berman 2011).

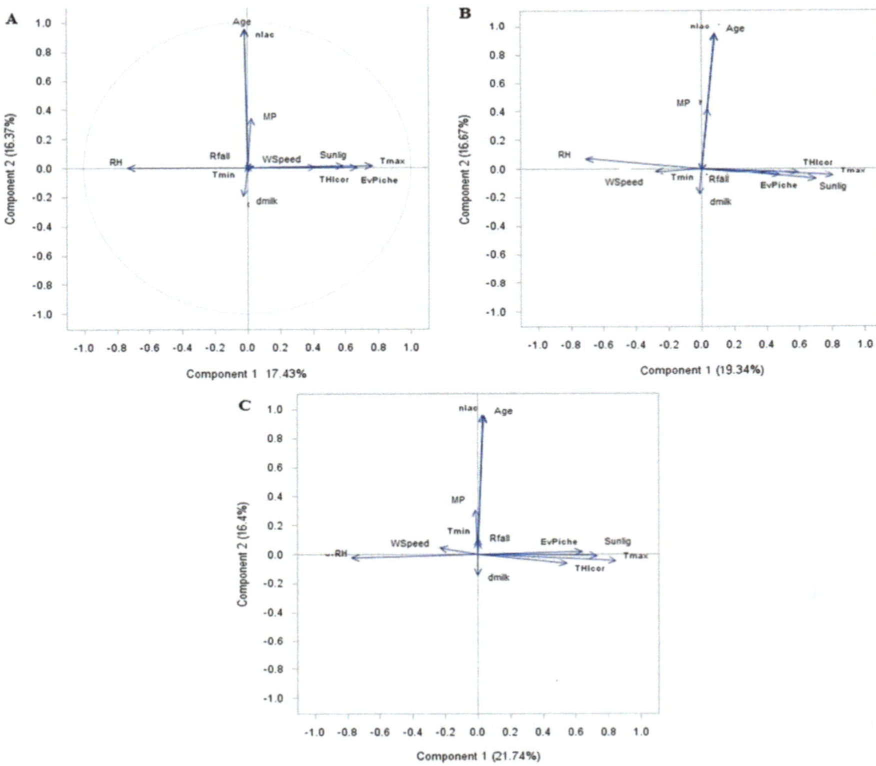

Fig. 24.4 Vector analysis of principal components for milk production, animal performance and meteorological variables in periods 1996–2000 (**a**), 2001–2005 (**b**) and 2006–2010 (**c**) in Castro, Paraná, Brazil. *MP* daily milk production, *nlac* number of lactation, *Age* cow's age, *dmilk* days in milking, *T min* minimum temperature, *T max* maximum temperature, *Rfall* rainfall, *WSpeed* wind speed, *RH* relative humidity, *Sunlig* solar radiation, *EvPiche* Piché's evaporation, *THIcor* temperature and humidity index corrected for solar radiation and wind speed

In all seasons evaluated (Fig. 24.5), meteorological variables presented larger eigenvectors in principal component 1. Maximum temperature showed the highest values, followed by solar radiation and Piché's evaporation. It is important to note that eigenvectors values for THI corrected are relatively high in principal component 1, which would indicate higher sensibility of milk production due to the cow's thermal comfort for each climate season, even though this effect have not show high significance for the animal comfort index and milk production, which means that THI for each season are important but do not alter milk production.

The relation between milk production and meteorological variables is more evident in summer and winter compared with autumn and spring. Three of four variables with importance in the determination of THI (temperature, relative humidity and solar radiation) presented higher extremes values, showing that seasons where extremes conditions occur may affect milk production or milk

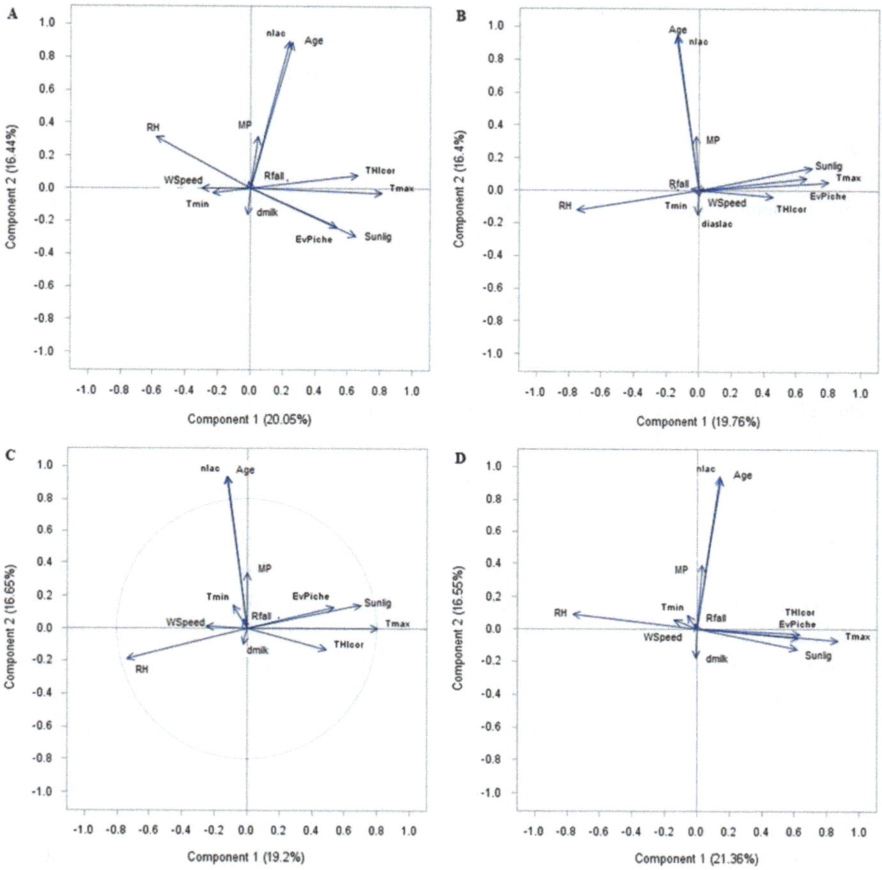

Fig. 24.5 Vector analysis of principal components for milk production, animal performance and meteorological variables in Summer (**a**), Fall (**b**), Winter (**c**) and Spring (**d**) in Castro, Paraná, Brazil. *MP* daily milk production, *nlac* number of lactation, *Age* cow's age, *dmilk* days in milking, *T min* minimum temperature, *T max* maximum temperature, *Rfall* rainfall, *WSpeed* wind speed, *RH* relative humidity, *Sunlig* solar radiation, *EvPiche* Piché's evaporation, *THIcor* temperature and humidity index corrected for solar radiation and wind speed

production is more sensitive to changes in these variables (Heck et al. 2009; Bertocchi et al. 2014).

Tables 24.4, 24.5 and 24.6 present discriminant analysis to period 1996–2010, for five-years periods and seasons, respectively. In these analyses, there were insert some management variables, as effect of comparison and to estimate the importance of each variable in dairy system of Cfb climate classification in Castro, Brazil.

All, variables included in the discriminant analysis were significants, but partial R^2 were low, which means that the variables utilized are important but are not enough to affect milk production in the Castro region, possibly due to high technological level of dairy production in the region and the use of mitigating

Table 24.4 Stepwise discriminant analysis of relation between milk production and meteorological variables in period 1996 to 2010 on Castro, Paraná, Brazil

Variables	Partial R^2	P > F	P > ASCC[a]
Days in milking	0.1837	<0.0001	<0.0001
Number of lactations	0.0568	<0.0001	<0.0001
Cow's age	0.0072	<0.0001	<0.0001
Minimum temperature	0.0028	0.0012	<0.0001
Wind speed	0.0027	0.0020	<0.0001
Maximum temperature	0.0027	0.0046	<0.0001
Rainfall	0.0024	0.1410	<0.0001

[a]Average square of canonical correlation

Table 24.5 Stepwise discriminant analysis of relation between milk production and meteorological variables evaluated from five-years periods during 1996 to 2010 on Castro, Paraná, Brazil

Variables	Partial R^2	P > F	P > ASCC[a]
Period 1996–2000			
Days in milking	0.2664	<0.0001	<0.0001
Number of lactations	0.0523	<0.0001	<0.0001
Relative humidity	0.0203	<0.0001	<0.0001
Cow's age	0.0155	<0.0001	<0.0001
Minimum temperature	0.0106	<0.0001	<0.0001
Solar radiation	0.0094	0.0099	<0.0001
Wind speed	0.0098	0.0018	<0.0001
Period 2001–2005			
Days in milking	0.2092	<0.0001	<0.0001
Number of lactations	0.0714	<0.0001	<0.0001
Cow's age	0.0081	0.0002	<0.0001
Minimum temperature	0.0078	0.0015	<0.0001
Maximum temperature	0.0078	0.0021	<0.0001
Period 2006–2010			
Days in milking	0.1791	<0.0001	<0.0001
Number of lactations	0.0466	<0.0001	<0.0001
Cow's age	0.0135	<0.0001	<0.0001
Relative humidity	0.0114	<0.0001	<0.0001
Solar radiation	0.0206	<0.0001	<0.0001
Rainfall	0.0070	0.0218	<0.0001

[a]Average square of canonical correlation

factors as described above (Silva 2006). It is important to highlight THI corrected in the discriminant analysis per season. When extremes are seen in temperature or humidity associated with solar radiation and wind speed, such as in summer and in winter, THI corrected takes on great importance, as an indicator of climatic welfare of lactating cows, even though this variable was not seen to influence milk production.

Table 24.6 Stepwise discriminant analysis of relation between milk production and meteorological variables evaluated from climate seasons periods during 1996 to 2010 on Castro, Paraná, Brazil

Variables	Partial R^2	P > F	P > ASCC[a]
Summer			
Days in milking	0.1798	<0.0001	<0.0001
THI[2] corrected	0.1588	<0.0001	<0.0001
Number of lactation	0.0564	<0.0001	<0.0001
Cow's age	0.0210	<0.0001	<0.0001
Maximum temperature	0.0139	<0.0001	<0.0001
Minimum temperature	0.0124	<0.0001	<0.0001
Rainfall	0.0114	0.0043	<0.0001
Fall			
Days in milking	0.1816	<0.0001	<0.0001
Number of lactation	0.0603	<0.0001	<0.0001
Cow's age	0.0178	<0.0001	<0.0001
Relative humidity	0.0131	<0.0001	<0.0001
Maximum temperature	0.0157	<0.0001	<0.0001
Minimum temperature	0.0121	0.0004	<0.0001
Rainfall	0.0103	0.1419	<0.0001
Winter			
Days in milking	0.1881	<0.0001	<0.0001
THI[b] corrected	0.0858	<0.0001	<0.0001
Number of lactation	0.0755	<0.0001	<0.0001
Maximum temperature	0.0148	<0.0001	<0.0001
Cow's age	0.0145	<0.0001	<0.0001
Minimum temperature	0.0134	<0.0001	<0.0001
Rainfall	0.0121	0.0018	<0.0001
Spring			
Days in milking	0.1986	<0.0001	<0.0001
Number of lactation	0.0661	<0.0001	<0.0001
Wind speed	0.0230	<0.0001	<0.0001
Maximum temperature	0.0140	<0.0001	<0.0001
Cow's age	0.0127	<0.0001	<0.0001

[a]Average square of canonical correlation
[b]Temperature and humidity index

Conclusions

There is not significant relationship between milk production from high-producing lactating Holstein cows and meteorological variables in a Cfb climate region of southern Brazil.

Acknowledgments Present work was supported by CNPq, Conselho Nacional de Desenvolvimento Científico e Tecnológico, CAPES and INCT- Pecuária (MCT/CNPq/FAPEMIG).

References

Berman A (2011) Are adaptations present to support dairy cattle productivity in warm climates? J Dairy Sci 94(5):2147–2158

Bertocchi L, Vitali A, Lacetera N, Nardone A, Varisco G, Bernabucci U (2014) Seasonal variations in the composition of Holstein cow's milk and temperature–humidity index relationship. Animal 8(4):667–674

Bohmanova J, Misztal I, Cole JB (2007) Temperature-Humidity Indices as indicators of milk production losses due to heat stress. J Dairy Sci 90(4):1947–1956

Costa NS, Hermuche P, Cobuci JA, Guimarães RF, Carvalho Júnior OA, Costa CN, Gomes RAT, Mcmanus CM (2014) Georeferenced evaluation of genetic breeding value patterns in Brazilian Holstein cattle. Genet Mol Res 13:9806–9816

Heck JML, Van Valenberg HJF, Dijkstra J, Van Hooijdonk ACM (2009) Seasonal variation in the Dutch bovine raw milk composition. J Dairy Sci 92(10):4745–4755

Hill DL, Wall E (2015) Dairy cattle in a temperate climate: the effects of weather on milk yield and composition depend on management. Animal 9(1):138–149

Kutner MH, Nachtsheim CJ, Neter J, Li W (2005) Applied linear models, 5th edn. McGraw-Hill Irwin, New York

Liang D, Wood CL, McQuerry KJ, Ray DL, Clark JD, Bewley JM (2013) Influence of breed, milk production, season, and ambient temperature on dairy cow reticulorumen temperature. J Dairy Sci 96(8):5072–5081

Mader T, Davis S, Gaughan J, Brown-Brandl T (2004) Wind speed and solar radiation adjustments for the temperature-humidity index. Meeting Abstract. In: 16th Conference on Biometeorology and Aerobiology, Vancouver, British Columbia, Canada

Martello LS, Savastano Júnior H, Silva SL, Titto EAL (2004) Respostas fisiológicas e produtivas de vacas Holandesas em lactação submetidas a diferentes ambientes [Physiologic and performance responses of holstein cows in milking under different environments]. Rev Bras Zootec 33(1):181–191

McManus C, Louvandini H, Paim TP, Martins RFS, Barcellos JOJ, Cardoso CC, Guimarães RF, Santana OA (2011a) The challenge of sheep farming in the tropics: aspects related to heat tolerance. Rev Bras Zootec 40(1):107–120

McManus C, Castanheira M, Paiva SR, Louvandini H, Fioravanti MCS, Paludo GR, Oliveira EMB, Corrêa PS (2011b) Use of multivariate analyses for determining heat tolerance in Brazilian cattle. Tropl Anim Health Prod 43:623–630

Peel MC, Finlayson BL, McMahon TA (2007) Updated world map of the Köppen-Geiger climate classification. Hydrol Earth Syst Sci 11:1633–1644

Pires MFA, Campos AT (2011) Relação do dados climáticos com o desempenho animal [Relation of climatic data with animal performance]. Embrapa Agrossilvopastoril, Sinop, pp 103–114

Silva HA (2006) Resultados econômicos de sistema de produção de leite com diferentes níveis tecnológicos na Cooperativa Castrolandia, Castro, PR [Economic results of dairy production system with different technological levels in Castrolanda Cooperative, Castro, Paraná, Brazil]. Emater, Curitiba, Brazil

Silva TGF, Moura MSB, Sá IIA, Zolnier S, Turco SHN, Justino F, Carmo JFA, Souza LSB (2009) Impactos das mudanças climáticas na produção leiteira do Estado de Pernambuco: análise para os cenários B2 e A2 do IPCC [Impacts of the climate change on regional variation of milk production in the Pernambuco state, Brazil: analysis for the B2 and A2 IPCC scenarios]. Rev Bras Meteorol 24(4):489–501

West JW (2003) Effects of heat-stress on production in dairy cattle. J Dairy Sci 86(6):2131–2144

Chapter 25
Climate Change Implications in Aviation and Tourism Market Equilibrium

Dimitrios J. Dimitriou

Abstract A significant proportion of capacity within the European air transport system is taken up by demand for leisure flying and it is particularly associated with summer holidays. Forecasts suggest that climate change has the potential to have a major impact upon levels and patterns of demand for leisure flying as seasons change, as some traditional locations become less attractive and as new markets emerge, either at different times of the year or in new geographical regions. This paper deals with the key challenges and issues for the aviation and tourism sectors towards a climate change adaptation strategy for attractive tourist destinations. Through a top-down analysis, the variables of climate change that impact on the supply and demand equilibrium for tourism and aviation are defined, and based on a gap analysis framework, the relationship of these variables to the aviation and tourism equilibrium is given. Conventional wisdom is to provide key messages to aviation authorities, decision makers and stakeholders regarding the expected changes in demand, the implications in airport operation and the effects in regional economic development, especially, for regions that are highly reliant upon income from tourism. The application includes the aviation and tourism in Greece, which is a very attractive tourist destination in southeast Mediterranean, highlighting the diverse impacts of a changing climate on aviation and tourism, that expected to have significant implications for air traffic flows and the economies across Europe.

Keywords Climate Change • Implications • Aviation • Tourism • Aviation-Tourism equilibrium

Introduction

A significant proportion of capacity within the European air transport system is taken up by demand for leisure flying. This is particularly associated with holidays taken in the Mediterranean during the summer months and ski holidays taken in the

---

D.J. Dimitriou (✉)
Department of Economics, Democritus University of Thrace, Panepistimioupoli, Komotini 69100, Greece
e-mail: ddimitri@econ.duth.gr

W.L. Filho et al. (eds.), *Climate Change Adaptation, Resilience and Hazards*, Climate Change Management, DOI 10.1007/978-3-319-39880-8_25

Alps and other parts of central Europe during the winter. Forecasts suggest that Climate Change (CC) has the potential to have a major impact upon levels and patterns of demand for leisure flying as seasons change, as some traditional locations become less attractive and as new markets emerge, either at different times of the year or in new geographical regions. Such changes could have significant implications for the economies of a number of regions across Europe that are highly reliant upon income from tourism.

A comprehensive assessment of these relationships would require significant investment so with this in mind, the content of this paper is designed merely to illustrate the nature of the phenomenon. This will be achieved through a literature review and based on the gap analysis methodological approach, the CC implications for aviation and tourism businesses are presented to illustrate the nature, timing and potential magnitude of changes, and their potential impact upon demand. The application deals with tourism and aviation in Greece, which is selected, for two reasons: firstly, the Greek economy is heavily dependent on aviation and tourism; and secondly, given the recent long term economic recession of the country, the additional income from tourism and aviation growth is seen as a key element of Greece's economy recovery plan (IMF 2014).

The Greek economy is heavily dependent on aviation for tourism which accounts for 20 % of Greek GDP. Approximately 73 % of tourists arrive in Greece by air while 80 % of international arrivals are tourists (Dimitriou et al. 2011; Eurocontrol 2010). CC not only has the potential to have significant impacts on the pattern of demand, it also has consequences for the ability of some tourism locations to meet demand when and where it arises (for example as a result of water shortages). Through a systemic approach this paper maps the CC variables which impact the aviation and tourism supply–demand equilibrium. The broad findings of this paper deals with the break down and the review of the CC implications on aviation and tourism supply–demand equilibrium, providing the framework for other similar applications.

The paper is divided in five sections. The first includes a short introduction to the topic, highlighting the key objectives of the paper. This is followed by the section that provides an overview of tourism and aviation trends. The third section deals with the CC implications for supply and demand equilibrium of tourism and aviation, providing a systemic approach to define the CC variables relevant to the tourism-aviation equilibrium and a gap analysis to define the relationship of these variables. The next section provides the results of a weighted SWOT analysis for aviation and tourism development in Greece and explaining the need for CC adaptation and mitigation policies to be incorporated into tourism and aviation strategies designed to support sustainable development. In the last section the conclusions and the references are cited.

Tourism and Aviation

Tourism Industry Trends

Over the last half century, tourism experienced continued expansion and diversification, becoming one of the largest and dynamically developing sectors of external economic activities. In 2013, Travel & Tourism's total contribution was 9.5 % of the global GDP (approx. US $7 trillion), not only outpacing the wider economy, but also growing faster than other significant sectors such as financial and business services, transport and manufacturing (WTTC 2014).

In Europe, the International Tourist Arrivals (ITA) present virtually uninterrupted growth—from 277 million in 1980, 528 million in 1995, and 940 million in 2010, and this trend is forecast to continue reaching the level of 1.8 billion ITA in 2030 (UNWTO 2011). Europe (EU28) achieves 54.1 % share of the global tourism market and the travel and tourism sector makes an increasingly large contribution to the overall economy, contributing US$1,512 billion to the European GDP, which represents 9 % of the overall GDP and creating over 21 mio jobs, which represents 10 % of the overall employment in 2013 (WTTC 2014). The Mediterranean region is one of the most attractive tourism destinations in the world, accounting more than a third of ITA (424 million ITA in 2013, UNWTO 2014). Spain, Portugal, Greece, Cyprus, Croatia, Turkey and Egypt are the leaders in attracting tourists in region of Mediterranean and tourism in these countries is one of the major source of national income (UNWTO 2011).

Aviation Growth

Aviation is a key driver for tourism development, accommodating the higher shares of ITA (Forsyth 2006). Whilst geography has meant that, in modern times, air travel has always been the dominant mode for long distance travel and ITA, moves towards deregulation, and in particular the emergence of the low cost carrier sector, has also increased aviation's significance for short and medium haul tourism trips (Burghouwt and Hakfoort 2001).

Due to the aviation industry liberalization along with falling costs of supply chain, the demand for air transport has increased exponentially over the years. Worldwide the aviation industry in 2013 shared 53 % of ITA, while the remainder 47 % travelled by all other transport modes—whether by road (40 %), rail (2 %) or over water (5 %) (UNWTO 2014). The vast majority of ITA are leisure travelers and for remote destinations, aviation is the key driver of economic development. ATAG (2014) estimates that 3.1 billion passengers were transported by aviation in 2012, which globally supports 58.1 mio jobs and contributes US$2.4 trillion that is equal with the 3.4 % of global GDP.

Aviation could be defined as providing "connectivity" or "accessibility" to a global market. Improving connectivity leads to more productivity, which in turn attracts more investment. Connectivity is an effective engine for increasing both competitiveness and economic growth. That is particularly true in Europe, which relies on aviation to provide the international transport links that make Europe a global hub of social and economic connectivity. Following the liberalization of aviation market in Europe (EU28), the number of flights within the EU has more than doubled, the destinations served by more than two airlines have quadrupled and low-cost carriers have boomed during last decade and they now account for almost half of the intra-European aviation market. ATAG (2014) present data showing that the European aviation industry (EU 28) supports 9.3 million jobs and generate US$658 billion (512 billion €) income, contributing essential in EU GDP.

The Link Between Tourism and Aviation

Many researchers review the relationship of tourism and economic development. Dwyer et al. (2004) underline that tourism benefits include investments in infrastructure, the development of management expertise and cultural exchange benefits which affect various sectors of the regional economy. Kim et al. (2006) provide evidence that for the tourist attractive destinations economic stability and financial sustainability related to tourism business development, while, Lee and Chang (2008) highlight that the tourism industry contribution is essential to regional economic development; and Dimitriou et al. (2011) provide the methodology to estimate the impact on the economic system for the regions heavily depended on aviation.

The passenger's decision for choosing the most suitable holiday option depends on a variety of factors such as the consumer profile, the distance of the final destination, the transport options and the price of services. (Mohamad and Jamil 2012) present the complexity of the tourist consumer's decision process and they support that the transportation accessibility and level of service are correlated to tourism market trends. Forsyth (2006) underline that tourism remains heavily dependent upon the aviation industry and any changes in its efficiency can have a significant impact on tourism development. In terms of prices, Tretheway and Mark (2006) demonstrate that tourist demand influenced from the supply chain cost and revenue management.

Institutions, associations and governmental bodies widely recognize the need for monitoring tourism demand and adopting strategies to exploit the economic benefits of tourism. According to WTTC (2014) many islands in the Mediterranean draw a considerable part of their income from the tourism industry, which in turn, is heavily dependent on the aviation industry. Dimitriou et al. (2011) estimate the contribution of aviation in regional development providing evidence that this relationship needs to be investigated not only to extrapolate the demand trends, but also to adopt policies, define strategies and support decisions towards tourism policies and investments in new infrastructure to accommodate additional demand.

CC Implications for Tourist Destinations

Different climatic changes may have a range of diverse impacts on aviation and tourism infrastructure and services. These may vary significant by region and depend on the local or regional circumstances and vulnerabilities, including those associated with the natural environment, as well as abroad range of socioeconomic factors (Dimitriou et al. 2011).

CC has the potential to have significant impacts on the pattern of demand for tourism and the ability of tourist destinations to supply their tourist product to international and domestic customers (Daley et al. 2008). According the concept of gap analysis, Fig. 25.1 illustrates how CC variables can impact on demand and supply side variables which produce the tourism equilibrium in tourist destinations (Eurocontrol 2010). The demand side consists primarily of mitigation and adaptation policies that will impact consumer choice (carbon pricing, climate change awareness and sensitivity). The orange middle arrows list ambient weather conditions (which are both demand and supply variables) in Northern and Southern Europe. Budgeting the CC implications could estimate the determinant of the tourism-aviation supply–demand equilibrium (Eurocontrol 2010; Thomas et al. 2010).

Aviation has become a major contributor to economic growth that requires operational productive and efficient infrastructures and services (Dimitriou et al. 2011). At the same time, CC is likely to have essential implication for aviation and tourism industry infrastructure (Daley et al. 2008). The need and the budget for

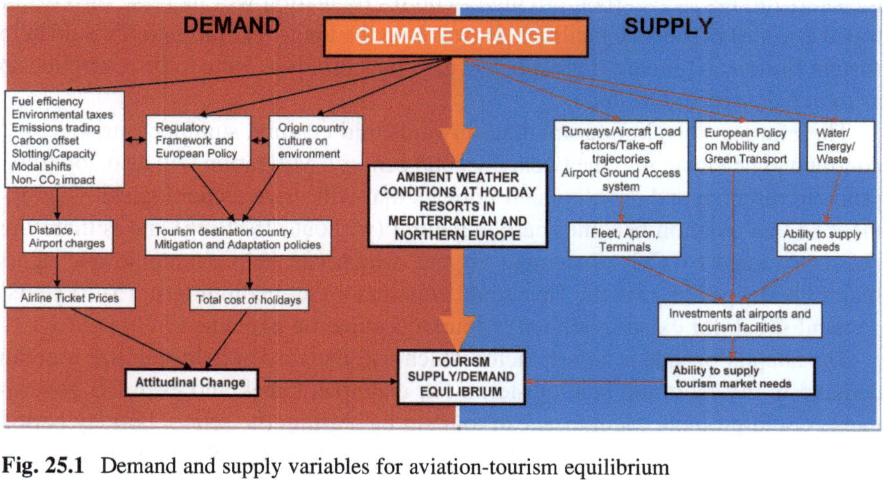

Fig. 25.1 Demand and supply variables for aviation-tourism equilibrium

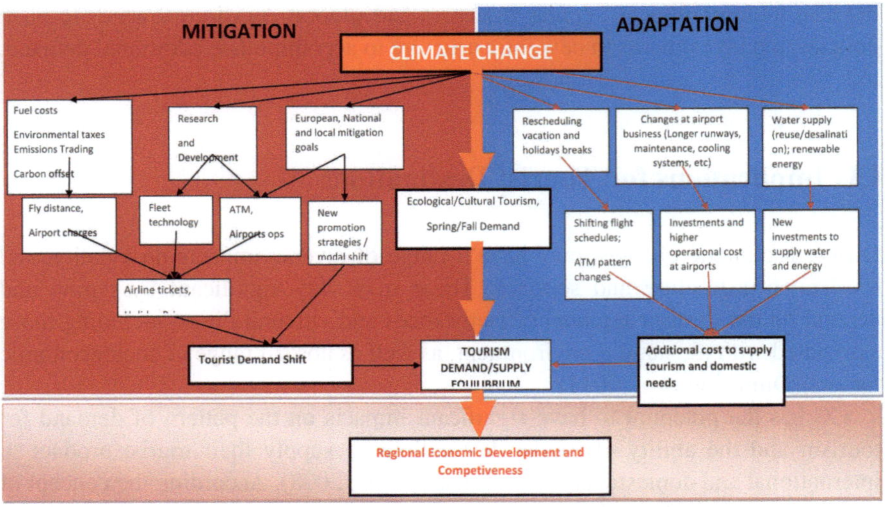

Fig. 25.2 CC mitigation and adaptation variables for aviation-tourism equilibrium

the inclusion of adaptation measures in the transport industry are highlighted by UNECE (2013). Dimitriou et al. (2014) review the CC adaptation strategies and environmental mitigation actions for 15 European airports, highlighting the low performance of Greece and South Europe.

Taking all these factors into account, a mitigation/adaptation chart (Fig. 25.2) is constructed to demonstrate that the supply/demand elements are two sides of the same coin (Eurocontrol 2010). This is because mitigation and adaptation will impact the supply/demand equilibrium differently in the short and long-run. The black arrows depict mitigation variables while the red arrows depict adaptation variables. The orange arrows list variables impacting both mitigation and adaptation. For example mitigation is pull factor, which could reduce tourism demand in the short-run because carbon pricing or control regulation will increase the cost of travel and cost of tourism facilities. However in the long-run, mitigation would help sustain European tourism if greenhouse gas concentrations were to be controlled so as to avoid adverse CC impacts.

Therefore, as the climate stabilizes towards mid-century, summer tourism could continue to be viable into the twenty-second century. Mitigation will also hopefully promote research and development policies that will drive technological development so that aviation becomes carbon neutral (without offsetting) and is therefore not constrained by CO_2 abatement targets, which can help overcome one of its major challenges to growth. This could counteract the medium-term (5–10 years) demand shock for existing attractive tourist destination explained in Fig. 25.1.

The adaptation side is a push factor which will increase or help stabilize tourism demand in the short and medium-term. For example, Greek and even EU governments could promote Mediterranean tourism by shifting the school holiday period to the months where comfort-indexes are appropriate for Northern Europeans.

Greece could do the same to encourage its citizens to continue holidaying domestically rather than travelling north to avoid the increasing summer heat (Dimitriou and Thomas 2008).

The CC Implications for the Greek Aviation and Tourism Markets

CC Implications for Tourist Demand

The purpose of the demand side analysis is to identify key variables from the literature and provide a qualitative assessment of climate change impacts on demand (Eurocontrol 2010; Daley et al. 2008; Dimitriou and Thomas 2008; Dimitriou et al. 2015). A substantial literature exists on tourism demand models for Greece using the classical variables listed in above figure. These forecasting approaches are based on quantitative methodologies using causal econometric models (Song and Li 2009). The operation of demand variables is intuitive; however their precise values in the models vary significantly. In general, income, stability and advertising are positively correlated with tourism demand, while prices, exchange rates and transportation costs are negatively correlated.

The following Fig. 25.3 divides the variables between classical (those which are used in existing tourism demand models in European destinations) and the new climate related variables which need to be included in the models.

CC variables will have significant overall effects on Greek tourism demand, because not only are new variables that should be internalized in existing causal models, but they will impact the classical variables in such a way that is likely to reduce overall demand. The branch of economics, economic geography, indicates that the values assigned to classical variables for the models that forecast tourism demand in Greece are unique to the country and Fig. 25.3 illustrates this (Dritsakis 2004). These variables are self-explanatory and interrelated. For example, although natural and climatic capital may exist in many countries, it is a combination of the existing tourism infrastructure and historical levels of tourism that together with the natural and climatic capital explain why volumes of tourism are higher in certain countries.

Indeed the costs of tourism trade, like trade in general, explains these patterns. A country's comparative advantage in tourism is a function of the costs associated with distance, infrastructure and the agency and transaction costs associated with industry and institutional structures. These accumulated comparative advantages are more complex than simply ambient temperature or even aviation infrastructure, they take time to develop and therefore the comparative advantages also take time to erode. In short, the classical variables assume existing assets and cannot be viewed as operating in a vacuum.

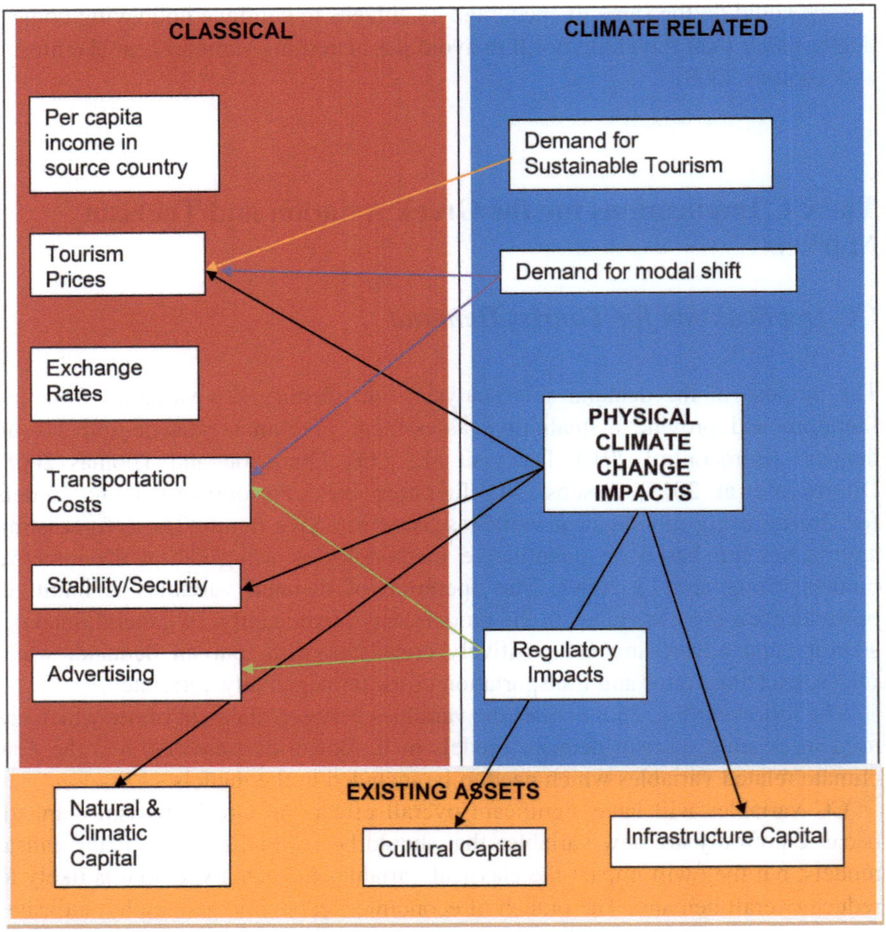

Fig. 25.3 Demand side variables for tourism

Demand for Sustainable Tourism and Transport

The sustainability of tourism resorts is an additional variable to the carbon intensity of transportation. The literature on sustainable tourism demand is not very developed. Although most studies have not found that sustainability ranks highly on factors affecting tourism destination choice (Brau 2008), it is possible that as climate change impacts become more severe and public awareness increases, sustainability will become a significant variable for tourism demand. This is a variable Greece will have to control. It should therefore develop a strategy to be ahead of the curve in sustainable tourism destinations to compensate for the carbon intensity of its reliance upon long distance transport. Another possibility is for Greece to pay for offsetting the emissions from tourism related transport, energy and water use. This may be contrary to EU state aid rules, but exceptions could be discussed for tourism dependent EU countries.

Literature reveals that passengers have a general understanding of climate change and the link between aviation and climate change (Daley et al. 2008). However, passengers seem to believe their aviation use is acceptable and that the responsibility for addressing aviation's impact on climate change lies with the government, airlines and manufacturers. This demonstrates that in Europe a high value is placed upon air travel to go on holiday, for weekend breaks, to visit friends and family and to meet religious and cultural demands. The multi-cultural nature of many European states, their relative geographic isolation and the poor quality of 'holiday weather' in home countries increases the perceived value of an "ability to fly". They therefore appear reluctant to give up the high-speed-long-distance services offered by the aviation industry unless a comparably priced alternative mode of transport, like high speed trains, becomes available. These generalizations can in principle be extended to the other major source countries for Greek tourism like Germany and Holland. Consequently passenger demand for leisure air travel is unlikely to diminish in the near future, without severe price or other forms of aviation rationing by government.

Regulatory Impacts

Policies that restrict aviation growth or price carbon will increase the cost of transportation to Greece. The exact value of demand decrease depends primarily on the estimated carbon price and price elasticity for each service, class and route as well as the ability of airlines to pass-through costs to passengers. If governments decide to constrain aviation growth—this is uncertain in the short to medium term—there are three regulatory options being discussed to deliver this objective:

- Personal carbon allowances (for domestic consumption (utility bills), travel (car, public transportation) and air travel (domestic and international aviation).
- Fiscal responses designed to constrain demand (see below).
- Slot pricing and availability controls.
- Airport infrastructure constraints to restrict the ability to respond to demand.

Policies (rationing forms of aviation, informational campaigns, etc) intended to promote local tourism as a way of meeting national carbon budgets and prioritizing 'necessary' aviation, could have a similar effect as carbon pricing. The need for this rationing mechanism is well recognized in Bank of Greece report (Tsartas and Papatheodorou 2014) where a cap on growth could imply a top-down form of government prioritization of different forms of aviation according to available substitutes and environmental impacts. For example, governments could prioritize long-haul over short haul, business travel over tourism. Control mechanisms existed in the past when traffic distribution rules were in place. These could be reinstituted and motivated less by who (which carriers) and more by what (what destinations for what purpose).

Demand for Modal Shifts

Emissions from aviation contribute the largest share of total Greek tourism industry emissions. If the aviation industry faces carbon pricing and caps on growth, there may be a significant demand for alternative forms of transportation to tourism destinations. Development plans for high speed rail lines are extensive in some parts of Europe (Committee on Climate Change 2009) and tourists could decide to use high speed rail links for Mediterranean or Atlantic destinations, particularly in Southern France, Northern Italy and Northern Spain. This combined with the heat stress induced by climate change could also shift tourism patterns away from the Southern Mediterranean and Aegean area (Eurocontrol 2010), with a consequential significant impact upon economies in that region.

Because Greece is isolated in South-East Europe, where no plans for high-speed rail exist, and 80 % of tourists arrive in Greece by air, its tourism is particularly vulnerable to policies that increase the price of aviation and decrease (in relative or absolute terms) the price of modes such as maritime and rail. Greek tourism is also vulnerable to changes in consumers' attitudes towards aviation for tourism transport (Daley et al. 2008). However, policies that promote and incentivize other modes of transportation like maritime and rail, combined with warmer dryer summers in Northern Europe could have a significant negative effect on Greek tourism demand. Subsidies or simply greater consumer demand for high-speed rail would bring down the cost of rail journeys on high density routes making transport and tourism packages for these routes cheaper compared to Greece which remains dependent on high carbon aviation.

SWOT Analysis for the Aviation and Tourism Development in Greece

The key objectives of the SWOT analysis are: (a) to identify key parameters regarding aviation and tourism development; (b) to evaluate the benefits and weaknesses, and (c) to support decisions at the level of strategic planning. The level of significance providing in a scoring scale from *−3* to *+3* (strongly positive: +3, medium: (−1,+1), significant negative: −3). The analysis key results are given in the following table (Table 25.1).

The next step is to analyse the key opportunities and threats from the aviation and tourism business, as briefly presented in the following Table 25.2.

Table 25.1 Aviation business strengths and weaknesses for Greece, by their level of significance

Level of significance			
Development issues	Positive (+3)	Medium (−1,+1)	Negative (−3)
Strengths (S)			
S1. Regional development	• Additional capacity and low pricing strategy at airports stimulate demand for air transport • New investments in airport and tourism infrastructures • Extension of the other trade sectors (e.g. agriculture) • Additional income to residents	• Improve airport facilities and services focused on leisure and business traffic • Increase of land prices in tourist destinations • Population growth • Control of unemployment • Improve quality of life and social services	
S2. National economy	• New traffic from European and long-haul destinations (e.g. Russia, Asia, etc.) • New jobs in aviation, tourism and other relative activities • Improve national GDP	• Improve accessibility to attractive tourism destinations • Additional business for the national air carriers • Establish Greece as the leading country in Balkans	
S3. European aviation and tourism industry	• More capacity and slots to the European air transport network • Stimulate demand to existing and new destinations • Additional income to tourism and aviation business	• Establishment of business oriented management culture • Improve knowledge regarding aviation and tourism business • Reduce business risk in aviation and tourism industry	
Weaknesses (W)			
W1. Aviation business		• Business focuses on holiday/seasonal traffic • Status of Greek national economy • Cost of the new investments • Competition in national aviation business	• Fare policy to meet future levels of demand • Taxation and accreditation of airports to meet national environmental targets • Uncertainty in tourism and aviation industry
W2. Cost of services and		• Economies of scale in aviation and tourism industry • Authorities	• Total transport chain costs for the users • No specific strategy towards mitigation

(continued)

Table 25.1 (continued)

Level of significance			
Development issues	Positive (+3)	Medium (−1,+1)	Negative (−3)
public sector productivity		performance and national regulation framework • Management performance of state owned operators in the most of the airports	environmental impacts • Price of holiday package
W3. Financial variables		• Government/public authorities procedures in decision making • Interest for investments in aviation and tourism business • Condition of private-public contracts to operate/manage airports	• Project cost and payback period for the new investments in aviation • Conditions in national economy • Financial risk in aviation and tourism business

The relationship between existing strengths/weaknesses and future opportunities/threats of the expected environmental changes in business for Greece is presented in the following Table 25.3.

Key Messages for the Aviation and Tourism in Greece

The central message for decision makers is that the positive impacts on regional development, the national economy, aviation and tourism businesses is greater than the negative impacts. This is because of the high potential of new investments and the changing environment in Greek airports management, lead to stimulate additional traffic. However, environment protection policies and the competitive environment in tourism and aviation are expected to be a significant threat for the further development of these industries in Greece. By mitigating or adapting to the environmental weaknesses and threats, the decision to promote aviation growth is strengthened in the SWOT analysis.

The above analysis indicates that the following issues will reduce the environment impacts associated with the development of both the air transport and tourism industries:

- Land use planning in tourism and airport areas;
- Improved ground transport;
- Reduced consumption of energy and water;

Table 25.2 Aviation business opportunities and threats for Greece, by their level of significance

	Level of significance		
Decision issues	Positive (+3)	Medium (−1,+1)	Negative (−3)
Opportunities (O)			
O1. National aviation business	• Add new airport capacity to meet demand growth • Budget carriers (LCC) growth • Attract long-haul carriers to/from Russia and Asia • Non-aeronautical revenues from the retail	• Reduce transportation cost to/from Greek domestic destinations/islands • More scheduled flights • Additional income from the growth of demand	
O2. Tourism business development	• New demand (Russia,, Middle East, Asia and Africa) • Provide advantages regarding competition in Mediterranean • New income from tourism	• New businesses in other tourism sectors (sports, education, health, etc) • Attract high income tourists • Extension of the tourism season	
O3. Aviation business in Europe	• New opportunities for European air carriers • Improve connectivity in European region • Stimulate demand in middle-long haul destinations in Asia	• Investment in aviation business • Stimulate additional traffic in regional—secondary—airports in south-east Mediterranean • Growth of leisure traffic in Europe	
Threats (T)			
T1. Business environment and economy		• Niche nature of the Greek market • Growth of European GDP and financial stability in Europe • Investments to other transport modes (sea, rail, etc.)	• Rise of oil and energy prices • Circularity and uncertainty in aviation and tourism market • Euro exchange policy • National economy and financial performance
T2. Environmental		• Growth of environment sensitivity in Europe (e.g. green travelers)	• Restrictions to aviation growth • EU green policies to

(continued)

Table 25.2 (continued)

	Level of significance		
Decision issues	Positive (+3)	Medium (−1,+1)	Negative (−3)
Impacts mitigation		• Environmental taxation • Water and energy resources	control carbon emissions • Operational constraints at European airports
T3. Competition		• Competition in aviation and transport industry • Advantages (e.g. infrastructures, quality of services) compared to the other destinations in Mediterranean	• Barriers to entry for new investors in aviation and tourism industry • National economy and financial statues in Europe • Euro exchange policy

Table 25.3 Analysis of the relationship between strengths/weaknesses and opportunities/threats

	Aviation and tourism business changes						Score	
Decision issues	Opportunities			Threats			(scale +3,−3)	
Strengths	O1	O2	O3	T1	T2	T3	(+)	(−)
S1	+3	+3	+3	+1	−1	+1	+11	−1
S2	+3	+3	+1	−3	−1	−1	+7	−5
S3	+3	+1	+3	−1	−3	+1	+8	−4
S. Score (+)	+9	+7	+7	+1	0	+2	**+26**	
(−)	0	0	0	−4	−5	−1		−**10**
Weaknesses								
W1	+1	+1	+1	−1	−1	−1	+3	**−3**
W2	−1	−1	−1	−1	**−3**	−1	0	−8
W3	−1	−1	−1	−1	−1	−1	0	**−3**
W. Score (+)	+1	+1	+1	0	0	0	**+3**	
(−)	−2	−2	−2	**−3**	−5	**−3**		−**17**
Aviation/tourism impacts score	+11	+9	+7	+2	0	+1	**+29**	
	−1	−1	−1	−4	−8	−5		**−27**

- Use of renewable energy;
- Introduction of specific regulation to address wastes;
- Re-development of the existing airports to reduce carbon intensity of their operations

Despite these, certain conclusions can be drawn with respect to demand for Greek tourism at least in the short-term (1–5 years) and medium term (5–10 years):

- It is likely that significant improvements in high speed rail combined with improving ambient summer temperature in Northern Europe will decrease demand for Greek tourism in the medium-term. In the long-term (after 2030) the interaction of these variables will be very significant. Maritime based tourism could help Greece fill the Northern European tourist gap, however this depends on the carbon price and the ability of Greece to attract tourists from nearby countries.
- Consumer behaviour is unlikely to be influenced by informational or promotional policies unless combined with high carbon prices and/or the awareness of increased severe physical climate change impacts.
- Certain adaptation measures are viable in Greece: (1) promoting off-season cultural tourism; (2) diversifying the regions offering beach tourism; (3) promoting sustainable tourism locations and offsetting aviation emissions.

Conclusions

Among the many effects of CC, the existing unforeseen strain on extend capacity by new infrastructure and business as usual model of growing tourism must be constantly revaluated. In this paper, the CC implications for key demand and supply variables impact the aviation-tourism equilibrium are described. The need for new actions and an updated strategic plan is highlighted. The findings of this research provide important messages for decision and policy makers for Greece, which is a very attractive tourist destination, messages that will be of equal importance and significance for many other regions in the same latitude.

The key findings provide evidence that the implications of CC for aviation and tourism require not only for air traffic accommodation activities but also national tourism planning activities. Ultimately, the costs of adaptation policies must be weighed against the economic benefits of tourism, which are very substantial in Greece. Consideration at the EU level must also be given, because tourism has been a strategic target for EU funds in promoting sustainable economic development in tourist destinations. While the costs of adaptation will likely be born in part by the EU, so too will the costs of inaction in the form of transfers to a weakened Greek economy; weighing the costs of action and inaction is an exercise worth pursuing through further research.

References

ATAG (2014) Aviation benefits beyond borders. Air Transport Action Group. Geneva, Switzerland, pp 10, 47

Brau R (2008) Demand-driven sustainable tourism? A choice modeling analysis. Tour Econ 14 (4):691–708

Burghouwt G, Hakfoort J (2001) The evolution of the European aviation network, 1990–1998. J Aviat Manag 7:311–318

Committee on Climate Change (CCC) (2009) Meeting the UK aviation target—options for reducing emissions to 2050, UK

Daley B, Dimitriou D, Thomas C (2008) Chapter 18: The environmental sustainability of aviation and tourism". Aviation and Tourism, Ashgate, pp 239–253

Dimitriou D, Thomas C (2008) Environmental constrains on airport development in tourist areas. In: Fifth international conference on transportation research in Greece, Hellenic Institute of Transportation Engineers, Hellenic Institute of Transports and National University of Athens, May 2008, Conference proceedings

Dimitriou D, Sartzetaki M, Voskaki A, Athanasiadis G (2011) An input–output model to quantify the benefits of tourist airports on regional economy. J Air Transp Stud 2(2):92–105

Dimitriou D, Voskaki A, Sartzetaki M (2014) Airports environmental management: results from the evaluation of European airports environmental plans. Int J Inform Syst Supply Chain Manag 7(1):1–14

Dimitriou D, Mourmouris J, Sartzetaki M (2015) Economic impact assessment of mega infrastructure projects. Int J Appl Econ 47(40):4310–4322

Dritsakis N (2004) Cointegration analysis of German and British tourism demand for Greece. Tour Manag 25:111–119

Dwyer L, Forsyth P, Madden J, Spurr R (2004) Evaluating tourism economic effects: new and old approaches. J Tour Manag 73:307–317

Eurocontrol (2010) Case Study 1: Changes to tourist activity in Greece. In: Dimitriou D (ed) MMU—OMEGA program, Brussels

Forsyth P (2006) Tourism benefits and aviation policy. J Air Transp Manag 12:3–13

International Monetary Fund (2014) Country Report No.14/151. Washington, DC, USA, p 43

Kim HJ, Chen M-H, Jang SC (2006) Tourism expansion and economic development: the case of Taiwan. J Tour Manag 27:925–933

Lee CC, Chang CP (2008) Tourism development and economic growth: a closer look at panels. Tour Manag J 29:180–192

Mohamad D, Jamil MR (2012) A preference analysis model for selecting tourist destinations based on motivational factors: a case study in Kedah, Malaysia. Proc Soc Behav Sci 65:20–25

Song H, Li G (2009) Tourism demand modelling and forecasting-a review of recent research. Tour Manag 29:203–220

Thomas C, McCarthy R, Lewis K, Boucher O, Hayward J, Owen B, Liggins F (2010) Challenges to growth environmental update study (EUROCONTROL). Case Study 1:4–5

Tretheway M, Mark D (2006) Emerging tourism markets: ageing and developing economies. J Air Transp Manag 12(1):21–27

Tsartas P, Papatheodorou A (2014) The pathogenicity of the Greek tourist product and the effort for upgrading. Bank of Greece, Report on the impacts of climate change on Greek tourism, pp 37–42. www.bankofgreece.gr/BoGDocuments/EMEKA_tourismos_2014_.pdf

UNECE (2013) Chapter 2: Climate Change Implications for Transport, Climate change impacts and adaptation for international transport networks. United Nations Economic Commission for Europe, Geneva, pp 19–46

UNWTO (2011) Tourism Towards 2030: Global overview. United Nations World Tourism Organization, Madrid, Spain, p 24

UNWTO (2014) Tourism Highlights Report. United Nations World Tourism Organization, Madrid, Spain, p 4

WTTC (2014) Travel and Tourism economic impact Report, European Union". World Travel and Tourism Council, London, UK, pp 1–4

Chapter 26
Canadian, Argentinean, and Colombian Programs Building Resiliency to Extreme Events

Margot A. Hurlbert, Paula Mussetta, and Sandra Turbay

Abstract An important determinant of adaptive capacity to climate change and extreme events of drought and flood is institutional capital, or the assistance provided by government, civil society and private companies through programs and policy. A strong institutional capital provides adaptive strategies for agricultural producers and rural communities that not only assist in prevention of disaster, but recovery and rebuilding from disaster. But what are the components of strong institutional capital?

This paper reports research studies drawing on comparative institutional governance studies of agricultural producers in river basins in Canada, Argentina, and Colombia, in relation to climate change and extreme events of drought and flood. An assessment is made comparing and contrasting the different suites of institutional capital (organizations, policy, and programs) in relation to drought and flood and their impact on different types, sizes, and sensitivities of agricultural producers. This comparative analysis provides useful insights into what specific policies and programs build resilience and how this institutional capital is distributed amongst agricultural producers. Recommendations for improving institutional capital and its equitable distribution are made. This paper will be informative for policy makers, civil society organizations, and government.

M.A. Hurlbert (✉)
Department of Justice Studies, University of Regina, 3737 Wascana Parkway, Regina, Canada, S4S 0A2

Department of Sociology and Social Studies CL 235, University of Regina, 3737 Wascana Parkway, Regina, Canada, S4S 0A2
e-mail: Margot.Hurlbert@uregina.ca

P. Mussetta
Human, Social, and Environmental Sciences Institute, Scientific and Technological National Research Institute, Av. Ruiz Leal. Parque San Martin, 5500 Mendoza, Argentina

S. Turbay
Antropología Social y Etnología, Departamento de Antropología, Universidad de Antioquia, Sede de Investigación Universitaria, torre 1, oficina 213, Medellín, Colombia

W.L. Filho et al. (eds.), *Climate Change Adaptation, Resilience and Hazards*,
Climate Change Management, DOI 10.1007/978-3-319-39880-8_26

Keywords Adaptive governance • Resilience • Institutional capital • Climate change

Introduction

In the future increasing climate variability and frequency and duration of weather events such as drought and flood are anticipated in many areas of the world. The study areas of Canada, Argentina, and Colombia are anticipated to face such a future (IPCC 2014). Impacts of drought and flood events are already occurring (ibid.). It is necessary to prepare for these events, but questions remain outstanding on how best to do this. The area of governance or societal decision-making is an important area of focus in this regard.

This paper explores studies of institutional governance in relation to extreme events in three case study areas in Argentina, Canada, and Colombia. In each, an institutional governance assessment was conducted through review of secondary sources and semi-structured qualitative interviews exploring dimensions of adaptive governance. This paper reports on these findings and makes recommendations for a framework for improved resilient governance based on the findings.

Institutional Capital

An important determinant of the ability of a community to adapt to future climate change impacts and current climate variability is its institutional setting and the degree to which this setting facilitates or hinders the community's adaptive capacity (Willems and Baumert 2003). As the IPCC argues, nations with "well developed institutional systems are considered to have greater adaptive capacity," and accordingly, developed countries have a better "institutional capacity to help deal with risks associated with future climate change" (2001: 896, 897). Institutions contribute to the management of a community's assets, the community members' interrelationships, and their relationships with natural resources. Formal institutions, like government, non-profit and civil society organizations, and informal institutions such as social norms, values and contexts all contribute to the relationships of people to each other and to natural resources (Hurlbert and Diaz 2013; Moser and Satterthwaite 2008).

Studying the institutional context of adaptive capacity can be done by examining the institutions involved in governance. Governance encompasses laws, regulations, and organizations, as well as governmental policies and actions, domestic activities and networks of influence, including international market forces, the private sector and civil society (Demetropoulou et al. 2010: 341). It entails the interactions among structures, processes, rules, and traditions that determine how people in societies make decisions and share power, exercise responsibility and ensure accountability (Lebel et al. 2006; Raik and Decker 2007; Cundhill and Fabricius 2010: 14). Thus, governance involves institutions through which citizens and groups articulate their interests, exercise their legal rights, meet their legal obligations and mediate their differences (Armitage et al. 2009; Kiparsky et al. 2012).

A rich literature has developed regarding adaptive governance and specifically how the wider institutional context of governance can facilitate adaptation and improve adaptive capacity of communities. This adaptive capacity is especially important in responding to extreme weather events, drought, and flood events. The governance framework surrounding extreme events (constituted by such things as water allocation laws, programs and policies allowing preparation for drought, and income stabilization in the event of drought) plays an important role. How do we recognize a system of governance as adaptive? Within the adaptive capacity literature, several dimensions have been identified as important characteristics called institutional design principles, or features of governance systems that define an institutional system as adaptive. These include such things as "availability of information," "openness for experimentation," "flexibility," "learning," and others. The discussion in some cases is generic and applies to institutions in general (Gupta et al. 2010; Folke et al. 2005; Gunderson and Holling 2002; Olsson et al. 2006). The literature refers to a proper understanding of the complexities of the phenomenon of climate change, which include the requirements imposed by different sectors, various levels of government, and diverse stakeholders, as well as the uncertainties surrounding the long-term time-frame of climate change (Gupta et al. 2010, Frohlich and Knieling 2013; Cook et al. 2011). Table 26.1 below outlines these various dimensions.

Adaptive governance entails a more flexible, participatory, experimental, collaborative and learning-based design and approach to policy making to increase the adaptive capacity of institutions and sustainability of natural resources (Pahl-Wostl 2010). Adaptive governance shifts focus from rule-based, fixed organizations to a view of institutions as dynamic, flexible, pluralistic, and adaptive in order to cope with present and future uncertain climatic conditions and the limits of predictability (Carpenter and Gunderson 2001). Adaptive governance is then a means to the achievement of adaptive capacity (Cook et al. 2011). Within this comparative study, the items in bold print in Table 26.1 focus on: responsiveness, flexibility, participation and learning, capacity building, and equity. Each of these themes will be discussed in relation to the case studies.

Table 26.1 Institutional design principles of adaptive governance

	Institutional design principle of adaptive governance	Related principles/Sub-principles	Explanation
1	**Responsiveness**		The ability of governance networks, organizations and actors to respond appropriately and in a timely manner to climate variability, hazards and extreme events in a manner that accounts for ecosystem dynamics
		Robust and Flexible Process	Institutions and policy processes that continue to work satisfactorily when confronted with social and physical challenges but which at the same time are capable of changing
2	Learning and institutional memory		Past experiences must be remembered, learned from, and routines improved.
		Participation	Participation by non-state actors
		Collective choice arrangements	To enhance participation of those involved in making decisions about the system in how to adapt
		Monitor and evaluate	Institutional evaluation processes must monitor and evaluate policy experiences
3	Trust		Institutional patterns must exist to promote mutual respect and trust such that participants continue involvement in the process of governance.
4	**Capacity Building**	Information Leadership Resources	Rigorous up to date information, sufficient and reliable Leadership must exist to act as a catalyst to change; and Appropriate resources (financial, political, human) must be available for this change
5	**Equity**	Legitimate Accountable Fair	The governance regime must be perceived as legitimate and accountable, as well as fair in its process and impact such that there is an equal and fair (re distribution of risks, benefits and costs
6	Political support		Responding to climate change is a long-term policy challenge which requires solid political support for plans longer than election cycles

Based on Hurlbert and Diaz (2013)

Methodology

In three case study areas (Southern Alberta and Saskatchewan, Canada; Mendoza, Argentina; and Manizales, Colombia) a multi-level institutional analysis was conducted exploring the institutions of governance relating to climate change and extreme weather events. In all three of the study areas future climate change of increasing variability and more frequent and extreme weather events are anticipated (IPCC 2014). In Argentina and Canada drought was explored; in Canada flood as well; and in Colombia, excessive moisture resulting in landslides and flood. First an inventory of organizations was prepared. Organizations were selected that are involved in assisting rural agricultural producers and their communities respond to climate change and extreme weather events of drought and flood. Semi-structured interviews with producers and people involved in governance were then conducted. These interviews explored the dimensions of adaptive governance outlined in Table 26.1. This paper provides the results of this research.

Institutional Governance in Canada, Argentina, and Colombia

Canada

In the Canadian case, national strategies have not been renewed surrounding climate change and adaptation. However, in relation to disaster and events of drought and flood, a host of institutions respond to these extreme weather events and (although not framed specifically in relation to climate adaptation) assist rural agricultural producers adapt to drought and flood (see Hurlbert 2016; Hurlbert et al. 2015; Hurlbert 2013). At the national level, Canada has established a Hyogo Platform for Disaster Risk Reduction that helps coordinate efforts in relation to disaster planning and response, knowledge exchange and dissemination. A federal Disaster Assistance program provides funds to provinces in times of disaster and the province will manage the payment of the disaster funds to impacted individuals. Although flood insurance is not available, these funds assist homeowners in rebuilding. Agricultural producers must rely on business continuity insurance or crop insurance (if they have purchased it) to cover their farm losses.

A certain amount of flexibility exists in the governance regimes of the case study areas (Alberta and Saskatchewan) in relation to water management and droughts and floods. Both systems are predominantly government administered licensing systems, with licenses containing certain restrictions and qualifications. Alberta provides for a limited water transfer system (or water market) which has been utilized in times of drought to facilitate adaptive responses via water transfers to allow some agricultural producers to produce a crop while those transferring their interests receive payment for their foregone water (Hurlbert 2009).

Local watershed planning has existed in Saskatchewan and Alberta for some time engaging in source water protection planning and education of communities about water. These groups have facilitated climate change and drought planning with rural agricultural producers using scenarios and discussion groups to build strategies at a local level (East et al. 2012; Alberta Government 2013). Province-wide consultations contributed to the development of long term water strategies in both Alberta (Alberta Environment 2003, 2008) and Saskatchewan (SWSA 2012).

Local watershed groups also contribute to capacity building in their local communities through their activities of source water protection planning and drought planning. Saskatchewan groups also facilitate a federal government program aimed at encouraging environmental farm practices through "best management practices' that can be done with groups of agricultural producers, or single producers (GS 2011). Although local drought plans exist in Saskatchewan, a provincial plan does not; a provincial drought plan exists in Alberta, but often local plans do not.

In Canada, local communities are tasked with responding to emergencies (which include flood). This study found in Canada that rural communities studied did have various emergency plans and services in place, but that these plans didn't include climate change per se considerations within their plans or planning processes (Hurlbert et al. 2015), but did have climate variability events in the short term. The federal government has withdrawn its support through funding the education of local community members in emergency planning in the last decade.

Considerations of equity in relation to the impacts of the institutions of governance were noted in the study. Agricultural producers demographics are changing. Producers are getting older, farm sizes are increasing, and debt levels are rising. Small agricultural producers are decreasing in number with many supporting the farm with off farm income in the oil and gas industry. Indigenous people were absent from local watershed groups; studies have concluded these groups are marginalized and especially vulnerable to the impacts of climate change (Corkal et al. 2016).

Argentina

Argentinian created the Argentinean Carbon Fund by National Decree 1070/05 in order to maximize Clean Development Mechanism projects (Pochat et al. 2006). Further, a National Program for Rational Use of Energy and Energy Efficiency was created in 2005 that encourages the use of bio-ethanol and biodiesel (Pochat et al. 2006; United Nations 2011). These programs do not relate to climate change adaptation. A Climate Change Agency at the provincial level (Mendoza) has had a minimal involvement in adaptation and no role in mitigation (Hurlbert et al. 2015). For the past four years Mendoza has had a Declaration of Drought by the Governor; a process of "turno" or water rationing has been implemented by inspectors charged with managing the irrigation canals. This process is highly responsive in managing the reduction in stream flow, but not highly responsive or flexible in meeting differing crop needs of agricultural producers. Water delivery is completely dependent on supply.

Water rights are based on the principle of inherence, or ownership of property. Water can't be transferred or sold separately from the land to which it pertains. This system results in a rigid allocation system incapable of adapting quickly to changing agricultural practices.

In Mendoza, Argentina, flood is unheard of; any additional water coming down the river is thought of favourably. Extreme events of hail figure more highly into the risk psyche of residents. Response to weather-related disasters centers more on drought. After four years of drought declarations, practices have changed and adapted. Some agricultural producers have diversified into producing olive oil; large producers have expanded at the headwaters and if they have enough financial capital, access groundwater. The regulatory regime does not have pricing or quantity restrictions relating to groundwater.

A very deep institutional structure of participation in water governance exists, for those with water rights. The riverbed inspectors and canal associations and general users assembly are only open to water rights holders for participation. Those without water, dry land farmers and ranchers, do not have a voice as they don't participate.

A Master Plan for Mendoza River Basin was developed in 2010 to 2012. However, this plan was developed in respect of five river basins after only one focus group meeting. Departamento General de Irrigacion (DGI) developed a water plan "H2020" which sets out its plans surrounding water infrastructure upgrades, legal and institutional changes; no mention is made of climate change and adaptation (DGI 2015). In the spring of 2014 the government of Mendoza was embarking on integrated land use planning (Planning Law 8051) through public consultations. There is potential to address issues of climate change, decreasing water supply, increasing aridity. However, the plan would be subject to the inflexible water laws of Mendoza.

In Argentina there are marked differences among farmers in horticulture and viticulture. Small grape producers with traditional vineyards exist generally downstream. The horticulturalists in some cases are of Bolivian origin and resort to social and family networks to organize their production and successfully develop their agricultural activities. A web of medium and small-sized towns spread over these agricultural lands away from the central Metropolitan Area of the capital city of Mendoza (Montaña and Boninsegna 2016). Campesinos or "guarpes" live in the dry lands raising goats. One interviewee identified that these producers, "never had water, and now have even less." These poorer people (dry land farmers and small irrigator/ horticulturalists) are described as having strong social capital within their local communities, but having no bridging connections with institutions outside their local communities. As a result, these voices would be in the audience of current participatory processes like integrated land planning, not at the table. Infrastructure development (such as the building of the Portrerillos Dam) occurs often at their expense. These small producers can't access financial tools as they generally don't qualify (with all taxes and fees being up to date). Access to services such as sanitation and living assistance occurs with the municipalities.

Twenty years ago, producers were all relatively small or medium sized; now there is a significant group of large producers. Today the DGI (the main water governance institution) and its system of managing water meets the interests of the most wealthy and powerful regional producers, perpetuating its development model established in 1884 in producing agricultural crops through irrigation systems for virtual water export (the trade of agricultural commodities with embedded water) (Hurlbert et al. 2015). This is implemented through the use of the tool of forfeiture only in two areas where the most expensive lands are (Uco Valley and to the right margin of the Luhan River). In these areas the small farmers' properties were appropriated. However, in other areas forfeiture is unheard of.

Colombia

In the Colombian case, a National Development Plan "Prosperity for All 2010–2014" was developed to promote a cultural change and anticipate and cope with the adverse effects of climate change and variability; one year later a policy to implement this was developed (Departamento Nacional de Planeación 2011). This policy takes an inter-sectorial focus on ensuring development projects account for climate change.

The national government has just moved from a service-oriented emergency response system (The National System of Disaster Prevention and Care created in 1989) to a disaster-response system focusing on risk prevention (National System For Disaster Risk Management) because of the La Nina event of 2010–2011. In the study area, the Chinchiná River basin, a Risk Management Unit coordinates with non-governmental organizations, universities, emergency response groups and private businesses. This model was found to promote flexibility and participation of many different actors in the community of Manizales.

Because Manizales is located near an active volcano on steep slopes, there is a culture of risk in relation to the environment. Mudslides from intense rainfalls (made worse with deforestation and increased livestock production) are the predominant concern. A warning system is in place that monitors rain and issues alerts in the event of excess accumulations. Unfortunately rural areas are not incorporated.

Local communities are starting to plan for climate change and further, some initiatives have been undertaken by the Regional Autonomous Corporation of Caldas. The Colombian Coffee Growers' Federation (CCGF) and the National Coffee Research Centre (NCRC) are important institutions providing research and outreach information (occasionally funding) to rural agricultural coffee farmers in relation to climate change, variability and agricultural practices. The main adaptation strategy to the changing climate of the CCGF is the renovation of coffee plantations with a new variety resistant to coffee rust (*Hemilea vastatrix*).

In Manizales, Colombia, drought is not the issue. The issues experienced by local agricultural producers relate to weather impacting production and falling coffee prices. In the highlands, small coffee farmers can't afford chemical inputs to respond to pests and diseases generated from excess moisture. Issues of equity arise as these small producers must recover the cultural memory of agricultural practices previous to the Green Revolution (the development of a coffee monoculture) in order to adapt to the changing climate. These farms remain small employing traditional practices to control weeds, properly handle vegetative cover, trench to drain water, and hold plots of coffee at various different stages (some of the traditional practices). In the lowlands coffee farmers are adapting by moving to livestock and fruit crops. As a result, hundreds of workers are unemployed and populating the informal sector in the urban areas, many of them living in precarious areas next to river banks prone to mudslides during torrential rain.

In the study region capacity is an issue. While the professionals and actors involved in the Risk Management Unit are highly educated, the local agricultural producers are not: the average is 5.1 years of schooling (Grajales Quintero 2013). Capacity of these producers is built through the CCGF and NCRC (organizations which are essential in communicating out best agricultural practices).

Two water-planning activities are currently under way in the study region. A Wathershed Council is governed by the environmental authority of the government and developing a Management Plan for the Chinchina River basin. Another initiative is the *Pactos por la Cuenca* (Pacts of the Basin), which is a voluntary initiative for the recovery and conservation of the river bringing together industrial companies, government officials and members of civil society. Although too early to tell, it is questionable if issues of equity and adaptation can be addressed with these two processes.

Comparative Analysis

There are some interesting differences between the case study areas. First the organization of water governance is markedly different. Both Mendoza and southern Alberta and Saskatchewan have a highly institutionalized system of water allocation and an irrigation industry. In Colombia irrigated agriculture and the highly institutionalized water allocation system doesn't exist. While Canada's systems are predominantly based on government-allocated licenses (albeit Alberta has a small water market (in the South Saskatchewan River Basin), Mendoza's water is governed on the principle of inherence. Thus far, the government allocation of water has provided some flexibility in Canada (and the water market an added feature, only used sparing thus far). Interviewees perceived that the inherence principle had favourable implications for equity as it ensured that water rights stayed with the land to which they were allocated thus preventing an entity purchase of water rights and market dominance. However, an element of inflexibility exists in an inability to adjust water supply based on demand factors.

In the Canadian case more institutional response has occurred in relation to the extreme event of flood. There is disaster assistance for homeowners and agricultural producers have some ability to access insurance. In Colombia, the institutional support in relation to mudslides and floods is predominantly local and regional.

Some initiatives (basin planning and *Pactos por la Cuenca*) hold promises to be successful, but these processes will require comprehensive participation.

These processes in Colombia have similar replications in Canada and Argentina. In Canada local watershed organizations have been active in source water protection and environmental planning. To date, these groups have not had significant involvement in land use planning (which significantly impacts source water protection). Integrated land planning by way of community plans is currently being undertaken in many local communities in Canada. Argentina currently is embarked in integrated land use planning. In all three study areas there is potential to address some of the major challenges of climate change and increasing extreme events within a context of a major driver of vulnerability, development and changes to land use. In Canada issues of building in flood plains and intensive production causing pollution could potentially be discussed; in Argentina the expansion of the oasis in the headwaters at the expense of downstream users; in Colombia the integration of different users along the river, the impacts of pollution, a building in flood and landslides zones.

Another striking similarity in these regions is the growing inequity occurring in relation to agricultural producers. In Argentina and Canada bigger agricultural production units are emerging and smaller units are experiencing increasing vulnerability. In Colombia, changes in agricultural production are resulting in more migration to urban areas (often residing in places of vulnerability to mud slides) and greater food insecurity. Current governance structures in all three countries have not effectively address this dynamic. A new governance framework for inclusive resilience governance is required.

Recommendations and Proposed Framework

The comparison of governance institutional frameworks doesn't offer simple solutions for improved governance; context and drivers operating within each case study have to be contemplated. Simple recommendations such as creating a government-based licensing framework such as Canada would not bode well in Mendoza, Argentina where the water rights of inherence have existed for centuries. Further, the limited but favourable experience of Alberta's water market might not be possible in some places depending on the water resource and built infrastructure. In order to trade water the physical reallocation must be possible. In Colombia, these solutions have no application to coffee growers depending on rain and moisture conditions.

The similarities existing within the case studies set the stage for the development of resilient governance for disaster. These similarities are the current forums for consultation and discussion of development, land use, and impacts on water. As water is ubiquitous and interconnected with so many other social processes (community drinking water, agriculture, industrial development) the current initiatives occurring within each case study offer opportunity to develop a robust resilient governance system.

The blueprint for this robust system rests on the dimensions of adaptive governance outlined in Section "Institutional Capital" and the adoption of several governance practices: continuous information flow, ongoing consultation, and meaningful participation. Each of these will be discussed in turn.

Often the purposes of public participation in governance decision-making are grouped into processes that seek to increase legitimacy of the process of knowledge generation (performing a normative function); those that seek to integrate more sources of knowledge and greater capacity for problem solving (performing a substantive function); and those seeking to build collaborative relationships (an instrumental function) (Jones et al. 2009: 1181). A normative function is fulfilled by increasing the accountability of the knowledge in existence through the scrutiny of stakeholder participants discussing, analyzing and critiquing such knowledge; the substantive function requires a commitment to listening to the knowledge of stakeholders in order to select an appropriate solution; lastly the instrumental function would appear to require a process of consultation specifically aimed at resolving disagreements about knowledge, enhancing individual and social learning, and assisting collective decision making. These purposes have been expressed in different manners by different authors. Table 26.2 organizes these functions of public participation into the normative, substantive, and instrumental functions.

Table 26.2 Functions of public participation

Function	Normative	Substantive	Instrumental
Purpose (Jones et al. 2009: 1181)	To increase legitimacy of the process of knowledge generation	To integrate more sources of knowledge and greater capacity for problem solving	To build collaborative relationships
Uncertainty (Bijlsma et al. 2011: 54)	To manage normative uncertainty (different societal values and goals contributing to the upcoming decision)	To manage informational or substantive uncertainty (improving the information basis of decisions)	To achieve procedural justice or obtain information about local acceptance of the proposal through illustrating how the decision was derived
Goal (Jackson 2001: 140)	Informing or public education, testing reactions to planners' ideas through consultation,	Seeking ideas and alternative solutions	Collaborative shared decision making
Methods (Lynam et al. 2007)	Diagnostic and informing methods that extract knowledge, values, or preferences from a target group to understand local issues	Co-learning methods in which perspectives of all groups change as a result of the process	Co-management in which actors involved are learning and are included in the decision making process

Table 26.3 Citizen power and degrees of involvement

Ladder of citizen participation Arnstein (1969)	Spectrum of public involvement Dorcey et al. (1994)	Information flows
Increasing citizen power	Increasing levels of involvement	
Citizen control Delegated power Partnership	Ongoing involvement Seek consensus	Ongoing, iterative information flow
Placation Consultation	Test ideas Seek advice Define Issues Consult on relationships	Two way information flow
Informing Therapy Manipulation	Gather information and perspectives Educate Inform	One way information flow

The direction of information flow correlates with the purpose of involving public in decision-making. Information flows can be one way (where government provides information to people or people engage in an activity like a demonstration and provide information to government), two way (where an exchange of information takes place), or iterative (wherein a process of ongoing dialogue occurs between government and people) (Barreteau et al. 2010). As the information flows increase, so does the degree of citizen participation in decision-making processes. Dorcey et al. (1994) built a ladder of participation from bottom rungs of lowest participation (where manipulation and placation of the public occurs) to the highest (where citizens are engaged in actual decision making). Congruence can be seen between the idea of citizen participation or power and levels of participation in Table 26.3.

In order for the higher rungs of the ladder to be engaged, there must be high levels of trust and expertise. At the bottom end of the ladder there would be lower levels of trust and expertise. Shared trust among stakeholders is key in reducing conflict in public participation processes and studies have found early public involvement in decision making processes decreases the conflict between people (Mackenzie and Krogman 2005). Figure 26.1 combines the concepts of function (normative, substantive, and instrumental), expertise, trust, levels of participation, and communication flows.

We argue that at the top end of the matrix, social learning can occur. Social learning is learning in and with social groups through interaction (Argyris and Schön 1978; Siebenhuner 2008). This is a process of iterative reflection that occurs when we share our experiences, ideas and environments with others. Pahl-Wost (2006) defines social learning as the process of model development where actors develop "their" system and their own behaviour as group model building or scenario development. By merging resilience thinking and action research/learning, a new form of learning, anticipatory learning has emerged in the literature (Tschakert and Dietrich 2010). In resilience literature, learning includes

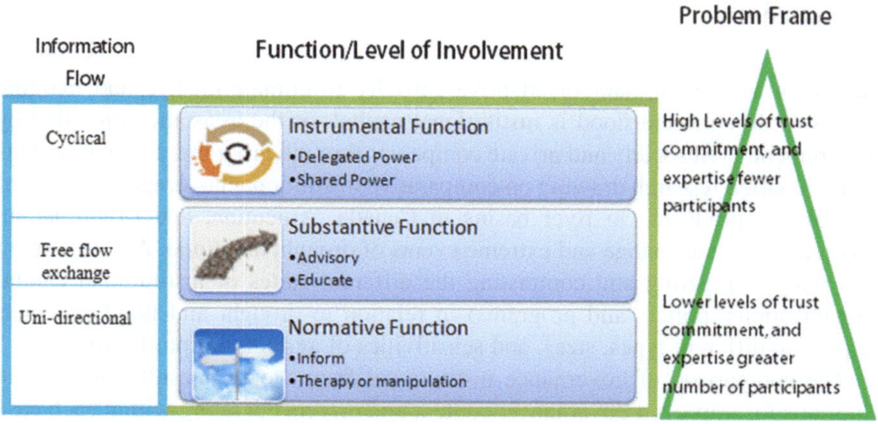

Fig. 26.1 Policy participation matrix (Hurlbert 2014)

incremental front – loop learning, spasmodic or profound back-loop learning, and transformational learning that can lead to innovative processes with high potential for transformability (Tschakert and Dietrich 2010). This learning occurs in small and fast cycles and larger and slower cycles over time. Action learning is a field of inquiry that has emerged with action research and future studies (Ramos 2006) that is a collaborative, democratic, and heuristic-reflexive process that links iterative questioning, anticipation, learning and creation with the ultimate purpose of crafting a different world (Stevenson 2002). As a critique of positivist research, it has emerged as a critical theory grounded in participatory worldview and is a particularly applicable to poor and vulnerable communities (Tschakert and Dietrich 2010).

Increasingly, iterative information flows and inclusive governance is recognized as necessary for both resilient planning for risk (Renn 2008) and engagement with uncertain science such as climate change (Weible et al. 2010; Darier et al. 1999). Including the public in the process of dialogue between and with policymakers and scientists is occurring (Carolan 2006) and the idea has emerged that formal science no longer speaks uniquely to determine policy, but that a plurality of knowledge claims exist (Lovebrand and Oberg 2005: 196). Learning is seen as an important part of this process and an integration of expert and non-expert or local knowledge (Weible et al. 2010, 522; Darier et al. 1999, 104) in a manner that co-produces knowledge at the front end of science (determining what is studied, how it is defined, and in what manner) and also the back end of science (determining how evidence is evaluated) (Morehouse et al. 2008: 280; Carolan 2006: 236). Buizer et al. (2011) sees this interface happening in relation to issues of scale such as where current events have long-term global consequences or the time-space compression issues. Others term this public ecology (Robertson and Hull 2003: 399). The reciprocating result is that the participation of the public facilitates the changes in values institutions and behaviours required to move people to recognition that variability and change, not stability is the more usual condition of life (Morehouse et al. 2008: 281).

Conclusions

An important determinant of adaptive capacity to climate change and extreme events of drought and flood is institutional capital, or the assistance provided by government, civil society and private companies through programs and policy. This paper reported research drawing on comparative institutional governance studies of agricultural producers in river basins in Canada, Argentina, and Colombia, in relation to climate change and extreme events of drought and flood. An assessment was made comparing and contrasting the different suites of institutional capital (organizations, policy, and programs) in relation to drought and flood and their impact on different types, sizes, and sensitivities of agricultural producers.

The comparison of governance institutional frameworks doesn't offer simple solutions for improved governance; context and drivers operating within each case study have to be contemplated. Simple recommendations such as creating a government-based licensing framework such as Canada would not bode well in Mendoza, Argentina where the water rights of inherence have existed for centuries. Further, the limited but favourable experience of Alberta's water market might not be possible in some places depending on the water resource and built infrastructure. In order to trade water the physical reallocation must be possible. In Colombia, these solutions have no application to coffee growers depending on rain and moisture conditions.

The review of these three case studies offered interesting insight into mechanisms for strengthening institutional capital and equity. These findings were contemplated within the literature describing what practices build a strong institutional capital. The practices of information flow, consultation, and meaningful participation are necessary.

This comparative analysis provided useful insights into what specific policies and programs build resilience and how this institutional capital is distributed amongst agricultural producers. In order to build more equitable distribution of institutional governance capital, it is first of all necessary to build iterative communication flows between the government, scientists and people. Within these iterative communication flows, consultation, and increased participation of people need and can occur. This participation must be meaningful. It must involve all people and allow for all opinions and ideas to be voiced. Through iterative communication flows issues and disagreements can be resolved.

Recommendations for improving institutional capital and its equitable distribution include the adoption of adaptive governance principles and these practices of iterative information flow, consultation, and meaningful participation. In order to establish this framework of resilient governance, government, policy makers, people, stakeholders and civil society groups must all engage in these practices.

References

Alberta Environment (2003) Water for Life. Alberta's strategy for sustainability. November, 2003. ISBN No. 0-7785-3058-2 Pub. No. 1/955. Available at: http://environment.gov.ab.ca/info/library/6190.pdf. Accessed 30 Jan 2015

Alberta Environment (2008) Water for life: a renewal. ISBN 978-0-7785-7670-9 November, 2008. Available at: http://environment.gov.ab.ca/info/library/8035.pdf Accessed 30 Jan 2015. Accessed 31 Mar 2015

Alberta Government (2013) Draft South Saskatchewan Regional Plan 2014-2024. Available at: https://landuse.alberta.ca/LandUse%20Documents/SSRP%20Draft%20SSRP%202014-2024_2013-10-10.pdf

Argyris C, Schön D (1978) Organizational learning: a theory of action perspective. Addison-Wesley, Reading, UK

Armitage DR, Plummer R, Berkes F, Arthur RI, Charles AT, Davidson-Hunt IJ, Diduck AP, Doubleday NC, Johnson DS, Marschke M, McConney P, Pinkerton EW, Wollenburg EK (2009) Adaptive Co-management for Social-ecological Complexity. Front Ecol Environ 7 (2):95–102

Arnstein SR (1969) A ladder of citizen participation. J Am Inst Plann 35(4):216–224

Barreteau O, Bots PWG, Daniell KA (2010) A framework for clarifying "Participation" in Participatory research to prevent its rejection for the wrong reasons. Ecol Soc 15(2):1–22

Bijlsma RM, Bots PWG, Wolters HA, Hoekstra AY (2011) An empirical analysis of stakeholders' influence on policy development: the role of uncertainty handling. Ecol Soc 16(1):51–67

Buizer M, Arts B, Kok K (2011) Governance, scale, and the environment: the importance of recognizing knowledge claims in transdisciplinary arenas [online]. Ecol Soc 16(1):21, http://www.ecologyandsociety.org/vol16/iss1/art21/

Carolan MS (2006) Sustainable agriculture, science and the coproduction of 'expert' knowledge: The value of interactional expertise. Local Environ 11(4):421–431

Carpenter SR, Gunderson LH (2001) Coping with collapse: ecological and social dynamics in ecosystem management. BioScience 51:451–57

Cook J, Freeman S, Levine E, Hill M (2011) Shifting course: climate adaptation for water management institutions. World Wildlife Fund, Washington. Available at www.adaptiveinstitutions.org

Corkal D, Morito B, Rojas A (2016). Values analysis as a decision-support tool to manage vulnerability and adaptation to drought. In: Diaz H, Hurlbert M, Warren J (eds) Vulnerability and adaptation to drought: the Canadian Prairies and South America. University of Calgary Press, Calgary, AB

Cundill G, Fabricius C (2010) Monitoring the Governance dimension of natural resource comanagement. Ecol Soc 15(1). Available at: www.ecologyand society.org/vol15/iss1/art15/

Darier E, Gough C, De Marchi B, Funtowicz S, Grove-White R, Kitchener D, Guimarães Pereira Â, Shackley S, Wynne B (1999) Between democracy and expertise? Citizens' participation and environmental integrated assessment in Venice (Italy) and St. Helens (UK). J Environ Policy Plann 1:103–120

Departamento Nacional de Planeacion (2011) Estrategia institucional para la articulacion de politicas y acciones en material de cambio climatico en Columbia. Documento CONPES 3700. Departmento Nacional de Planeacion. Available at: http://oab.ambientebogota.gov.co/es/con-la-comunidad/ES/conpres-3700-estrategia-institucional-para-la-articulacion-de-politicas-y-acciones-en-materia-de-cambio-climatico-en-colombia. Accessed 27 Nov 2014

Demetropoulou L, Nikolaidis N, Papadoulakis V, Tsakiris K, Koussouris T, Kalogerakis N, Koukaras K (2010) Water framework directive implementation in Greece: introducing participation in water Governance—The case of the Evrotas river basin management plan. Environ Policy Gov 20:336–49

DGI (2015) Plan H2020. Available at: www.aqua.gov.ar/2020/. Accessed 21 Jan 2015

Dorcey A, Doney L, Rueggeberg H (1994) Public involvement in government decision-making: choosing the right model (Victoria, Round Table on the Enviornment and the Economy)
East V, Pittman J, McRae K, Soulodre E (2012) Drought and excessive moisture pre-paredness plan. Upper Souris River Watershed. Available at: http://www.parc.ca/rac/fileManagement/upload/Upper_Souris_DEM_plan_July%209_2012.pdf
Folke C, Hahn T, Olsson P, Norberg J (2005) Adaptive governance of social-ecological systems. Annu Rev Environ Resour 30:411–73
Frohlich J, Knieling J (2013) Conceptualizing climate change. In: Knieling J, Leal W (eds) Climate change governance. Hamburg University, Springer, pp 9–26
Grajales Quintero A (2013) Desarrollo de una metodología para el diagnóstico rural: la pobreza rural en el Departamento de Caldas. Tesis doctoral en Ciencias Agrarias. Universidad de Caldas, Manizales
Gunderson LH, Holling CS (eds) (2002) Panarchy: understanding transformations in human and natural systems. Island Press, Washington
(GS) Government of Saskatchewan (2011) Farm and Ranch Water Infrastructure. Available at: http://www.agriculture.gov.sk.ca/FRWIP_2009. Accessed 29 Sep 2011
Gupta J et al (2010) The adaptive capacity wheel; a method to assess the inherent characteristics of institutions to enable the adaptive capacity of society. Environ Sci Policy 13:459–471
Hurlbert M (2009) The adaptation of water law to climate change. Int J Clim Change Strat Manag 1(3):230–240
Hurlbert M (2013) Learning, participation, and adaptation: exploring agri-environmental programs. J Environ Plann Manag. doi:10.1080/09640568.2013.847823
Hurlbert M (2014) Evaluating public consultation in nuclear energy: the importance of problem structuring and scale. Int J Energy Sector Manag 8(1):56–75
Hurlbert M (2016) Water governance in the Prairie Provinces. In: Diaz H, Hurlbert M, Warren J (eds) Vulnerability and adaptation to drought: the Canadian Prairies and South America. University of Calgary Press, Calgary, AB
Hurlbert M, Diaz H (2013) Water Governance in Chile and Canada- A comparison of adaptive characteristics. Ecol Soc 18(4):61–76
Hurlbert M, Mussetta P, Ivars J (2015) A multi-level analysis and comparison of climate change policies in Argentina and Canada. In: Leale W (ed) Policy and climate change, vol 2. Springer, Berlin, Heidelberg
IPCC (2001) Climate Change 2001: impacts, adaptation, and vulnerability technical summary. A Report of Working Group II of the Intergovernmental Panel on Climate Change 2001. Intergovernmental Panel on Climate Change, WMO and UNEP, Geneva
IPCC (2014) Summary for policymakers. In Edenhofer O, Pichs-Madruga R, Sokona Y, Farahani E, Kadner S, Seyboth K, Adler A, Baum I, Brunner S, Eickemeier P, Kriemann B, Savolainen J, Schlömer S, von Stechow C, Zwickel T, Minx JC (eds) Climate change 2014, Mitigation of climate change. Contribution of Working Group III to the Fifth Assessment Report of the Intergovernmental Panel on Climate Change. Cambridge University Press, Cambridge, United Kingdom and New York, NY, USA
Jackson LS (2001) Contemporary public involvement: toward a strategic approach. Local Environ 6(2):135–147
Jones N, Perez P, Measham TG, Kelly GJ, D'Aquino P, Daniell KA, Dray A, Ferrand N (2009) Evaluating participatory modeling: Developing a framework for cross-case analysis. Socio-Economics and the Environment in Discussion (SEED). CSIRO Working Paper Series, 44 (6), 1180–1195
Kiparsky M, Milman A, Vicuna S (2012) Climate and water: knowledge of impacts to actions on adaptation. Annu Rev Environ Resour 37:163–94
Lebel L, Anderies JM, Cambell C, Folke S, Hatfield-Dodds S, Hughes TP, Wilson J (2006) Governance and the capacity to manage resilience in regional social-ecological systems. Ecol Soc 11(1):19

Lovebrand E, Oberg G (2005) Comment on "How science makes environmental controversies worse" by Daniel Sarewitz, Environmental Science and Policy, 7, 385–403 and "When Scientists politicise science: making sense of the controversy over The Skeptical Environmentalist" by Roger A. Pielke Jr., Environmental Science and Policy, 7, 405–417. Environ Sci Policy 8:195–197

Lynam T, De Jong W, Sheil D, Kusumanto T, Evans K (2007) A review of tools for incorporating community knowledge, preferences, and values into decision making in natural resources management [online]. Ecol Soc 12(1):5, http://www.ecologyandsociety.org/vol12/iss1/art5/

Mackenzie J, Krogman N (2005) Public involvement processes, conflict, and challenges forrural residents near intensive hog farms. Local Environment. 10(5): 513-524. Environ Sci Policy 8:195–197

Montaña E, Boninsegna JA (2016) Drought in the oases of Central Western Argentina. Chapter 14. In: Diaz H, Warren J, Hurlbert M (eds) Vulnerability and adaptation to drought: the Canadian Prairies and South America. University of Calgary Press, Calgary, AB, pp 327–348, ISSN 1919-7144

Morehouse B, Ferguson DB, Owen G, Browning-Aiken A, Wong-Gonzalez P, Pineda N, Varady R (2008) Science and Socio-ecological resilience: examples from the Arizona-Sonora Border. Environ Sci Policy 11(3):272–284

Moser C, Satterthwaite D (2008) "Towards pro-poor adaptation to climate change in the urban centres of low-and middle-income countries" Climate Change and Cities Discussion Paper 3, in Climate change and Cities Discussion Paper 3. Global Urban Research Centre, IIED, Washington, DC

Olsson P, Gunderson LH, Carpenter SR, Ryan P, Lebel L, Folke C, Holling CS (2006) Shooting the rapids: navigating transitions to adaptive governance of socio-ecological systems. Ecol Soc 11(1):1–18

Pahl-Wost C (2006) The Importance of Social Learning in Restoring the Multifunctionality of Rivers and Floodplains. Ecol Soc 11(1):10–24

Pahl-Wostl C (2010) Water governance in times of change. Environ Sci Policy 13:567–70

Pochat V, Natenzon CE, Murgida AM (2006) Argentina, Country case study on domestic policy frameworks for adaptation in the water sector. Background information for presentation given by Victor Pochat at the Annex I Expert Group Seminar in Conjunction with the OECD Global Forum on Sustainable Development, March 28, 2006. Available at: www.oecd. orgéenvironmentéccé36448827.pdf. Accessed 30 Sep 2014

Raik DB, Decker DJ (2007) A multisector framework for assessing community-based forest management: lessons from Madagascar. Ecol Soc 12(1): 14. Available online: http://www. ecologyandsociety.org/ vol12/iss1/art14/

Ramos JM (2006) Action research and futures studies. Futures 31:639–641

Renn O (2008) Risk governance: coping with uncertainty in a complex world. Earthscan, London

Robertson DP, Hull RB (2003) Public ecology: an environmental science and policy for global society. Environ Sci Policy 6:399–410

Siebenhuner B (2008) Learning in international organizations in global environmental governance. Glob Environ Pol 8(4):92–116

Stevenson T (2002) Anticipatory action learning: conversations about the future. Futures 34:417–425

SWSA (Saskatchewan Water Security Agency) (2012) 25 Year Saskatchewan Water Security Plan. Queen's Printer, Regina. Available at: http://www.gov.sk.ca/adx/aspx/adxGetMedia. aspx?mediaId=1798&PN=Shared. Accessed 17 Sep 2014

Tschakert P, Dietrich KA (2010) Anticipatory learning for climate change adaptation and resilience. Ecol Soc 15(2):11–34

United Nations (2011) Ad Hoc Working Group on Long-term Cooperative Action under the Convention. Compilation of information on nationally appropriate mitigation actions to be implemented by Parties not included in Annex I to the Convention. Framework Convention on Climate change

Weible CM, Pattison A, Sabatier PA (2010) Harnessing expert-based information for learning and the sustainable management of complex socio-ecological systems. Environl Sci Policy 13:522–534

Willems S, Baumert K (2003) Institutional Capacity and climate actions: organization for economic co-operation and development. Environment Directorate. International Energy Agency, Publication. com/env/epok/iea/slt(200305)

Chapter 27
Mobilizing Private Sector Funds for Climate Change Adaptation: Nordic Climate Facility (NCF) as a Case Study

Kari Haemekoski and Heli Sinkko

Abstract Although estimates of the global climate finance flows vary, bulk of the funding has been targeted at climate change mitigation and only a fraction to adaptation to the impacts of climate change. Mitigation actions have also received more private sector funds as compared to adaptation. The current financial flows targeted at adaptation fall short of the current estimates for funding needs and call for increased private sector involvement and finance. This article discusses the mobilisation of private sector funds for climate change adaptation using experiences from the Nordic Climate Facility (NCF), a competitive partial grant facility, as a case study. Since its launch in 2009 NCF has approved financing for 49 climate change mitigation and adaptation projects in selected developing countries in Asia, Africa and Latin America. The aim of this paper is to showcase, using selected NCF projects as examples, how adapting to climate change can also have business linkages, especially when combined with mitigation activities, and how public sector funding can be used to leverage private sector funds through local business development for climate change adaptation.

Keywords Climate change • Adaptation • Climate finance • Private sector • Business

Introduction

For a long time adaptation to climate change was absent from the international climate change debate. Calling attention on the need to adapt to climate change was regarded almost as admitting defeat, which meant that mitigation actions dominated

---

Any views or opinions expressed are those of the authors acting in their personal capacity and do not necessarily reflect those of the Nordic Environment Finance Corporation or of the Nordic Development Fund.

K. Haemekoski, MSc • H. Sinkko, MSSc (✉)
Nordic Environment Finance Corporation (NEFCO), Fabianinkatu 34, P.O. Box 241, 00171 Helsinki, Finland
e-mail: heli.sinkko@nefco.fi

W.L. Filho et al. (eds.), *Climate Change Adaptation, Resilience and Hazards*, Climate Change Management, DOI 10.1007/978-3-319-39880-8_27

the international discussions (Tanner and Horn-Phathanothai 2014). Developing countries brought adaptation on the agenda of international climate debate in late 1990s and today there is international consensus that adaptation is necessary especially in the developing world which is deemed most vulnerable to the impacts of climate change.

Despite of the acknowledged need for adaptation actions, the volume of climate finance targeted at adaptation has not yet reached levels that effective adaptation is estimated to require. Parties to the United Nations Framework Convention on Climate Change (UNFCCC) have committed to set the limit of global warming to 2 °C but even the 2 °C warming is likely to bear significant adaptation costs in particular for developing countries that are most vulnerable to climate change. The World Bank's Economics of Adaptation to Climate Change (EACC) study estimated for example that the cost of adapting to a 2 °C warmer world by 2050 comes with a price tag of 70–100 billion US dollars per year (Narain et al. 2011). To meet the adaptation costs entails thus massive financial resources for developing countries.

Any estimate of global climate finance flows depends on the very definition of climate finance. While there is no uniform definition for "climate finance", a key requirement is that climate finance has the objective of either greenhouse gas mitigation or adaptation, i.e. reduction of vulnerability to climate change (Gupta et al. 2014). For UNFCCC "Climate finance aims at reducing emissions, and enhancing sinks of greenhouse gases and aims at reducing vulnerability of, and maintaining and increasing the resilience of, human and ecological systems to negative climate impacts" (UNFCCC 2014). In the fifth assessment report of the Inter-governmental Panel of Climate Change (IPCC) climate finance has been defined more concretely as capital flows that are targeted at low-carbon or climate resilient development. The sources of climate finance vary but can derive either from public or private, international or domestic sources (Gupta et al. 2014).

The current climate finance available for adaptation does not yet quite fill the funding gap. The latest Global Landscape of Climate Finance 2014 report published by the Climate Policy Initiative (CPI) estimated that in 2013 the total climate finance flows amounted to 331 billion US dollars. Out of this, 302 billion dollars were utilized for mitigation, 25 billion for adaptation, and only four billion for projects with multiple, i.e. both mitigation and adaptation objectives. In addition, according to the CPI report, the entire 25 billion US dollars directed at adaptation activities originated exclusively from public sector sources (Buchner et al. 2014).

To estimate the financing available for adaptation is a challenge, which also the CPI's report rightly acknowledges. Data sources are in general unreliable but CPI estimates that adaptation funding by development finance institutions and multilateral development banks has without doubt leveraged some private sector funding because of the dual relationship between adaptation and development (Buchner et al. 2014). The decision on how to secure funding to strengthen especially developing countries' resilience will be one of the key topics in the upcoming COP meeting, to be held in Paris in December 2015. One of the key mechanisms will be the Green Climate Fund (GCF). While not yet operational, the GCF with current pledges of US 10.2 billion[1] is expected to provide considerable adaptation funding. GCF's board of directors has already decided that 50 % of the funding will go to adaptation and 50 % of the funding will be allocated to Least Developed Countries (LDCs) (Green Climate Fund 2014, 2015). It has however already now become evident that in order to fill the "funding gap", private sector's involvement will be paramount (UNEP 2015).

Approach and Methodology

The object of this case study is the Nordic Climate Facility and the purpose is, using selected NCF projects as material, to discuss tested ways to increase the private sector's involvement in adaptation efforts. The purpose of this paper is twofold: it showcases how public funding can be used to leverage private sector funding for adaptation efforts especially when combined with mitigation actions. Secondly, the paper argues that interventions with adaptation benefits to climate change can create business opportunities also in low-income countries.

The analysis is based on a qualitative assessment of selected NCF projects representing three project categories in use in the NCF—mitigation, adaptation and combination. The analysis has focused on the stated objectives and achieved results of the projects. A key question guiding the analysis has been to what extent has the project contributed to climate change adaptation and private sector development. In addition to the presentation and analysis of selected projects, the paper also presents a simple quantitative analysis of the sources and amounts of co-financing across the project categories and calls for proposals. The analysis and results presented in this paper are based on progress reports submitted by the projects selected to be showcased in this paper and on a simple quantitative analysis of co-financing and grant amounts for the whole NCF portfolio.

The following sections will present a quantitative analysis of co-financing for the whole NCF portfolio after which some projects the selected NCF projects representing the three project categories will be presented and analysed. Finally, the last section contains conclusions and implications for other similar initiatives. There are certain limitations to the conclusions that will be presented. First of all, as in all case studies, the results are not easily generalized or replicated. Secondly, the NCF has targeted applicants originating from the Nordic countries and replication of a similar programme in another geographical context may generate different outcomes.

[1]As of June 2015.

Nordic Climate Facility (NCF)

The Nordic Climate Facility (NCF) is a competitive financing mechanism that provides grants with co-financing requirements to encourage and promote innovations in areas susceptible to climate change in low-income countries. NCF is financed by the Nordic Development Fund (NDF) and administered by the Nordic Environment Finance Corporation (NEFCO). NDF is the joint development finance institution of the five Nordic countries whereas NEFCO is an international financial institution established by the Nordic countries that finances green growth investments and projects primarily in Russia, Ukraine and Belarus as well as climate projects across the world. Both institutions have their headquarters in Helsinki, Finland.

The NCF is targeted at private as well as public Nordic organizations with relevant experience that in partnership with local partners and, in some cases, with other partners undertake to implement projects in one or more developing countries eligible for NDF funding to generate both climate and development benefits. The main objectives of NCF are to: (i) facilitate the exchange of technology, knowledge, know-how and innovative ideas between the Nordic countries and low-income countries in the field of climate change; (ii) increase the low-income countries' capacity to mitigate and adapt to climate change; and (iii) to contribute to sustainable development and the reduction of poverty. NCF's purpose and objective is also to encourage testing of concrete concepts relating to climate change and, especially, to facilitate partnerships.

To date, NCF has launched five calls for proposals for innovative project concepts. Each call has focused on a specific theme in relation to climate change and development, and applicants with their partners are free to propose any relevant project that fits in to the theme. The first call, NCF1, launched in 2009 focused on *water resources* and *energy efficiency*. The second call, NCF2, launched in 2010 had two focus themes: *renewable energy* and *urban adaptation*. Since the third call, NCF3, launched in 2011 with the theme of *Innovative low-cost climate solutions with focus on local business development* more focus has been shifted towards various direct and indirect ways of supporting private sector development, promoting economic activity and facilitating private sector's participation in climate-related development efforts. The fourth call, NCF4, launched in 2013 looked for *Inclusive green growth projects contributing to private sector development*. The currently ongoing NCF5 has invited project proposals under the theme of *Climate Resilience in Urban and Private Sector Contexts*. The call also increased focus on gender aspects.

To date, 49 projects have been selected via two phase evaluation process out of altogether 580 applications for funding through the four calls for proposals and further projects are expected to be included in the project pipeline from the fifth call. At the time of writing, the cumulative NDF funding for five calls amounted to 19.2 million euros and, when also co-financing is accounted for, the total value of the programme was approximately 32.6 million euros. One project can receive a grant between 250,000 and 500,000 € and the maximum support period is in principle from 2 to 2.5 years but in practise the average implementation period of projects can be slightly longer. NCF is a results-based instrument, and disbursements are linked to achieved milestones and realized co-financing. Only an agreed percentage of the total incurred and audited costs can be covered by NCF.

As the design feature of NCF is to support, as cost efficiently as possible, small-scale projects, longer term monitoring is currently not required or possible for NCF projects. The main results and impacts of individual projects are, however, captured in the final project reporting. In addition, the evaluation of the Nordic Climate Facility conducted in 2013 concluded that NCF has international added value as a quite rare mechanism combining innovation, leverage and partnership (Sigvaldsen et al. 2013). A recent independent assessment of selected NCF projects states that: "*All* [NCF] *projects also have a strong development agenda, highlighting the need not to separate development and climate projects into silos of their own, but rather merge these two financing opportunities in developing countries. Development co-benefits range from reducing the amount of water-borne diseases in communities, which have a direct effect on household income levels and children's school attendance. The projects also result in increased local level employment and new types of income generating activities, to name a few.*" (Brüning and Hamro-Drotz 2015).

Private Sector's Role in Adaptation: The NCF Case Study

As discussed above, any estimates of total funding directed at climate change adaptation, whether originating from the public or the private sector, are challenged by the very nature of those activities since most forms of adaptation interventions are difficult to distinguish from standard development interventions. The dual relationship is further complicated by the fact that the spectrum of adaptation activities varies from traditional development interventions to activities that are targeted to address a specific climate change impact and that do not coincide with any type of activity usually understood as development (McGray et al. 2007). In addition, adaptation activities usually focus on the technical aspects of climate impacts and fail to address the reasons for which people are vulnerable to climate change in the first place (Eriksen et al. 2015). The debate related to the dual relationship between adaptation and development is not new and falls outside the scope of this paper, but it should be remembered that the difficulties in having a concrete operational definition of adaptation and its relationship to development means that also the amount of funding targeted at adaptation activities is, at best, an informed estimate (Narain et al. 2011; see also McGray et al. 2007).

Contrary to adaptation actions and impacts, assessment of mitigation impacts is fairly straightforward. Methodologies to define baseline emissions as well as monitoring of greenhouse gas emissions reductions are well developed and considerable amount of work to measure mitigation impacts has been conducted especially via the development of detailed Clean Development Mechanism (CDM) methodologies (UNFCCC 2015). In the discussion on climate change and development the relationship between mitigation and development remains often overlooked. Developing countries are not big emitters and mitigation actions traditionally require large investments in industry and infrastructure that concern more the developed world. The truth lies somewhere in between since for example Africa's soil carbon is estimated to form a considerable share of world's carbon stock. What comes to mitigation and development, there can also be significant development co-benefits from some mitigation actions. Clean cookstoves are a typical example of such actions as they reduce greenhouse gas emissions but also deliver strong development impacts such as improved indoor air quality and reduced household energy costs (see for example Tanner and Horn-Phathanothai 2014).

As mentioned above, NCF projects are both climate and development projects. Projects are categorised either as mitigation, adaptation or the combination of both. The portfolio of 49 projects includes 14 pure adaptation projects, 16 projects that are classified as combination projects and 18 pure mitigation projects. In many projects the climate impacts do not always follow the labelling given but, as will be discussed below, mitigation projects often generate clear adaptation benefits and vice-versa. In the first and second calls for proposals most adaptation projects were oriented towards capacity building or feasibility studies, which sometimes were combined with small-scale yet tangible pilot activities. Since NCF3, only concrete projects or projects that had feasible direct linkages to subsequent concrete activities, have been granted funding.

During selection phases, scoring of project proposals has followed the same principles for both categories, i.e. both types could score equally well in evaluation. Assessment of adaptation impacts is, however, still somewhat more challenging especially in quantitative terms and due to the higher variety of project types as compared to mitigation projects, which has been clearly noted in the evaluation of projects proposals as well as in the subsequent monitoring of NCF projects.[2] Despite of these challenges and the competitive selection method of projects, the NCF portfolio is balanced between mitigation and adaptation projects.

Adaptation Funding in the NCF Portfolio

The Table **27.1** shows the division of funding sources across adaptation and combination project categories for four NCF call for proposals. The total value of the current NCF portfolio is EUR 32.6 million of which NCF grant funding covers 59 %. The value is calculated on the basis of actual total costs for 19 fully

[2]For more information, please see NCF Annual Review for 2014 available at http://www.ndf.fi/project/nordic-climate-facility-ncf

Table 27.1 Division of funding sources across project categories in NCF adaptation and combination projects

	NCF Call	Total project budget (EUR million)	Total co-finance (excl. NCF grants, EUR million)	% of total project budget	Private sector co-finance (EUR million)	% of total project budget	% of total co-finance (excl. NCF grant)
Adaptation projects	NCF1	3.20	1.43	45	0.74	23	52
	NCF2	3.04	0.79	26	0.16	5	20
	NCF3	1.49	0.37	25	0.27	18	74
	NCF4	1.00	0.50	50	–	0	0
		8.72	**3.09**	**35**	**1.17**	**13**	**38**
Combination projects	NCF1	3.20	1.33	42	1.29	40	97
	NCF2	0.51	0.17	33	0.13	25	75
	NCF3	4.09	1.49	37	1.49	37	100
	NCF4	3.45	1.36	39	0.75	22	55
		11.25	**4.36**	**39**	**3.66**	**33**	**84**

completed projects and on the budgeted total value of on-going and recently approved projects. NCF also allows limited in-kind contributions, as long as they are justified and recorded e.g. using timesheets to assess the actual value of all contributions.

The total amount of co-financing originating from private and public sectors alike covers 39 % of the costs in combination projects and slightly less, 35 %, in adaptation only projects. The share of private sector co-finance is 33 % in combination projects and somewhat less, 13 %, in pure adaptation projects. These findings are in line with the common perception that there is less private sector involvement in adaptation only projects. It is, however, important to note that in only four out of 30 NCF adaptation/combination projects there is no private sector finance at all. The Fig. 27.1 shows also which proportion of the co-financing (NCF grants excluded) has originated from the private sector.

As mentioned above, the estimated funding of 25 billion US dollars targeted at adaptation activities originated almost exclusively from the public sector (Buchner et al. 2014). While this is not the whole truth due to the challenges discussed in the introduction to this paper, there is growing awareness that in order to reach the funding target of 100 billion US dollars per year requires more active participation of the private sector in adaptation efforts. Based on experiences from the Nordic Climate Facility, arguments in favour of involving the private sector in adaptation activities should not focus on the private sector only as a source of funding but as a key actor.

To back this argument, an OECD working paper published in 2011 called for more attention to private sector's role in adaptation to climate change (Agrawala et al. 2011). Despite of its focus on European companies, the paper concluded that the private sector is crucial for adaptation not only as a provider of financing but

Fig. 27.1 Share of private sector funding in adaptation and combination projects

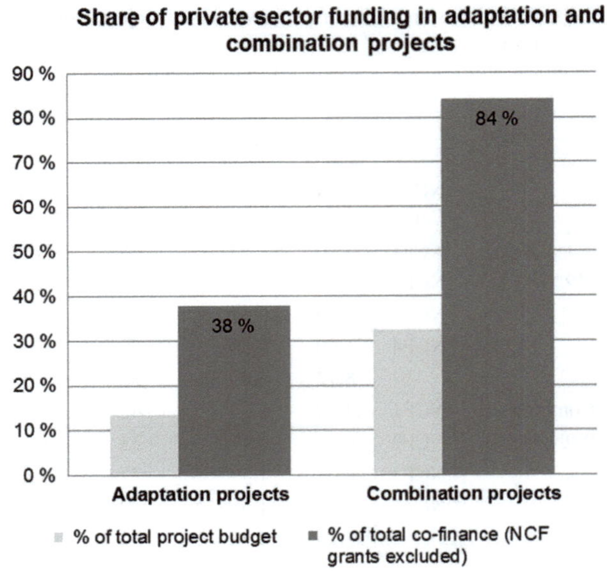

because a country's success in adapting to climate change depends on the capacity of its private sector in responding to climate change impacts. Private sector's involvement in adaptation to climate change should be therefore seen as an opportunity for the private sector to generate new business and new sources of income (Agrawala et al. 2011).

In this context it is noteworthy to point out that the economic growth rate in many low-income countries has exceeded that of high-income countries (World Bank 2015). Growth in these countries was 4.9 % in 2013 while being 1.4 % for high-income countries. As the economic growth is forecasted to follow the same path in numerous low-income countries and translate into positive development outcomes, further discussion is needed on how linkages with future adaptation actions could be strengthened and synergies identified.

Mitigation Projects with Adaptation Impacts

There are several examples of NCF projects classified as mitigation only projects due to their primary focus but which also have adaptation co-benefits. An example of this kind of project is the *Scaling up low carbon household water purification technologies in the Mekong Sub Region* that is being implemented in Cambodia and Laos. In this case, NCF is mainly supporting the scaling-up and new product development to ensure the long term sustainability of the business, i.e. sufficiently high production volumes, increased coverage for sales network, and to develop Gold Standard emission reductions project to further support the activities via carbon credits. Local companies are providing 85 % of the total budget of some 3.5 million euros.

The main target of the project is to achieve sales of 154,000 Ceramic Water Purifiers (CWP) that should lead to a total of 200,000 tonnes of monitored CO_{2e} emission reductions by 2015 as mitigation benefits. At the same time, the project has generated adaptation impacts since households get access to purified drinking water in the likely case that climate change has further adverse impacts on the availability of drinking water. As development co-benefits, household members encounter better health due to reduced indoor air pollution and unsanitary drinking water. Households will also save time and money, as they do not have to obtain wood for boiling water, either. While not subject to monitoring as per NCF Grant Agreement requirements for this project, reduced usage of fuel wood from unsustainable sources for water boiling will undoubtedly reduce deforestation with linkages to various positive adaptation impacts, such as protection of land from floods.

Many efficient cook stove and sustainable charcoal projects, also present in the NCF portfolio, are similar to the project described above: they are typically designed to generate mainly mitigation impacts (e.g. as carbon finance projects) as well as direct development and health impacts but they also relate to adaptation co-benefits.

Adaptation Projects with Business Linkages

This section will take a look at three adaptation projects, where NCF funding has worked as seed money in support of local business development while at the same time increasing local communities' resilience and adaptive capacity. The NCF portfolio includes now 14 "adaptation only" projects out of which 10 have direct private sector and/or business linkages many of which are agricultural projects. The business and private sector linkages with adaptation are typically related to increased yields in agricultural sector and increased resilience of current businesses, i.e. also closely linked with development co-benefits. Climate change can have direct impacts especially on agricultural productivity and hence food security through a diversity of effects. Climate change is likely to have a direct impact on water resources causing more frequent droughts or floods or changing the rainfall patterns altogether as discussed recently e.g. by Tanner and Horn-Phathanothai (2014). The United Nations' Food and Agriculture Organization (FAO) has estimated that 40 % of world population, mostly poor, are reliant on agriculture on arid and semi-arid lands (ASALs). A key question with regard to food security and development, especially in Africa but also elsewhere, is therefore related to the question on how effectively agricultural systems are able to adapt to climate change for example through innovative ways of collecting and using scarce water resources or through the use of more climate resilient crops.

In addition to increasing the resilience of rural populations, adaptation actions in the agricultural sector can also create additional income and form a basis for commercial, albeit small-scale, business activity. The *Providing Assistance for Design and Management of Appropriate Water Harvesting Technologies in Arid Lands* project, completed in 2012, provided assistance for the design and management of water harvesting technologies in arid lands of Kenya in collaboration with the local communities, authorities and Nordic and Kenyan consulting companies. The objective of the project was to increase resilience of the local communities through improved water utilization techniques, mainly trapezoidal bunds (i.e. small dams that harvest rainwater) to contribute to increased long-term food security, health and income diversification for 15,000 beneficiaries.

This project was a typical example of an adaptation project where adaptation impacts went hand in hand with development impacts. The monitoring report from the first growing season in 2011–2012 showed that rainfall was low being well below the "design rainfall" for the design of the water harvesting structures. During the first growing season, the trapezoidal bunds produced crops valued at on average EUR 100 per bund. Under good management, there were strong indications that the bunds could yield crops valued at EUR 470 per successful season. These figures can be compared with the average estimated cost of 375 EUR per hand-built bund and approximately 820 EUR per bund built by mechanized means in the project. The returns were higher than the returns per acre under rain-fed conditions in the arid and semi-arid lands of Kenya. The project therefore contributed positively to income generation and food security for the households involved.

The *Mount Elgon Integrated Watershed Management Project*, implemented in 2013 in Kenya, addressed land degradation, forest resource conservation and improved livelihoods for approximately 7000 households in the Mount Elgon water catchment area through improved crop and livestock production methods. Farmers in the project area faced challenges in producing food for the entire year and were forced to buy food instead of being self-sufficient in food production. This has resulted in increasing poverty levels. Since the introduction of sustainable soil and land management systems (SALM),[3] together with improved crop and livestock production methods, farmers can produce sufficient food not only for their own use, but also to sell to the markets. One of the sub-projects, a quarter of an acre onion farm, generated baseline income of Kenyan shillings (KES) 12,000–15,000 for the local farmer per season.[4] After practicing SALM, the income increased to KES 40,000. In this case the SALM practises exercised included the building of trenches to protect the farm from river bursts, the use of compost, and increasing the distance of the planting rows. Once completed, the project had increased farmers' cash income and can pave the way for the establishment of a co-operative type of

---

[3]For a complete account of the SALM methodology, please refer to Wekesa, A. and Jönsson, M. 2014.

[4]One EUR equals approximately to 100 KES.

small-scale business activity in the future. This case project has also some likely mitigation impacts via sequestration, but they are not monitored.

A third example of an agricultural adaptation project is the currently on-going project in Bolivia, *Promoting cañahua in the Andean highland: a highly nutritive crop with a great market potential, adapted to extreme climate conditions.* Expected changes in precipitation patterns create a need for crops with short growing cycles and good tolerance for extreme weather conditions such as drought and frost.

Cañahua is a native but highly underutilised goosefoot plant of Bolivia and it is characterised by its high nutritive value and good resilience to extreme weather conditions. It is more tolerant to drought and frost, and mature in a shorter time, allowing farmers to obtain locally produced food and cash product for market devising successful adaptation process. The project is expected to generate local business opportunities among poor including households mainly headed by women. The total project costs are EUR 348,936 with NCF financing of EUR 269,952 with the rest covered by local private sources.

Combination of Adaptation and Mitigation with Business Linkages

Prolonged droughts in the Isiolo district in Kenya put great stress on the communities' water supplies, reduced livestock production and thus creating food insecurity and increased the incidence of water-borne diseases. The *Community based adaptation to climate change through environmentally sustainable water resource management project* implemented in the Isiolo District addressed these issues by increasing access to safe water and promoting hygiene awareness for selected vulnerable communities. The project ensured that more than 15,000 people living in the target area have improved access to safe water and 90 % of them have improved knowledge on hygiene and sanitation issues.

At the same time, through the use of modern and innovative technology of 9 Grundfos LIFELINK water systems, the project also reduced emissions of greenhouse gases by replacing previously used diesel pumps. The heart of the innovative systems is a submersible pump driven by solar panels. The water pump includes a satellite link to a computer-based system with an integrated communication and surveillance module. The operating performance of the community water system can be monitored remotely and in case it breaks down a local service partner will provide the necessary maintenance. The communities pay for their water via mobile telephones, which are already widely used throughout the country, and the user fees are utilised to cover the operating and maintenance costs. In this NCF case, the private sector business incentive combined with a clear technology transfer component, has led to adaptation impacts alongside mitigation and development benefits, generating also scaling-up activities elsewhere (Grundfos 2015).

Another example of a successful combination project is *The Cambodian Farmland Carbon (CAFACA)* tree planting project, which is expected to plant 300,000 trees as the key outcome. The total project costs are EUR 526,054 with NCF financing of EUR 386,130. The rest of the costs are covered by Nordic and local private sources as well as by revenues generated by the project.

The project will disseminate practical approaches to climate-resilient agriculture, including creative low-cost ways to organise tree planting in the farming landscape. The project has established a local company that connects the existing farmers' associations and small-holder farmers and supports them to increase their incomes from carbon sequestration via voluntary carbon credits and corporate social responsibility markets. The relevance of the project in terms of adaptation is related to climate adaptive functions of trees in the farming landscape, introduction of climate-resilient agricultural practices and increased climate change awareness of the farmers. Adaptation benefits are also expected to occur through decreased erosion and improved groundwater availability.

The ongoing pilot project of *Efficiency Enhancement and Entrepreneurship Development in Sustainable Biomass Charcoaling* is expected to reduce up to approximately 20,000 tonnes of CO_{2e} annually by reducing deforestation in Ghana. The local population is being trained in plantation management and in efficient charcoal production utilizing efficient kilns. Development benefits are linked to poverty reduction impacts through improved efficiency in charcoal production and creation of local entrepreneurship. The adaptation benefits relate to improvement of energy security and reduced reforestation but the development components have also adaptation linkages. Out of the budget of EUR 848,000, EUR 308,000 in provided by private sector and EUR 30,000 by the local communities.

Conclusions

As mentioned above, the CPI estimated that in 2013, 91.2 % of the global climate finance flows were directed at mitigations projects (Buchner et al. 2014). The estimate is based on a categorisation of climate actions, which, despite of being understandably inevitable, fails to depict the variety of linkages mitigation projects have with adaptation impacts as discussed above. As far as climate benefits are concerned, several NCF projects indicate that the distinction between mitigation and adaptation is partially artificial. Based on the lessons learned from NCF projects, the division of mitigation and adaptation seems partially to be linked to the fact that adaptation impacts, especially when quantified, are more challenging to conceptualize and monitor whereas mitigation impacts are readily monitorable. Quantification of adaptation impacts needs therefore further development and longer term monitoring in general.

Business initiatives geared towards mitigation can also have notable adaptation impacts. Typical examples in NCF portfolio are sustainable charcoal, efficient cook-stoves, and water filters. These project types are rather common among mitigation-oriented project types in general. Also many adaptation projects can have mitigation (i.e. sequestration) impacts, especially when aimed at improving agricultural productivity and conserving the soil.

Lessons learned so far from 49 NCF projects indicate that it is possible to attract co-financing also from the private sector to adaptation projects, especially when combined with mitigation actions. While adaptation projects attracted 13 % private sector co-financing, in combination projects the private sector's share of funding is considerably higher, 33 %. Based on the NCF experience, the volume and value of current adaptation activities may be larger than currently captured by the climate finance flow estimates. NCF projects suggest that adaptation co-benefits may not be accounted for in activities labelled as mitigation.

Even if re-classifying some mitigation projects as adaptation or multi-purpose would not increase the actual adaptation co-benefits, it can be argued that re-classification, when applicable, could further help to conceptualize the still challenging adaptation concept and encourage the consideration of adaptation needs, actions and impacts also in the context of mitigation projects. In the long run, this could function in support of the general adaptation agenda and possibly also result in an increase of the much-needed adaptation co-benefits and funding. In addition, a more thorough understanding of the interlinkages between mitigation and adaptation impacts could help to improve project designs and lead to additional adaptation co-benefits, especially when taking into account the projected growth in many low-income countries and possibilities for synergies.

The NCF projects thus indicate that adaptation activities can attract private sector interest, co-funding and create business linkages—especially when combined with mitigation components.

References

Agrawala S et al (2011) Private sector engagement in adaptation to climate change: approaches to managing climate risk. OECD Environment Working Papers, No. 39, OECD Publishing, Paris, France. http://www.oecd-ilibrary.org/docserver/download/5kg221jkf1g7.pdf?expires=1426583832&id=id&accname=guest&checksum=0D0C86D7C36E97D206B05F819632CA0C. Accessed 17 Mar 2015

Brüning K, Hamro-Drotz D (2015) Nordic climate finance opportunities—the NCF case study. Nordic Working Group for Climate Change Negotiations NOAK, Nordic Council of Ministers (Nordiske Arbejdspapirer, 2015:910, Nordisk Ministerråd, 2015, Copenhagen)

Buchner B, Stadelmann M, Wilkinson J, Mazza F, Rosenberg A, Abramskiehn D (2014) Global landscape of climate finance, CPI Report. The Climate Policy Initiative. http://climatepolicyinitiative.org/wp-content/uploads/2014/11/The-Global-Landscape-of-Climate-Finance-2014.pdf. Accessed 24 Mar 2015

Eriksen S, Inderberg TH, O'Brien K, Sygna L (2015) Introduction: development as usual is not enough. In: Inderberg TH, Eriksen S, O'Brien K, Sygna L (eds) Climate change adaptation and development. Transforming paradigms and practices. Routledge, Abingdon

Green Climate Fund (GCF) (2014) Green Climate Fund Board takes key decisions on operations and makes progress on 'Essential Eight'. Press release, Green Climate Fund, Songdo, South Korea 22.2.2014. http://gcfund.net/fileadmin/00_customer/documents/pdf/GCF_Press_Release_fin_20140222.pdf. Accessed 30 Mar 2015

Green Climate Fund (GCF) (2015) GCF set to allocate resources before Paris Climate Change Conference. Press Release, Green Climate Fund, Songdo, South Korea http://www.gcfund.org/fileadmin/00_customer/documents/Press/GCF_Press_Release_2015_03_26_accreditation_entities.pdf. Accessed 29 Mar 2015

Grundfos (2015) Grundfos Lifelink projects in Kenya—connecting the link to sustainable water supply. http://www.grundfos.com/cases/find-case/grundfos-lifelink-projects-in-kenya.html. Accessed 31 Mar 2015

Gupta S, Harnisch J, Barua DC, Chingambo L, Frankel P, Garrido Vázquez RJ, Gómez-Echeverri L, Haites E, Huang Y, Kopp R, Lefèvre B, Machado-Filho H, Massetti E (2014) Cross-cutting investment and finance issues. In: Edenhofer O, Pichs-Madruga R, Sokona Y, Farahani E, Kadner S, Seyboth K, Adler A, Baum I, Brunner S, Eickemeier P, Kriemann B, Savolainen J, Schlömer S, von Stechow C, Zwickel T, Minx JC (eds) Climate change 2014: mitigation of climate change. Contribution of working group III to the fifth assessment report of the intergovernmental panel on climate change. Cambridge University Press, Cambridge. http://www.ipcc.ch/pdf/assessment-report/ar5/wg3/ipcc_wg3_ar5_chapter16.pdf. Accessed 9 June 2015

McGray H, Hammill A, Bradley R (2007) Weathering the storm, options for framing adaptation and development. WRI Report, World Resources Institute, Washington. http://pdf.wri.org/weathering_the_storm.pdf. Accessed 28 Mar 2015

Narain U, Margulis S, Essam T (2011) Estimating costs of adaptation to climate change. Clim Policy 11(3):1001–1019

Sigvaldsen E, Hansen S, Laugerud T (2013) NCG evaluation of nordic climate facility. Nordic Development Fund, Nordic Consulting Group, Oslo. http://www.ndf.fi/sites/ndf.fi/files/attach/ncf_evaluation_report_june_2013.pdf. Accessed 10 June 2015

Tanner T, Horn-Phathanothai L (2014) Climate change and development. Routledge, Abingdon

United Nations Environment Programme (UNEP) (2015) Africa's adaptation gap 2. Technical Report, UNEP, Nairobi. http://apps.unep.org/publications/pmtdocuments/-Africa%E2%80%99s_Adaptation_Gap_2__.pdf. Accessed 30 Mac 2015

United Nations Framework Convention on Climate Change (UNFCCC) (2014) 2014 Biennial assessment and overview of climate finance flows report. UNFCCC, Bonn. https://unfccc.int/files/cooperation_and_support/financial_mechanism/standing_committee/application/pdf/2014_biennial_assessment_and_overview_of_climate_finance_flows_report_web.pdf. Accessed 8 June 2015

United Nations Framework Convention on Climate Change (UNFCCC) (2015) CDM Methodologies. UNFCCC, Bonn. http://cdm.unfccc.int/methodologies/index.html. Accessed 24 Mar 2015

World Bank (2015) Global economic prospects. Global and Country and region specific forecasts and data. World Bank, Washington. http://www.worldbank.org/en/publication/global-economic-prospects/data. Accessed 24 Mar 2015

MIX
Papier aus verantwortungsvollen Quellen
Paper from responsible sources
FSC® C105338

If you have any concerns about our products,
you can contact us on
ProductSafety@springernature.com

In case Publisher is established outside the EU,
the EU authorized representative is:
Springer Nature Customer Service Center GmbH
Europaplatz 3, 69115 Heidelberg, Germany

Printed by Libri Plureos GmbH
in Hamburg, Germany